Remarkable recent developments in the field of quantum optics have given rise to experimental techniques, of unprecedented sensitivity, and theoretical tools which are being used to investigate the fundamental concepts of quantum mechanics, and to perform extremely sophisticated measurements of important physical parameters. This book provides an introduction to this exciting area of physics by giving a comprehensive account of the basic theory of the interaction between atoms and electromagnetic fields.

The first four chapters describe the different forms of the interaction between atoms and radiation fields. The rest of the book deals with how these interactions lead to the formation of dressed states, in the presence of vacuum fluctuations, as well as in the presence of external fields. Also covered are the role of dressed atoms in quantum measurement theory, and the physical interpretation of vacuum radiative effects.

Treating a key field on the boundary between quantum optics and quantum electrodynamics, the book will be of great use to graduate students, as well as to established experimentalists and theorists, in either of these areas.

T0275569

CAMBRIDGE STUDIES IN MODERN OPTICS

Series Editors

P. L. KNIGHT
Department of Physics, Imperial College of Science, Technology and Medicine

A. MILLER
Department of Physics and Astronomy, University of St Andrews

Atom-Field Interactions and Dressed Atoms

TITLES IN PRINT IN THIS SERIES

Optical Holography – Principles, Techniques and Applications
P. Hariharan

Fabry-Perot Interferometers
G. Hernandez

Holographic and Speckle Interferometry (Second edition)
R. Jones and C. Wykes

Laser Chemical Processing for Microelectronics
edited by K. G. Ibbs and R. M. Osgood

The Elements of Nonlinear Optics
P. N. Butcher and D. Cotter

Optical Solitons – Theory and Experiment
edited by J. R. Taylor

Particle Field Holography
C. S. Vikram

Ultrafast Fiber Switching Devices and Systems
M. N. Islam

Optical Effects of Ion Implantation
P. D. Townsend, P. J. Chandler and L. Zhang

Diode-Laser Arrays
edited by D. Botez and D. R. Scifres

The Ray and Wave Theory of Lenses
A. Walther

Design Issues in Optical Processing
edited by J. N. Lee

Atom-Field Interactions and Dressed Atoms
G. Compagno, R. Passante and F. Persico

Atom-Field Interactions and Dressed Atoms

G. COMPAGNO,[*] R. PASSANTE[†] and F. PERSICO[*][†]

[*] *Istituto di Fisica, Università di Palermo*

[†] *Istituto per le Applicazioni Interdisciplinari della Fisica, CNR Palermo*

CAMBRIDGE
UNIVERSITY PRESS

CAMBRIDGE UNIVERSITY PRESS
Cambridge, New York, Melbourne, Madrid, Cape Town, Singapore, São Paulo

Cambridge University Press
The Edinburgh Building, Cambridge CB2 2RU, UK

Published in the United States of America by Cambridge University Press, New York

www.cambridge.org
Information on this title: www.cambridge.org/9780521419482

© Cambridge University Press 1995

This publication is in copyright. Subject to statutory exception
and to the provisions of relevant collective licensing agreements,
no reproduction of any part may take place without
the written permission of Cambridge University Press.

First published 1995
This digitally printed first paperback version 2005

A catalogue record for this publication is available from the British Library

Library of Congress Cataloguing in Publication data
Compagno, G.
Atom-field interactions and dressed atoms/G. Compagno, R.
Passante, and F. Persico.
p. cm. – (Cambridge studies in modern optics ; 17)
includes bibliographical references and index.
ISBN 0 521 41948 4
1. Quantum optics. 2. Quantum electrodynamics. 3. Quantum field theory.
I. Passante, R. II. Persico, F. (Franco) III. Title. IV. Series.
QC446.2.C65 1995
537.6'7 – dc20 94 – 31627 CIP

ISBN-13 978-0-521-41948-2 hardback
ISBN-10 0-521-41948-4 hardback

ISBN-13 978-0-521-01972-9 paperback
ISBN-10 0-521-01972-9 paperback

Contents

Preface

Quantum Optics is a branch of physics which has developed recently in different directions relevant to fundamental physics as well as to highly sophisticated technological applications. The scientific roots of quantum optics, however, originate from the broader subject of Quantum Electrodynamics and, more generally, from quantum field theory. Thus the boundary between quantum optics and quantum field theory is a particularly delicate conceptual ground which should be properly mastered by any prospective quantum optician, theorist or experimentalist alike. This book is intended to foster understanding and knowledge of this boundary region by presenting in a pedagogical fashion the basic theory of dressed atoms, which has been established as a concept of central importance in quantum optics, since it is capable of shedding light on such diverse physical phenomena as resonance fluorescence, the Lamb shift and van der Waals forces.

Coherently with the aims outlined above, the first part of this book, consisting of the first four chapters, is dedicated to the foundations of atom-field interactions. Both radiation and matter are treated from the quantum field theory point of view, and the coupling between matter and the electromagnetic field is derived using the principle of gauge invariance. The atom-photon Hamiltonian is obtained by specializing this general treatment to a nonrelativistic electron field describing the electrons around an atomic nucleus. It should be noted that these first four chapters are specifically aimed at deriving the atom-photon interaction from general principles of quantum field theory, and are not in the form of a balanced compendium of QED.

The impossibility of separating atoms and radiation leads naturally to the second part of the book and to the concept of atoms dressed by the radiation field. Actually, the expression "dressed atom" in nonrelativistic

QED is used with reference to two physical situations which, broadly speaking, are different with respect to the photon number of occupancy. In the first situation an atom is in the presence of real photons, such as those produced by an external source, coherent or incoherent. Then the atom-photon coupling admixes, shifts and splits the levels of the system constituted by the atom and the field. The resulting atom-field energy levels display correlations between bare atom and field states, which are interpreted as states of a new composite object, namely the dressed atom. This kind of dressed atom is the subject of the fifth chapter. In the second situation one has a ground-state bare atom interacting with the vacuum electromagnetic field. Taking the total atom-field system to be in the lowest possible energy state, the zero-point quantum fluctuations of the field induce virtual absorption and re-emission processes of photons by the atom. Since these processes take place continuously they create a cloud of virtual photons around the bare atom. The complex object, bare atom plus cloud of virtual photons, is what one calls a dressed atom in this second physical situation. Naturally this concept of dressed atom is a specialization of the much more general concept of dressed source in quantum field theory, which finds application in diverse branches of physics as solid state or elementary particles, and it is discussed from this general point of view in Chapters 6 and 7. The final chapter of the book is dedicated to some general features of the theory of dressed atoms relevant for the quantum theory of measurement and for the theory of self-reaction. Chapters 6 to 8 summarize investigations by different authors and attempt to present them in a systematic way, with an emphasis on the conceptual QED foundations. On the whole, the last four chapters of the book are aimed at deriving and describing the properties of dressed atoms and are inevitably more specific than the first four. However, the presentation of the subject in the more general context of quantum fields, using models taken from the theory of the electron-phonon and of the nucleon-meson interactions, is intended to smooth out discontinuities in the treatment as far as possible. The appendices are intended to complement or to generalize the topics discussed in the text.

In view of the rather broad scope of the book, no attempt has been made to provide the reader with a complete set of references. The general criterion has been to refer to an easily available book wherever possible, or to a review article. Reference to normal papers has been done only in the case of historic and fundamental papers or when the information contained in the paper has been deemed necessary and not available in a book or in a review article. Furthermore, in order to avoid divergence of citation lists we have also made recourse to the well-known formula "and papers cited therein". Thus the reference lists at the end of each chapter cannot be used to assess priority of a discovery, of a new idea or of a suggestion.

The book is a theoretical text, but it is not intended for theoreticians only. It has been assumed that the reader is a postgraduate student specializing in quantum optics or in nonrelativistic QED possessing a working knowledge of quantum mechanics and classical electrodynamics. The derivation of the main results, however, is displayed in a rather complete fashion and the mathematics involved should be well within the grasp of theorists and experimentalists alike. We also hope that the book may interest more senior scientists working in quantum optics, in QED and in neighbouring fields where a knowledge of the basis of quantum optics is useful or even necessary.

The system of units throughout the book is the Gauss system. A recipe for translating all expressions and formulae quickly and efficiently into SI units is presented in Appendix E.

This book owes much to our collaboration with E. A. Power, from whom we have learned most of the QED we know. We are particularly grateful to P. L. Knight for suggesting and encouraging the project. P. L. Knight, E. A. Power and T. Thirunamachandran have read the complete manuscript and have spent long hours with us in discussions relating to this book. S. M. Barnett has made useful comments on Chapters 5 and 6. L. de la Pen̄a has volunteered a thorough debugging of the manuscript as well as interesting remarks on the structure of the book. We feel indebted to these friends and colleagues who have generously dedicated a substantial part of their time to improve our book and to purify it from a large number of what pudic authors usually call misprints. We thank P. W. Milonni for reading the manuscript, for suggesting changes and for encouragement, and we gratefully acknowledge a much appreciated conversation on the contents of the book with C. Cohen-Tannoudji, who in addition has generously sent us useful material on dressed atoms.

Most of the discussions on the manuscript with S. M. Barnett, P. L. Knight, E. A. Power and T. Thirunamachandran have taken place in the pleasantly stimulating environment of the International Centre for Theoretical Physics in Trieste, and we are grateful to the ICTP organizers for liberal hospitality and collaboration. Finally we wish to thank Mr. S. Pappalardo for helpful assistance with the figures.

1

The classical electromagnetic field in the absence of sources

Introduction. The principal aim of this chapter is to familiarize the reader with the notation adopted in the text, as well as to introduce some concepts, such as the energy-momentum tensor of the electromagnetic field, the partition of its total angular momentum into an orbital and a spin contribution and its expansion in vector spherical harmonics, which are not usually included in an undergraduate course on electrodynamics. The chapter is entirely dedicated to the classical electromagnetic field in the absence of charges and currents. In the first two sections we present Maxwell's equations, the vector potential and different forms of the Lagrangian density of the free field from which Maxwell's equations can be obtained as Euler-Lagrange equations. In Section 1.3 we discuss briefly the properties of the field under pure Lorentz transformation and tensor notation. Then we introduce the concept of local gauge invariance and of gauge transformation, and we define the constraints leading to the Lorentz and to the Coulomb gauge. Using a canonical formalism, in Section 1.5 we obtain the Hamiltonian density of the field in the Coulomb gauge. The energy-momentum tensor of the field, the momentum and the angular momentum, along with their important conservation properties, are discussed in Section 1.6. The attention is focused on the angular momentum in Section 1.7, with a discussion of the partition into orbital and spin contributions. The mathematical properties of a general vector field are described in Section 1.8, with particular reference to the partition into longitudinal and transverse fields and to the definition of the longitudinal and transverse δ-functions and of their Fourier transform. These mathematical tools are used in Section 1.9 where the properties of the vector potential and its changes under gauge transformations are succinctly summarized. In the next section the solutions of Maxwell's equation, in the absence of charges and currents, are expanded in plane and spherical waves. This gives the opportunity for a short discussion of boundary conditions imposed on the electromagnetic field on surfaces of different shape, which is done in Section 1.11. The chapter is concluded by Section 1.12, where the plane wave field amplitude expansion, introduced in Section 1.10, is used to express various quantities relevant for the description of the dynamics of the electromagnetic field, such as the momentum, the angular momentum and the Hamiltonian density.

1.1 Maxwell's equations in the absence of sources

In the absence of charges and currents and in empty space Maxwell's equations, which describe the propagation of the electromagnetic field, take the form

$$\nabla \cdot \mathbf{E} = 0 \,; \nabla \times \mathbf{H} - \frac{1}{c}\dot{\mathbf{E}} = 0 \,;\; \nabla \cdot \mathbf{H} = 0 \,;\; \nabla \times \mathbf{E} + \frac{1}{c}\dot{\mathbf{H}} = 0 \qquad (1.1)$$

In these equations **E** and **H** are the electric and magnetic field respectively, both functions of space coordinates $\mathbf{r} = (x_1, x_2, x_3)$ and time t, and c is the speed of light in vacuo. The Gaussian system of units is used throughout this book. Conversion tables between Gaussian and SI are given in Appendix E. It should be noted that in the present book no use is made of the field auxiliary to the magnetic field. Thus the magnetic field has been denoted by **H** rather than **B**. This choice is similar to that of Heitler (1960) and of Landau and Lifshitz (1975). On the other hand, Power (1964) and Craig and Thirunamachandran (1984) denote the magnetic field by **B**.

As is well known and as we will discuss briefly in Section 1.3, the properties of the electromagnetic field under a coordinate transformation are most conveniently expressed by the mathematical properties of the field tensor

$$F_{\mu\nu} = \begin{pmatrix} 0 & H_3 & -H_2 & -iE_1 \\ -H_3 & 0 & H_1 & -iE_2 \\ H_2 & -H_1 & 0 & -iE_3 \\ iE_1 & iE_2 & iE_3 & 0 \end{pmatrix} \qquad (1.2)$$

In terms of $F_{\mu\nu}$, the first and the second pair of Equations (1.1) take either of the two particularly compact forms

$$\frac{\partial F_{\mu\nu}}{\partial x_\nu} = 0 \,;\; e_{\kappa\lambda\mu\nu}\frac{\partial F_{\mu\nu}}{\partial x_\lambda} = 0 \text{ or } \frac{\partial F_{\lambda\mu}}{\partial x_\nu} + \frac{\partial F_{\mu\nu}}{\partial x_\lambda} + \frac{F_{\nu\lambda}}{\partial x_\mu} = 0 \qquad (1.3)$$

where $e_{\kappa\lambda\mu\nu}$ is the completely antisymmetric four-tensor whose components change sign under the interchange of any pair of indices and which has $e_{1234} = 1$ (see e.g. Landau and Lifshitz 1975). In (1.3) the convention of implicit summation over repeated indices is used, and the four-vector x_μ is defined as (x_1, x_2, x_3, x_4) with $x_4 = ict$. Since we shall limit our discussion to the domain of the special theory of relativity, no distinction is necessary between covariant and contravariant components and no metric tensor is introduced (see Sakurai 1982). Note also that all Greek indices run from 1 to 4; in the future we shall also use Latin indices which run from 1 to 3. It is evident that form (1.3) of Maxwell's equations is covariant under Lorentz transformations.

Another form of Maxwell's equations can be obtained in terms of the four-vector A_μ, which is related to the field tensor by (see e.g. Sakurai 1982)

$$\frac{\partial A_\nu}{\partial x_\mu} - \frac{\partial A_\mu}{\partial x_\nu} = F_{\mu\nu} \tag{1.4}$$

From (1.2) we obtain

$$\nabla \times \mathbf{A} = \mathbf{H}\ ;\ \nabla A_4 = -i(\mathbf{E} + \frac{1}{c}\dot{\mathbf{A}}) \tag{1.5}$$

Clearly A_i are the components of the vector potential whereas A_4 coincides with iV, where V is the scalar potential familiar from elementary electromagnetism (see e.g. Bleaney and Bleaney 1985). Substituting (1.5) into the second pair of (1.1) yields two trivial identities. The first two of (1.1), however, yield

$$\nabla^2\mathbf{A} - \nabla\left(\nabla\cdot\mathbf{A} + \frac{1}{c}\dot{V}\right) - \frac{1}{c^2}\ddot{\mathbf{A}} = 0\ ;\ \nabla^2 V + \frac{1}{c}\nabla\cdot\dot{\mathbf{A}} = 0 \tag{1.6}$$

or in tensor notation

$$\frac{\partial^2 A_\mu}{\partial x_\nu^2} - \frac{\partial}{\partial x_\mu}\left(\frac{\partial A_\nu}{\partial x_\nu}\right) = 0 \tag{1.7}$$

Equations (1.6) or, equivalently, (1.7) are the third form of Maxwell's equations *in vacuo* considered here. The slight abuse of language should be noted. Strictly speaking in fact (1.7) are the equations of field A_μ rather than Maxwell's equations which are normally expressed in terms of **E** and **H**.

1.2 Lagrangian of the free field

The equations of motion of any field should be derivable from an appropriate Lagrangian density using the Euler-Lagrange equations. It is easy to show that the following Lagrangian density

$$\mathcal{L}_0 = -\frac{1}{8\pi}(\mathbf{H}^2 - \mathbf{E}^2) = -\frac{1}{16\pi}F_{\mu\nu}F_{\mu\nu}$$
$$= -\frac{1}{16\pi}\left(\frac{\partial A_\nu}{\partial x_\mu} - \frac{\partial A_\mu}{\partial x_\nu}\right)\left(\frac{\partial A_\nu}{\partial x_\mu} - \frac{\partial A_\mu}{\partial x_\nu}\right) \tag{1.8}$$

is suitable for our needs. It should be noted that in a canonical formalism the Lagrangian density of any field should be expressed in terms of the field amplitude and of its derivatives, whereas its Hamiltonian density should be expressed in terms of the field amplitudes and of their canonically conjugate momenta. In this book we shall occasionally infringe this rule when extreme precision of language is superfluous. In (1.8), for example, the first two forms of $\mathcal{L}_0$ should not, strictly speaking, be considered as Lagrangian density, although they do coincide with the value taken by the Lagrangian density when expressed in terms of **E** and **H**. It is evident, however, that no confusion should arise in the case at hand, where it is natural to adopt the third form of (1.8) to obtain the Euler-Lagrange equations. This form for $\mathcal{L}_0$ in (1.8) is a function of the first derivatives of A_μ only, and not of A_μ. Therefore the Euler-Lagrange equations, valid for a general Lagrangian density $\mathcal{L}$ function of $\frac{\partial A_\mu}{\partial x_\nu}$ and of A_μ,

$$\frac{\partial}{\partial x_\nu}\frac{\partial \mathcal{L}}{\partial\left(\frac{\partial A_\mu}{\partial x_\nu}\right)} - \frac{\partial \mathcal{L}}{\partial A_\mu} = 0 \tag{1.9}$$

reduce for $\mathcal{L}_0$ to

$$\frac{\partial}{\partial x_\nu}\frac{\partial \mathcal{L}_0}{\partial\left(\frac{\partial A_\mu}{\partial x_\nu}\right)} = 0 \tag{1.10}$$

Moreover, from (1.8)

$$\frac{\partial \mathcal{L}_0}{\partial\left(\frac{\partial A_\mu}{\partial x_\nu}\right)} = -\frac{1}{16\pi}\left\{\frac{\partial}{\partial\left(\frac{\partial A_\mu}{\partial x_\nu}\right)}\left[\left(\frac{\partial A_\sigma}{\partial x_\lambda} - \frac{\partial A_\lambda}{\partial x_\sigma}\right)\left(\frac{\partial A_\sigma}{\partial x_\lambda} - \frac{\partial A_\lambda}{\partial x_\sigma}\right)\right]\right\}$$
$$= -\frac{1}{8\pi}\left[\frac{\partial}{\partial\left(\frac{\partial A_\mu}{\partial x_\nu}\right)}\left(\frac{\partial A_\sigma}{\partial x_\lambda}\frac{\partial A_\sigma}{\partial x_\lambda} - \frac{\partial A_\sigma}{\partial x_\lambda}\frac{\partial A_\lambda}{\partial x_\sigma}\right)\right]$$
$$= \frac{1}{4\pi}\left(\frac{\partial A_\nu}{\partial x_\mu} - \frac{\partial A_\mu}{\partial x_\nu}\right) \tag{1.11}$$

Substitution of result (1.11) into the Euler-Lagrange equation (1.10) immediately yields Maxwell's equations in the form (1.7).

It should be remarked that $\mathcal{L}_0$ is not the only Lagrangian density giving rise to Maxwell's equations. For example, if we add to $\mathcal{L}_0$ a scalar which is the four-divergence of a four-vector Γ_μ, function of A_μ and x_μ, the equations of motion for the new Lagrangian density $\mathcal{L}_0 + \frac{\partial \Gamma_\mu}{\partial x_\mu}$ will coincide with the equations of motion for $\mathcal{L}_0$. This is due to the fact that the Euler-Lagrange equations are obtained by varying the action $\int \mathcal{L} dx$, and the four-divergence term introduced above contributes a vanishing boundary term to the variation of the action, as discussed by Barut (1980). In fact, more generally but for the same reason, two Lagrangian densities which differ by a term which vanishes upon space-time integration and possibly upon application of additional appropriate constraints, yield the same Euler-Lagrange equations (Bogoliubov and Shirkov (1960). Examples of such equivalent Lagrangian densities are (see e.g. Schweber 1964)

$$\mathcal{L}_1 = \mathcal{L}_0 - \frac{1}{8\pi}\left(\frac{\partial A_\mu}{\partial x_\mu}\right)^2 ; \; \mathcal{L}_2 = -\frac{1}{8\pi}\left(\frac{\partial A_\mu}{\partial x_\nu}\right)^2 \tag{1.12}$$

1.3 Pure Lorentz transformations

Any four-vector, such as x_μ, under a homogeneous Lorentz transformation yields a new four vector such that

$$x'_\mu = \Lambda_{\mu\nu} x_\nu \tag{1.13}$$

where $\Lambda_{\mu\nu}$ are matrix elements of a 4×4 matrix Λ representing the effects of the Lorentz transformation (see e.g. Ugarov 1982). The four-vector transformation property specified by (1.13) is called "Lorentz covariance". If Λ is a pure Lorentz transformation (Goldstein 1980) the matrix elements are given by

$$\begin{aligned} &\Lambda_{ij} = \delta_{ij} + \beta^{-2}\beta_i\beta_j(\gamma - 1) \; ; \; \Lambda_{i4} = i\beta_i\gamma \; ; \\ &\Lambda_{4j} = -i\beta_j\gamma \; ; \; \Lambda_{44} = \gamma \end{aligned} \tag{1.14}$$

where $\boldsymbol{\beta} = \mathbf{v}/c$, $\gamma = (1 - \beta^2)^{-1/2}$ and $\mathbf{v}$ is the relative velocity of the transformation. Lorentz transformations are orthogonal and their matrix elements satisfy

$$\Lambda_{\mu\nu}\Lambda_{\mu\lambda} = \delta_{\nu\lambda} \; ; \; \left(\Lambda^{-1}\right)_{\mu\nu} = \Lambda_{\nu\mu} \tag{1.15}$$

The latter property can be exploited to invert (1.13) as

$$x_\mu = \left(\Lambda^{-1}\right)_{\mu\nu} x'_\nu = \Lambda_{\nu\mu} x'_\nu \tag{1.16}$$

Consequently from (1.16)

$$\frac{\partial}{\partial x'_\mu} = \frac{\partial x_\nu}{\partial x'_\mu}\frac{\partial}{\partial x_\nu} = \Lambda_{\mu\nu}\frac{\partial}{\partial x_\nu} \tag{1.17}$$

which shows that the operator $\frac{\partial}{\partial x_\nu}$, sometimes called the four-gradient, is also a four-vector.

The scalar product of two four-vectors, such as x_μ and y_μ, is invariant under a Lorentz transformation, in view of (1.15). In fact

$$x'_\mu y'_\mu = \Lambda_{\mu\nu} x_\nu \Lambda_{\mu\lambda} y_\lambda = \delta_{\nu\lambda} x_\nu y_\lambda = x_\mu y_\mu \tag{1.18}$$

A quantity which is invariant under a Lorentz transformation is called a "Lorentz scalar". A tensor with two indices, such as $F_{\mu\nu}$, is a "tensor of second rank". A tensor of second rank transforms according to the rule

$$F'_{\mu\nu} = \Lambda_{\mu\lambda}\Lambda_{\nu\rho} F_{\lambda\rho} \tag{1.19}$$

Hence the quantity $F_{\mu\nu}F_{\mu\nu}$ transforms as

$$F'_{\mu\nu}F'_{\mu\nu} = \Lambda_{\mu\lambda}\Lambda_{\nu\rho}F_{\lambda\rho}\Lambda_{\mu\sigma}\Lambda_{\nu\tau}F_{\sigma\tau} = \delta_{\lambda\sigma}\delta_{\rho\tau}F_{\lambda\rho}F_{\sigma\tau} = F_{\mu\nu}F_{\mu\nu} \tag{1.20}$$

and thus is a Lorentz scalar.

Thus the three Lagrangian densities $\mathcal{L}_i$ appearing in (1.8) and (1.12) are Lorentz scalars and invariant under Lorentz transformations. It is also easy to check that Maxwell's equations are Lorentz-covariant, as it is particularly evident when they are expressed in the form (1.3). This means that they are of the same form in reference frames related by Lorentz transformations, in accord with the principles of special relativity. Indeed, it is possible to show in general that the Lagrangian density of any field must be a Lorentz scalar if the Euler-Lagrange equations are to be Lorentz-covariant.

1.4 Gauge transformations

Suppose we have obtained a four-vector A_μ satisfying relation (1.4) for a given field tensor $F_{\mu\nu}$. We now add to A_μ the four-gradient of a scalar field $\chi(x_\mu)$, a function of position and time, and obtain a new four-vector

$A_\mu + \partial\chi/\partial x_\mu$. It is easy to see that this new four-vector, like A_μ, satisfies (1.4), for the same field tensor $F_{\mu\nu}$. In fact

$$\frac{\partial A_\nu}{\partial x_\mu} + \frac{\partial^2 \chi}{\partial x_\mu x_\nu} - \frac{\partial A_\mu}{\partial x_\nu} - \frac{\partial^2 \chi}{\partial x_\nu x_\mu} = \frac{\partial A_\nu}{\partial x_\mu} - \frac{\partial A_\mu}{\partial x_\nu} = F_{\mu\nu} \tag{1.21}$$

Indeed, until the scalar $\chi(\mathbf{r}, t)$ is subjected to some condition which determines it, there are an infinite number of four-vectors $A_\mu + \partial\chi/\partial x_\mu$ which correspond to the same electromagnetic field $F_{\mu\nu}$. Clearly the Lagrangian density $\mathcal{L}_0$ given by (1.8) is invariant with respect to the transformation A_μ to $A_\mu + \partial\chi/\partial x_\mu$. Such an invariance is called local gauge invariance, or gauge invariance of the second kind, where χ is a function of the space-time point x_μ. Gauge invariance of the second kind is discussed by Doughty (1990) and Mandl and Shaw (1984) among others. On the other hand $\mathcal{L}_1$ and $\mathcal{L}_2$ in (1.12) are not gauge invariant.

As for Maxwell's equations, clearly forms (1.1) and (1.3) are gauge invariant, but form (1.6) in terms of the four-potential, or equivalently (1.7), is not. Clearly the four-potential A_μ cannot be regarded as representing a physical observable, because the form of the equation of motion of such an observable cannot depend on an arbitrary mathematical function such as $\partial\chi/\partial x_\mu$. It should be emphasized, however, that once A_μ has been chosen according to some criteria, the electromagnetic field is uniquely determined on the basis of (1.4). The choice of these criteria is a procedure which is called "fixing the gauge". This absence of a unique correspondence between the description in terms of **E** and **H** and the description in terms of A_μ is to be expected on the basis of the fact that Maxwell's equations (1.1) are first-order differential equations, whose solution necessitates knowledge of six functions of **r** as initial conditions. These are the three components of **E** and **H** for each point of space. On the other hand, Maxwell's equations (1.6) are of second order, and their solution implies knowledge of eight functions, which are the components A_μ and their time derivatives. Thus the A_μ description of the field displays some redundancy of dynamical variables, which calls for the introduction of constraints between some of them. This point is discussed in clear terms by Cohen-Tannoudji *et al.* (1989). These constraints are the conditions which fix the gauge.

Consider, for example, the following constraint

$$\frac{\partial A_\mu}{\partial x_\mu} = 0 \text{ or } \nabla \cdot \mathbf{A} + \frac{1}{c}\dot{V} = 0 \tag{1.22}$$

which fixes a gauge called the Lorentz gauge. The scalar nature of the quantity on the LHS of (1.22), which is a four-divergence and Lorentz invariant, makes this gauge particularly useful for treatments which fully exploit the relativistic nature of electrodynamics. Then the second term on the LHS of (1.7) vanishes, and Maxwell's equations (1.6) take the form

$$\nabla^2\mathbf{A} - \frac{1}{c^2}\ddot{\mathbf{A}} = 0 \; ; \; \nabla^2 V - \frac{1}{c^2}\ddot{V} = 0 \tag{1.23}$$

It is interesting to note that constraint (1.22) does not completely lead to a unique relation between A_μ and $(\mathbf{E}, \mathbf{H})$. In fact, we see that any new four-vector $A_\mu + \partial\chi/\partial x_\mu$, where A_μ satisfies (1.22) and $\chi(\mathbf{r}, t)$ is a scalar subject to the condition

$$\nabla^2\chi - \frac{1}{c^2}\ddot{\chi} = 0 \tag{1.24}$$

satisfies the same constraint (1.22). This shows that the Lorentz gauge is really a class of subgauges determined by the Lorentz condition. The arbitrariness of χ can be exploited to eliminate V from Maxwell's equation, by choosing among the various χ solutions of (1.24) that particular χ for which $V + \dot{\chi}/c = 0$ (see e.g. Heitler 1960). In this case the Lorentz constraint (1.22) reduces to

$$\nabla \cdot \mathbf{A} = 0 \; , \; V = 0 \tag{1.25}$$

and Maxwell's equations are simply given by the first of (1.23).

Another gauge, the so-called "transverse" or "Coulomb" gauge, is obtained by imposing a constraint only on the first three components of A_μ, in the form

$$\nabla \cdot \mathbf{A} = 0 \tag{1.26}$$

and not on A_4. Thus the Coulomb gauge (1.26) should not be confused with the Lorentz subgauge (1.25) where V is constrained to vanish. In the absence of charges and currents and in infinite unbounded space, however, these two gauges coincide, since substitution of (1.26) in (1.6) yields

$$\nabla^2\mathbf{A} - \frac{1}{c}\nabla\dot{V} - \frac{1}{c^2}\ddot{\mathbf{A}} = 0 \; ; \; \nabla^2 V = 0 \tag{1.27}$$

and since the only nondiverging solution of Laplace's equation for V, which vanishes at infinity, is the null $V = 0$ solution. This point is discussed e.g. by Morse and Feshbach (1953).

1.5 Hamiltonian density of the free field in the Coulomb gauge

The usual Hamiltonian formalism can be applied to the free electromagnetic field. We start from the gauge invariant Lagrangian density $\mathcal{L}_0$ in (1.8). With A_μ as the generalized coordinates of the field one should be able to obtain conjugate momenta Π_μ

$$\Pi_\mu = \frac{1}{ic} \frac{\partial \mathcal{L}_0}{\partial \left(\frac{\partial A_\mu}{\partial x_4} \right)} \tag{1.28}$$

and the Hamiltonian density as

$$\mathcal{H}_0 = ic\Pi_\mu \frac{\partial A_\mu}{\partial x_4} - \mathcal{L}_0 \tag{1.29}$$

Such a general approach, however, is not as simple as it looks. In fact, from (1.11) we immediately see that Π_4, the momentum conjugate to the scalar potential V, vanishes identically. This is going to cause some difficulties, for example in connection with quantization where one is expected to impose noncommutation of Π_4 and V. The relativistic field theorist's way out of this difficulty is to change the Lagrangian while keeping the four generalized coordinates A_μ for the field. This permits us to preserve the manifest relativistic features of the theory at each step of the calculations, but it gives rise to a series of formal difficulties which are unnecessary for our purposes, as discussed by Schweber (1964).

We shall take a different approach, and exploit the gauge invariance of $\mathcal{L}_0$ in order to reduce the number of field coordinates. This will give rise to a theory which is not manifestly relativistically covariant, but it is much more simple formally. Thus, remembering we are discussing the case of no charges and currents, we choose the Lorentz subgauge with constraints (1.25), or equivalently the Coulomb gauge, and set $A_4 = 0$. Then (1.8) becomes (Craig and Thirunamachandran 1984)

$$\mathcal{L}_0 = \frac{1}{8\pi} \left\{ \frac{1}{c^2} \dot{\mathbf{A}}^2 - \left(\frac{\partial A_i}{\partial x_j} - \frac{\partial A_j}{\partial x_i} \right)^2 \right\} \tag{1.30}$$

Expression (1.30) depends on three generalized coordinates only, since A_4 does not appear. This coordinate will reappear, however, if we subject $\mathcal{L}_0$ to a pure Lorentz transformation, thereby spoiling manifest Lorentz invariance. This means that our approach is not really suited to describing processes which involve Lorentz transformations where non-Galilean features are important. On the other hand the formalism should handle low-energy processes properly, which are those of interest for the present book. This point is discussed by Cohen-Tannoudji *et al.* (1989).

The equations of motion of the field are now given by the first of (1.23) alone or, equivalently, by the first of (1.27) with $V = 0$. The momentum components conjugate to A_i are, from (1.28)

$$\Pi_i = \frac{\partial \mathcal{L}_0}{\partial \dot{A}_i} = \frac{1}{4\pi c^2}\dot{A}_i = -\frac{1}{4\pi c}E_i \tag{1.31}$$

where (1.5) with $A_4 = 0$ has been used. The Hamiltonian density is obtained from (1.29) and

$$\begin{aligned} \mathcal{H}_0 &= \Pi_i \frac{\partial A_i}{\partial t} - \mathcal{L}_0 \\ &= \frac{1}{8\pi}\left\{\frac{1}{c^2}\dot{A}_i^2 + \left(\frac{\partial A_i}{\partial x_j} - \frac{\partial A_j}{\partial x_i}\right)^2\right\} = \frac{1}{8\pi}(\mathbf{E}^2 + \mathbf{H}^2) \end{aligned} \tag{1.32}$$

1.6 Energy-momentum tensor and conservation laws

Following Heitler's notation (1960), we define the energy-momentum tensor

$$T_{\mu\nu} = \frac{1}{4\pi}\left[F_{\mu\lambda}F_{\lambda\nu} + \frac{1}{4}\delta_{\mu\nu}F_{\lambda\rho}F_{\lambda\rho}\right] \tag{1.33}$$

From (1.2) it is easy to obtain the following expression for the elements of $T_{\mu\nu}$

$$\begin{aligned} &T_{ii} = \frac{1}{4\pi}(E_i^2 + H_i^2) - \mathcal{H}_0; \; T_{44} = \frac{1}{8\pi}(\mathbf{E}^2 + \mathbf{H}^2) \equiv \mathcal{H}_0; \\ &T_{ij} = T_{ji} = \frac{1}{4\pi}(E_iE_j + H_iH_j) \;\; (i \neq j); \\ &T_{i4} = T_{4i} = -\frac{i}{4\pi}(\mathbf{E} \times \mathbf{H})_i \end{aligned} \tag{1.34}$$

where $\mathcal{H}_0$ is the energy density. T_{44} coincides with the Hamiltonian density (1.32) which was obtained in the Coulomb gauge. So the Hamiltonian density, contrary to the Lagrangian density of the free field, is not a Lorentz scalar, but transforms like the time-time component of a tensor of second rank. We remark that $T_{\mu\nu}$ in (1.33) is gauge invariant, symmetric ($T_{\mu\nu} = T_{\nu\mu}$) and traceless ($T_{\mu\mu} = 0$). We also remark that the three components T_{i4} are related to the momentum density of the field, according to the relation (see e.g. Jackson 1988)

$$p_i = \frac{1}{4\pi c}(\mathbf{E} \times \mathbf{H})_i = \frac{i}{c} T_{i4} \tag{1.35}$$

In fact, integrating over a volume outside which the field is assumed to vanish, one can show that

$$P_\mu = \frac{i}{c} \int T_{\mu 4} d^3\mathbf{x} \tag{1.36}$$

is a four-vector which is the momentum of the field, analogous to the relativistic mechanical four-momentum of a particle. In general, however, if the field is not of finite extension the four integrals in (1.36) do not transform like the components of a four-vector, which emphasizes the limits of some of the analogies between the electromagnetic field and massive classical matter (Heitler 1960).

Further, we define the tensor of third rank (see e.g. Jackson 1988)

$$M_{\mu\nu\lambda} = T_{\mu\nu} x_\lambda - T_{\mu\lambda} x_\nu \tag{1.37}$$

which is called the angular momentum density tensor. This is antisymmetric with respect to exchange of λ and ν. It is easy to see from (1.34) that the components M_{4ij} are related to the angular momentum density of the field as

$$\ell_i = \frac{1}{4\pi c}[\mathbf{x} \times (\mathbf{E} \times \mathbf{H})]_i = -\frac{1}{2}\frac{i}{c} M_{4jk} e_{ijk} \tag{1.38}$$

where, as in Landau and Lifshitz (1975), e_{ijk} is the completely antisymmetric unit pseudotensor with null components except for the following

$$e_{123} = e_{312} = e_{231} = 1; \; e_{321} = e_{132} = e_{213} = -1 \tag{1.39}$$

Differentiating (1.33) term by term, one obtains

$$\frac{\partial T_{\mu\nu}}{\partial x_\mu} = \frac{1}{4\pi}\left(\frac{\partial F_{\mu\lambda}}{\partial x_\mu} F_{\lambda\nu} + F_{\mu\lambda}\frac{\partial F_{\lambda\nu}}{\partial x_\mu} + \frac{1}{2}\frac{\partial F_{\lambda\rho}}{\partial x_\nu} F_{\lambda\rho}\right) \tag{1.40}$$

Using Maxwell's equations (1.3), the first term within brackets in (1.40) vanishes, whereas the third can be written as

$$\begin{aligned}
\frac{1}{2}\frac{\partial F_{\lambda\rho}}{\partial x_\nu} F_{\lambda\rho} &= \frac{1}{2}\left(-\frac{\partial F_{\rho\nu}}{\partial x_\lambda} - \frac{\partial F_{\nu\lambda}}{\partial x_\rho}\right) F_{\lambda\rho} \\
&= \frac{1}{2}\left(\frac{\partial F_{\nu\rho}}{\partial x_\lambda} F_{\lambda\rho} - \frac{\partial F_{\nu\lambda}}{\partial x_\rho} F_{\lambda\rho}\right) = \frac{1}{2}\left(\frac{\partial F_{\nu\lambda}}{\partial x_\rho} F_{\rho\lambda} - \frac{\partial F_{\nu\lambda}}{\partial x_\rho} F_{\lambda\rho}\right) \\
&= -\frac{\partial F_{\nu\lambda}}{\partial x_\rho} F_{\lambda\rho} = -\frac{\partial F_{\nu\lambda}}{\partial x_\mu} F_{\lambda\mu} = -\frac{\partial F_{\lambda\nu}}{\partial x_\mu} F_{\mu\lambda}
\end{aligned}$$

where the symmetry of $F_{\mu\nu}$ has been used. The latter expression cancels with the second term in (1.40) and one obtains

$$\frac{\partial T_{\mu\nu}}{\partial x_\mu} = 0 \tag{1.41}$$

This is an important relation, which encompasses several conservation laws. Thus, taking $\nu = 4$, we obtain from (1.41) and (1.34) (see Jackson 1988)

$$\frac{\partial T_{\mu 4}}{\partial x_\mu} = 0 \quad \text{or} \quad -\frac{i}{4\pi}\nabla \cdot (\mathbf{E} \times \mathbf{H}) - \frac{i}{c}\frac{\partial \mathcal{H}_0}{\partial t} = 0 \tag{1.42}$$

which can be put in the form

$$\nabla \cdot \mathbf{S} + \frac{\partial \mathcal{H}_0}{\partial t} = 0\ ;\ \mathbf{S} = \frac{c}{4\pi}(\mathbf{E} \times \mathbf{H}) \tag{1.43}$$

where $\mathbf{S}$ is the Poynting vector, and which expresses conservation of energy. On the other hand, taking $\mu = j$, we have from (1.41)

$$\frac{\partial T_{\mu j}}{\partial x_\mu} = \frac{\partial T_{ij}}{\partial x_i} + \frac{\partial T_{4j}}{\partial x_4} = \frac{\partial T^M_{ij}}{\partial x_i} - \frac{1}{4\pi c}\frac{\partial}{\partial t}(\mathbf{E} \times \mathbf{H})_j$$

$$\text{or} \quad \frac{\partial T^M_{ij}}{\partial x_i} - \dot{p}_j = 0 \tag{1.44}$$

which expresses conservation of impulse, since T^M is the Maxwell stress tensor as defined by Heitler (1960) and p_j is the momentum density of the field defined in (1.35).

Conservation of angular momentum of the free field also follows from (1.41) and from the symmetry of $T_{\mu\nu}$. In fact, differentiating (1.37) yields (see Barut 1980)

$$\frac{M_{\mu\nu\lambda}}{\partial x_\mu} = \frac{\partial T_{\mu\nu}}{\partial x_\mu} x_\lambda - \frac{T_{\mu\lambda}}{\partial x_\mu} x_\nu + T_{\mu\nu}\delta_{\lambda\mu} - T_{\mu\lambda}\delta_{\nu\mu} = 0 \tag{1.45}$$

since the first two terms on the RHS of (1.45) vanish in view of (1.41) and the other two terms cancel because $T_{\lambda\nu} = T_{\nu\lambda}$. Thus (1.45) implies in particular

$$\frac{\partial M_{ijk}}{\partial x_i} + \frac{\partial M_{4jk}}{\partial x_4} = \frac{\partial}{\partial x_i}\left(T^M_{ij} x_k - T^M_{ik} x_j\right) + \frac{\partial M_{4jk}}{\partial x_4} = 0 \tag{1.46}$$

Multiplication of (1.46) by e_{njk} (and summation over dummy indices) yields (see Healy 1982)

$$\frac{\partial}{\partial x_i} e_{njk}\left(x_k T^M_{ij} - x_j T^M_{ik}\right) + \frac{1}{ic}\frac{\partial}{\partial t} e_{njk} M_{4jk} = 0 \tag{1.47}$$

Finally, from (1.38), one obtains

$$\frac{\partial}{\partial x_i}\left(e_{njk}x_k T_{ij}^{M}\right) + \dot{\ell}_n = 0 \tag{1.48}$$

which expresses conservation of angular momentum for the free field.

The energy-momentum tensor can be defined for any field considered in this book and even for a point particle (see e.g. Doughty 1990). Indeed, it can be derived from the Lagrangian density by exploiting the symmetry properties of the field and Noether's theorem as pointed out by Bogoliubov and Shirkov (1960). Since such a derivation is beyond the scope of this book, for each field we shall simply give the appropriate definition of $T_{\mu\nu}$ and we shall discuss the related conservation laws, using the fact that T_{44} is directly related to the energy density of the field.

1.7 Angular momentum of the free field

Integrating the angular momentum density ℓ_i over a volume, one obtains the well-known expression for the angular momentum of the free field

$$\mathbf{L} = \frac{1}{4\pi c}\int \mathbf{x}\times(\mathbf{E}\times\mathbf{H})d^3\mathbf{x} \tag{1.49}$$

Using (1.5) and a theorem of elementary vector analysis (see e.g. Spiegel 1959), one finds

$$\begin{aligned}\mathbf{E}\times\mathbf{H} &= \mathbf{E}\times(\nabla\times\mathbf{A})\\ &= \nabla(\mathbf{E}\cdot\mathbf{A}) - (\mathbf{A}\cdot\nabla)\mathbf{E} - \mathbf{A}\times(\nabla\times\mathbf{E}) - (\mathbf{E}\cdot\nabla)\mathbf{A}\\ &= E_i\nabla A_i - (\mathbf{E}\cdot\nabla)\mathbf{A}\end{aligned} \tag{1.50}$$

Substitution of (1.50) in (1.49) yields

$$\mathbf{L} = \frac{1}{4\pi c}\int E_i(\mathbf{x}\times\nabla)A_i d^3\mathbf{x} - \frac{1}{4\pi c}\int \mathbf{x}\times(\mathbf{E}\cdot\nabla)\mathbf{A}d^3\mathbf{x} \tag{1.51}$$

Moreover, using as in Cohen-Tannoudji *et al.* (1989)

$$\begin{aligned}(\mathbf{E}\cdot\nabla)(\mathbf{x}\times\mathbf{A}) &= \mathbf{x}\times(\mathbf{E}\cdot\nabla)\mathbf{A} + (\mathbf{E}\cdot\nabla)\mathbf{x}\times\mathbf{A}\\ &= \mathbf{x}\times(\mathbf{E}\cdot\nabla)\mathbf{A} + \mathbf{E}\times\mathbf{A}\end{aligned}$$

the second integral in (1.51) becomes

$$-\frac{1}{4\pi c}\int \mathbf{x}\times(\mathbf{E}\cdot\nabla)\mathbf{A}d^3\mathbf{x} = \frac{1}{4\pi c}\int \mathbf{E}\times\mathbf{A}d^3\mathbf{x} - \frac{1}{4\pi c}\int(\mathbf{E}\cdot\nabla)(\mathbf{x}\times\mathbf{A})d^3\mathbf{x} \tag{1.52}$$

If the field vanishes outside the integration volume (Haus and Pan 1993), the second integration on the RHS of (1.52) is zero, and from (1.51) one has

$$\mathbf{L} = \frac{1}{4\pi c}\int E_i(\mathbf{x}\times\nabla)A_i d^3\mathbf{x} + \frac{1}{4\pi c}\int \mathbf{E}\times\mathbf{A}d^3\mathbf{x} \tag{1.53}$$

This may be compared with the analogous expression for the angular momentum of a multicomponent scalar field ψ_i

$$\mathbf{L} = -\int \Pi_i(\mathbf{x}\times\nabla)\psi_i d^3x \tag{1.54}$$

where Π_i is the canonical momentum of the field ψ_i (Henley and Thirring 1962). In the electromagnetic case one might assume by analogy that $\psi_i = A_i$, $\Pi_i = -E_i/4\pi c$ on the basis of (1.31), which accounts for the first of the two contributions to **L** in (1.53). The second contribution in (1.53) is due to the vector character of the e.m. field, which shows that a vector field is intrinsically different from a multicomponent scalar field. This second contribution is sometimes called the "spin" contribution (see e.g. Jackson 1988) whereas the first is the "orbital" contribution. It is appropriate to emphasize, however, that no new degrees of freedom are involved in the definition of the spin contribution, which is completely expressed in terms of the usual dynamical variables of the free field.

1.8 Transverse and longitudinal vector fields

The following discussion is based mainly on the treatment given by Craig and Thirunamachandran (1984). Any three-vector field $\mathbf{f}(\mathbf{x})$ can be decomposed as a sum of an irrotational vector field $\mathbf{f}_\parallel(\mathbf{x})$, called the longitudinal component, and a solenoidal vector field $\mathbf{f}_\perp(\mathbf{x})$, called the transverse component, such that

$$\mathbf{f}(\mathbf{x}) = \mathbf{f}_\parallel(\mathbf{x}) + \mathbf{f}_\perp(\mathbf{x});\ \nabla\times\mathbf{f}_\parallel(\mathbf{x}) = 0;\ \nabla\cdot\mathbf{f}_\perp(\mathbf{x}) = 0 \tag{1.55}$$

at each point of space. The Fourier transform $\mathbf{g}(\mathbf{k})$ of the total field with respect to $\mathbf{x}$ can be introduced as

$$\mathbf{f}(\mathbf{x}) = \int \mathbf{g}(\mathbf{k})e^{i\mathbf{k}\cdot\mathbf{x}}d^3\mathbf{k};\ \mathbf{g}(\mathbf{k}) = \frac{1}{(2\pi)^3}\int \mathbf{f}(\mathbf{x})e^{-i\mathbf{k}\cdot\mathbf{x}}d^3\mathbf{x} \tag{1.56}$$

Expressions of the same form are valid for $\mathbf{f}_\parallel(\mathbf{x})$ and $\mathbf{f}_\perp(\mathbf{x})$, defining $\mathbf{g}(\mathbf{k}) = \mathbf{g}_\parallel(\mathbf{k}) + \mathbf{g}_\perp(\mathbf{k})$, in view of the linearity of the Fourier transform. In particular the transverse component gives

$$\nabla\cdot\mathbf{f}_\perp(\mathbf{x}) = i\int \mathbf{k}\cdot\mathbf{g}_\perp(\mathbf{k})e^{i\mathbf{k}\cdot\mathbf{x}}d^3\mathbf{k} = 0 \tag{1.57}$$

Since (1.57) is valid for any **x**, we have for all **k**

$$\mathbf{k}\cdot\mathbf{g}_{\perp}(\mathbf{k})=0 \tag{1.58}$$

which means that $\mathbf{g}_{\perp}(\mathbf{k})$ is perpendicular to **k** everywhere in **k** space. This feature of the $\mathbf{f}_{\perp}(\mathbf{x})$ field, that is of having its Fourier transform transverse to **k**, is called transversality. On the other hand

$$\nabla\times\mathbf{f}_{\parallel}(\mathbf{x})=i\int\mathbf{k}\times\mathbf{g}_{\parallel}(\mathbf{k})e^{i\mathbf{k}\cdot\mathbf{x}}d^3\mathbf{k}=0 \tag{1.59}$$

so that for all **k**

$$\mathbf{k}\times\mathbf{g}_{\parallel}(\mathbf{k})=0 \tag{1.60}$$

This implies that $\mathbf{g}_{\parallel}(\mathbf{k})$ is parallel to **k** everywhere in **k** space. The latter condition can be expressed as $\mathbf{g}_{\parallel}(\mathbf{k})=(\mathbf{g}\cdot\hat{\mathbf{k}})\hat{\mathbf{k}}$, where $\hat{\mathbf{k}}$ is the unit vector of **k**. Hence, in terms of components,

$$g_{\parallel i}(\mathbf{k})=\hat{k}_i\hat{k}_j g_j(\mathbf{k});\ g_{\perp i}(\mathbf{k})=\left(\delta_{ij}-\hat{k}_i\hat{k}_j\right)g_j(\mathbf{k}) \tag{1.61}$$

since $\mathbf{g}_{\parallel}(\mathbf{k})+\mathbf{g}_{\perp}(\mathbf{k})=\mathbf{g}(\mathbf{k})$. Taking the Fourier transform of both Equations (1.61) yields

$$\begin{aligned}
f_{\parallel i}(\mathbf{x})&=\int\hat{k}_i\hat{k}_j g_j(\mathbf{k})e^{i\mathbf{k}\cdot\mathbf{x}}d^3\mathbf{k}\\
&=\frac{1}{(2\pi)^3}\int\int\hat{k}_i\hat{k}_j f_j(\mathbf{x}')e^{i\mathbf{k}\cdot(\mathbf{x}-\mathbf{x}')}d^3\mathbf{k}d^3\mathbf{x}'\\
&\equiv\int f_j(\mathbf{x}')\delta_{\parallel ij}(\mathbf{x}-\mathbf{x}')d^3\mathbf{x}';\\
f_{\perp i}(\mathbf{x})&=\int\left(\delta_{ij}-\hat{k}_i\hat{k}_j\right)g_j(\mathbf{k})e^{i\mathbf{k}\cdot\mathbf{x}}d^3\mathbf{k}\\
&=\frac{1}{(2\pi)^3}\int\int\left(\delta_{ij}-\hat{k}_i\hat{k}_j\right)f_j(\mathbf{x}')e^{i\mathbf{k}\cdot(\mathbf{x}-\mathbf{x}')}d^3\mathbf{k}d^3\mathbf{x}'\\
&\equiv\int f_j(\mathbf{x}')\delta_{\perp ij}(\mathbf{x}-\mathbf{x}')d^3\mathbf{x}'
\end{aligned} \tag{1.62}$$

where the longitudinal and transverse δ-tensors have been introduced as

$$\begin{aligned}
\delta_{\parallel ij}(\mathbf{x})&=\frac{1}{(2\pi)^3}\int\hat{k}_i\hat{k}_j e^{i\mathbf{k}\cdot\mathbf{x}}d^3\mathbf{k}\,;\\
\delta_{\perp ij}(\mathbf{x})&=\frac{1}{(2\pi)^3}\int\left(\delta_{ij}-\hat{k}_i\hat{k}_j\right)e^{i\mathbf{k}\cdot\mathbf{x}}d^3\mathbf{k}
\end{aligned} \tag{1.63}$$

From (1.63) one has

$$\delta_{\parallel ij}(\mathbf{x}) + \delta_{\perp ij}(\mathbf{x}) = \delta_{ij}\frac{1}{(2\pi)^3}\int e^{i\mathbf{k}\cdot\mathbf{x}} d^3\mathbf{k} = \delta_{ij}\delta(\mathbf{x}) \tag{1.64}$$

The integrals in (1.63) can be calculated explicitly starting from the Fourier representation of the Green's function of the Poisson equation (see e.g. Barton 1989)

$$\frac{1}{4\pi x} = \frac{1}{(2\pi)^3}\int \frac{1}{k^2} e^{i\mathbf{k}\cdot\mathbf{x}} d^3\mathbf{k}\ ;$$
$$\nabla_i\nabla_j\left(\frac{1}{4\pi x}\right) = -\frac{1}{(2\pi)^3}\int \hat{k}_i\hat{k}_j e^{i\mathbf{k}\cdot\mathbf{x}} d^3\mathbf{k} \tag{1.65}$$

and we find (see also Cohen-Tannoudji *et al.* 1989)

$$\delta_{\parallel ij}(\mathbf{x}) = -\nabla_i\nabla_j\left(\frac{1}{4\pi x}\right) = \frac{1}{3}\delta_{ij}\delta(\mathbf{x}) + \frac{1}{4\pi x^3}\left(\delta_{ij} - 3\hat{x}_i\hat{x}_j\right);$$
$$\delta_{\perp ij}(\mathbf{x}) = \delta_{ij}\delta(\mathbf{x}) + \nabla_i\nabla_j\left(\frac{1}{4\pi x}\right)$$
$$= \frac{2}{3}\delta_{ij}\delta(\mathbf{x}) - \frac{1}{4\pi x^3}\left(\delta_{ij} - 3\hat{x}_i\hat{x}_j\right) \tag{1.66}$$

1.9 Vector properties of the free field

It is immediate that **E** and **H** for a free field are purely transverse on the basis of Maxwell's equations and of (1.55). The same is not true for the vector potential **A** except in the Coulomb gauge or in the Lorentz gauge with $V = 0$. In the Coulomb gauge we have

$$\nabla\cdot\mathbf{A}^{\text{coul}} = 0\ ;\ \mathbf{A}^{\text{coul}} = \mathbf{A}_{\perp} \tag{1.67}$$

In general a gauge transformation leading from $\mathbf{A}^{\text{old}}$ to $\mathbf{A}^{\text{new}}$

$$A_i^{\text{new}} = A_i^{\text{old}} + \nabla_i\chi;\ V^{\text{new}} = V^{\text{old}} - \frac{1}{c}\dot{\chi} \tag{1.68}$$

adds to $\mathbf{A}^{\text{old}}$ a field $\nabla\chi$ which is pure longitudinal since

$$\nabla\times\nabla\chi = 0 \tag{1.69}$$

Thus two **A** fields related by a gauge transformation have the same $\mathbf{A}_{\perp}$ at any point in space (see Figure 1.1). Moreover, this $\mathbf{A}_{\perp}(\mathbf{x}, t)$, which is gauge invariant, must coincide with $\mathbf{A}^{\text{coul}}(\mathbf{x}, t)$. In conclusion, in any

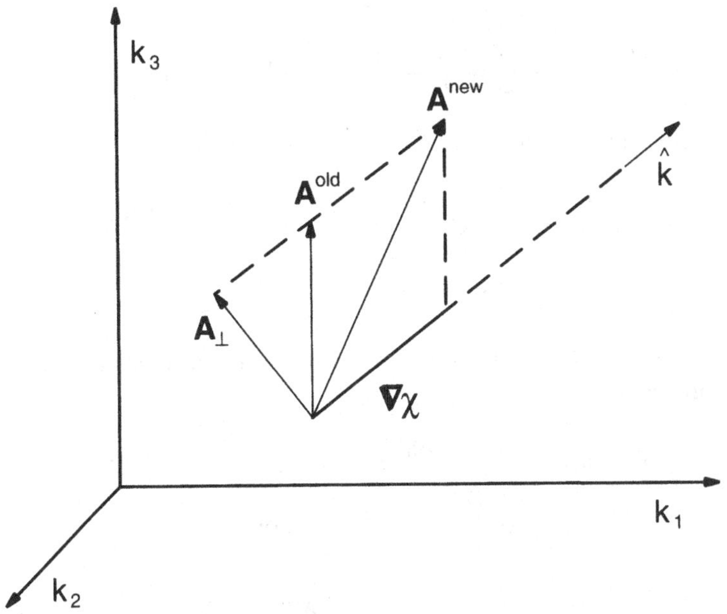

Fig. 1.1 The effect of a gauge transformation on the vector potential $\mathbf{A}^{\text{old}}$ is to add $\nabla\chi$, which gives a new vector potential $\mathbf{A}^{\text{new}}$. In k-space $\nabla\chi$ is along $\mathbf{k}$ and orthogonal to $\mathbf{A}_{\perp}$, which is unchanged by the transformation.

gauge the component $\mathbf{A}_{\perp}$ coincides with $\mathbf{A}^{\text{coul}}$. Consider a gauge transformation leading from any arbitrary $\mathbf{A}^{\text{old}}$ to $\mathbf{A}^{\text{coul}}$. On the basis of (1.68) and (1.67) one has

$$\mathbf{A}^{\text{coul}} = \mathbf{A}^{\text{old}} + \nabla\chi;\ \nabla\cdot\mathbf{A}^{\text{coul}} = \nabla\cdot\mathbf{A}^{\text{old}} + \nabla^2\chi = 0;$$
$$\nabla^2\chi = -\nabla\cdot\mathbf{A}^{\text{old}} \tag{1.70}$$

The last equation in (1.70) can be regarded as a Poisson equation for the field χ, which is the scalar field connecting $\mathbf{A}^{\text{old}}$ to $\mathbf{A}^{\text{coul}}$. If we assume that all fields involved vanish at infinity, the solution of such a Poisson equation is

$$\chi(\mathbf{x},t) = \frac{1}{4\pi}\int \frac{1}{|\mathbf{x}-\mathbf{x}'|}\nabla'\cdot\mathbf{A}^{\text{old}}(\mathbf{x}',t)d^3\mathbf{x}' \tag{1.71}$$

where ∇' is the grad with respect to $\mathbf{x}'$. On the basis of the first equation of (1.70), the ∇ of expression (1.71) is the vector field which should be added to $\mathbf{A}^{\mathrm{old}}$ in order to obtain $\mathbf{A}^{\mathrm{coul}} \equiv \mathbf{A}_{\perp}^{\mathrm{old}}$. Thus

$$\mathbf{A}_{\parallel}^{\mathrm{old}}(\mathbf{x}, t) = -\frac{1}{4\pi}\nabla \int \frac{1}{|\mathbf{x}-\mathbf{x}'|}\nabla'\cdot\mathbf{A}^{\mathrm{old}}(\mathbf{x}', t)d^3\mathbf{x}' \tag{1.72}$$

On the other hand a slight modification of (1.65) yields

$$\nabla_i\nabla'_j\left(\frac{1}{4\pi\,|\mathbf{x}-\mathbf{x}'|}\right) = \frac{1}{(2\pi)^3}\int \hat{k}_i\hat{k}_j e^{i\mathbf{k}\cdot(\mathbf{x}-\mathbf{x}')}d^3\mathbf{k}\,;$$

$$\delta_{\parallel ij}(\mathbf{x}-\mathbf{x}') = \nabla_i\nabla'_j\left(\frac{1}{4\pi\,|\mathbf{x}-\mathbf{x}'|}\right) \tag{1.73}$$

Substituting in (1.62) and using elementary vector analysis (e.g. Spiegel 1959) one gets

$$\begin{aligned} A_{\parallel i}^{\mathrm{old}}(\mathbf{x}, t) &= \nabla_i \int A_j^{\mathrm{old}}(\mathbf{x}', t)\nabla'_j\left(\frac{1}{4\pi\,|\mathbf{x}-\mathbf{x}'|}\right)d^3\mathbf{x}' \\ &= \nabla_i\left\{\int_{S'} \frac{1}{4\pi\,|\mathbf{x}-\mathbf{x}'|}\mathbf{A}^{\mathrm{old}}(\mathbf{x}', t)\cdot\mathbf{n}dS' \right. \\ &\quad \left. - \int \frac{1}{4\pi\,|\mathbf{x}-\mathbf{x}'|}\nabla'\cdot\mathbf{A}^{\mathrm{old}}(\mathbf{x}', t)d^3\mathbf{x}'\right\} \end{aligned} \tag{1.74}$$

In (1.74) the surface integral is taken at infinity where it has been assumed to vanish. Thus (1.74) coincides with (1.72), and it is an explicit expression for the longitudinal component of **A** in an arbitrary gauge.

1.10 Solutions of Maxwell's equations in the Coulomb gauge

We now restrict our discussion to the Coulomb gauge, which is characterized by the constraint (1.26) on the vector potential. As we have seen, in the absence of charges and currents, this condition implies $V = 0$ and pure transversality of **A**. For simplicity we shall drop the transversality index in what follows, except when appropriate in order to avoid possible ambiguities. In terms of **A**, Maxwell's equations take the form (1.27), or

$$\nabla^2\mathbf{A} - \frac{1}{c^2}\ddot{\mathbf{A}} = 0 \tag{1.75}$$

According to (1.56), $\mathbf{A}(\mathbf{x}, t)$ can be Fourier-analysed as

$$\mathbf{A}(\mathbf{x}, t) = \int \mathbf{A}(\mathbf{k}, t)e^{i\mathbf{k}\cdot\mathbf{x}}d^3\mathbf{k}\,;\ \mathbf{A}(\mathbf{k}, t) = \frac{1}{(2\pi)^3}\int \mathbf{A}(\mathbf{x}, t)e^{-i\mathbf{k}\cdot\mathbf{x}}d^3\mathbf{x} \tag{1.76}$$

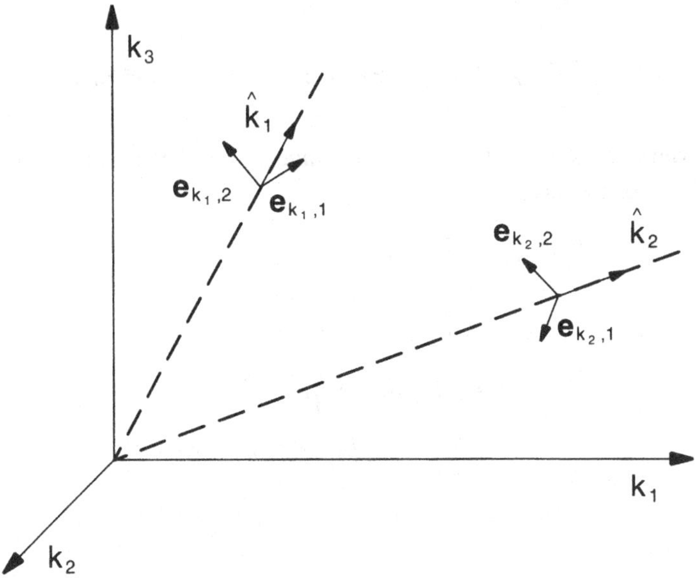

Fig. 1.2 Unit vectors $\mathbf{e}_{\mathbf{k}j}$, $\hat{\mathbf{k}}$ for two different directions of $\mathbf{k}$. $\mathbf{e}_{\mathbf{k}j}$ and $\hat{\mathbf{k}}$ are mutually perpendicular.

Substitution of (1.76) into (1.75) yields an equation for the field amplitude $\mathbf{A}(\mathbf{k}, t)$ of the Helmholtz form

$$\ddot{\mathbf{A}}(\mathbf{k}, t) + \omega_k^2 \mathbf{A}(\mathbf{k}, t) = 0 \; ; \; \omega_k = ck \tag{1.77}$$

which can be immediately solved as

$$\mathbf{A}(\mathbf{k}, t) = \mathbf{A}^{(+)}(\mathbf{k}) e^{-i\omega_k t} + \mathbf{A}^{(-)}(\mathbf{k}) e^{i\omega_k t} \tag{1.78}$$

We remark that due to the transversality condition in $\mathbf{A}(\mathbf{x})$, for any $\mathbf{k}$ one has $\mathbf{k} \cdot \mathbf{A}(\mathbf{k}, t) = 0$, which implies that each vector $\mathbf{A}^{\pm}(\mathbf{k})$ is perpendicular to $\mathbf{k}$. Thus for each $\mathbf{k}$ one can expand $\mathbf{A}^{\pm}(\mathbf{k})$ in terms of two unit vectors $\mathbf{e}_{\mathbf{k}j} (j = 1, 2)$ both perpendicular to $\mathbf{k}$ and to each other as in Figure 1.2 for real vectors. Thus one has

$$\mathbf{A}(\mathbf{k}, t) = \sum_j \left\{ A_j^{(+)}(\mathbf{k}) \mathbf{e}_{\mathbf{k}j} e^{-i\omega_k t} + A_j^{(-)}(\mathbf{k}) \mathbf{e}_{\mathbf{k}j} e^{i\omega_k t} \right\} \tag{1.79}$$

Vectors $\mathbf{e}_{\mathbf{k}j}$ are called polarization vectors, and they may be complex. In this case their orthogonality properties are expressed as

$$\mathbf{e}_{\mathbf{k}j} \cdot \mathbf{e}^*_{\mathbf{k}j'} = \delta_{jj'} \; ; \; \mathbf{e}_{\mathbf{k}j} \cdot \hat{\mathbf{k}} = 0 \tag{1.80}$$

Substitution of (1.79) in (1.76) yields the general solution of (1.75) in the form

$$\mathbf{A}(\mathbf{x}, t) = \sum_j \int \left\{ A_j^{(+)}(\mathbf{k}) \mathbf{e}_{\mathbf{k}j} e^{-i\omega_k t} + A_j^{(-)}(\mathbf{k}) \mathbf{e}_{\mathbf{k}j} e^{i\omega_k t} \right\} e^{i\mathbf{k}\cdot\mathbf{x}} d^3\mathbf{k} \tag{1.81}$$

Now from (1.5), since in the Coulomb gauge and in the absence of charges $A_4 = iV = 0$, we have

$$\begin{aligned}
\mathbf{E}(\mathbf{x}, t) &= -\frac{1}{c}\dot{\mathbf{A}}(\mathbf{x}, t) \\
&= \frac{i}{c}\sum_j \int \omega_k \Big\{ \mathbf{e}_{\mathbf{k}j} A_j^{(+)}(\mathbf{k}) e^{-i\omega_k t} \\
&\quad - \mathbf{e}_{\mathbf{k}j} A_j^{(-)}(\mathbf{k}) e^{i\omega_k t} \Big\} e^{i\mathbf{k}\cdot\mathbf{x}} d^3\mathbf{k} \\
&= \frac{i}{c}\sum_j \int \omega_k \Big\{ \mathbf{e}_{\mathbf{k}j} A_j^{(+)}(\mathbf{k}) e^{-i\omega_k t} e^{i\mathbf{k}\cdot\mathbf{x}} \\
&\quad - \mathbf{e}_{-\mathbf{k}j} A_j^{(-)}(-\mathbf{k}) e^{i\omega_k t} e^{-i\mathbf{k}\cdot\mathbf{x}} \Big\} d^3\mathbf{k}
\end{aligned} \tag{1.82}$$

Requiring that $\mathbf{E}$ should be a real field leads to

$$\mathbf{e}_{-\mathbf{k}j} A_j^{(-)}(-\mathbf{k}) = \mathbf{e}_{\mathbf{k}j}^{*} A_j^{(+)*}(\mathbf{k}) \tag{1.83}$$

which enables us to eliminate the $A^{(-)}$ coefficients in favour of $A^{(+)}$. Thus (1.81) and (1.82) can be put in the form

$$\mathbf{A}(\mathbf{x}, t) = \sum_j \int \left\{ \mathbf{e}_{\mathbf{k}j} A_j(\mathbf{k}) e^{-i\omega_k t} e^{i\mathbf{k}\cdot\mathbf{x}} + \mathbf{e}_{\mathbf{k}j}^{*} A_j^{*}(\mathbf{k}) e^{i\omega_k t} e^{-i\mathbf{k}\cdot\mathbf{x}} \right\} d^3\mathbf{k} \tag{1.84}$$

$$\begin{aligned}
\mathbf{E}(\mathbf{x}, t) &= -\frac{1}{c}\dot{\mathbf{A}}(\mathbf{x}, t) \\
&= i\sum_j \int k \Big\{ \mathbf{e}_{\mathbf{k}j} A_j(\mathbf{k}) e^{-i\omega_k t} e^{i\mathbf{k}\cdot\mathbf{x}} \\
&\quad - \mathbf{e}_{\mathbf{k}j}^{*} A_j^{*}(\mathbf{k}) e^{i\omega_k t} e^{-i\mathbf{k}\cdot\mathbf{x}} \Big\} d^3\mathbf{k}
\end{aligned} \tag{1.85}$$

where the now superfluous superscript (+) has been dropped. The solution for $\mathbf{H}$ can be obtained from the first of (1.5) as

$$\begin{aligned}
\mathbf{H}(\mathbf{x}, t) &= \nabla \times \mathbf{A}(\mathbf{x}, t) \\
&= i\sum_j \int k \Big\{ (\hat{\mathbf{k}} \times \mathbf{e}_{\mathbf{k}j}) A_j(\mathbf{k}) e^{-i\omega_k t} e^{i\mathbf{k}\cdot\mathbf{x}} \\
&\quad - (\hat{\mathbf{k}} \times \mathbf{e}_{\mathbf{k}j}^{*}) A_j^{*}(\mathbf{k}) e^{i\omega_k t} e^{-i\mathbf{k}\cdot\mathbf{x}} \Big\} d^3\mathbf{k}
\end{aligned} \tag{1.86}$$

Expressions (1.84, 1.85, 1.86) are solutions of Maxwell's equations expressed in terms of an expansion in plane waves. The analogous expansion in terms of spherical waves is also very useful. It is called the multipolar expansion and it is obtained in Appendix A for the vector potential starting from (1.84). It yields

$$\begin{aligned}\mathbf{A}(\mathbf{x},t) = 4\pi\sum_{\ell m}\int_0^\infty \Big\{ & A(k,\mathcal{M},\ell,m)j_\ell(kx)\mathbf{Y}_{\ell m0}(\vartheta,\varphi) \\ & + A(k,\mathcal{E},\ell,m)\frac{1}{\sqrt{2\ell+1}}\Big[\sqrt{\ell}j_{\ell+1}(kx)\mathbf{Y}_{\ell m+}(\vartheta,\varphi) \\ & - \sqrt{\ell+1}j_{\ell-1}(kx)\mathbf{Y}_{\ell m-}(\vartheta,\varphi)\Big]\Big\}e^{-i\omega_k t}kdk + \text{c.c.}\end{aligned} \tag{1.87}$$

$$\begin{aligned}\mathbf{E}(\mathbf{x},t) = 4\pi i\sum_{\ell m}\int_0^\infty \Big\{ & A(k,\mathcal{M},\ell,m)j_\ell(kx)\mathbf{Y}_{\ell m0}(\vartheta,\varphi) \\ & + A(k,\mathcal{E},\ell,m)\frac{1}{\sqrt{2\ell+1}}\Big[\sqrt{\ell}j_{\ell+1}(kx)\mathbf{Y}_{\ell m+}(\vartheta,\varphi) \\ & - \sqrt{\ell+1}j_{\ell-1}(kx)\mathbf{Y}_{\ell m-}(\vartheta,\varphi)\Big]\Big\}e^{-i\omega_k t}k^2dk + \text{c.c.}\end{aligned} \tag{1.88}$$

$$\begin{aligned}\mathbf{H}(\mathbf{x},t) = -4\pi i\sum_{\ell m}\int_0^\infty \Big\{ & A(k,\mathcal{E},\ell,m)j_\ell(kx)\mathbf{Y}_{\ell m0}(\vartheta,\varphi) \\ & - A(k,\mathcal{M},\ell,m)\frac{1}{\sqrt{2\ell+1}}\Big[\sqrt{\ell}j_{\ell+1}(kx)\mathbf{Y}_{\ell m+}(\vartheta,\varphi) \\ & - \sqrt{\ell+1}j_{\ell-1}(kx)\mathbf{Y}_{\ell m-}(\vartheta,\varphi)\Big]\Big\}e^{-i\omega_k t}k^2dk + \text{c.c.}\end{aligned} \tag{1.89}$$

In (1.87, 1.88, 1.89), $j_\ell(kx)$ are spherical Bessel functions as defined in Abramovitz and Stegun (1965), $\mathbf{Y}_{\ell mi}$ $(i = \pm, 0)$ are vector spherical harmonics defined in Appendix A, and the sums run over all ℓ from 1 to ∞ since the terms with $\ell = 0$ can be shown to vanish, and, for each ℓ, over integer m from $-\ell$ to $+\ell$. The multipolar field amplitudes $A(k,\lambda,\ell,m)$ $(\lambda = \mathcal{E},\mathcal{M})$ can be expressed in terms of the plane wave field amplitude $A_j(\mathbf{k})$ as in Appendix A. The relative weight of the electric multipole wave amplitudes $A(k,\mathcal{E},\ell,m)$ and of the magnetic multipole wave amplitude $A(k,\mathcal{M},\ell,m)$ determine the physical characteristics of the e.m. field solution of Maxwell's equations. For example, for $A(k,\mathcal{M},\ell,m) = 0$, expressions (1.87, 1.88, 189) can be shown to give rise to an electric field which dominates the magnetic one near the origin $x = 0$. For this reason they are called electric multipole waves. On the contrary, the magnetic field dominates the electric field near the origin for magnetic multipole waves with $A(k,\mathcal{E},\ell,m) = 0$, as discussed by Cohen-Tannoudji *et al.* (1989).

1.11 Boundary conditions

In the solutions of Maxwell's equations discussed in Section 1.10 the field amplitudes $A_j(\mathbf{k})$ or $A(k, \lambda, \ell, m)$ depend on different variables, some of which are discrete (j, λ, ℓ, m) and others which are continuous (k_i, k). The presence of continuous variables tends to introduce algebraic complications in the quantization procedure which in many cases are unnecessary. One can in fact introduce appropriate constraints on the system which turn a continuous range into a discrete, albeit infinite, one. These constraints are called boundary conditions. Under some circumstances, boundary conditions are imposed on the basis of physical considerations. For example, on the surface of a perfectly conducting wall the component of the electric field $\mathbf{E}(\mathbf{x}, t)$ parallel to the wall must vanish together with the component of the magnetic field $\mathbf{H}(\mathbf{x}, t)$ perpendicular to the wall: fields which do not satisfy these conditions would induce currents inside the wall which would soon dissipate the field energy. Imposing these constraints on the solution of Maxwell's equations eliminates those solutions which do not conform to the same constraints and limits the number of acceptable solutions. Under other circumstances it may be convenient to limit the solutions by imposing more abstract conditions, such as periodicity in normal three-dimensional space of the acceptable solutions, which does not constrain the field inside a limited region of space, but which again selects a certain class of solutions.

The shape of the surface on which boundary conditions are to be imposed is naturally strictly related to the physical nature of the problem at hand. In turn, this shape determines the kind of space coordinates which is more appropriate for the description of the field in mathematical terms. For example, if the boundary is made of plane surfaces, then one would expect rectangular coordinates to be more apt to render the problem mathematically manageable, whereas if the boundary is a sphere, spherical coordinates should be used. In this section we shall limit our considerations to these two coordinate systems although a limited number of other coordinate systems exists, appropriate to their related shapes of boundary surfaces, in which the problem of solving Maxwell's equations can be handled (Morse and Feshbach 1953).

Consider, for example, a field between two infinite plane surfaces parallel to the (x_2, x_3) plane as in Figure 1.3. Then the geometry of the boundaries is determined by the two x_1 values of the two planes. Take the two planes to be situated at $x_1 = 0$ and at $x_1 = L$ respectively. In a case like this the obvious choice is rectangular coordinates. Expansion (1.86)

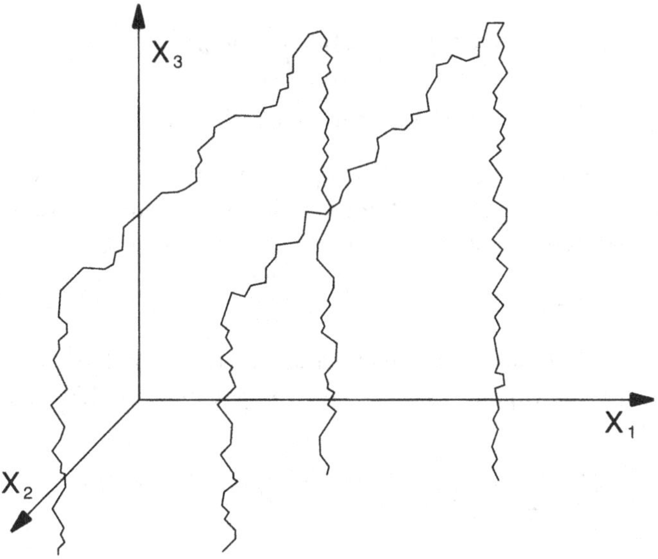

Fig. 1.3 Two infinite parallel planes perpendicular to $\hat{x}_1$. If boundary conditions are set on these planes, the natural choice to treat a problem is rectangular coordinates.

for the magnetic field can be used to obtain its component normal to the two surfaces in the form

$$\begin{aligned}
\hat{x}_1 \cdot \mathbf{H}(\mathbf{x}, t) &= i \sum_j \int k(\mathbf{b}_{\mathbf{k}j})_1 A_j(k_1\hat{x}_1 + k_2\hat{x}_2 + k_3\hat{x}_3) \\
&\quad \times e^{i(k_1x_1+k_2x_2+k_3x_3)} e^{-i\omega_k t} d^3\mathbf{k} + \text{c.c.} \\
&= i \sum_j \int_{-\infty}^{\infty} dk_2 e^{ik_2x_2} \int_{-\infty}^{\infty} dk_3 e^{ik_3x_3} \int_0^{\infty} dk_1 k\{(\mathbf{b}_{\mathbf{k}j})_1 \\
&\quad \times A_j(k_1\hat{x}_1 + k_2\hat{x}_2 + k_3\hat{x}_3) e^{ik_1x_1} - (\mathbf{b}'_{\mathbf{k}_j})_1 \\
&\quad \times A_j(-k_1\hat{x}_1 + k_2\hat{x}_2 + k_3\hat{x}_3) e^{-ik_1x_1}\} e^{-i\omega_k t} + \text{c.c.}
\end{aligned} \tag{1.90}$$

where $\mathbf{b}_{\mathbf{k}j} = \hat{\mathbf{k}} \times \mathbf{e}_{\mathbf{k}j}$ and $\mathbf{b}'_{\mathbf{k}j} = \hat{\mathbf{k}} \times \mathbf{e}'_{\mathbf{k}j}$ with $\mathbf{k}' = -k_1\hat{x}_1 + k_2\hat{x}_2 + k_3\hat{x}_3$. Adopting the convention $(\mathbf{b}_{\mathbf{k}j})_1 = (\mathbf{b}_{\mathbf{k}'j})_1$ and constraining the amplitudes in such a way that

$$A_j(k_1\hat{x}_1 + k_2\hat{x}_2 + k_3\hat{x}_3) = A_j(-k_1\hat{x}_1 + k_2\hat{x}_2 + k_3\hat{x}_3) \tag{1.91}$$

yields

$$\hat{x}_1 \cdot \mathbf{H}(\mathbf{x}, t) = -2 \sum_j \int_{-\infty}^{\infty} dk_2 e^{ik_2 x_2} \int_{-\infty}^{\infty} dk_3 e^{ik_3 x_3} \int_0^{\infty} dk_1$$
$$\times k(\mathbf{b_{k}}_j)_1 A_j(\mathbf{k}) \sin k_1 x_1 e^{-i\omega_k t} + \text{c.c.} \tag{1.92}$$

In view of the arbitrariness of $A_j(k)$ appearing in the k_1 integration, (1.92) vanishes at $x_1 = 0$ and at $x_1 = L$ for any time only if the integration over k_1 can be reduced to a sum over particular values such that

$$\sin k_1 L = 0 \text{ or } k_1 = k_1^{(n)} \equiv \frac{n\pi}{L} \quad (n = 1, 2, ...) \tag{1.93}$$

All waves with $k_1 \neq k_1^{(n)}$ are eliminated by the boundary conditions. Consequently the expansion for the vector potential in (1.84) takes the form

$$\mathbf{A}(\mathbf{x}, t) = \sum_j \sum_{n=-\infty}^{\infty} \int \{\mathbf{e_{k}}_j A_j(\mathbf{k}) e^{-i\omega_k t} e^{i\mathbf{k}\cdot\mathbf{x}}\} \delta(k_1 - k_1^{(n)}) d^3\mathbf{k} + \text{c.c.} \tag{1.94}$$

together with constraint (1.91). The fields **E** and **H** are easily obtained from (1.94). We see that the waves do not propagate in the $\hat{x}_1$ direction, where they become standing waves, but no constraint is posed by the two surfaces along the $\hat{x}_2$ and $\hat{x}_3$ directions, where the waves keep their travelling character.

Another set of important boundary conditions is obtained by imposing that the vector potential $\mathbf{A}(\mathbf{x})$ takes the same values on opposite faces of a parallelepiped. If the parallelepiped is a cube of side L this constraint is easily seen to transform (1.84) into

$$\mathbf{A}(\mathbf{x}, t) = \sum_j \sum_{\mathbf{k}} \{\mathbf{e_{k}}_j A_{\mathbf{k}j} e^{-i\omega_k t} e^{i\mathbf{k}\cdot\mathbf{x}} + \text{c.c.}\} \tag{1.95}$$

where the sum runs over all **k** vectors subject to

$$k_1 = \frac{2\pi n_1}{L}; \; k_2 = \frac{2\pi n_2}{L}; \; k_3 = \frac{2\pi n_3}{L} \tag{1.96}$$

and where, in the presence of a discrete set of **k** values, we have set $A_j(\mathbf{k}) \equiv A_{\mathbf{k}j}$. The boundary conditions leading to (1.95), as well as to the analogous expansion for **E** and **H**, are called periodic boundary conditions.

As previously mentioned, if the boundary is a spherical surface it is more convenient to adopt spherical coordinates, and to use expansions (1.87, 1.88, 1.89) in terms of vector spherical harmonics. Thus, if the field is constrained within a metallic spherical enclosure of radius R, the condition that the component of **H** normal to the sphere vanishes on the boundary is written, after some algebra, as

$$\hat{k}\cdot\mathbf{H}(\mathbf{x},t)=\frac{4\pi}{R}\sum_{\ell m} i^{\ell}\sqrt{\ell(\ell+1)}\,Y_{\ell}^{m}(\vartheta,\varphi)$$
$$\times\int_{0}^{\infty}A(k,\mathcal{M},\ell,m)j_{\ell}(kR)e^{-i\omega_{k}t}kdk=0 \qquad (1.97)$$

In view of the arbitrariness of A as a function of k, the integral in (1.97) vanishes only if it can be reduced to a sum over particular values of k such that $j_{\ell}(kR)=0$. This condition selects for each ℓ a discrete number of k-values in correspondence with the zeros of the spherical Bessel function (given e.g. in Abramowitz and Stegun 1965). Thus the expansion for the vector potential in (1.87) takes the form

$$\mathbf{A}(\mathbf{x},t)=4\pi\sum_{k\ell m}\Big\{A_{k\ell m}(\mathcal{M})j_{\ell}(kx)\mathbf{Y}_{\ell m0}(\vartheta,\varphi)+A_{k\ell m}(\mathcal{E})\frac{1}{\sqrt{2\ell+1}}$$
$$\times\Big[\sqrt{\ell}j_{\ell+1}(kx)\mathbf{Y}_{\ell m+}(\vartheta,\varphi)-\sqrt{\ell+1}j_{\ell-1}(kx)\mathbf{Y}_{\ell m-}(\vartheta,\varphi)\Big]\Big\}$$
$$\times e^{-i\omega_{k}t}+\text{c.c.} \qquad (1.98)$$

The sum over k runs over $k_{\ell n}$ where $k_{\ell n}R$ is the nth zero of the spherical Bessel function of order ℓ and where, in the presence of a discrete set of k-values, we have set $A(k,\lambda,\ell,m)\equiv A_{k\ell m}(\lambda)$.

1.12 Free field in terms of field amplitudes

We specialize our discussion to periodic boundary conditions on the surface of a cube of side L. The electric and magnetic fields are obtained from (1.95) as

$$\mathbf{E}(\mathbf{x},t)=-\frac{1}{c}\dot{\mathbf{A}}=i\sum_{\mathbf{k}j}k\Big\{\mathbf{e}_{\mathbf{k}j}A_{\mathbf{k}j}e^{-i\omega_{k}t}e^{i\mathbf{k}\cdot\mathbf{x}}-\mathbf{e}^{*}_{\mathbf{k}j}A^{*}_{\mathbf{k}j}e^{i\omega_{k}t}e^{-i\mathbf{k}\cdot\mathbf{x}}\Big\};$$
$$\mathbf{H}(\mathbf{x},t)=\nabla\times\mathbf{A}$$
$$=i\sum_{\mathbf{k}j}k\Big\{\mathbf{b}_{\mathbf{k}j}A_{\mathbf{k}j}e^{-i\omega_{k}t}e^{i\mathbf{k}\cdot\mathbf{x}}-\mathbf{b}^{*}_{\mathbf{k}j}A^{*}_{\mathbf{k}j}e^{i\omega_{k}t}e^{i\mathbf{k}\cdot\mathbf{x}}\Big\}$$
$$(\mathbf{b}_{\mathbf{k}j}=\hat{\mathbf{k}}\times\mathbf{e}_{\mathbf{k}j}) \qquad (1.99)$$

Consider for each **k** three real orthogonal unit vectors as in Figure 1.4. Then

$$\hat{\mathbf{k}}\times\hat{\epsilon}_{1}=\hat{\epsilon}_{2};\ \hat{\mathbf{k}}\times\hat{\epsilon}_{2}=-\hat{\epsilon}_{1};\ \hat{\epsilon}_{1}\times\hat{\epsilon}_{2}=\hat{\mathbf{k}} \qquad (1.100)$$

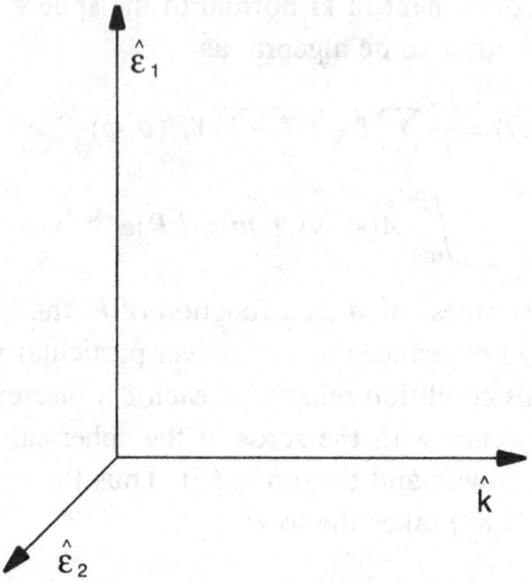

Fig. 1.4 The three real unit vectors $\hat{\epsilon}_1, \hat{\epsilon}_2$ and $\hat{k}$ are mutually perpendicular. Complex polarization vectors $\mathbf{e}_{\mathbf{k}1}$ and $\mathbf{e}_{\mathbf{k}2}$ can be obtained by linear combinations of $\hat{\epsilon}_1$ and $\hat{\epsilon}_2$ with **k**-dependent complex coefficients.

The two complex vectors

$$\mathbf{e}_{\mathbf{k}1} = \hat{\epsilon}_1 \cos\theta + i\hat{\epsilon}_2 \sin\theta \; ; \; \mathbf{e}_{\mathbf{k}2} = \hat{\epsilon}_1 \sin\theta - i\hat{\epsilon}_2 \cos\theta \tag{1.101}$$

satisfy conditions (1.80) and can be taken as polarization vectors in expansions (1.99). Naturally the directions of $\hat{\epsilon}_1$ and $\hat{\epsilon}_2$, as well as the parameter θ, may be different for different $\hat{\mathbf{k}}$. A component in the expansion for **E** corresponding to a particular **k** and j can be written as

$$\begin{aligned} &ik\left\{\mathbf{e}_{\mathbf{k}1} A_{\mathbf{k}1} e^{i(\mathbf{k}\cdot\mathbf{x}-\omega_k t)} - e^*_{\mathbf{k}1} A^*_{\mathbf{k}1} e^{-i(\mathbf{k}\cdot\mathbf{x}-\omega_k t)}\right\} \\ &= -2k \mid A_{\mathbf{k}1} \mid \{\hat{\epsilon}_1 \cos\theta \sin(\mathbf{k}\cdot\mathbf{x} - \omega_k t + \delta) \\ &+\hat{\epsilon}_2 \sin\theta \cos(\mathbf{k}\cdot\mathbf{x} - \omega_k t + \delta)\} \end{aligned} \tag{1.102}$$

where we have set $A_{\mathbf{k}1} =\mid A_{\mathbf{k}1} \mid e^{i\delta}$. Clearly, expression (1.102) describes a travelling wave of wavevector **k**, frequency ω_k, phase δ and amplitude proportional to $\mid A_{\mathbf{k}1} \mid$, which is elliptically polarized in the plane orthogonal to $\hat{\mathbf{k}}$. This wave becomes linearly polarized for $\theta = 0$ and circularly polarized for $\theta = \pi/4$. This example illustrates the important role played by the polarization vectors in determining the physical nature of expressions (1.99) for the electric and magnetic fields.

The following useful properties of the polarization vectors can be easily obtained from (1.101)

$$\begin{aligned}
&\mathbf{e}_{\mathbf{k}j} \cdot \mathbf{e}_{\mathbf{k}j} = -(-1)^j \cos 2\theta; \ \mathbf{e}_{\mathbf{k}1} \cdot \mathbf{e}_{\mathbf{k}2} = \sin 2\theta; \\
&\hat{\mathbf{k}} \times \mathbf{e}_{\mathbf{k}1} = -i\mathbf{e}^*_{\mathbf{k}2}; \ \hat{\mathbf{k}} \times \mathbf{e}_{\mathbf{k}2} = i\mathbf{e}^*_{\mathbf{k}1}; \\
&\mathbf{e}_{\mathbf{k}j} \times \mathbf{e}_{\mathbf{k}j} = 0; \ \mathbf{e}^*_{\mathbf{k}j} \times \mathbf{e}_{\mathbf{k}j} = (-1)^{j+1} i\hat{\mathbf{k}} \sin 2\theta; \\
&\mathbf{e}_{\mathbf{k}1} \times \mathbf{e}_{\mathbf{k}2} = -i\hat{\mathbf{k}}; \ \mathbf{e}^*_{\mathbf{k}1} \times \mathbf{e}_{\mathbf{k}2} = -i\hat{\mathbf{k}} \cos 2\theta
\end{aligned} \tag{1.103}$$

Thus, taking $\mathbf{e}_{-\mathbf{k}j} = \mathbf{e}_{\mathbf{k}j}$, the following equalities are also obtained

$$\begin{aligned}
&\mathbf{e}_{\mathbf{k}j} \cdot \mathbf{e}_{-\mathbf{k}j'} + \mathbf{b}_{\mathbf{k}j} \cdot \mathbf{b}_{-\mathbf{k}j'} \\
&= \mathbf{e}_{\mathbf{k}j} \cdot \mathbf{e}_{-\mathbf{k}j'} - (\hat{\mathbf{k}} \times \mathbf{e}_{\mathbf{k}j}) \cdot (\hat{\mathbf{k}} \times \mathbf{e}_{-\mathbf{k}j'}) = 0; \\
&\mathbf{b}_{\mathbf{k}j} \cdot \mathbf{b}^*_{\mathbf{k}j'} = (\hat{\mathbf{k}} \times \mathbf{e}_{\mathbf{k}j}) \cdot (\hat{\mathbf{k}} \times \mathbf{e}^*_{-\mathbf{k}j}) = \delta_{jj'};
\end{aligned} \tag{1.104}$$

$$\begin{aligned}
&\mathbf{e}_{\mathbf{k}j} \times \mathbf{b}_{-\mathbf{k}j'} = -\hat{\mathbf{k}}(\mathbf{e}_{\mathbf{k}j} \cdot \mathbf{e}_{-\mathbf{k}j'}); \\
&\mathbf{e}_{\mathbf{k}1} \times \mathbf{b}_{-\mathbf{k}1} = -\mathbf{e}_{\mathbf{k}2} \times \mathbf{b}_{-\mathbf{k}2} = -\hat{\mathbf{k}} \cos 2\theta; \\
&\mathbf{e}_{\mathbf{k}1} \times \mathbf{b}_{-\mathbf{k}2} = \mathbf{e}_{\mathbf{k}2} \times \mathbf{b}_{-\mathbf{k}1} = -\hat{\mathbf{k}} \sin 2\theta; \\
&\mathbf{e}_{\mathbf{k}j} \times \mathbf{b}^*_{\mathbf{k}j'} = \hat{\mathbf{k}}(\mathbf{e}_{\mathbf{k}j} \cdot \mathbf{e}^*_{\mathbf{k}j'}) = \hat{\mathbf{k}}\delta_{jj'}
\end{aligned} \tag{1.105}$$

$$\begin{aligned}
&\mathbf{e}_{\mathbf{k}j} \times \mathbf{e}_{-\mathbf{k}j} = 0; \ \mathbf{e}_{\mathbf{k}1} \times \mathbf{e}_{-\mathbf{k}2} = -\mathbf{e}_{\mathbf{k}2} \times \mathbf{e}_{-\mathbf{k}1} = -i\hat{\mathbf{k}}; \\
&\mathbf{e}_{\mathbf{k}1} \times \mathbf{e}^*_{\mathbf{k}1} = -\mathbf{e}_{\mathbf{k}2} \times \mathbf{e}^*_{\mathbf{k}2} = i\hat{\mathbf{k}} \sin 2\theta; \\
&\mathbf{e}_{\mathbf{k}1} \times \mathbf{e}^*_{\mathbf{k}2} = \mathbf{e}_{\mathbf{k}2} \times \mathbf{e}^*_{\mathbf{k}1} = i\hat{\mathbf{k}} \cos 2\theta
\end{aligned} \tag{1.106}$$

Real or imaginary polarization vectors can be obtained from (1.101) by adopting the pair $\mathbf{e}_{\mathbf{k}1}$, $i\mathbf{e}_{\mathbf{k}2}$ for $\theta = 0$ and $\pi/2$ respectively. Expansions (1.99) can now be used to express the Hamiltonian density (1.32) as

$$\begin{aligned}
\mathcal{H}_0 = -\frac{1}{8\pi} \sum_{\mathbf{k}\mathbf{k}'} \sum_{jj'} kk' \Big\{ &(\mathbf{e}_{\mathbf{k}j} \cdot \mathbf{e}_{\mathbf{k}'j'} + \mathbf{b}_{\mathbf{k}j} \cdot \mathbf{b}_{\mathbf{k}'j'}) A_{\mathbf{k}j} A_{\mathbf{k}'j'} \cdot \\
&\times e^{-i(\omega_k + \omega_{k'})t} e^{i(\mathbf{k}+\mathbf{k}')\cdot\mathbf{x}} - \left(\mathbf{e}_{\mathbf{k}j} \cdot \mathbf{e}^*_{\mathbf{k}'j'} + \mathbf{b}_{\mathbf{k}j} \cdot \mathbf{b}^*_{\mathbf{k}'j'}\right) \\
&\times A_{\mathbf{k}j} A^*_{\mathbf{k}'j'} e^{-i(\omega_k - \omega_{k'})t} e^{i(\mathbf{k}-\mathbf{k}')\cdot\mathbf{x}} \Big\} + \text{c.c.}
\end{aligned} \tag{1.107}$$

Using

$$\frac{1}{V} \int e^{\pm i(\mathbf{k}-\mathbf{k}')\cdot\mathbf{x}} d^3\mathbf{x} = \delta_{\mathbf{k}\mathbf{k}'} \tag{1.108}$$

where $V = L^3$, and integrating (1.107) over V one gets the free-field Hamiltonian

$$\begin{aligned} H_F &= \int \mathcal{H}_0(\mathbf{x}, t) d^3\mathbf{x} \\ &= -\frac{V}{8\pi} \sum_{\mathbf{k}} \sum_{jj'} k^2 \Big\{ \left(\mathbf{e}_{\mathbf{k}j} \cdot \mathbf{e}_{-\mathbf{k}j'} + \mathbf{b}_{\mathbf{k}j} \cdot \mathbf{b}_{-\mathbf{k}j'} \right) A_{\mathbf{k}j} A_{-\mathbf{k}j'} e^{-2i\omega_k t}. \\ &- \left(\mathbf{e}_{\mathbf{k}j} \cdot \mathbf{e}^*_{\mathbf{k}j'} + \mathbf{b}_{\mathbf{k}j} \cdot \mathbf{b}^*_{\mathbf{k}j'} \right) A_{\mathbf{k}j} A^*_{\mathbf{k}j'} \Big\} + \text{c.c.} \\ &= \frac{V}{4\pi} \sum_{\mathbf{k}j} k^2 \left(A_{\mathbf{k}j} A^*_{\mathbf{k}j} + \text{c.c.} \right) \end{aligned} \tag{1.109}$$

where (1.104) has been used.

The momentum density of the field defined in (1.35) can also be expanded in the field amplitudes as

$$\begin{aligned} \mathbf{p} &= \frac{1}{4\pi c} \mathbf{E} \times \mathbf{H} \\ &= -\frac{1}{4\pi c} \sum_{\mathbf{k}\mathbf{k}'} \sum_{jj'} kk' \Big\{ \mathbf{e}_{\mathbf{k}j} \times \mathbf{b}_{\mathbf{k}'j'} A_{\mathbf{k}j} A_{\mathbf{k}'j'} e^{-i(\omega_k + \omega_{k'})t} e^{i(\mathbf{k}+\mathbf{k}')\cdot\mathbf{x}} \\ &- \mathbf{e}_{\mathbf{k}j} \times \mathbf{b}^*_{\mathbf{k}'j'} A_{\mathbf{k}j} A^*_{\mathbf{k}'j'} e^{-i(\omega_k - \omega_{k'})t} e^{i(\mathbf{k}-\mathbf{k}')\cdot\mathbf{x}} \Big\} + \text{c.c.} \end{aligned} \tag{1.110}$$

Integration over V, use of (1.108) and of (1.105) yields the momentum of the field

$$\begin{aligned} \mathbf{P} &= \int \mathbf{p} d^3\mathbf{x} = -\frac{V}{4\pi c} \sum_{\mathbf{k}} \sum_{jj'} k^2 \{ \mathbf{e}_{\mathbf{k}j} \times \mathbf{b}_{-\mathbf{k}j'} A_{\mathbf{k}j} A_{-\mathbf{k}j'} e^{-2i\omega_k t} \\ &- \mathbf{e}_{\mathbf{k}j} \times \mathbf{b}^*_{\mathbf{k}j'} A_{\mathbf{k}j} A^*_{\mathbf{k}j'} \} + \text{c.c.} \\ &= \frac{V}{4\pi c} \sum_{\mathbf{k}j} k^2 \hat{\mathbf{k}} \left(A_{\mathbf{k}j} A^*_{\mathbf{k}j} + \text{c.c.} \right) \end{aligned} \tag{1.111}$$

where the time-dependent part of the expression vanishes because terms with $\mathbf{k}$ balance exactly terms with $-\mathbf{k}$ in the sum over $\mathbf{k}$.

As for angular momentum, we have shown in Section 1.7 that the total $\mathbf{L}$ can be separated into two parts according to (1.53). In some cases orbital and spin contributions to $\mathbf{L}$ can be approximately decoupled (Allen *et al.* 1992). This separation is suggestive since the first term on the RHS of (1.53) is the same as if the electromagnetic field could be decomposed into a set of three scalar fields, bearing no vectorial relation to each other. The second term on the RHS of (1.53) has a structure similar to what one would expect of a spin 1 particle in quantum mechanics, and it is often called the spin contribution, as in Cohen-Tannoudji *et al.* (1989). We shall concentrate on this spin contribution, which in terms of field amplitudes takes the form

$$\frac{1}{4\pi c}\int \mathbf{E}\times\mathbf{A}d^3\mathbf{x} = \frac{iV}{4\pi c}\sum_{\mathbf{k}}\sum_{jj'} k\Big\{\mathbf{e}_{\mathbf{k}j}\times\mathbf{e}_{-\mathbf{k}j'}A_{\mathbf{k}j}A_{-\mathbf{k}j'}e^{-2i\omega_k t}.$$

$$+\,\mathbf{e}_{\mathbf{k}j}\times\mathbf{e}^*_{\mathbf{k}j'}A_{\mathbf{k}j}A^*_{\mathbf{k}j'}\Big\} + \text{c.c.} \qquad (1.112)$$

Contrarily to the case of the linear momentum, in (1.112) the terms in $\mathbf{k}$ do not balance the terms in $-\mathbf{k}$. The time average of these terms, however, vanishes and one is left with

$$\frac{1}{4\pi c}\overline{\int \mathbf{E}\times\mathbf{A}d^3\mathbf{x}} = -\frac{V}{4\pi c}\sum_{\mathbf{k}}\mathbf{k}\Big\{\sin 2\theta\left[A_{\mathbf{k}1}A^*_{\mathbf{k}1} - A_{\mathbf{k}2}A^*_{\mathbf{k}2}\right].$$

$$+\cos 2\theta\left[A_{\mathbf{k}1}A^*_{\mathbf{k}2} - A_{\mathbf{k}2}A^*_{\mathbf{k}1}\right] + \text{c.c.}\Big\} \qquad (1.113)$$

where the bar over the integral indicates time average. For circularly polarized modes $\theta = \pi/4$, $\cos 2\theta = 0$ and the cross terms in the quadratic form for the amplitudes do not contribute to the time average (1.113). In this case the time-averaged spin contribution to the field angular momentum corresponds to the difference in intensity of the left-polarized mode and of the right-polarized mode, the resultant contribution being along the direction $\hat{\mathbf{k}}$ of wave propagation. In this sense the use of the term "spin part" for the contribution (1.112) to the total angular momentum of the field can be partially justified. This point is also discussed by Jackson (1988).

References

M. Abramovitz, I.A. Stegun (eds.) (1965). *Handbook of Mathematical Functions* (Dover Publications Inc., New York)

L. Allen, M.W. Beijersbergen, R.J.C. Spreeuw, J.P. Woerdman (1992). *Phys. Rev. A* **45**, 8185

G. Barton (1989). *Elements of Green's Functions and Propagation* (Oxford University Press, Oxford)

A.O. Barut (1980). *Electrodynamics and Classical Theory of Fields and Particles* (Dover Publications Inc., New York)

B.I. Bleaney, B. Bleaney (1985). *Electricity and Magnetism* (Oxford University Press, Hong Kong)

N.N. Bogoliubov, D.V. Shirkov (1960). *Introduction to the Theory of Quantized Fields* (Interscience Publishers Inc., New York)

C. Cohen-Tannoudji, J. Dupont-Roc, G. Grynberg (1989). *Photons and Atoms* (John Wiley and Sons, New York)

D.P. Craig, T. Thirunamachandran (1984). *Molecular Quantum Electrodynamics* (Academic Press Inc., London)
N.A. Doughty (1990). *Lagrangian Interaction* (Addison-Wesley Publishing Co., Sydney)
H. Goldstein (1980). *Classical Mechanics* (Addison-Wesley Publishing Co., Reading, Ma.)
H.A. Haus, J.L. Pan (1993). *Am. J. Phys.* **61**, 818
W.P. Healy (1982). *Non-relativistic Quantum Electrodynamics* (Academic Press Inc., London)
W. Heitler (1960). *The Quantum Theory of Radiation* (Oxford University Press, London)
E.M. Henley, W. Thirring (1962). *Elementary Quantum Field Theory* (McGraw-Hill Book Co., New York)
J.D. Jackson (1988). *Elettrodinamica Classica* (Zanichelli, Bologna)
L.D. Landau, E.M. Lifshitz (1975). *The Classical Theory of Fields* (Pergamon Press, London)
F. Mandl, G. Shaw (1984). *Quantum Field Theory* (John Wiley and Sons, Norwich)
P.M. Morse, H. Feshbach (1953). *Methods of Theoretical Physics*, vol. 1 (McGraw-Hill Book Co., New York)
E.A. Power (1964). *Introductory Quantum Electrodynamics* (Longmans, Green and Co. Ltd., London)
J.J. Sakurai (1982). *Advanced Quantum Mechanics* (Addison-Wesley Publishing Co., Reading, Ma.)
S.S. Schweber (1964). *An Introduction to Relativistic Quantum Field Theory* (John Weatherhill Inc., Tokyo)
M.R. Spiegel (1959). *Vector Analysis* (Schaum Publishing Co., New York)
V.A. Ugarov (1982). *Teoria della Relatività Ristretta* (Edizioni MIR)

Further reading

The reader interested in the theory of the classical electromagnetic field in the optical range of frequency must be referred to the outstanding textbook
M. Born, E. Wolf, *Principles of Optics* (Pergamon Press, Exeter, 1980).

2

The quantum electromagnetic field in the absence of sources

Introduction. The purpose of this chapter is to present the quantum theory of the electromagnetic field in the absence of charges and currents. Thus the classical field discussed in the previous chapter is subjected to canonical quantization in Section 2.1, where creation and annihilation operators for plane-wave and spherical wave modes, as well as their commutation relations, are derived along with various field-field commutators related to field propagators. In the next section we introduce the concept of the photon as an elementary excitation of the electromagnetic field. The attention is focused on the ground state of the quantized electromagnetic field in the absence of sources, which is the photon vacuum. The amplitude fluctuations, or zero-point fluctuations of this vacuum, are evaluated. Excited states of the field are examined in the next sections. In particular, Section 2.3 is concerned with number states and coherent states, the latter being obtained by a Glauber transformation of the vacuum, and with their statistical properties. Squeezed states of the field are introduced in Section 2.4 by a unitary transformation leading from the normal to the squeezed vacuum, whose statistical properties are compared with those of a coherent state. Section 2.5 is devoted to a brief discussion of thermal states. The final section of this chapter is dedicated to a discussion of the nonlocalizability of the photon.

2.1 Canonical quantization in the Coulomb gauge

As discussed in Section 1.5, renouncing manifest Lorentz covariance of the theory makes it convenient to adopt the Coulomb gauge to discuss the dynamics of the free e.m. field in the absence of charges and currents. In this way A_4 is not regarded as a dynamical variable, and one is left only with the three components of the transverse potential $\mathbf{A}(\mathbf{x}, t)$ which play the role of generalized coordinates of the field. The canonical momenta conjugate to these coordinates are given by (1.31) as

$$\Pi_i = -\frac{1}{4\pi c} E_i \tag{2.1}$$

At this point one might be tempted to apply the canonical quantization rules to the set of A_i and Π_i in the form

$$\begin{aligned}&\left[\Pi_i(\mathbf{x}, t), A_j(\mathbf{x}', t)\right] \\ &= -\frac{1}{4\pi c}\left[E_i(\mathbf{x}, t), A_j(\mathbf{x}', t)\right] = -i\hbar\delta_{ij}\delta(\mathbf{x}-\mathbf{x}')\end{aligned} \tag{2.2}$$

which is perfectly legitimate, for example, in the case of a multicomponent scalar field. This procedure, however, does not take adequately into account the vector nature of the electromagnetic field, that is the fact that its components are not independent of each other. In fact, it is immediately seen that (2.2) is incompatible with Maxwell's equation $\nabla \cdot \mathbf{E} = 0$: taking the divergence with respect to $\mathbf{x}$ of both sides of (2.2), considered as a relation between vector components for each i, the LHS vanishes but the RHS does not, since

$$\begin{aligned}-i\frac{\partial}{\partial x_i}\delta_{ij}\delta(\mathbf{x}-\mathbf{x}') &= \delta_{ij}\frac{1}{(2\pi)^3}\int k_i e^{i\mathbf{k}\cdot(\mathbf{x}-\mathbf{x}')}d^3\mathbf{k} \\ &= \frac{1}{(2\pi)^3}\int k_j e^{i\mathbf{k}\cdot(\mathbf{x}-\mathbf{x}')}d^3\mathbf{k}\end{aligned} \tag{2.3}$$

where the Fourier transform (1.64) has been used. Thus (2.2) must be modified to remove the divergence of the RHS. The most natural choice is to replace $\delta_{ij}\delta(\mathbf{x}-\mathbf{x}')$ in (2.2) with $\delta_{\perp ij}(\mathbf{x}-\mathbf{x}')$, since

$$\begin{aligned}&-i\frac{\partial}{\partial x_i}\delta_{\perp ij}\delta(\mathbf{x}-\mathbf{x}') = \frac{1}{(2\pi)^3}\int\left(\delta_{ij}-\hat{k}_i\hat{k}_j\right)k_i e^{i\mathbf{k}\cdot(\mathbf{x}-\mathbf{x}')}d^3\mathbf{k} \\ &= \frac{1}{(2\pi)^3}\int\left(1-\hat{k}_i^2\right)\hat{k}_j e^{i\mathbf{k}\cdot(\mathbf{x}-\mathbf{x}')}d^3\mathbf{k} = 0\end{aligned} \tag{2.4}$$

where (1.63) has been used. Then the canonical quantization rules to be adopted are the following (Bjorken and Drell 1964)

$$\begin{aligned}&\left[\Pi_i(\mathbf{x}, t), A_j(\mathbf{x}', t)\right] \\ &= -\frac{1}{4\pi c}\left[E_i(\mathbf{x}, t), A_j(\mathbf{x}', t)\right] = -i\hbar\delta_{\perp ij}(\mathbf{x}-\mathbf{x}')\end{aligned} \tag{2.5}$$

Since we have shown that the fields $\mathbf{E}$ and $\mathbf{A}$ can be expressed in terms of the complex field amplitudes $A_{\mathbf{k}j}$, (2.5) implies a set of nontrivial commutation relations among these field amplitudes which become operators upon quantization. Using (1.95), (1.99) and the set of polarization sum rules (Craig and Thirunamachandran 1984)

$$\sum_j (\mathbf{e}_{\mathbf{k}j})_m(\mathbf{e}^*_{\mathbf{k}j})_n = \sum_j (\mathbf{b}_{\mathbf{k}j})_m(\mathbf{b}^*_{\mathbf{k}j})_n = \delta_{mn} - \hat{k}_m\hat{k}_n\ ;$$
$$\sum_j (\mathbf{e}_{\mathbf{k}j})_m(\mathbf{b}^*_{\mathbf{k}j})_n = e_{mnl}\hat{k}_l \tag{2.6}$$

where e_{mnl} is the completely antisymmetric pseudotensor defined in (1.38), it is easy to show that (2.5) (with $A^\dagger_{\mathbf{k}_j}$ Hermitian conjugate of $A_{\mathbf{k}_j}$) implies

$$\left[A_{\mathbf{k}j}, A_{\mathbf{k}'j'}\right] = 0;\ \left[A_{\mathbf{k}j}, A^\dagger_{\mathbf{k}'j'}\right] = \frac{2\pi\hbar c^2}{V\omega_k}\delta_{\mathbf{k}\mathbf{k}'}\delta_{jj'} \tag{2.7}$$

Thus, defining scaled creation and annihilation operators $a_{\mathbf{k}_j}$ and $a^\dagger_{\mathbf{k}_j}$

$$a_{\mathbf{k}_j} = \sqrt{\frac{V\omega_k}{2\pi\hbar c^2}}A_{\mathbf{k}j};\ \left[a_{\mathbf{k}_j}, a_{\mathbf{k}'j'}\right] = 0;\ \left[a_{\mathbf{k}_j}, a^\dagger_{\mathbf{k}'j'}\right] = \delta_{\mathbf{k}\mathbf{k}'}\delta_{jj'} \tag{2.8}$$

one can express the quantized fields as

$$\mathbf{A}(\mathbf{x},t) = \sum_{\mathbf{k}j}\sqrt{\frac{2\pi\hbar c^2}{V\omega_k}}\left\{\mathbf{e}_{\mathbf{k}j}a_{\mathbf{k}_j}(t)e^{i\mathbf{k}\cdot\mathbf{x}} + \mathbf{e}^*_{\mathbf{k}j}a^\dagger_{\mathbf{k}_j}(t)e^{-i\mathbf{k}\cdot\mathbf{x}}\right\};$$
$$\mathbf{E}(\mathbf{x},t) = i\sum_{\mathbf{k}j}\sqrt{\frac{2\pi\hbar\omega_k}{V}}\left\{\mathbf{e}_{\mathbf{k}j}a_{\mathbf{k}_j}(t)e^{i\mathbf{k}\cdot\mathbf{x}} - \mathbf{e}^*_{\mathbf{k}j}a^\dagger_{\mathbf{k}_j}(t)e^{-i\mathbf{k}\cdot\mathbf{x}}\right\};$$
$$\mathbf{H}(\mathbf{x},t) = i\sum_{\mathbf{k}j}\sqrt{\frac{2\pi\hbar\omega_k}{V}}\left\{\mathbf{b}_{\mathbf{k}j}a_{\mathbf{k}_j}(t)e^{i\mathbf{k}\cdot\mathbf{x}} - \mathbf{b}^*_{\mathbf{k}j}a^\dagger_{\mathbf{k}_j}(t)e^{-i\mathbf{k}\cdot\mathbf{x}}\right\} \tag{2.9}$$

which satisfy the commutation relations (2.5) in view of (2.8). It should be emphasized that expressions (2.9) are now operators and not c-numbers.

Following the same procedure as in Section 1.12, one obtains the quantized expression for the operators corresponding to the various field observables. In particular, the field Hamiltonian can be obtained as in (1.109) as

$$H_F = \frac{1}{2}\sum_{\mathbf{k}j}\hbar\omega_k\left(a_{\mathbf{k}_j}a^\dagger_{\mathbf{k}_j} + a^\dagger_{\mathbf{k}_j}a_{\mathbf{k}_j}\right) = \sum_{\mathbf{k}j}\hbar\omega_k\left(a^\dagger_{\mathbf{k}_j}a_{\mathbf{k}_j} + \frac{1}{2}\right) \tag{2.10}$$

The time-dependence of the quantized field amplitudes $a_{\mathbf{k}_j}$ is implicit in (2.9) and can be found by solving the Heisenberg equations

$$\dot{a}_{\mathbf{k}j} = \frac{i}{\hbar}\left[H_F, a_{\mathbf{k}_j}\right] = -i\omega_k a_{\mathbf{k}_j}\ ;\ \text{ or }\ a_{\mathbf{k}_j}(t) = a_{\mathbf{k}_j}(0)e^{-\omega_k t} \tag{2.11}$$

Inserting the solutions (2.11) for $a_{\mathbf{k}_j}(t)$ into the field operators (2.9) one obtains expressions for **A**, **E** and **H** which are formally identical to the classical expressions (1.95) and (1.99). On the other hand, the latter are solutions of Maxwell's equations (1.1). Thus it follows that the Heisenberg equations of motion for operators **A**, **E** and **H** must be formally identical to Maxwell's equations.

It is now interesting to evaluate the commutators of the field at two different points in space-time, because this commutator is related to the possibility of performing correlated measurements (see e.g. Heitler 1960). Using (2.9) and (2.11) one obtains

$$\begin{aligned}
&[E_m(\mathbf{x},t), E_n(\mathbf{x}',t')] \\
&= \sum_{\mathbf{k}j} \frac{2\pi\hbar\omega_k}{V} \left\{ (\mathbf{e}_{\mathbf{k}j})_m \left(\mathbf{e}^*_{\mathbf{k}j}\right)_n e^{i[\mathbf{k}\cdot(\mathbf{x}-\mathbf{x}')-\omega_k(t-t')]} - \text{c.c.} \right\} \\
&= 4\pi i\hbar c \frac{1}{V} \sum_{\mathbf{k}} k(\delta_{mn} - \hat{k}_m\hat{k}_n) \sin[\mathbf{k}\cdot(\mathbf{x}-\mathbf{x}') - ck(t-t')] \\
&= -4\pi i\hbar c \left(\delta_{mn} \frac{1}{c^2}\frac{\partial^2}{\partial t^2} - \frac{\partial^2}{\partial x_m \partial x_n} \right) \\
&\times \frac{1}{V} \sum_{\mathbf{k}} \frac{1}{k} \sin[\mathbf{k}\cdot(\mathbf{x}-\mathbf{x}') - ck(t-t')] \qquad (2.12)
\end{aligned}$$

The last summation over **k** vanishes for $t = t'$, as it is easily seen by associating terms with **k** and $-$**k** in the sum. For $t \neq t'$ this trick is not useful and the sum over **k** must be transformed into an integral which yields the well-known result (Heitler 1960)

$$\begin{aligned}
&\frac{1}{V} \sum_{\mathbf{k}} \frac{1}{k} \sin[\mathbf{k}\cdot(\mathbf{x}-\mathbf{x}') - ck(t-t')] \\
&= \frac{1}{(2\pi)^3} \int \frac{1}{k} \sin[\mathbf{k}\cdot(\mathbf{x}-\mathbf{x}') - ck(t-t')] d^3\mathbf{k} \\
&= -\frac{\delta[\mid \mathbf{x}-\mathbf{x}' \mid - c(t-t')] - \delta[\mid \mathbf{x}-\mathbf{x}' \mid + c(t-t')]}{4\pi \mid \mathbf{x}-\mathbf{x}' \mid} \qquad (2.13)
\end{aligned}$$

Substituting (2.13) into (2.12) yields

$$\begin{aligned}
&[E_m(\mathbf{x},t), E_n(\mathbf{x}',t')] = i\hbar c \left(\delta_{mn} \frac{1}{c^2}\frac{\partial^2}{\partial t^2} - \frac{\partial^2}{\partial x_m \partial x_n} \right) \\
&\times \frac{\delta[\mid \mathbf{x}-\mathbf{x}' \mid - c(t-t')] - \delta[\mid \mathbf{x}-\mathbf{x}' \mid + c(t-t')]}{\mid \mathbf{x}-\mathbf{x}' \mid} \qquad (2.14)
\end{aligned}$$

Proceeding along the same lines one obtains

$$[E_m(\mathbf{x},t), H_n(\mathbf{x}',t')] = -i\hbar e_{mn\ell} \frac{\partial^2}{\partial x_\ell \partial t} \times \frac{\delta[\mid \mathbf{x}-\mathbf{x}' \mid -c(t-t')] - \delta[\mid \mathbf{x}-\mathbf{x}' \mid +c(t-t')]}{\mid \mathbf{x}-\mathbf{x}' \mid} \tag{2.15}$$

From (2.14) and (2.15) we conclude that any two components of the electromagnetic field can be measured at two different points in space-time with absolute precision, unless these two points can be connected by a light signal. In the latter case the operators corresponding to the field components at $(\mathbf{x}, t)$ generally fail to commute with those at $(\mathbf{x}', t')$ and the measurement performed at $(\mathbf{x}, t)$ perturbs the measurement performed at $(\mathbf{x}', t')$ in a complementary fashion. The discussion above leads quite naturally to the concept of propagator of the electromagnetic field, for which we refer the reader to the textbook by Bjorken and Drell (1964).

Quantization of the free field can also be attained in the absence of walls affecting the density of allowed k values. Since such an expression is, the continuum limit one should expect that the Kronecker δ appearing in the Bose commutation relations of the form (2.8) is substituted by a δ-function. This leads for plane wave operators to

$$\left[a(\mathbf{k},j), a^\dagger(\mathbf{k}',j')\right] = \delta(\mathbf{k}-\mathbf{k}')\delta_{jj'} \ ; \tag{2.16}$$

$$\left[a(\mathbf{k},j), a(\mathbf{k}',j')\right] = 0 \tag{2.17}$$

and to the continuum plane wave expansions corresponding to (2.9)

$$\mathbf{A}(\mathbf{x},t) = \frac{1}{2\pi}\hbar^{1/2} c \sum_j \int \frac{1}{\sqrt{\omega_k}} \{\mathbf{e}_{\mathbf{k}j} a(\mathbf{k},j) e^{i\mathbf{k}\cdot\mathbf{x}} + \text{h.c.}\} d^3\mathbf{k} \tag{2.18}$$

$$\mathbf{E}(\mathbf{x},t) = \frac{i}{2\pi}\hbar^{1/2} \sum_j \int \sqrt{\omega_k} \{\mathbf{e}_{\mathbf{k}j} a(\mathbf{k},j) e^{i\mathbf{k}\cdot\mathbf{x}} - \text{h.c.}\} d^3\mathbf{k} \tag{2.19}$$

$$\mathbf{H}(\mathbf{x},t) = \frac{i}{2\pi}\hbar^{1/2} \sum_j \int \sqrt{\omega_k} \{\mathbf{b}_{\mathbf{k}j} a(\mathbf{k},j) e^{i\mathbf{k}\cdot\mathbf{x}} - \text{h.c.}\} d^3\mathbf{k} \tag{2.20}$$

On the other hand, for a spherical expansion starting from (1.87, 1.88, 1.89) and introducing the scaled amplitudes

$$a(k,\lambda,\ell,m) = \frac{2\pi}{\hbar c}\sqrt{\hbar\omega_k} A(k,\lambda,\ell,m) \tag{2.21}$$

one obtains (see e.g. Passante *et al.* 1985)

$$\mathbf{A}(\mathbf{x},t) = 2\sqrt{\hbar c}\sum_{\ell m}\int_0^\infty \Big\{a(k,\mathcal{M},\ell,m)j_\ell(kx)\mathbf{Y}_{\ell m0}(\vartheta,\varphi)$$
$$+a(k,\mathcal{E},\ell,m)\frac{1}{\sqrt{2\ell+1}}\Big[\sqrt{\ell}j_{\ell+1}(kx)\mathbf{Y}_{\ell m+}(\vartheta,\varphi)$$
$$-\sqrt{\ell+1}j_{\ell-1}(kx)\mathbf{Y}_{\ell m-}(\vartheta,\varphi)\Big]\Big\}k^{1/2}dk + \text{h.c.}$$
$$\mathbf{E}(\mathbf{x},t) = i2\sqrt{\hbar c}\sum_{\ell m}\int_0^\infty \Big\{a(k,\mathcal{M},\ell,m)j_\ell(kx)\mathbf{Y}_{\ell m0}(\vartheta,\varphi)$$
$$+a(k,\mathcal{E},\ell,m)\frac{1}{\sqrt{2\ell+1}}\Big[\sqrt{\ell}j_{\ell+1}(kx)\mathbf{Y}_{\ell m+}(\vartheta,\varphi)$$
$$-\sqrt{\ell+1}j_{\ell-1}(kx)\mathbf{Y}_{\ell m-}(\vartheta,\varphi)\Big]\Big\}k^{3/2}dk + \text{h.c.}$$
$$\mathbf{H}(\mathbf{x},t) = -i2\sqrt{\hbar c}\sum_{\ell m}\int_0^\infty \Big\{a(k,\mathcal{E},\ell,m)j_\ell(kx)\mathbf{Y}_{\ell m0}(\vartheta,\varphi)$$
$$-a(k,\mathcal{M},\ell,m)\frac{1}{\sqrt{2\ell+1}}\Big[\sqrt{\ell}j_{\ell+1}(kx)\mathbf{Y}_{\ell m+}(\vartheta,\varphi)$$
$$-\sqrt{\ell+1}j_{\ell-1}(kx)\mathbf{Y}_{\ell m-}(\vartheta,\varphi)\Big]\Big\}k^{3/2}dk + \text{h.c.} \quad (2.22)$$

where

$$[a(k,\lambda,\ell,m), a^\dagger(k',\lambda',\ell',m')] = \delta(k-k')\delta_{\lambda\lambda'}\delta_{\ell\ell'}\delta_{mm'}\ ; \quad (2.23)$$

$$[a(k,\lambda,\ell,m), a(k',\lambda',\ell',m')] = 0 \quad (2.24)$$

Substituting (2.22) into (1.32) and integrating over the volume, one obtains the field Hamiltonian in the multipolar expansion

$$H_F = \sum_{\lambda\ell m}\int \hbar\omega_k a^\dagger(k,\lambda,\ell,m)a(k,\lambda,\ell,m)dk \quad (2.25)$$

2.2 Photons and the vacuum state

The reader will immediately realize that our canonical quantization procedure has led to a description of the quantum electromagnetic field, in the Coulomb gauge and in the absence of sources, in terms of an ensemble of Bose particles, since the commutation rules of the scaled field amplitude operators $a_{\mathbf{k}j}$, as given by (2.8), are exactly those of the creation and annihilation operators for a set of bosons. Moreover, the field

Hamiltonian in (2.10) contains the form $a^\dagger_{\mathbf{k}_j} a_{\mathbf{k}_j}$ which is to be interpreted as the boson number operator for the $\mathbf{k}j$ mode of the field. These bosons are called photons. As it is well known (see e.g. Roman 1965), the operator algebra (2.8) of the photon creation and annihilation operators leads to a representation of the state of the field in terms of a set of normalized kets $| \{n_{\mathbf{k}j}\}\rangle$ in Fock space, where the curly brackets indicate a distribution of photons over the various modes of the field, each mode being characterized by a wavevector $\mathbf{k}$ and a polarization index j. The action of operators $a_{\mathbf{k}_j}(0)$ and $a^\dagger_{\mathbf{k}_j}(0)$ on one of such states can be expressed as

$$a_{\mathbf{k}_j}(0) \mid \{n_{\mathbf{k}j}\}\rangle = \sqrt{n_{\mathbf{k}j}} \mid \{n_{\mathbf{k}j}\}'\rangle \; ;$$
$$a^\dagger_{\mathbf{k}_j}(0) \mid \{n_{\mathbf{k}j}\}\rangle = \sqrt{n_{\mathbf{k}j}+1} \mid \{n_{\mathbf{k}j}\}''\rangle \tag{2.26}$$

where $\{n_{\mathbf{k}j}\}'$ is the same photon distribution as $\{n_{\mathbf{k}j}\}$ except in the particular mode $\mathbf{k}j$, in which $\{n_{\mathbf{k}j}\}'$ has one photon less. Analogously, $\{n_{\mathbf{k}j}\}''$ is the same as $\{n_{\mathbf{k}j}\}$ except for having one extra photon in mode $\mathbf{k}j$. Clearly (2.26) justify the terminology of annihilation and creation operators for $a_{\mathbf{k}_j}$ and $a^\dagger_{\mathbf{k}_j}$, as well as of number operator for $a^\dagger_{\mathbf{k}_j} a_{\mathbf{k}_j}$, since

$$a^\dagger_{\mathbf{k}_j} a_{\mathbf{k}_j} \mid \{n_{\mathbf{k}j}\}\rangle = n_{\mathbf{k}j} \mid \{n_{\mathbf{k}j}\}\rangle \tag{2.27}$$

Of special interest for our treatment is the state $| \{0_{\mathbf{k}j}\}\rangle$ where no photons are present in the field. In this state $n_{\mathbf{k}j} = 0$ for each $\mathbf{k}$ and j. For this reason $| \{0_{\mathbf{k}j}\}\rangle$ is usually called the vacuum state. Since after (2.26)

$$a_{\mathbf{k}_j}(0) \mid \{0_{\mathbf{k}_j}\}\rangle = 0, \; a^\dagger_{\mathbf{k}_j}(0) \mid \{0_{\mathbf{k}_j}\}\rangle = \mid 1_{\mathbf{k}j}\rangle \tag{2.28}$$

where $| 1_{\mathbf{k}j}\rangle$ is the state with one photon in mode $\mathbf{k}j$ and zero photons in all other modes, we find

$$\begin{aligned}\langle\{0_{\mathbf{k}j}\} \mid \mathbf{A}(\mathbf{x},t) \mid \{0_{\mathbf{k}j}\}\rangle &= \langle\{0_{\mathbf{k}j}\} \mid \mathbf{E}(\mathbf{x},t) \mid \{0_{\mathbf{k}j}\}\rangle \\ &= \langle\{0_{\mathbf{k}j}\} \mid \mathbf{H}(\mathbf{x},t) \mid \{0_{\mathbf{k}_j}\}\rangle = 0\end{aligned} \tag{2.29}$$

Thus the quantum average of the field on the vacuum state vanishes at every point in space and at any time. The quantum average of the square of the field on the same state, however, does not vanish. In fact we have, for example, from (2.9)

$$\langle\{0_{\mathbf{k}j}\} \mid \mathbf{E}^2(\mathbf{x},t) \mid \{0_{\mathbf{k}j}\}\rangle = \langle\{0_{\mathbf{k}j}\} \mid \mathbf{H}^2(\mathbf{x},t) \mid \{0_{\mathbf{k}j}\}\rangle$$

$$= \sum_{\mathbf{k}j} \frac{2\pi\hbar\omega_k}{V} \langle\{0_{\mathbf{k}j}\} \mid a_{\mathbf{k}_j} a^\dagger_{\mathbf{k}_j} \mid \{0_{\mathbf{k}_j}\}\rangle = \frac{4\pi\hbar c}{V} \sum_{\mathbf{k}} k = \frac{2}{\pi}\hbar c \int_0^\infty k^3 dk \tag{2.30}$$

where (1.80) and (1.104) have been used and the sum over discrete $\mathbf{k}$ has been converted into an integral in the continuum limit $V \to \infty$ according to

$$\frac{1}{V}\sum_{\mathbf{k}} \to \frac{1}{(2\pi)^3}\int d^3\mathbf{k} \tag{2.31}$$

From (2.30) we see not only that the quantum average of the energy density of the field in the vacuum state does not vanish, but that it actually diverges at any $\mathbf{x}, t$. However, if instead of evaluating the quantum average of the square of the fields at $\mathbf{x}, t$ one evaluates the quantum average of operators $\overline{\mathbf{E}^2}(\mathbf{x}, t)$ or $\overline{\mathbf{H}^2}(\mathbf{x}, t)$, which are defined here as operators averaged over a finite region of linear dimensions r_0, then naturally the integration over k in (2.30) would extend only up to $k_M \sim r_0^{-1}$. In this case the quantum averages on the vacuum state yield

$$\langle\{0_{\mathbf{k}j}\} \mid \overline{\mathbf{E}^2}(\mathbf{x}, t) \mid \{0_{\mathbf{k}j}\}\rangle = \langle\{0_{\mathbf{k}j}\} \mid \overline{\mathbf{H}^2}(\mathbf{x}, t) \mid \{0_{\mathbf{k}j}\}\rangle \sim \frac{1}{2\pi}\hbar c k_M^4 \tag{2.32}$$

which shows that the quantum average of the energy density of the vacuum, as measured by a probe of linear dimensions r_0, is proportional to r_0^{-4}. This leads to the concept of vacuum fluctuations of the electromagnetic field, since for example

$$\begin{aligned} &\langle\{0_{\mathbf{k}j}\} \mid \mathbf{E}^2(\mathbf{x}, t) \mid \{0_{\mathbf{k}j}\}\rangle \\ &= \langle\{0_{\mathbf{k}j}\} \mid \mathbf{E}^2(\mathbf{x}, t) \mid \{0_{\mathbf{k}j}\}\rangle - (\langle\{0_{\mathbf{k}j}\} \mid \mathbf{E}(\mathbf{x}, t) \mid \{0_{\mathbf{k}j}\}\rangle)^2 \\ &\equiv (\Delta\mathbf{E})^2 \end{aligned} \tag{2.33}$$

is the variance of the electric field at the point $\mathbf{x}, t$. Our results mean that the fields in the vacuum state fluctuate around their mean value which is zero. These fluctuations are a typical quantum effect, as is also indicated by the appearance of $\hbar$ in the expression of $(\Delta\mathbf{E})^2$, and they should not be confused with thermal fluctuations, which obviously vanish at $T = 0K$. For this reason these fluctuations are called zero-point fluctuations. They are clearly related to the 1/2 term appearing in the expression (2.10) for the field Hamiltonian, since

$$\begin{aligned} \langle\{0_{\mathbf{k}j}\} \mid H_F \mid \{0_{\mathbf{k}j}\}\rangle &= \frac{1}{2}\hbar\sum_{\mathbf{k}j}\omega_k = \frac{V}{2\pi^2}\hbar c\int_0^\infty k^3 dk \\ &= \frac{V}{8\pi}\langle\{0_{\mathbf{k}j}\} \mid \{\mathbf{E}^2(\mathbf{x}, t) + \mathbf{H}^2(\mathbf{x}, t)\} \mid \{0_{\mathbf{k}j}\}\rangle \end{aligned} \tag{2.34}$$

Thus the 1/2 term in H_F can be interpreted as the contribution to the energy of the field arising from the zero-point fluctuations.

2.3 Number states and coherent states

It should also be noted that expression (2.10) for the field Hamiltonian H_F is formally equivalent to the Hamiltonian of an infinite set of harmonic oscillators, one for each mode $\mathbf{k}j$ of the field. Since each of the sub-Hamiltonians $\hbar\omega_k(a^\dagger_{\mathbf{k}_j}a_{\mathbf{k}_j}+1/2)$ commutes with the others, the oscillators are independent, and their dynamics can be discussed separately. Thus we shall concentrate on a single mode $\mathbf{k}j$, which we shall assume to be populated by photons ($n_{\mathbf{k}j}\neq 0$), the other modes being assumed to be completely depleted except for the unavoidable zero-point photons.

Suppose first that the number of photons in the $\mathbf{k}j$ mode is well-defined. This means the state of that particular mode can be described by a ket $|\,n_{\mathbf{k}j}\rangle$ which is an eigenstate of H_F corresponding to the eigenvalue $\hbar\omega_k(n_{\mathbf{k}j}+1/2)$. This state of the field is called a number state. The electric and magnetic contributions of mode $\mathbf{k}j$ to the energy density of the field, assuming periodic boundary conditions, are easily obtained using (2.9) as

$$\begin{aligned}\frac{1}{8\pi}\langle n_{\mathbf{k}j}\,|\,\mathbf{E}^2(\mathbf{x},t)\,|\,n_{\mathbf{k}j}\rangle &= \frac{1}{8\pi}\langle n_{\mathbf{k}j}\,|\,\mathbf{H}^2(\mathbf{x},t)\,|\,n_{\mathbf{k}j}\rangle\\ &= \frac{1}{V}\frac{1}{2}\left(n_{\mathbf{k}j}+\frac{1}{2}\right)\hbar\omega_k \end{aligned}\qquad (2.35)$$

which shows that the two contributions are equal at any point in space and that the energy of the $n_{\mathbf{k}j}$ photons is time-independent and distributed uniformly over all space. The latter feature is of course strictly related to the mathematical nature of the normal modes chosen to expand the field, and it depends on the assumptions of periodic boundary conditions and of absence of sources or dielectric boundaries. Moreover, since the quantum averages of the field amplitudes over $|\,n_{\mathbf{k}j}\rangle$ vanish, the variance of these field amplitudes due to the excited mode are immediately obtained from (2.35). For example

$$\begin{aligned}(\Delta\mathbf{E})^2 &= \langle n_{\mathbf{k}j}\,|\,\mathbf{E}^2(\mathbf{x},t)\,|\,n_{\mathbf{k}j}\rangle - \left(\langle n_{\mathbf{k}j}\,|\,\mathbf{E}(\mathbf{x},t)\,|\,n_{\mathbf{k}j}\rangle\right)^2\\ &= \frac{4\pi}{V}\left(n_{\mathbf{k}j}+\frac{1}{2}\right)\hbar\omega_k\end{aligned}\qquad (2.36)$$

Thus the fluctuations of the electric field amplitude of a single mode in a number state increase with increasing excitation of the mode, and are a minimum for the vacuum state with $n_{\mathbf{k}j}=0$.

Suppose now that the **k***j* field mode is in a state $| \alpha\rangle$ given by (Glauber 1963, 1970, Titulaer and Glauber 1965)

$$\begin{aligned} | \alpha\rangle = D(\alpha) | 0_{\mathbf{k}j}\rangle &= \exp\left(\alpha a^\dagger_{\mathbf{k}_j} - \alpha^* a_{\mathbf{k}_j}\right) | 0_{\mathbf{k}j}\rangle \\ &= e^{-\frac{1}{2}|\alpha|^2} \sum_{n_{\mathbf{k}j}} \frac{(\alpha a^\dagger_{\mathbf{k}_j})^{n_{\mathbf{k}j}}}{n_{\mathbf{k}j}!} | 0_{\mathbf{k}j}\rangle \\ &= e^{-\frac{1}{2}|\alpha|^2} \sum_{n_{\mathbf{k}j}} \frac{\alpha^{n_{\mathbf{k}j}}}{\sqrt{n_{\mathbf{k}j}!}} | n_{\mathbf{k}j}\rangle \end{aligned} \tag{2.37}$$

where α is a c-number, $D(\alpha)$ is the coherent displacement operator and all operators are taken at $t = 0$. This state is called coherent (see e.g. Sargent *et al.* 1974). As we will see, $| \alpha\rangle$ is not an eigenstate of H_F, but it is an eigenstate of $a_{\mathbf{k}_j}$. In fact, using (see e.g. Louisell 1990)

$$\begin{aligned} \left[a_{\mathbf{k}_j}, D(\alpha)\right] &= D(\alpha)\Big\{\left[a_{\mathbf{k}_j}, \alpha a^\dagger_{\mathbf{k}_j} - \alpha^* a_{\mathbf{k}_j}\right] \\ &\quad + \frac{1}{2!}\left[\left[a_{\mathbf{k}_j}, \alpha a^\dagger_{\mathbf{k}_j} - \alpha^* a_{\mathbf{k}_j}\right], \alpha a^\dagger_{\mathbf{k}_j} - \alpha^* a_{\mathbf{k}_j}\right] + \ldots\Big\} \\ &= \alpha D(\alpha) \end{aligned} \tag{2.38}$$

one obtains immediately

$$a_{\mathbf{k}_j} | \alpha\rangle = a_{\mathbf{k}_j} D(\alpha) | 0_{\mathbf{k}j}\rangle = \alpha D(\alpha) | 0_{\mathbf{k}j}\rangle = \alpha | \alpha\rangle \tag{2.39}$$

which shows that $|\alpha\rangle$ is an eigenvector of $a_{\mathbf{k}_j}$ corresponding to the eigenvalue α. The latter is complex since $a_{\mathbf{k}_j}$ is non-Hermitian; therefore $| \alpha\rangle$ is not an eigenvector of $a^\dagger_{\mathbf{k}_j}$.

The contribution of the **k***j* mode to the quantum average of H_F on state $| \alpha\rangle$ is immediately obtained in the form

$$\langle\alpha | \hbar\omega_k\left(a^\dagger_{\mathbf{k}_j} a_{\mathbf{k}_j} + 1/2\right) | \alpha\rangle = \hbar\omega_k \overline{(n_{\mathbf{k}j} + 1/2)} \; ; \tag{2.40}$$

$$\begin{aligned} \overline{(n_{\mathbf{k}j} + 1/2)} &= \sum_{n_{\mathbf{k}j}=0}^{\infty} P(n_{\mathbf{k}j})(n_{\mathbf{k}j} + 1/2) \; ; \\ \overline{n_{\mathbf{k}j}} &\equiv \langle\alpha | a^\dagger_{\mathbf{k}_j} a_{\mathbf{k}_j} | \alpha\rangle = | \alpha |^2 \end{aligned} \tag{2.41}$$

where $P(n)$ is the Poisson distribution defined as

$$P(n) = \frac{e^{-|\alpha|^2} | \alpha |^{2n}}{n!} \tag{2.42}$$

The Poisson distribution $P(n_{\mathbf{k}j})$, in view of the last equality in (2.37), represents the probability of finding in a coherent state $| \alpha\rangle$ the $\mathbf{k}j$ mode populated by $n_{\mathbf{k}j}$ photons, and it peaks at $n_{\mathbf{k}j} \sim | \alpha |^2$. Moreover,

$$\begin{aligned} &\langle\alpha \mid (a^\dagger_{\mathbf{k}_j} a_{\mathbf{k}_j})^2 \mid \alpha\rangle \\ &= \langle\alpha \mid a^\dagger_{\mathbf{k}_j}(a^\dagger_{\mathbf{k}_j} a_{\mathbf{k}_j} + 1) a_{\mathbf{k}_j} \mid \alpha\rangle = | \alpha |^4 + | \alpha |^2 \,; \\ &\langle\alpha \mid a^\dagger_{\mathbf{k}_j} a_{\mathbf{k}_j} \mid \alpha\rangle = | \alpha |^2 \end{aligned} \tag{2.43}$$

Thus the width of the photon population distribution of the $\mathbf{k}j$ mode in a coherent state $| \alpha\rangle$ is given by

$$\sqrt{(\Delta n_{\mathbf{k}j})^2} = | \alpha | = \sqrt{\overline{n_{\mathbf{k}j}}} \tag{2.44}$$

This width increases more slowly than the average value of photon number $\overline{n_{\mathbf{k}j}}$, which shows that the peak of the Poisson distribution at $\overline{n_{\mathbf{k}j}}$ looks sharper and sharper as $\overline{n_{\mathbf{k}j}}$ increases so that the fractional width decreases. Thus one might be tempted to conclude that for large $\overline{n_{\mathbf{k}j}}$ the physical properties of a coherent state tend to those of a number state. That this is at best a misleading statement can be appreciated by evaluating the expectation value of the field amplitude operators on the coherent state $\mathbf{k}j$ with all other modes empty. For example,

$$\langle\alpha \mid \mathbf{E}(\mathbf{x}, t) \mid \alpha\rangle = i\sqrt{\frac{2\pi\hbar\omega_k}{V}}\left\{\mathbf{e}_{\mathbf{k}j}\alpha e^{-i\omega_k t} e^{i\mathbf{k}\cdot\mathbf{x}} - \mathbf{e}^*_{\mathbf{k}j}\alpha^* e^{i\omega_k t} e^{-i\mathbf{k}\cdot\mathbf{x}}\right\} \tag{2.45}$$

Taking $\mathbf{e}_{\mathbf{k}j}$ and α real in (2.45) yields (Loudon 1981)

$$\langle\alpha \mid \mathbf{E}(\mathbf{x}, t) \mid \alpha\rangle = -\sqrt{\frac{8\pi\hbar\omega_k}{V}}\,\alpha\mathbf{e}_{\mathbf{k}j}\sin(\mathbf{k}\cdot\mathbf{x} - \omega_k t) \tag{2.46}$$

which represents a travelling wave along $\hat{\mathbf{k}}$ linearly polarized along $\mathbf{e}_{\mathbf{k}j}$. The amplitude of this wave is given by

$$\sqrt{\frac{8\pi\overline{n_{\mathbf{k}j}}\,\hbar\omega_k}{V}} \tag{2.47}$$

and it does not vanish, in contrast with the case of the number state. Further, a straightforward calculation yields

$$\langle\alpha \mid \mathbf{E}^2(\mathbf{x}, t) \mid \alpha\rangle = \frac{8\pi\overline{n_{\mathbf{k}j}}\,\hbar\omega_k}{V}\sin^2(\mathbf{k}\cdot\mathbf{x} - \omega_k t) + \frac{2\pi\hbar\omega_k}{V} + ZPT' \tag{2.48}$$

where ZPT' represents the contribution from zero-point terms of all modes different from $\mathbf{k}j$. Thus the variance of the electric field amplitude

$$(\Delta \mathbf{E})^2 = \langle \alpha \mid \mathbf{E}^2(\mathbf{x}, t) \mid \alpha \rangle - (\langle \alpha \mid \mathbf{E}(\mathbf{x}, t) \mid \alpha \rangle)^2$$
$$= \frac{2\pi\hbar\omega_k}{V} + ZPT' = ZPT \tag{2.49}$$

One can see that the variance of the electric field amplitude for any coherent state is the same as that for the vacuum state: the contribution of the coherent excitation of any mode vanishes entirely. This behaviour is very different from that of a number state, where according to (2.36) the variance of the electric field amplitude depends on the degree of excitation of the mode. Analogous results can be obtained for the magnetic field amplitude.

In conclusion, the behaviour of the coherent state for a field in the absence of sources resembles that of a classical field, with the addition of the zero-point terms. This property is mirrored by the behaviour of the field amplitude operators. For example, it is possible to define a new quantum field amplitude operator $\mathbf{E}^{\text{new}}(\mathbf{x}, t)$ related to $\mathbf{E}(\mathbf{x}, t)$ by a unitary transformation induced by $D(\alpha)$. Thus, using (2.38), we find (see e.g. Cohen-Tannoudji *et al.* 1992)

$$\mathbf{E}^{\text{new}}(\mathbf{x}, t) \equiv D(\alpha)\mathbf{E}(\mathbf{x}, t)D^{-1}(\alpha)$$
$$= i\sum_{\mathbf{k}'j'} \sqrt{\frac{2\pi\hbar\omega_{k'}}{V}} \Big\{ \mathbf{e}_{\mathbf{k}'j'} a_{\mathbf{k}'j'}(t) e^{i\mathbf{k}'\cdot\mathbf{x}}$$
$$- \mathbf{e}^*_{\mathbf{k}'j'} a^\dagger_{\mathbf{k}'j'}(t) e^{-i\mathbf{k}'\cdot\mathbf{x}} \Big\}$$
$$- i\sqrt{\frac{2\pi\hbar\omega_k}{V}} \Big\{ \mathbf{e}_{\mathbf{k}j} \alpha e^{-i\omega_k t} e^{i\mathbf{k}\cdot\mathbf{x}} - \mathbf{e}^*_{\mathbf{k}j} \alpha^* e^{i\omega_k t} e^{-i\mathbf{k}\cdot\mathbf{x}} \Big\}$$
$$= \mathbf{E}(\mathbf{x}, t) - \langle \alpha \mid \mathbf{E}(\mathbf{x}, t) \mid \alpha \rangle \tag{2.50}$$

This shows that $\mathbf{E}$ can be partitioned into a classical part $\mathbf{E}_{\text{cl}} \equiv \langle \alpha \mid \mathbf{E} \mid \alpha \rangle$ and a quantum part $\mathbf{E}_Q \equiv \mathbf{E}^{\text{new}}$, such that

$$\mathbf{E}(\mathbf{x}, t) = \mathbf{E}_{\text{cl}}(\mathbf{x}, t) + \mathbf{E}_Q(\mathbf{x}, t) \tag{2.51}$$

The classical features of the field mode in state $\mid \alpha \rangle$ are included in $\mathbf{E}_{\text{cl}}$, whereas the quantum features, such as zero-point fluctuations, are included in $\mathbf{E}_Q$. In fact, in the limit of high-intensity field, the zero-point contribution can be neglected with respect to the intensity of the coherent field, which does not contribute at all to the quantum fluctuations. It is true, however, that the relationship between quantum and classical electrodynamics is not a straightforward problem and that it presents rather subtle conceptual complications (Bialynicki-Birula 1977, Stenholm 1985).

2.4 Squeezed states of the field

Suppose now that the **k**j field mode is in a quantum state defined as (see e.g. Loudon and Knight 1987)

$$S(\zeta) \mid 0_{\mathbf{k}j}\rangle \ ; \ S(\zeta) = \exp\left(\frac{1}{2}(\zeta^* a_{\mathbf{k}_j}^2 - \zeta a_{\mathbf{k}_j}^{\dagger 2})\right) \tag{2.52}$$

A state of the form (2.52) is called a squeezed state, and it is endowed with peculiar statistical properties which, apart from their undoubted scientific interests, are likely to be very useful from a technological point of view (see e.g. Walls 1983). The unitary squeezing operator S can be generalized to cover multimode squeezing (see e.g. Collett *et al.* 1987), but such an extension is not of interest to us at this stage in our discussion. Further, $S(\zeta)$ is usually applied in conjunction with the coherent displacement operator $D(\alpha)$ defined in the previous section. Here we shall limit ourselves to a brief discussion of some of the statistical properties of (2.52) which is called a squeezed vacuum state and which should be sufficient to indicate the potentialities of squeezed states. Using (see e.g. Louisell 1990)

$$e^{-A}Be^{A} = B + [B, A] + \frac{1}{2!}[[B, A], A] + \frac{1}{3!}[[[B, A], A], A] + \ldots.. \tag{2.53}$$

one easily obtains, through a simple Bogoliubov transformation (see e.g. Kaempffer 1965),

$$\begin{aligned}
&S^{-1}(\zeta)a_{\mathbf{k}_j}S(\zeta) = a_{\mathbf{k}_j}\cosh|\zeta| - a_{\mathbf{k}_j}^{\dagger}\frac{\zeta}{|\zeta|}\sinh|\zeta|;\\
&S^{-1}(\zeta)a_{\mathbf{k}_j}a_{\mathbf{k}_j}^{\dagger}S(\zeta) = a_{\mathbf{k}_j}a_{\mathbf{k}_j}^{\dagger}\cosh^2|\zeta|\\
&-\left(a_{\mathbf{k}_j}^2\frac{\zeta^*}{|\zeta|} + a_{\mathbf{k}_j}^{\dagger 2}\frac{\zeta}{|\zeta|}\right)\sinh|\zeta|\cosh|\zeta| + a_{\mathbf{k}_j}^{\dagger}a_{\mathbf{k}_j}\sinh^2|\zeta|;\\
&S^{-1}(\zeta)a_{\mathbf{k}_j}^2S(\zeta) = a_{\mathbf{k}_j}^2\cosh^2|\zeta|\\
&-\left(2a_{\mathbf{k}_j}^{\dagger}a_{\mathbf{k}_j} + 1\right)\frac{\zeta}{|\zeta|}\sinh|\zeta|\cosh|\zeta|\\
&+a_{\mathbf{k}_j}^{\dagger\,2}\frac{\zeta^2}{|\zeta|^2}\sinh^2|\zeta|
\end{aligned} \tag{2.54}$$

Assuming that no real photon exists in all modes $\mathbf{k}'j' \neq \mathbf{k}j$ of the field, one can evaluate the quantum average of various field operators. In particular, using $|0\rangle \equiv |\{0_{\mathbf{k}j}\}\rangle$,

$$\langle 0 \mid S^{-1}(\zeta)\mathbf{E}(\mathbf{x}, t)S(\zeta) \mid 0\rangle = 0 \tag{2.55}$$

differently from a coherent state field configuration. Moreover, using (2.54) in addition to (1.80) and (1.103), one has

$$\begin{aligned}\langle 0 \mid S^{-1}(\zeta)\mathbf{E}^2(\mathbf{x}, t)S(\zeta) \mid 0\rangle &= \sum_{\mathbf{k}'j'}{}' \frac{2\pi\hbar\omega_{k'}}{V} \\ &+ \frac{2\pi\hbar\omega_k}{V}\Big\{\cosh 2 \mid \zeta \mid -(-1)^j \cos 2\theta \Big[\zeta e^{2i(\mathbf{k}\cdot\mathbf{x}-\omega_k t)} \\ &+ \zeta^* e^{-2i(\mathbf{k}\cdot\mathbf{x}-\omega_k t)}\Big]\frac{1}{2 \mid \zeta \mid}\sinh 2 \mid \zeta \mid\Big\}\end{aligned} \tag{2.56}$$

where the dashed sum runs over all field modes except $\mathbf{k}j$. Assuming for example $j = 1$, $\theta = 0$ (linear polarization of $\mathbf{k}j$ mode) and taking the squeezing parameter ζ as real, (2.55) and (2.56) yield

$$\begin{aligned}&\langle 0 \mid S^{-1}(\zeta)\mathbf{E}^2(\mathbf{x}, t)S(\zeta) \mid 0\rangle - (\langle 0 \mid S^{-1}(\zeta)\mathbf{E}(\mathbf{x}, t)S(\zeta) \mid 0\rangle)^2 \\ &\equiv (\Delta\mathbf{E})^2 \\ &= \sum_{\mathbf{k}'j'}{}' \frac{2\pi\hbar\omega_{k'}}{V} + \frac{2\pi\hbar\omega_k}{V} \\ &\times \{\cosh 2\zeta + \cos 2(\mathbf{k}\cdot\mathbf{x} - \omega_k t)\sinh 2\zeta\}\end{aligned} \tag{2.57}$$

The summation over $\mathbf{k}'j'$ on the RHS of (2.57) represents the contribution to the variance of $\mathbf{E}$ arising from the zero-point fluctuation of the empty modes. The other term is the contribution of the squeezed $\mathbf{k}j$ mode. For $\zeta = 0$, that is in the absence of squeezing, the contribution of the $\mathbf{k}j$ mode reduces to that of any other empty mode. For $\zeta > 0$, however, since

$$\cosh 2\zeta - \sinh 2\zeta < 1 \tag{2.58}$$

there must be values of the total phase of the wave $\varphi = \mathbf{k}\cdot\mathbf{x} - \omega_k t$ for which this contribution of the $\mathbf{k}j$ mode is smaller than if the mode were empty. These values are graphically displayed in Figure 2.1. Thus it is possible, albeit in limited zones of space-time, to reduce the quantum mechanical noise well below the zero-point value. This idea forms the basis for many conceptual developments in quantum optics and in related fields.

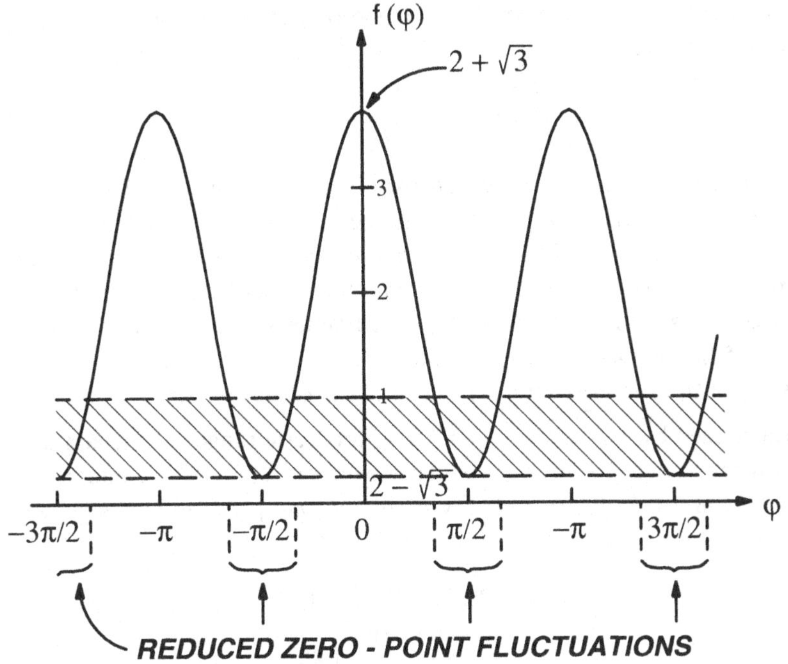

Fig. 2.1 The regions in which fluctuations of the squeezed $\mathbf{k}j$ mode are smaller than normal zero-point fluctuations are those for which $f(\varphi) = \cosh 2\zeta + \cos 2\varphi \sinh 2\zeta < 1$. In this figure we have taken $\cosh 2\zeta = 2$. The space-time region of reduced fluctuations corresponds to values of $\varphi = \mathbf{k} \cdot \mathbf{x} - \omega_k t$ for which $f(\varphi)$ lies inside the shaded area.

Turning to the statistical properties of the photon distribution in a vacuum squeezed state, from (2.54) one has

$$\begin{aligned} S^{-1}(\zeta) a_{\mathbf{k}_j}^\dagger a_{\mathbf{k}_j} S(\zeta) &= S^{-1}(\zeta) a_{\mathbf{k}_j} a_{\mathbf{k}_j}^\dagger S(\zeta) - 1 \\ &= a_{\mathbf{k}_j}^\dagger a_{\mathbf{k}_j} \cosh 2 \mid \zeta \mid - \left(a_{\mathbf{k}_j}^2 \zeta^* + a_{\mathbf{k}_j}^{\dagger 2} \zeta \right) \frac{1}{2 \mid \zeta \mid} \sinh 2 \mid \zeta \mid \\ &+ \sinh^2 \mid \zeta \mid \end{aligned} \tag{2.59}$$

Consequently, the average number of photons in a vacuum squeezed state (2.52) is

$$\overline{n_{\mathbf{k}j}} = \langle \{0_{\mathbf{k}j}\} \mid S^{-1}(\zeta) a_{\mathbf{k}_j}^\dagger a_{\mathbf{k}_j} S(\zeta) \mid \{0_{\mathbf{k}j}\} \rangle = \sinh^2 \mid \zeta \mid \tag{2.60}$$

Moreover, a simple calculation based on (2.59) yields

$$\langle \{0_{\mathbf{k}j}\} \mid S^{-1}(\zeta) \left(a_{\mathbf{k}_j}^\dagger a_{\mathbf{k}_j} \right)^2 S(\zeta) \mid \{0_{\mathbf{k}j}\} \rangle = \frac{1}{2} \sinh^2 2 \mid \zeta \mid + \sinh^4 \mid \zeta \mid \tag{2.61}$$

Hence one obtains for the width of the photon population distribution in a squeezed state (2.52)

$$\sqrt{(\Delta n_{\mathbf{k}j})^2} = \frac{1}{\sqrt{2}} \sinh 2 \mid \zeta \mid \tag{2.62}$$

From (2.62) and (2.60) we see that for a squeezed state (2.52) $\sqrt{(\Delta n_{\mathbf{k}j})^2} > \sqrt{\overline{n}_{\mathbf{k}j}}$. Comparison with (2.44) hence shows that for a squeezed vacuum, as state (2.52) is sometimes called, fluctuations in photon numbers are larger than for a coherent state. This behaviour is usually referred to as super-Poissonian. Squeezed states of a form more general than (2.52), obtained as $D(\alpha)S(\zeta) \mid 0_{\mathbf{k}j}\rangle$ (Caves 1981, Loudon and Knight 1987) with $D(\alpha)$ defined by (2.37), can also display sub-Poissonian behaviour characterized by $\sqrt{(\Delta n_{\mathbf{k}j})^2} < \sqrt{\overline{n}_{\mathbf{k}j}}$ in appropriate ranges of the complex parameters α and ζ. Finally, it can be shown (see Yuen 1976, Loudon and Knight 1987) that the probability of finding $n_{\mathbf{k}j}$ photons in a squeezed state (2.52) is

$$\begin{aligned} P(n_{\mathbf{k}j}) &= \mid \langle n_{\mathbf{k}j} \mid S(\zeta) \mid \{0_{\mathbf{k}j}\}\rangle \mid^2 \\ &= \frac{\left(\frac{1}{2}\tanh \mid \zeta \mid\right)^{n_{\mathbf{k}j}}}{n_{\mathbf{k}j}! \cosh \mid \zeta \mid} \mid H_{n_{\mathbf{k}j}}(0) \mid^2 \end{aligned} \tag{2.63}$$

where H_n is the Hermite polynomial of degree n.

From the experimental point of view, squeezed states have been generated by Slusher *et al.* (1985) and by Shelby *et al.* (1986) using four wave mixing techniques, and by Wu *et al.* (1986) in the output of an optical parametric oscillator. Squeezing in photon number fluctuations has also been observed in the output of semiconductor lasers (see e.g. Machida *et al.* 1987). From a theoretical point of view, we just mention an interesting development based on thermofield analysis (Takahashi and Umezawa 1975). This analysis introduces a representation in which the ensemble averages of statistical mechanics are replaced by expectation values in a temperature-dependent vacuum state. Thermofield analysis has been used by Barnett and Knight (1985) to discuss two-mode squeezing of light fluctuations.

2.5 Thermal states of the field

The states discussed so far are pure quantum states, that is, states which can be described by a well-defined ket in the Hilbert space of the system. Other states exist, however, which do not possess such a simple description, but which correspond to a statistical ensemble. In the case of the electromagnetic field, one must imagine that the object to be described consists of a statistical ensemble of a large number of subsystems described by identical free field Hamiltonians H_F. Each subsystem is in a pure state of its own Hamiltonian H_F, but this pure state is not the same for different subsystems. One has no way of knowing on which subsystem a given observation is being performed. The quantum average of a physical field observable O must be averaged over all subsystems of the ensemble in order to obtain a result which can be compared with observation (see e.g. Kaempffer 1965).

The situation described above arises in particular when a field is in thermal equilibrium with a reservoir at finite temperature. The interaction of the field with the reservoir leads to a state of the field which is a superposition of pure states, but due to the statistical nature of this reservoir the relative phases of the superposition are completely lost. The state attained by the field in these circumstances is called a thermal state, and it can be described mathematically in terms of a density matrix, but the discussion of such a description goes beyond the scope of our treatment. We shall limit our considerations to observing that a given mode $\mathbf{k}j$ of the field has, in principle, a different population $n_{\mathbf{k}j}$ for each subsystem of the statistical ensemble, with a probability $P(n_{\mathbf{k}j})$ which depends on the temperature of the reservoir and possibly on the nature of the field-reservoir interaction. In thermal equilibrium at temperature T, the probability of finding $n_{\mathbf{k}j}$ photons in the mode $\mathbf{k}j$ is given by Boltzmann statistics as

$$\begin{aligned} P(n_{\mathbf{k}j}) &= \frac{e^{-n_{\mathbf{k}j}\hbar\omega_k/KT}}{\sum_{n'_{\mathbf{k}j}=0}^{\infty} e^{-n'_{\mathbf{k}j}\hbar\omega_k/KT}} \\ &= \left(1 - e^{-\hbar\omega_k/KT}\right) e^{-n_{\mathbf{k}j}\hbar\omega_k/KT} \end{aligned} \tag{2.64}$$

where K is the Boltzmann constant. Then one can obtain the mean value of $n_{\mathbf{k}j}$ by evaluating

$$\begin{aligned} \overline{n_{\mathbf{k}j}} &= \left(1 - e^{-\hbar\omega_k/KT}\right) \sum_{n'_{\mathbf{k}j}=0}^{\infty} n'_{\mathbf{k}j} e^{-n'_{\mathbf{k}j}\hbar\omega_k/KT} \\ &= \frac{1}{e^{\hbar\omega_k/KT} - 1} \end{aligned} \tag{2.65}$$

which is the well-known Bose-Einstein distribution for vanishing chemical potential.

As for the variance of the photon number distribution, one has (Loudon 1981)

$$\overline{n_{\mathbf{k}j}^2} = \sum_{n'_{\mathbf{k}j}=0}^{\infty} n'^2_{\mathbf{k}j} P(n'_{\mathbf{k}j})$$

$$= \left(1 - e^{\hbar\omega_k/KT}\right)\left(\frac{\partial}{\partial x_k}\right)^2 \sum_{n'_{\mathbf{k}j}} e^{-n'_{\mathbf{k}j}x_k}$$

$$= \frac{e^{\hbar\omega_k/KT}+1}{\left(e^{\hbar\omega_k/KT}-1\right)^2} = \overline{n_{\mathbf{k}j}} + 2\left(\overline{n_{\mathbf{k}j}}\right)^2 \tag{2.66}$$

where $x_k = \hbar\omega_k/KT$. Then the variance follows

$$\left(\Delta n_{\mathbf{k}j}\right)^2 = \overline{n_{\mathbf{k}j}}^2 + \overline{n_{\mathbf{k}j}} \tag{2.67}$$

from which we see that the width of the distribution for a thermal state increases almost linearly with the average occupation number. The difference in the statistical properties of a thermal and of a coherent state of the field has been emphasized in an early experiment by Arecchi *et al.* (1966).

2.6 Nonlocalizability of the photon

The previous discussion of the quantized e.m. field in terms of photons might induce us to think that photons possess particle-like properties to the extent that it is possible to localize them at a point in space, at least at a given instant. That the situation is not so simple can be realized by noting that it is in fact impossible to define a position operator for the photon (Newton and Wigner 1949) with commuting components (Mourad 1993) in the same way as for a massive particle, e.g. an electron. We shall not follow the rather abstract treatment leading to this conclusion, but will show that serious difficulties arise when one tries to build a photon wavepacket localized at one point in space.

Consider the general one-photon state

$$|\psi\rangle = \sum_{\mathbf{k}j} c_{\mathbf{k}j} a^\dagger_{\mathbf{k}_j} \mid \{0_{\mathbf{k}j}\}\rangle \tag{2.68}$$

where $c_{\mathbf{k}j}$ are arbitrary c-number coefficients. The quantum average of the electric part of the energy density on a state of the form (2.68), after some straightforward algebra, can be put in the form

$$\frac{1}{8\pi}\langle\psi \mid \mathbf{E}^2(\mathbf{x},t) \mid \psi\rangle$$

$$= \frac{1}{8\pi} \mid \sum_{\mathbf{k}j} \sqrt{\frac{2\pi\hbar\omega_k}{V}} \mathbf{e}_{\mathbf{k}j} c_{\mathbf{k}j} e^{-i\mathbf{k}\cdot\mathbf{x}} \mid^2 + \frac{1}{2V}\sum_{\mathbf{k}j}\frac{\hbar\omega_k}{2} \tag{2.69}$$

Clearly, the last term on the RHS of (2.69) represents the electric contribution to the zero-point energy of the field. Hence the first term on the RHS of the same expression represents the contribution of the single photon present in the field. The question one could ask is: is it possible by a suitable choice of the coefficients $c_{\mathbf{k}j}$ to localize the latter contribution at a single point in space, e.g. at $\mathbf{x} = 0$?

In order to answer this question we remark first that the quantity

$$\mathbf{F}(\mathbf{x}) = \sum_{\mathbf{k}j} \sqrt{\frac{2\pi\hbar\omega_k}{V}} \mathbf{e}_{\mathbf{k}j} c_{\mathbf{k}j} e^{-i\mathbf{k}\cdot\mathbf{x}} \tag{2.70}$$

whose squared modulus appears on the RHS of (2.69) is actually a transverse vector field. Moreover, for $\mid \mathbf{F}(\mathbf{x}) \mid^2$ to be localized at the origin, it is sufficient to require that $\mathbf{F}(\mathbf{x})$ be localized at $\mathbf{x} = 0$. One can write these two conditions as

$$\mathbf{F}(\mathbf{x}) = \mathbf{F}(0)\delta(\mathbf{x})\ ;\ \nabla\cdot\mathbf{F}(\mathbf{x}) = 0 \tag{2.71}$$

which are incompatible, since

$$\nabla\cdot[\mathbf{F}(0)\delta(\mathbf{x})] = F_i(0)\frac{\partial}{\partial x_i}\frac{1}{(2\pi)^3}\int e^{i\mathbf{k}\cdot\mathbf{x}} d^3\mathbf{k}$$

$$= F_i(0)\frac{i}{(2\pi)^3}\int k_i e^{i\mathbf{k}\cdot\mathbf{x}} d^3\mathbf{k} \tag{2.72}$$

which does not vanish everywhere, as discussed in Section 2.1. In fact, it has been shown that the best localization possible for a one-photon state around $\mathbf{x} = 0$ yields an energy density of the form $\mid x \mid^{-7}$ (as discussed by Amrein 1969 and by Pike and Sarkar 1987).

References

W.O. Amrein (1969). *Helv. Phys. Acta* **42**, 149
F.T. Arecchi, E. Gatti, A. Sona (1966). *Phys. Lett.* **20**, 27
S.M. Barnett, P.L. Knight (1985). *J. Opt. Soc. Am. B* **2**, 467
I. Bialynicki-Birula (1977) *Acta Phys. Austr.*, Suppl. XVIII, 111

J.D. Bjorken, S. Drell (1964). *Relativistic Quantum Mechanics*, Vol. I (McGraw-Hill Book Co., New York)
C.M. Caves (1981). *Phys. Rev. D* **23**, 1693
C. Cohen-Tannoudji, J. Dupont-Roc, G. Grynberg (1992). *Atom-Photon Interactions* (John Wiley and Sons, New York)
M.J. Collett, R. Loudon, C.W. Gardiner (1987). *J. Mod. Opt.* **34**, 881
D.P. Craig, T. Thirunamachandran (1984). *Molecular Quantum Electrodynamics* (Academic Press Inc., London)
R.J. Glauber (1963). *Phys. Rev.* **130**, 2529
R.J. Glauber (1963) *Phys. Rev.* **131**, 2766
R.J. Glauber (1970), in *Quantum Optics*, M. Kay, A. Maitland (eds.) (Academic Press Inc., London), p. 53
W. Heitler (1960). *The Quantum Theory of Radiation* (Oxford University Press, London)
F.A. Kaempffer (1965). *Concepts in Quantum Mechanics* (Academic Press Inc., New York)
R. Loudon (1981). *The Quantum Theory of Light* (Oxford University Press, London)
R. Loudon, P.L. Knight (1987). *J. Mod. Opt.* **34**, 709
W.H. Louisell (1990). *Quantum Statistical Properties of Radiation* (John Wiley and Sons, New York)
S. Machida, Y. Yamamoto, Y. Itaya (1987). *Phys. Rev. Lett.* **58**, 1000
J. Mourad (1993). *Phys. Lett. A* **182**, 319
T.D. Newton, E.P. Wigner (1949). *Rev. Mod. Phys.* **21**, 400
R. Passante, G. Compagno, F. Persico (1985). *Phys. Rev. A* **31**, 2827
E.R. Pike, S. Sarkar (1987). *Phys. Rev. A* **35**, 926
P. Roman (1965). *Advanced Quantum Theory* (Addison-Wesley Publishing Co., Reading, Ma.)
M. Sargent III, M.O. Scully, W.E. Lamb Jr. (1974). *Laser Physics* (Addison-Wesley Publishing Co., Reading, Ma.)
R.M. Shelby, M.D. Levenson, S.M. Perlmutter, R.G. DeVoe, D.F. Walls (1986). *Phys. Rev. Lett.* **57**, 691
R.E. Slusher, L.W. Hollberg, B. Yurke, J.C. Mertz, J.F. Valley (1985). *Phys. Rev. Lett.* **55**, 2409
S. Stenholm (1985) *Ann. de Phys.* **10**, 817
Y. Takahashi, H. Umezawa (1975). *Collect. Phenom.* **2**, 55
U.M. Titulaer, R.J. Glauber (1965). *Phys. Rev. B* **140**, 676
D.F. Walls (1983). *Nature* **306**, 141
L.A. Wu, H.J. Kimble, J.L. Hall, H. Wu (1986). *Phys. Rev. Lett.* **57**, 2520
H.P. Yuen (1976). *Phys. Rev. A* **13**, 2226

Further reading

A discussion on the localizability of the photon can be found in
C. Cohen-Tannoudji, J. Dupont-Roc, G. Grynberg, *Photons and Atoms* (John Wiley and Sons, New York 1989).

An excellent review of modern concepts of vacuum in quantum field theory is given in the collection of essays
S. Saunders, H.R. Brown, *The Philosophy of Vacuum* (Oxford University Press, Oxford 1991).

The concept of vacuum in nonrelativistic QED has been reviewed in the following collection of papers

F. Persico, E.A. Power (eds.) *Vacuum in Nonrelativistic Matter-Radiation Systems, Physica Scripta* **T21**, (1988)

and by the recently published book

P.W. Milonni *The Quantum Vacuum* (Academic Press, Inc., San Diego 1994).

For recent reviews of nonclassical states of the field, see

D.F. Walls, Nonclassical Optical Phenomena, *Sci. Progress Oxford* **74**, 291 (1990)
W.M. Zhang, D.H. Feng, R. Gilmore, *Rev. Mod. Phys.* **62**, 867 (1990)
M.C. Teich, B.E.A. Saleh, in *Progress in Optics* XXVI, E. Wolf (ed.) (North-Holland, Amsterdam 1988), p. 1.

Squeezed states of light have been reviewed by

Y. Yamamoto, S. Machida, N. Imoto, M.K. Tagawa, G. Björk, *J. Opt. Soc. Am. B* **4**, 1645 (1987)
M.C. Teich, B.E.A. Saleh, *Quantum Opt.* **1**, 153 (1989)
D.F. Walls, P.D. Drummond, A. Lane, M. Marte, M.D. Reid, H. Ritsch, in *Squeezed and Nonclassical Light*, P. Tombesi, E.R. Pike (eds.) (Plenum Publishing Co., 1989), p. 1
R. Muñor-Tapica, *Am. J. Phys.* **61**, 1005 (1993).

An interesting space-time description of the squeezed states of the e.m. field has been produced by

Z. Bialynicka-Birula, I. Bialynicki-Birula, *J. Opt. Soc. Am. B* **4**, 1621 (1987).

A collection of early papers on the statistical properties of classical and quantum electromagnetic fields is the following

L. Mandel, E. Wolf, *Selected Papers on Coherence and Fluctuations of Light*, Vols. I,II (Dover Publications Inc., New York 1980).

An introduction to quantum optics together with a selection of early papers on this subject can be found in

P.L. Knight, L. Allen, *Concepts of Quantum Optics* (Pergamon Press, Oxford 1983).

3
The quantum matter field

Introduction. In the previous two chapters we have discussed electrodynamics in the absence of charges and currents. We are now ready to investigate the nature of these charges and currents. Thus in this chapter we introduce the concept of matter field, both classical and quantized, which as we will see acts as a source of the electromagnetic field. The difficulties encountered in the definition of convenient wave equations (Klein-Gordon and Dirac) for a relativistic particle are examined in Section 3.1, and they lead naturally to consider these equations as equations of motion of a field, obtainable from an appropriate field Lagrangian. Thus the probabilistic single-particle interpretation of the wave equations is abandoned, and Section 3.2 is dedicated to the Klein-Gordon field, which is introduced by an appropriate Klein-Gordon Lagrangian, yielding the Klein-Gordon equation. The Klein-Gordon field is then second-quantized, both in its real and complex versions. The eigenstates of the Hamiltonian of this second-quantized field are shown to correspond to many-particle states satisfying Bose-Einstein statistics. An analogous procedure is followed in Section 3.3 for the Dirac equation, leading to the definition of a Dirac field which upon second quantization yields a field Hamiltonian whose eigenstates correspond to many-particle states satisfying Fermi-Dirac statistics. For both fields the energy-momentum tensors are defined and various conservation properties are obtained. The difficulties from the relativistic single-particle wave equation interpretation attempted in Section 3.1 are shown to disappear from the second-quantized field point of view in terms of many-particle states. In view of this success, the idea of the second-quantized matter field is extended in Section 3.4 to the nonrelativistic Schrödinger field. Introducing a Lagrangian density, whose Euler-Lagrange equation coincides with the Schrödinger equation, leads to a Hamiltonian which is second-quantized. Its eigenstates are many-particle states whose creation and annihilation operators can satisfy Bose or Fermi commutation relations. In Section 3.5 we show that if the Schrödinger field is subjected to a local gauge transformation, its Lagrangian density is gauge-dependent, unless an appropriate term is added. This new term which makes the total Lagrangian gauge-independent contains both the Schrödinger field and a new field which is called a gauge field. This gauge field turns out to be the electromagnetic field, and the additional term in the Lagrangian density indicates the presence of an interaction between the electromagnetic and the Schrödinger field. Thus the interaction between matter and electromagnetic field is obtained as a consequence of the assumption of local gauge invariance of the Lagrangian density.

3.1 The wave equation for the free particle

As is well known from elementary quantum mechanics, the dynamics of a free particle of mass m, in the context of nonrelativistic theory, is governed by the Schrödinger equation

$$\frac{\hbar}{i}\frac{\partial\psi(\mathbf{x},t)}{\partial t}=\frac{\hbar^2}{2m}\frac{\partial^2\psi(\mathbf{x},t)}{\partial x_i^2} \tag{3.1}$$

where $\psi(\mathbf{x},t)$ is the wavefunction of the particle, which in general is a complex-valued function of position and time, whose squared modulus $|\psi(\mathbf{x},t)|^2$ yields the probability density of finding, upon a position measurement performed at time t, the particle at point $\mathbf{x}$. Thus the physical interpretation of the Schrödinger equation rests on at least two basic assumptions: the possibility of performing a position measurement at a definite space-time point and the possibility of interpreting $|\psi(\mathbf{x},t)|^2$ as a probability density. The first assumption is substantiated by the possibility of defining an Hermitian position operator $\mathbf{x}$, which in the coordinate representation is simply multiplication by the eigenvalue $\mathbf{x}$, with a complete set of eigenstates $|\mathbf{x}\rangle$. The second assumption is supported by the continuity equation

$$\frac{\partial\rho}{\partial t}+\nabla\cdot\mathbf{j}=0 \tag{3.2}$$

$$\rho=|\psi(\mathbf{x},t)|^2\,;\ \mathbf{j}=\frac{i\hbar}{2m}[\psi\nabla\psi^*-\psi^*\nabla\psi] \tag{3.3}$$

which can be derived from (3.1) and which ensures conservation of probability through the existence of a probability density current $\mathbf{j}(\mathbf{x},t)$.

The two basic assumptions above make it difficult to generalize wave equation (3.1) to the relativistic case. Perhaps the most obvious relativistic generalization of (3.1) is obtained by observing that the Schrödinger equation is related to the classical expression, valid for a free particle,

$$E = \frac{1}{2m} p_i^2 \tag{3.4}$$

by the substitution

$$E = -\frac{\hbar}{i}\frac{\partial}{\partial t}\ ;\ p_i = \frac{\hbar}{i}\frac{\partial}{\partial x_i} \tag{3.5}$$

leading to operator equation

$$\frac{\hbar}{i}\frac{\partial}{\partial t} = \frac{\hbar^2}{2m}\frac{\partial^2}{\partial x_i^2} \tag{3.6}$$

and hence to (3.1). Since within classical relativistic mechanics the generalization of (3.4) is

$$E^2 = c^2 p_i^2 + m^2 c^4 \tag{3.7}$$

one is naturally led to consider the Klein-Gordon equation

$$-\hbar^2 \frac{\partial^2 \phi}{\partial t^2} = -\hbar^2 c^2 \frac{\partial^2 \phi}{\partial x_i^2} + m^2 c^4 \phi$$

$$\text{or}\quad \frac{\partial^2 \phi}{\partial x_\mu^2} = \frac{m^2 c^2}{\hbar^2}\phi \tag{3.8}$$

as the relativistic generalization of the Schrödinger equation, which is manifestly Lorentz-covariant. This equation has plane wave solutions of the form (see e.g. Roman 1965)

$$\phi(\mathbf{x}, t) \sim e^{-i/\hbar(\mathbf{p}\cdot\mathbf{x} - Et)}\ ;\ E = \pm\sqrt{p^2 c^2 + m^2 c^4} \tag{3.9}$$

which is immediately obtained by substituting (3.9) in (3.8). Thus we find an energy spectrum which is unbounded from below. Moreover, the continuity equation which is derived from (3.8) is indeed in the right form (3.2), provided one takes

$$\rho = \frac{i\hbar}{2mc^2}\left(\phi^* \frac{\partial \phi}{\partial t} - \phi \frac{\partial \phi^*}{\partial t}\right); \tag{3.10}$$

$$\mathbf{j} = \frac{i\hbar}{2m}(\phi \nabla \phi^* - \phi^* \nabla \phi) \tag{3.11}$$

Since (3.8) is of second order in time, however, one has to assign arbitrarily both ϕ and $\frac{\partial\phi}{\partial t}$ to select a solution. Hence $\rho(\mathbf{x}, t)$ can in principle be negative for complex ϕ, whereas it vanishes identically for real ϕ. Thus $\rho(\mathbf{x}, t)$ can hardly be interpreted as a probability density as in the case of the Schrödinger equation. Both difficulties (negative energy eigenvalues and negative values of $\rho(\mathbf{x}, t)$) can be overcome, at least for a free particle, by limiting the Hilbert space arbitrarily to the positive energy half. However, with this limitation it becomes impossible to localize the particle at a particular point in space by linear superposition of the positive energy plane wave eigenfunctions: the best that one can obtain is a localization within a region of the order of $\hbar/mc$ (the Compton radius of the particle). In fact the particle position operator within the positive energy Hilbert space is no longer $\mathbf{x}$, as in the nonrelativistic case, but it takes a more complicated form (Schweber 1964). One has to conclude that the physical interpretation of the Klein-Gordon equation as a single-particle wave equation is unclear even for the simplest case of a free particle.

Another possible relativistic generalization of (3.1) was obtained by Dirac. It has the advantage of being of the first order in time, as is the Schrödinger equation (3.1). The starting point for the Dirac equation is again (3.7) in the form

$$E = \sqrt{c^2 p_i^2 + m^2 c^4} \tag{3.12}$$

followed by linearization of the RHS

$$E = \sqrt{c^2 p_i^2 + m^2 c^4} = c\alpha_i p_i + \beta mc^2 \tag{3.13}$$

Clearly the α_i and the β cannot be ordinary numbers, and it can be shown that they must satisfy

$$\alpha_i^2 = \beta_i^2 = 1; \ \{\alpha_i, \alpha_k\} = 2\delta_{ik}; \ \{\alpha_i, \beta\} = 0 \tag{3.14}$$

where curly brackets indicate anticommutators. Use of (3.5) and (3.13) thus leads to

$$\frac{\hbar}{i}\frac{\partial\phi}{\partial t} = -\left(\frac{\hbar c}{i}\alpha_i \frac{\partial}{\partial x_i} + \beta mc^2\right)\phi \tag{3.15}$$

which by means of appropriate algebraic manipulations can be cast in the form (see e.g. Roman 1965)

$$\left(\gamma_\mu \frac{\partial}{\partial x_\mu} + \kappa\right)\phi = 0\,;$$
$$\gamma_i = -i\beta\alpha_i\,;\ \gamma_4 = \beta\,;\ \kappa = mc/\hbar \tag{3.16}$$

Hence, from (3.14),

$$\{\gamma_\mu, \gamma_\nu\} = 2\delta_{\mu\nu} \tag{3.17}$$

The simplest possible set of four objects satisfying (3.17) can be shown to be a set of 4 × 4 matrices. Since 16 of such independent 4 × 4 matrices exist, there is a certain latitude in picking up only four of them. A possible choice is (Sakurai 1982)

$$\gamma_1 = \begin{pmatrix} 0 & 0 & 0 & -i \\ 0 & 0 & -i & 0 \\ 0 & i & 0 & 0 \\ i & 0 & 0 & 0 \end{pmatrix}; \ \gamma_2 = \begin{pmatrix} 0 & 0 & 0 & -1 \\ 0 & 0 & 1 & 0 \\ 0 & 1 & 0 & 0 \\ -1 & 0 & 0 & 0 \end{pmatrix};$$
$$\gamma_3 = \begin{pmatrix} 0 & 0 & -i & 0 \\ 0 & 0 & 0 & i \\ i & 0 & 0 & 0 \\ 0 & -i & 0 & 0 \end{pmatrix}; \ \gamma_4 = \begin{pmatrix} 1 & 0 & 0 & 0 \\ 0 & 1 & 0 & 0 \\ 0 & 0 & -1 & 0 \\ 0 & 0 & 0 & -1 \end{pmatrix} \tag{3.18}$$

With the γ_μ chosen to be 4 × 4 matrices, ϕ in (3.16) cannot be a scalar, as in the Klein-Gordon or in the Schrödinger equations, but must have four components. We shall indicate such a four-component object by the symbol

$$\phi(\mathbf{x}, t) = \begin{vmatrix} \phi_1(\mathbf{x}, t) \\ \phi_2(\mathbf{x}, t) \\ \phi_3(\mathbf{x}, t) \\ \phi_4(\mathbf{x}, t) \end{vmatrix} \tag{3.19}$$

The four-fold increase of the Hilbert space compared with the nonrelativistic case in (3.1) is explained in terms of two spin degrees of freedom and two types of energy states. The former feature is apparent from the fact that the orbital angular momentum $\mathbf{L} = \frac{\hbar}{i}(\mathbf{x} \times \nabla)$ of the particle does not commute with the Dirac Hamiltonian (see e.g. Roman 1965)

$$H = \frac{\hbar c}{i} \alpha_i \frac{\partial}{\partial x_i} + \beta mc^2 \tag{3.20}$$

which follows from (3.15) as the generator of infinitesimal time shifts. However the total angular momentum $\mathbf{J} = \mathbf{L} + \mathbf{S}$ with $\mathbf{S}$ given by

$$\mathbf{S} = \frac{1}{2}\hbar\boldsymbol{\sigma}; \ \sigma_1 = -i\gamma_2\gamma_3; \ \sigma_2 = -i\gamma_3\gamma_1; \ \sigma_3 = -i\gamma_1\gamma_2 \tag{3.21}$$

commutes with H. Negative energy states appear for the Dirac as well as for the Klein-Gordon free particles. Eigensolutions of (3.15) can be written in the form (see e.g. Sakurai 1982)

$$\phi(\mathbf{x},t) = \begin{cases} \sqrt{\frac{mc^2}{E_p V}}\, u^{(r)}(\mathbf{p}) e^{(i/\hbar)(\mathbf{p}\cdot\mathbf{x}-E_p t)}; \; r=1,2; E_p = \sqrt{p^2c^2+m^2c^4} \\ \sqrt{\frac{mc^2}{-E_p V}}\, u^{(r)}(\mathbf{p}) e^{(i/\hbar)(\mathbf{p}\cdot\mathbf{x}-E_p t)}; \; r=3,4; E_p = -\sqrt{p^2c^2+m^2c^4} \end{cases} \tag{3.22}$$

where $u^{(r)}(\mathbf{p})$ are four-component objects called spinors. These are orthogonal and normalized in the sense that

$$u^{(r)\dagger}(\mathbf{p})u^{(r')}(\mathbf{p}) = \frac{|\, E_p \,|}{mc^2}\delta_{rr'} \tag{3.23}$$

A representation for the γs can be shown to exist (the Foldy-Wouthuysen representation) in which states of positive energy have the two lower elements of the spinor part equal to zero and states of negative energy have the two upper elements of the spinor equal to zero. Thus one can reduce the dynamics of a free particle of positive energy to have a form similar to that of a nonrelativistic particle of spin $S = 1/2$. The action of the position operator of the particle in the Foldy-Wouthuysen representation is much more complicated than multiplication by $\mathbf{x}$ (Schweber 1964), so that, as in the Klein-Gordon case, it becomes impossible to localize the free particle by any linear combination of positive-energy solutions within a region of linear dimensions smaller than the Compton radius κ^{-1} of the particle. Thus, although from the Dirac equation (3.16) it is possible to derive a continuity equation (3.3) with

$$\rho = \phi^*\phi \; ; \; \mathbf{j}_i = ic\phi^*\gamma_4\gamma_i\phi \tag{3.24}$$

where ρ is positive definite, the difficulties of the probabilistic interpretation of the relativistic wave equation seem to persist in the Dirac theory as well as in the Klein-Gordon case.

These difficulties can be better appreciated from two other points of view. The first difficulty is connected with the appearance of negative energy states in both the Klein-Gordon and the Dirac relativistic extensions of (3.1). As long as one considers a free particle, one can argue that since the particle is isolated it will remain with positive energy indefinitely and that the negative energy solutions do not play any role. However, as soon as there is a force acting on the particle whose matrix elements connect positive and negative energy states, this force induces downwards transitions with the consequent emission of an unlimited amount of energy, which is contrary to experience. If the particle under study is a boson, this is a completely hopeless situation. If, on the other hand, the particle is a fermion, by exploiting the Pauli exclusion principle it is possible to imagine that all negative energy states are full with other particles of the same kind, so that transitions of the particle under study to negative energy states cannot take place. Even in this case the overall picture is far from satisfactory, since the wave equations under consideration are supposed to describe single-particle behaviour, whereas in this scheme one is introducing, albeit surreptitiously, an infinite number of other particles of the same kind. Second, as is evident from the form of the energy eigenvalues (3.9) and (3.22), mc^2 plays the role of a threshold energy which is the minimum required to create a particle of zero velocity. From the Heisenberg uncertainty relations, $\hbar/mc$ corresponds to the uncertainty in the position of a particle which has this threshold energy. Therefore the mere localization of a particle within a volume of linear dimensions $\hbar/mc$, corresponding to the Compton radius of the particle, establishes the energetic conditions necessary for the creation of other particles out of the vacuum. Thus we see that in relativistic quantum mechanics there exists a tendency towards a many-body reformulation of the dynamics.

The difficulties connected with the physical interpretation of the relativistic wave equations discussed above can only be resolved by a radical change of viewpoint. In fact, these difficulties arise when one insists that these equations should be considered as governing the motion of a single particle. On the other hand, there is no general principle that compels one to adopt such an interpretation. In the next sections we explore the consequences of a different assumption, namely that the relativistic wave equations actually describe the motion of a field $\phi(\mathbf{x}, t)$. The way in which this field should be related to the dynamics of a single particle is unclear at this stage, but an answer to this question will be obtained toward the end of the exploration of this new fruitful point of view.

3.2 The Klein-Gordon field

In the case of the Klein-Gordon equation one can define a Lagrangian density (Corinaldesi and Strocchi 1963)

$$\mathcal{L}(\phi, \phi^*, \partial\phi/\partial x_\mu, \partial\phi^*/\partial x_\mu) = -\frac{1}{2}\left(\frac{\partial\phi^*}{\partial x_\mu}\frac{\partial\phi}{\partial x_\mu} + \kappa^2\phi^*\phi\right)$$

$$= \frac{1}{2}\{\dot{\phi}\dot{\phi}^* - c^2(\nabla\phi)\cdot(\nabla\phi^*) - \kappa^2c^2\phi^*\phi\} \; ; \; \kappa = mc/\hbar \qquad (3.25)$$

from which the Euler-Lagrange equations take the form, analogous to (1.9) (see e.g. Henley and Thirring 1962)

$$\frac{\partial}{\partial t}\frac{\partial \mathcal{L}}{\partial \dot{\phi}} + \nabla \cdot \frac{\partial \mathcal{L}}{\partial (\nabla \phi)} = \frac{\partial \mathcal{L}}{\partial \phi} ;$$

$$\frac{1}{c^2}\ddot{\phi}^* - \nabla^2 \phi^* = -\kappa^2 \phi^* ; \quad \frac{\partial^2 \phi^*}{\partial x_\mu^2} = \kappa^2 \phi^* \tag{3.26}$$

and its complex conjugate. This is exactly the Klein-Gordon equation (3.8). Since we have taken $\phi(\mathbf{x})$ and $\phi^*(\mathbf{x})$ as the generalized coordinates, the conjugate momenta are

$$\Pi(\mathbf{x}) = \frac{\partial \mathcal{L}}{\partial \dot{\phi}} = \frac{1}{2}\dot{\phi}^*(\mathbf{x}) ;$$

$$\Pi^*(\mathbf{x}) = \frac{\partial \mathcal{L}}{\partial \dot{\phi}^*} = \frac{1}{2}\dot{\phi}(\mathbf{x}) \tag{3.27}$$

and the Hamiltonian density is

$$\mathcal{H}(\mathbf{x}) = \Pi(\mathbf{x})\dot{\phi}(\mathbf{x}) + \Pi^*(\mathbf{x})\dot{\phi}^*(\mathbf{x}) - \mathcal{L}$$

$$= \frac{1}{2}\{\dot{\phi}\dot{\phi}^* + c^2(\nabla\phi)\cdot(\nabla\phi^*) + \kappa^2 c^2 \phi\phi^*\} \tag{3.28}$$

The energy-momentum tensor can be obtained as (Corinaldesi and Strocchi 1963)

$$T_{\mu\nu} = \frac{1}{2}\left\{\frac{\partial \phi^*}{\partial x_\mu}\frac{\partial \phi}{\partial x_\nu} + \frac{\partial \phi^*}{\partial x_\nu}\frac{\partial \phi}{\partial x_\mu} - \left(\frac{\partial \phi^*}{\partial x_\lambda}\frac{\partial \phi}{\partial x_\lambda} + \kappa^2 \phi^* \phi\right)\delta_{\mu\nu}\right\} \tag{3.29}$$

A straightforward calculation shows that $\mathcal{H} = -T_{44}$. Thus again we see the Hamiltonian density transforms like the $4-4$ component of a symmetric tensor, whereas the Lagrangian density (3.25) is obviously a scalar. A little algebra and use of (3.26) shows that

$$\frac{\partial T_{\mu\nu}}{\partial x_\mu} = 0 \tag{3.30}$$

The various conservation laws follow from (3.30), in a way analogous to that discussed in Section 1.6 for the classical electromagnetic field.

The Klein-Gordon field can be expressed in terms of field amplitudes, imposing periodic boundary conditions on the surface of a cube of side L. If the field is real $\phi^* = \phi$ and

$$\phi(\mathbf{x}, t) = \frac{1}{V^{1/2}}\sum_{\mathbf{k}}(A_{\mathbf{k}}e^{i\mathbf{k}\cdot\mathbf{x}} + A_{\mathbf{k}}^* e^{-i\mathbf{k}\cdot\mathbf{x}}) \tag{3.31}$$

$(V = L^3, k_i = \frac{2\pi}{L} n_i \ (n_i = 1, 2 \ldots))$. Substituting (3.31) into (3.26) yields

$$\ddot{A}_{\mathbf{k}} + \omega_k^2 A_{\mathbf{k}} = 0 \ ; \ A_{\mathbf{k}}(t) = A_{\mathbf{k}}(0)e^{-i\omega_k t} \tag{3.32}$$

where $\omega_k = [c^2k^2 + m^2c^4/\hbar^2]^{1/2}$. Thus

$$\dot{\phi}(\mathbf{x}) = -\frac{i}{V^{1/2}} \sum_{\mathbf{k}} \omega_k \left(A_{\mathbf{k}} e^{i\mathbf{k}\cdot\mathbf{x}} - A_{\mathbf{k}}^* e^{-i\mathbf{k}\cdot\mathbf{x}}\right);$$

$$\nabla\phi(\mathbf{x}) = \frac{i}{V^{1/2}} \sum_{\mathbf{k}} \mathbf{k} \left(A_{\mathbf{k}} e^{i\mathbf{k}\cdot\mathbf{x}} - A_{\mathbf{k}}^* e^{-i\mathbf{k}\cdot\mathbf{x}}\right) \tag{3.33}$$

Using (3.31) and (3.33) the Hamiltonian density (3.28) as well as the total Hamiltonian can be expanded in terms of the field amplitudes. A simple calculation yields

$$\Pi(\mathbf{x}) = \frac{\partial \mathcal{L}}{\partial \dot{\phi}} = \dot{\phi}(\mathbf{x}) \ ;$$

$$\mathcal{H}(\mathbf{x}) = \Pi\dot{\phi} - \mathcal{L} = \frac{1}{2}\left\{\dot{\phi}^2 + c^2(\nabla\phi)^2 + \kappa^2 c^2 \phi^2\right\} \ ;$$

$$H = \int_V \mathcal{H}(\mathbf{x}) d^3\mathbf{x} = \sum_{\mathbf{k}} \omega_k^2 \left(A_{\mathbf{k}} A_{\mathbf{k}}^* + A_{\mathbf{k}}^* A_{\mathbf{k}}\right) \tag{3.34}$$

where for future purposes the order of the amplitudes has been preserved throughout the calculation in spite of their c-number nature, and where

$$\int_V e^{i(\mathbf{k}-\mathbf{k}')\cdot\mathbf{x}} d^3\mathbf{x} = V\delta_{\mathbf{k}\mathbf{k}'} \tag{3.35}$$

has been used.

If the field is complex, it is possible to introduce two real fields ϕ_1 and ϕ_2 such that

$$\phi = \phi_1 + i\phi_2 \ ; \phi^* = \phi_1 - i\phi_2 \tag{3.36}$$

Substitution of (3.36) into (3.25) yields immediately

$$\mathcal{L} = \frac{1}{2} \sum_{j=1}^{2} \left\{\dot{\phi}_j^2 - c^2 (\nabla\phi_j)^2 - c^2\kappa^2\phi_j^2\right\} \tag{3.37}$$

which shows that the complex field can be analysed in terms of two independent real fields. Each of them can be expanded in complex amplitudes

$$\phi_j(\mathbf{x}) = \frac{1}{V^{1/2}} \sum_{\mathbf{k}} \left(A_{j\mathbf{k}} e^{i\mathbf{k}\cdot\mathbf{x}} + A_{j\mathbf{k}}^* e^{-i\mathbf{k}\cdot\mathbf{x}}\right) \tag{3.38}$$

and the total Hamiltonian can be expressed as

$$H = \sum_{j=1}^{2} \sum_{\mathbf{k}} \omega_k^2 \left(A_{j\mathbf{k}} A_{j\mathbf{k}}^* + A_{j\mathbf{k}}^* A_{j\mathbf{k}} \right) \tag{3.39}$$

The Klein-Gordon matter field is now quantized, proceeding in the canonical fashion. For the real field, on the basis of (3.34), one imposes

$$\begin{aligned} [\phi(\mathbf{x}), \Pi(\mathbf{x}')] &= i\hbar\delta(\mathbf{x} - \mathbf{x}') \\ \text{or } \left[\phi(\mathbf{x}), \dot{\phi}(\mathbf{x}')\right] &= i\hbar\delta(\mathbf{x} - \mathbf{x}') \end{aligned} \tag{3.40}$$

When the Klein-Gordon field (3.31) is expressed in terms of the scaled complex amplitudes $a_{\mathbf{k}}$ given by

$$A_{\mathbf{k}} = \sqrt{\frac{\hbar}{2\omega_k}} a_{\mathbf{k}} \tag{3.41}$$

the quantization condition (3.40) follows immediately if one takes the Bose commutation relations

$$\left[a_{\mathbf{k}}, a_{\mathbf{k}'}^{\dagger}\right] = \delta_{\mathbf{k}\mathbf{k}'} \; ; \; [a_{\mathbf{k}}, a_{\mathbf{k}'}] = 0 \tag{3.42}$$

Thus the Hamiltonian operator of the real Klein-Gordon field is obtained from (3.34) as

$$H = \sum_{\mathbf{k}} \hbar\omega_k \left(a_{\mathbf{k}}^{\dagger} a_{\mathbf{k}} + \frac{1}{2} \right) \tag{3.43}$$

which describes a set of bosons of energy $\hbar c[k^2 + \kappa^2]^{1/2}$, rather than a single particle as in the naive interpretation of the Klein-Gordon equation (3.8). The particle number operator is given by

$$N = \sum_{\mathbf{k}} a_{\mathbf{k}}^{\dagger} a_{\mathbf{k}} \tag{3.44}$$

The following operator

$$\begin{aligned} \rho(\mathbf{x}) &= -\frac{i}{\hbar} [\dot{\phi}_-(\mathbf{x})\phi_+(\mathbf{x}) - \phi_-(\mathbf{x})\dot{\phi}_+(\mathbf{x})] \; ; \\ \phi_+(\mathbf{x}) &= \frac{1}{V^{1/2}} \sum_{\mathbf{k}} \sqrt{\frac{\hbar}{2\omega_k}} a_{\mathbf{k}} e^{i\mathbf{k}\cdot\mathbf{x}} \; ; \; \phi_-(\mathbf{x}) = [\phi_+(\mathbf{x})]^{\dagger} \end{aligned} \tag{3.45}$$

yields N upon integration over V, and can thus be taken as the operator for the particle density of the Klein-Gordon field. We see that it has a different form from what one would expect on the basis of (3.10), which would in any case vanish identically for a real field. $\rho(\mathbf{x})$, however, has the interesting property that it does not commute at different points. In fact, a little algebra leads to (Henley and Thirring 1962)

$$[\rho(\mathbf{x}),\rho(\mathbf{x}')] = \frac{1}{2}\left[\dot{\phi}_-(\mathbf{x})\dot{\phi}_+(\mathbf{x}') - \dot{\phi}_-(\mathbf{x}')\dot{\phi}_+(\mathbf{x})\right]\Delta(\mathbf{x}-\mathbf{x}')$$
$$+\frac{1}{2}[\phi_-(\mathbf{x})\phi_+(\mathbf{x}') - \phi_-(\mathbf{x}')\phi_+(\mathbf{x})]\Gamma(\mathbf{x}-\mathbf{x}')\,;$$
$$\Delta(\mathbf{x}) = \frac{\hbar}{V}\sum_{\mathbf{k}}\frac{1}{\omega_k}e^{i\mathbf{k}\cdot\mathbf{x}} = -\frac{m}{4\pi x}H_1^{(1)}(i\kappa x)\,;$$
$$\Gamma(\mathbf{x}) = \frac{\hbar}{V}\sum_{\mathbf{k}}\omega_k e^{i\mathbf{k}\cdot\mathbf{x}} = i\frac{\kappa mc^2}{4\pi x^2}H_2^{(1)}(i\kappa x) \tag{3.46}$$

where $H_i^{(1)}$ is a Haenkel function of order i (see e.g. Abramowitz and Stegun 1965). For $\kappa x >> 1$ both Δ and Γ behave as $e^{-\kappa x}$ and the commutator in (3.46) can be safely neglected, which, however, is not the case for $x < \kappa^{-1}$. What (3.46) tells us is that it is impossible to measure the number of particles within a volume of linear dimensions smaller than κ^{-1}, since one does not know the number of particles in an adjacent volume of the same size. One can see what happens, since localizing a particle within κ^{-1} implies giving the particle an energy $\sim (\hbar\kappa)^2/m = mc^2$ which is the minimum energy required to create a second particle. Thus any attempt to measure the position of a particle with precision larger than κ^{-1} results in the production of other particles. This displays clearly the limits of the interpretation of the Klein-Gordon equation as a single particle wave equation. On the contrary, the formalism ensuing from the interpretation of (3.8) as the equation of the field leads to a mathematically coherent incorporation of the many-particle features into the dynamics of the system.

If the field is complex, on the basis of (3.27) the quantization condition is

$$[\phi(\mathbf{x}),\dot{\phi}^\dagger(\mathbf{x}')] = 2i\hbar\delta(\mathbf{x}-\mathbf{x}') \tag{3.47}$$

Since

$$\phi(\mathbf{x}) = \frac{1}{V^{1/2}}\sum_{\mathbf{k}}\{(A_{1\mathbf{k}} + iA_{2\mathbf{k}})e^{i\mathbf{k}\cdot\mathbf{x}} + (A_{1\mathbf{k}}^* + iA_{2\mathbf{k}}^*)e^{-i\mathbf{k}\cdot\mathbf{x}}\}\,;$$
$$\dot{\phi}^*(\mathbf{x}) = -\frac{i}{V^{1/2}}\sum_{\mathbf{k}}\{(A_{1\mathbf{k}} - iA_{2\mathbf{k}})e^{i\mathbf{k}\cdot\mathbf{x}}$$
$$-(A_{1\mathbf{k}}^* - iA_{2\mathbf{k}}^*)e^{-i\mathbf{k}\cdot\mathbf{x}}\}\omega_k \tag{3.48}$$

and rescaling the field amplitudes as

$$A_{j\mathbf{k}} = \sqrt{\frac{\hbar}{2\omega_k}}\, a_{j\mathbf{k}} \tag{3.49}$$

it is easy to see that condition (3.47) is met by imposing

$$\left[a_{j\mathbf{k}}, a^{\dagger}_{j'\mathbf{k}'}\right] = \delta_{jj'}\delta_{\mathbf{k}\mathbf{k}'} \; ; \; \left[a_{j\mathbf{k}}, a_{j'\mathbf{k}'}\right] = 0 \tag{3.50}$$

When (3.50) are used in the Hamiltonian (3.39), one has

$$H = \sum_{j=1}^{2}\sum_{\mathbf{k}} \hbar\omega_k \left(a^{\dagger}_{j\mathbf{k}} a_{j\mathbf{k}} + \frac{1}{2}\right) \tag{3.51}$$

from which one sees that the complex Klein-Gordon field is described in terms of two different kinds of Bose particles with the same mass. We note that the real field Hamiltonian (3.43) as well as that for the complex field (3.51) do not possess a negative energy spectrum. This resolves the paradox of negative energy states of the Klein-Gordon equation, whose status is now degraded to the equation of motion for a field amplitude, whose negative eigenvalues do not have any particular physical meaning. Also, the other paradox, of a negative probability such as could be obtained from (3.10) by volume integration, can be resolved in the new approach. First we note that extension of the wave equation to the relativistic case, as obtained in Section 3.1, involved assuming the dimension of $\phi(\mathbf{x})$ to be $\ell^{-3/2}$ like in the Schrödinger case, whereas in the field approach developed in this section it has been more convenient to take the dimensions of $\phi(\mathbf{x})$ to be $m^{1/2}\ell^{-1/2}$. Taking this into account, the expression for the density ρ in (3.10) in terms of the quantized field is

$$\rho = \frac{i}{2\hbar}\left(\phi^{\dagger}\dot{\phi} - \dot{\phi}^{\dagger}\phi\right) \tag{3.52}$$

Next we write the field and its derivative from (3.48) as

$$\begin{aligned} \phi(\mathbf{x}) &= \frac{1}{V^{1/2}}\sum_{\mathbf{k}} \sqrt{\frac{\hbar}{\omega_k}}\left(\alpha_{\mathbf{k}} e^{i\mathbf{k}\cdot\mathbf{x}} + \beta^{\dagger}_{\mathbf{k}} e^{-i\mathbf{k}\cdot\mathbf{x}}\right) ; \\ \dot{\phi}^{\dagger}(\mathbf{x}) &= -\frac{i}{V^{1/2}}\sum_{\mathbf{k}} \sqrt{\hbar\omega_k}\left(\beta_{\mathbf{k}} e^{i\mathbf{k}\cdot\mathbf{x}} - \alpha^{\dagger}_{\mathbf{k}} e^{-i\mathbf{k}\cdot\mathbf{x}}\right) ; \end{aligned} \tag{3.53}$$

where we have introduced the new Bose operators

$$\alpha_{\mathbf{k}} = \frac{1}{\sqrt{2}}\left(a_{1\mathbf{k}} + i a_{2\mathbf{k}}\right) \; ; \; \beta_{\mathbf{k}} = \frac{1}{\sqrt{2}}\left(a_{1\mathbf{k}} - i a_{2\mathbf{k}}\right) \tag{3.54}$$

such that

$$\left[\alpha_{\mathbf{k}}, \alpha^{\dagger}_{\mathbf{k}'}\right] = \left[\beta_{\mathbf{k}}, \beta^{\dagger}_{\mathbf{k}'}\right] = \delta_{\mathbf{k}\mathbf{k}'} \tag{3.55}$$

$$[\alpha_{\mathbf{k}}, \alpha_{\mathbf{k}'}] = [\beta_{\mathbf{k}}, \beta_{\mathbf{k}'}] = [\alpha_{\mathbf{k}}, \beta_{\mathbf{k}'}] = \left[\alpha_{\mathbf{k}}, \beta^{\dagger}_{\mathbf{k}'}\right] = 0 \tag{3.56}$$

It is then easy to substitute (3.53) into (3.52) and to perform the volume integration, which yields (Berestetskii *et al.* 1982)

$$\int \rho(\mathbf{x}) d^3\mathbf{x} = \sum_{\mathbf{k}} \left(\alpha^{\dagger}_{\mathbf{k}}\alpha_{\mathbf{k}} - \beta_{\mathbf{k}}\beta^{\dagger}_{\mathbf{k}}\right) = \sum_{\mathbf{k}} \left(\alpha^{\dagger}_{\mathbf{k}}\alpha_{\mathbf{k}} - \beta^{\dagger}_{\mathbf{k}}\beta_{\mathbf{k}} - 1\right) \tag{3.57}$$

One can see from (3.57) that it is not surprising, from the field point of view, that the integral on the LHS can be negative, since it has now essentially the meaning of the difference in the number between two kinds of particles α and β. Obviously this number can be positive or negative. These two sets of particles, which appear in the case of a complex Klein-Gordon field, are called the particle and the antiparticle respectively. In terms of the particle and antiparticle picture one obtains the Hamiltonian by inverting (3.54) and substituting in (3.51), which yields

$$H = \sum_{\mathbf{k}} \hbar\omega_k \left(\alpha^{\dagger}_{\mathbf{k}}\alpha_{\mathbf{k}} + \beta^{\dagger}_{\mathbf{k}}\beta_{\mathbf{k}} + 1\right) \tag{3.58}$$

3.3 The Dirac field

One takes the following Lagrangian density (Corinaldesi and Strocchi 1963)

$$\mathcal{L} = -\frac{1}{2}\hbar c\left\{\bar{\phi}\left(\gamma_\mu \frac{\partial}{\partial x_\mu} + \kappa\right)\phi - \left(\frac{\partial\bar{\phi}}{\partial x_\mu}\gamma_\mu - \kappa\bar{\phi}\right)\phi\right\} \tag{3.59}$$

where ϕ and $\bar{\phi}$ are independent coordinates. Other Lagrangian densities for the Dirac field are of course in use. The main advantage of (3.59) for our purposes is its symmetry with respect to ϕ and $\bar{\phi}$. Thus the Euler-Lagrange equations are

$$\frac{\partial\mathcal{L}}{\partial\phi} - \frac{\partial}{\partial x_\mu}\frac{\partial\mathcal{L}}{\partial(\partial\phi/\partial x_\mu)} = 0\,;$$
$$\frac{\partial\mathcal{L}}{\partial\bar{\phi}} - \frac{\partial}{\partial x_\mu}\frac{\partial\mathcal{L}}{\partial(\partial\bar{\phi}/\partial x_\mu)} = 0 \tag{3.60}$$

Since

$$\frac{\partial \mathcal{L}}{\partial \phi} = -\frac{1}{2}\hbar c\left(2\kappa\bar{\phi} - \frac{\partial \bar{\phi}}{\partial x_\mu}\gamma_\mu\right);$$
$$\frac{\partial \mathcal{L}}{\partial(\partial\phi/\partial x_\mu)} = -\frac{1}{2}\hbar c\bar{\phi}\gamma_\mu \tag{3.61}$$

the two equations in (3.60) yield respectively

$$\frac{\partial \bar{\phi}}{\partial x_\mu}\gamma_\mu - \kappa\bar{\phi} = 0\,;\ \gamma_\mu\frac{\partial \phi}{\partial x_\mu} + \kappa\phi = 0 \tag{3.62}$$

which are the Dirac equation and its adjoint, as required. The energy-momentum tensor can be obtained (Corinaldesi and Strocchi 1963) as

$$T_{\mu\nu} = \frac{1}{4}\hbar c\left(\bar{\phi}\gamma_\mu\frac{\partial \phi}{\partial x_\nu} - \frac{\partial \bar{\phi}}{\partial x_\nu}\gamma_\mu\phi + \bar{\phi}\gamma_\nu\frac{\partial \phi}{\partial x_\mu} - \frac{\partial \bar{\phi}}{\partial x_\mu}\gamma_\nu\phi\right) \tag{3.63}$$

Using Dirac's equations (3.62) as well as

$$\frac{\partial^2 \phi}{\partial x_\mu^2} - \kappa^2\phi = 0\,;\ \frac{\partial^2 \bar{\phi}}{\partial x_\mu^2} - \kappa^2\bar{\phi} = 0 \tag{3.64}$$

it is an easy matter to show that

$$\frac{\partial T_{\mu\nu}}{\partial x_\mu} = 0 \tag{3.65}$$

We note, *en passant*, that (3.64), which are obtainable from the second of (3.62) by application of operators $\gamma_\mu \partial/\partial x_\mu$, are Klein-Gordon equations. Thus the Dirac equation implies the Klein-Gordon equation, although vice versa does not hold true. The various conservation laws relative to the Dirac field follow from (3.65). Thus for $\nu = 4$ we obtain conservation of energy analogous to (1.43) for the e.m. field, since

$$\mathcal{H} = -T_{44} = \frac{i}{2}\hbar(\phi^*\dot{\phi} - \dot{\phi}^*\phi) \tag{3.66}$$

is defined as the energy density of the Dirac field. We note that this may be negative, contrary to the Klein-Gordon case, where (3.28) is positive definite. This inconvenience will be eliminated by quantizing the field in such a way that the negative energy densities are eliminated (Ryder 1986).

The angular momentum density tensor, in analogy to the e.m. case of Section 1.6, is defined as

$$M_{\mu\nu\lambda} = x_\nu T_{\mu\lambda} - x_\lambda T_{\mu\nu} \tag{3.67}$$

It is easy to show, using (3.65) and the symmetry of $T_{\mu\nu}$ with respect to interchange of μ and ν, that

$$\frac{\partial}{\partial x_\mu} M_{\mu\nu\lambda} = 0 \tag{3.68}$$

From this, multiplication by the previously introduced tensor e_{njk} yields

$$\frac{\partial}{\partial x_i}(e_{njk}x_jT_{ik}) + \dot{\ell}_n = 0\,;\; \ell_n = \frac{1}{2ic}e_{njk}M_{4jk} \tag{3.69}$$

where ℓ_n is the nth component of the angular momentum of the field. (3.69) expresses conservation of angular momentum for the free Dirac field. This leads us to discuss more closely the expression for

$$\begin{aligned}\ell_n &= \frac{1}{2ic}e_{njk}\left(x_jT_{4k} - x_kT_{4j}\right) = \frac{1}{ic}e_{njk}x_jT_{4k}\\ &= \frac{\hbar}{4i}e_{njk}x_j\left(\phi^*\frac{\partial\phi}{\partial x_k} - \frac{\partial\phi^*}{\partial x_k}\phi + \phi^*\gamma_4\gamma_k\frac{\partial\phi}{\partial x_4}\right.\\ &\quad\left. - \frac{\partial\phi^*}{\partial x_4}\gamma_4\gamma_k\phi\right)\end{aligned} \tag{3.70}$$

Use of the Dirac equations in order to eliminate the x_4 derivative yields, after some manipulation,

$$\begin{aligned}\ell_n &= \frac{\hbar}{2i}e_{njk}x_j\left\{\phi^*\frac{\partial\phi}{\partial x_k} - \frac{\partial\phi^*}{\partial x_k}\phi + \frac{1}{2}\sum_{\ell\neq k}\left(\phi^*\gamma_k\gamma_\ell\frac{\partial\phi}{\partial x_l}\right.\right.\\ &\quad\left.\left. - \frac{\partial\phi^*}{\partial x_l}\gamma_\ell\gamma_k\phi\right)\right\}\\ &= \frac{\hbar}{2i}e_{njk}x_j\left\{2\phi^*\frac{\partial\phi}{\partial x_k} - \frac{\partial}{\partial x_k}(\phi^*\phi) - \frac{1}{2}\sum_{\ell\neq k}\frac{\partial}{\partial x_\ell}(\phi^*\gamma_\ell\gamma_k\phi)\right\}\\ &= \frac{\hbar}{2i}e_{njk}\left\{2x_j\phi^*\frac{\partial\phi}{\partial x_k} - \frac{\partial}{\partial x_k}(x_j\phi^*\phi) - \frac{1}{2}\sum_{\ell\neq k}\frac{\partial}{\partial x_k}(x_j\phi^*\gamma_\ell\gamma_k\phi)\right.\\ &\quad\left. + \frac{1}{2}\phi^*\gamma_j\gamma_k\phi\right\}\end{aligned} \tag{3.71}$$

From (3.21)

$$e_{njk}\gamma_j\gamma_k = 2i\sigma_n \tag{3.72}$$

and consequently

$$\ell_n = \phi^* e_{njk} x_j \frac{\hbar}{i} \frac{\partial}{\partial x_k} \phi + \frac{\hbar}{2} \phi^* \sigma_n \phi$$
$$- \frac{\hbar}{2i} e_{njk} \left\{ \frac{\partial}{\partial x_k} (x_j \phi^* \phi) + \frac{1}{2} \sum_{\ell \neq k} \frac{\partial}{\partial x_l} (x_j \phi^* \gamma_\ell \gamma_k \phi) \right\} \tag{3.73}$$

The last term on the RHS of (3.73), upon integration over a large volume V outside which the Dirac field can be taken to vanish, yields surface terms which do not contribute to the total expression, whereas the contribution of the first two terms can be put into the form

$$L_n = \int \phi^* \left(\mathbf{x} \times \frac{\hbar}{i} \nabla \right)_n \phi d^3\mathbf{x} + \int \phi^* \mathbf{S}_n \phi d^3\mathbf{x} \tag{3.74}$$

which displays clearly the orbital and spin contribution to the angular momentum of the Dirac field.

Like the fields previously introduced, the Dirac field can also be expressed in terms of field amplitudes by expanding ϕ in series of the plane wave eigensolutions of the Dirac equation (3.22). Thus, using periodic boundary conditions on the surface of a cube of volume V, we put

$$\phi(\mathbf{x}, t) = \frac{1}{\sqrt{V}} \sum_{s=1}^{2} \sum_{\mathbf{p}} \sqrt{\frac{mc^2}{|E_p|}} \left\{ b_{\mathbf{p}}^{(s)}(0) u^{(s)}(\mathbf{p}) e^{(i/\hbar)(\mathbf{p}\cdot\mathbf{x} - |E_p| t)} \right.$$
$$\left. + d_{\mathbf{p}}^{(s)\,*}(0) v^{(s)}(\mathbf{p}) e^{-(i/\hbar)(\mathbf{p}\cdot\mathbf{x} - |E_p| t)} \right\} \tag{3.75}$$

where the negative energy spinors have been relabelled according to

$$v^{(1,2)}(\mathbf{p}) = (-1)^{r+1} u^{(3,4)}(-\mathbf{p}) \tag{3.76}$$

and where $b_{\mathbf{p}}^{(s)}(0)$ and $d_{\mathbf{p}}^{(s)}(0)$ are the classical amplitudes of plane wave of momentum $\mathbf{p}$ and positive energy and that of plane wave of momentum $-\mathbf{p}$ and negative energy respectively, taken at $t = 0$. From (3.75) it is easy to obtain

$$\dot{\phi}(\mathbf{x}, t) = \frac{1}{\sqrt{V}} \frac{i}{\hbar} \sum_{s=1}^{2} \sum_{\mathbf{p}} \sqrt{mc^2 |E_p|} \left\{ -b_{\mathbf{p}}^{(s)}(0) u^{(s)}(\mathbf{p}) e^{(i/\hbar)(\mathbf{p}\cdot\mathbf{x} - |E_p| t)} \right.$$
$$\left. + d_{\mathbf{p}}^{(s)*}(0) v^{(s)}(\mathbf{p}) e^{-(i/\hbar)(\mathbf{p}\cdot\mathbf{x} - |E_p| t)} \right\} \tag{3.77}$$

Substitution of (3.75) and (3.77) into the expression of the field Hamiltonian, which can be obtained from (3.66) by volume integration, yields after some algebra

$$H = \int_V \mathcal{H}(\mathbf{x}) d^3\mathbf{x} = \frac{i\hbar}{2} \int_V (\phi^* \dot{\phi} - \dot{\phi}^* \phi) d^3\mathbf{x}$$
$$= mc^2 \sum_{ss'} \sum_{\mathbf{p}} \left\{ b_{\mathbf{p}}^{(s)*} b_{\mathbf{p}}^{(s')} u^{(s)\dagger}(\mathbf{p}) u^{(s')}(\mathbf{p}) - d_{\mathbf{p}}^{(s)} d_{\mathbf{p}}^{(s')*} v^{(s)\dagger}(\mathbf{p}) v^{(s')}(\mathbf{p}) \right\} \tag{3.78}$$

where the order of the amplitudes has been preserved throughout the calculation in spite of their c-number nature. Finally, through use of the orthogonality condition (3.23), expression (3.78) takes the form

$$H = \sum_{s=1}^{2} \sum_{\mathbf{p}} | E_p | \left(b_{\mathbf{p}}^{(s)*} b_{\mathbf{p}}^{(s)} - d_{\mathbf{p}}^{(s)} d_{\mathbf{p}}^{(s)*} \right) \tag{3.79}$$

We see that as long as the amplitudes b and d are classical, positivity of H is not ensured, which cannot be regarded as reasonable. This is different from the analogous expression (3.34) for the Hamiltonian of the Klein-Gordon field, which is always positive definite. Moreover, imposing Bose commutation relations for b and d in (3.79) would not really solve this problem, since the spectrum of H would then be proportional to that of the operator $b^\dagger b - dd^\dagger = b^\dagger b - d^\dagger d - 1$ which has intrinsically negative eigenvalues. Positivity of H is on the contrary ensured by assuming Fermi commutation relations among b and d, and precisely

$$\{d_{\mathbf{p}}^{(s)}, d_{\mathbf{p}'}^{(s')\dagger}\} = \{b_{\mathbf{p}}^{(s)}, b_{\mathbf{p}'}^{(s')\dagger}\} = \delta_{ss'} \delta_{\mathbf{pp}'} \; ;$$
$$\{d_{\mathbf{p}}^{(s)}, d_{\mathbf{p}'}^{(s')}\} = \{b_{\mathbf{p}}^{(s)}, b_{\mathbf{p}'}^{(s')}\} = 0 \tag{3.80}$$

with any d operator anticommuting with any b. In this case, in fact, the quantized version of (3.79) takes the form

$$H = \sum_{s=1}^{2} \sum_{\mathbf{p}} | E_p | \left(b_{\mathbf{p}}^{(s)\dagger} b_{\mathbf{p}}^{(s)} + d_{\mathbf{p}}^{(s)\dagger} d_{\mathbf{p}}^{(s)} \right) + \mathrm{C} \; ; \; \mathrm{C} = - \sum_{s=1}^{2} \sum_{\mathbf{p}} | E_p | \tag{3.81}$$

in which the infinite negative constant C formally plays the same role as the zero-point energy for the electromagnetic or the Klein-Gordon field. Thus (3.81) has a well-defined ground state associated with zero occupation number of the particles of kind b and d. It should be remarked at this point that if we had applied Fermi commutation rules to the amplitudes of the complex Klein-Gordon field in Hamiltonian (3.34) we would have obtained a constant c-number, which does not make any sense since one would expect a classical dynamical variable such as H to correspond to a quantum mechanical operator with a nontrivial spectrum. It should also be noted, however, that in analogy with the Klein-Gordon case, the solution of the problem of the negative energy spectrum in relativistic quantum mechanics entails the appearance of two kinds of field excitations instead of only one, pertaining to particles and antiparticles.

It is instructive to investigate the meaning of the two kinds of particles by studying the second-quantized expression for the quantity ρ defined in (3.24) and appearing in the continuity equation. On the basis of (3.77) and after some algebra we obtain

$$\begin{aligned}\int_V \rho d^3\mathbf{x} = mc^2 \sum_{ss'} \sum_{\mathbf{p}} \frac{1}{|E_p|} \Big\{ & b_{\mathbf{p}}^{(s)\dagger} b_{\mathbf{p}}^{(s')} u^{(s)\dagger}(\mathbf{p}) u^{(s')}(\mathbf{p}) \\ & + b_{\mathbf{p}}^{(s)\dagger} d_{-\mathbf{p}}^{(s')\dagger} u^{(s)\dagger}(\mathbf{p}) v^{(s')}(-\mathbf{p}) + d_{\mathbf{p}}^{(s)} b_{-\mathbf{p}}^{(s')} v^{(s)\dagger}(\mathbf{p}) u^{(s')}(-\mathbf{p}) \\ & + d_{\mathbf{p}}^{(s)} d_{\mathbf{p}}^{(s')\dagger} v^{(s)\dagger}(\mathbf{p}) v^{(s')}(\mathbf{p}) \Big\}\end{aligned} \tag{3.82}$$

In view of orthogonality between any u and v, stemming from (3.23) and (3.76), expression (3.82) reduces to

$$\begin{aligned}\int_V \rho d^3\mathbf{x} &= \sum_{s=1}^{2} \sum_{\mathbf{p}} \left(b_{\mathbf{p}}^{(s)\dagger} b_{\mathbf{p}}^{(s)} + d_{\mathbf{p}}^{(s)} d_{\mathbf{p}}^{(s)\dagger} \right) \\ &= \sum_{s=1}^{2} \sum_{\mathbf{p}} \left(b_{\mathbf{p}}^{(s)\dagger} b_{\mathbf{p}}^{(s)} - d_{\mathbf{p}}^{(s)\dagger} d_{\mathbf{p}}^{(s)} + 1 \right)\end{aligned} \tag{3.83}$$

which shows that, although the integral on the RHS is always positive, the difference between (3.57) and (3.83) is more apparent than real, since at the core of both expressions there is the difference between the numbers of the two kinds of particles populating each field.

3.4 The spinless Schrödinger field

As previously discussed, the necessity of taking a matter field point of view really exists only within the domain of relativistic quantum mechanics, where it is impossible to interpret the square modulus of ψ in terms of a probability density. Thus there is no reason in principle to adopt the field point of view in nonrelativistic quantum theory. The success of the relativistic theory outlined in the previous sections, however, suggests the opportunity to explore the consequences of the application of the matter field idea to the Schrödinger equation. This leads us to look for a Lagrangian density from which one can derive the ordinary Schrödinger equation (3.1) for a free particle by means of a field formalism. Here and in the following, when dealing with nonrelativistic situations, we shall limit our considerations to spinless particles. Generalization of the Lagrangian approach to obtain the Schrödinger-Pauli equation from an appropriate Lagrangian density should not lead to conceptual difficulties, but it would constitute an unnecessary complication in the context of the present book. Along these lines one introduces the following Lagrangian density (Corinaldesi and Strocchi 1963)

$$\mathcal{L}(\psi, \psi^*, \nabla\psi, \nabla\psi^* \dot{\psi}, \dot{\psi}^*) = -\frac{\hbar^2}{2m}\nabla\psi^* \cdot \nabla\psi - \frac{i\hbar}{2}(\dot{\psi}^*\psi - \psi^*\dot{\psi}) \tag{3.84}$$

From the Euler-Lagrange equations

$$\begin{aligned} \frac{\partial}{\partial t}\frac{\partial \mathcal{L}}{\partial \dot{\psi}^*} + \nabla \cdot \frac{\partial \mathcal{L}}{\partial \nabla\psi^*} &= \frac{\partial \mathcal{L}}{\partial \psi^*} \\ \frac{\partial}{\partial t}\frac{\partial \mathcal{L}}{\partial \dot{\psi}} + \nabla \cdot \frac{\partial \mathcal{L}}{\partial \nabla\psi} &= \frac{\partial \mathcal{L}}{\partial \psi} \end{aligned} \tag{3.85}$$

one immediately obtains (3.1) and its complex conjugate. The energy-momentum tensor can be obtained as

$$\begin{aligned} T_{\mu\nu} = & \frac{\hbar^2}{2m}\left(\frac{\partial\psi^*}{\partial x_\mu}\frac{\partial\psi}{\partial x_\nu} + \frac{\partial\psi^*}{\partial x_\nu}\frac{\partial\psi}{\partial x_\mu}\right)(1 - \delta_{\mu 4}) \\ & + \frac{1}{2}\hbar c\left(\psi^*\frac{\partial\psi}{\partial x_\nu} - \frac{\partial\psi^*}{\partial x_\nu}\psi\right)\delta_{\mu 4} - \mathcal{L}\delta_{\mu\nu} \end{aligned} \tag{3.86}$$

Since $T_{4i} \neq T_{i4}$, this tensor is not symmetric, but $T_{ik} = T_{ki}$. Moreover, using (3.1) it is easy to prove that

$$\frac{\partial T_{\mu\nu}}{\partial x_\nu} = 0 \tag{3.87}$$

All conservation properties of the Schrödinger field follow from (3.87). The energy density of the field is given by

$$\mathcal{H} = -T_{44} = -\frac{i\hbar}{2}(\dot{\psi}^*\psi - \psi^*\dot{\psi}) - \mathcal{L} = \frac{\hbar^2}{2m}\nabla\psi^* \cdot \nabla\psi \tag{3.88}$$

from which the Hamiltonian is obtained by partial integration in the form

$$H = -\int T_{44} d^3\mathbf{x} = \frac{\hbar^2}{2m}\int \nabla\psi^* \cdot \nabla\psi d^3\mathbf{x} = -\int \psi^* \frac{\hbar^2}{2m}\nabla^2\psi d^3\mathbf{x} \tag{3.89}$$

The momentum density is

$$p_i = \frac{1}{ic} T_{4i} = \frac{\hbar}{2i} \left(\psi^* \frac{\partial \psi}{\partial x_i} - \frac{\partial \psi^*}{\partial x_i} \psi \right) = m j_i \tag{3.90}$$

where we have used the definition of probability density current given in (3.3). The total momentum of the Schrödinger field is obtained from (3.90) by partial integration as

$$\mathbf{P} = \frac{\hbar}{2i} \int (\psi^* \nabla \psi - \psi \nabla \psi^*) d^3\mathbf{x} = \frac{\hbar}{i} \int \psi^* \nabla \psi d^3\mathbf{x} \tag{3.91}$$

Also the position of the Schrödinger field can be defined as

$$\mathbf{X} = \int \psi^* \mathbf{x} \psi d^3\mathbf{x} \tag{3.92}$$

A slightly more involved procedure, along the same lines as in the previous sections, leads also from the energy momentum tensor to the angular momentum density and finally to the following expression for the angular momentum of the field

$$\mathbf{L} = \frac{\hbar}{i} \int \psi^* (\mathbf{x} \times \nabla) \psi d^3\mathbf{x} \tag{3.93}$$

In the face of results (3.89), (3.91), (3.92), (3.93) nothing new would seem to have resulted from the field approach to the Schrödinger equation, since these expressions are exactly the same as those well known from elementary quantum theory. To show that this is not really the case and that the field point of view opens a new perspective on the problem we now push the field procedure one step further and apply second quantization as we did in the relativistic case. First we expand ψ in terms of field amplitudes $a_k(t)$ as (see e.g. Roman 1965)

$$\psi(\mathbf{x}, t) = \sum_k a_k(t) u_k(\mathbf{x}) \tag{3.94}$$

where $u_k(\mathbf{x})$ are eigensolutions of the free-particle Hamiltonian

$$H = -\frac{\hbar^2}{2m} \nabla^2 \; ; \; H u_k(\mathbf{x}) = E_k u_k(\mathbf{x}) \tag{3.95}$$

subject to appropriate boundary conditions. These eigensolutions form a complete set of eigenfunctions and can thus be used for the expansion (3.94). Second quantization is performed by transforming the field amplitudes a_k into operators. On the basis of our previous experience, this can be done imposing either Bose or Fermi commutation rules. In the first case one has, as in (3.42),

$$\left[a_k, a_{k'}^\dagger\right] = \delta_{kk'} \; ; \; [a_k, a_{k'}] = 0 \tag{3.96}$$

Consequently one has

$$\begin{aligned} [\psi(\mathbf{x},t), \psi^\dagger(\mathbf{x}',t)] &= \sum_{kk'} \left[a_k(t), a_{k'}^\dagger(t)\right] u_k(\mathbf{x}) u_{k'}^*(\mathbf{x}') \\ &= \sum_k u_k(\mathbf{x}) u_k^*(\mathbf{x}') = \delta(\mathbf{x}-\mathbf{x}') \\ [\psi(\mathbf{x},t), \psi(\mathbf{x}',t)] &= \sum_{kk'} [a_k(t), a_{k'}(t)] u_k(\mathbf{x}) u_{k'}(\mathbf{x}') = 0 \end{aligned} \tag{3.97}$$

We see that $\psi(\mathbf{x})$ and $\psi^\dagger(\mathbf{x}')$ have been transformed into noncommuting operators, and that a_k and $a_k^\dagger$ are now annihilation and creation operators of particles in an eigenstate $u_k(\mathbf{x})$ of the single-particle Hamiltonian H. A general state of the field with well-defined number of particles in each u_k can be simply described in Fock space by a ket

$$| \, n_{k_1,k_2,\ldots} \rangle \equiv | \, \{n_k\} \rangle \tag{3.98}$$

which indicates the number of particles in each single-particle state. It can be shown that the probability amplitude for finding the $N = \sum_k n_k$ particles in state (3.98) distributed at space points $\mathbf{x}_1, \mathbf{x}_2, \ldots, \mathbf{x}_N$ is simply the wavefunction of N Bose particles symmetrized with respect to exchange of any two bosons (Roman 1965). Moreover, substituting expansion (3.94) into the field Hamiltonian (3.89), the latter can be put in the form

$$H = \sum_k E_k a_k^\dagger a_k \; ; \; E_k = -\int u_k^*(\mathbf{x}) \frac{\hbar^2 \nabla^2}{2m} u_k(\mathbf{x}) d^3\mathbf{x} \tag{3.99}$$

where E_k are the eigenvalues of the single-particle Hamiltonian, which one should not confuse with the field Hamiltonian as given by (3.89) or (3.99). Clearly the number state (3.98) is an eigenstate of the field Hamiltonian, as is apparent from its second quantized form (3.99).

The second possibility is to choose Fermi anticommutation for the field amplitude operators, as in (3.80)

$$\left\{a_k, a_{k'}^\dagger\right\} = \delta_{kk'} \; ; \; \{a_k, a_{k'}\} = 0 \tag{3.100}$$

which yields, for field operators,

$$\{\psi(\mathbf{x},t), \psi^\dagger(\mathbf{x}',t)\} = \delta(\mathbf{x}-\mathbf{x}') \; ; \; \{\psi(\mathbf{x},t), \psi(\mathbf{x}',t)\} = 0 \tag{3.101}$$

Since, from (3.100), $a_k^{\dagger 2} \mid \{0_k\}\rangle = 0$, where $\mid \{0_k\}\rangle$ is the vacuum field state where no fermions are present in the field, a state of the form (3.98) in the Fermi case can have $n_k = 0, 1$ only for any k. This expresses the Pauli exclusion principle, which thus becomes a consequence of the form of commutation relations (3.100). Moreover, the probability amplitude for finding the $N = \sum_k n_k$ fermions in a number state of the form (3.98) distributed at space points $\mathbf{x}_1, \mathbf{x}_2, \ldots, \mathbf{x}_N$ is the wavefunction of N Fermi particles antisymmetrized with respect to exchange of any two of them (Roman 1965). Finally, substitution of (3.94) into (3.89) yields again expression (3.99) similarly to the Bose case.

In conclusion, assuming the matter-field point of view for the Schrödinger case and proceeding on to second quantization leads from a single-particle to a system of many identical particles, as in the relativistic case. In the nonrelativistic case, however, no compelling reason seems to exist for second-quantizing the field according to Fermi or to Bose commutation relations. We recall that in the relativistic case, using the "wrong" quantization rule yielded unacceptable properties for the second-quantized Hamiltonian, whereas for the Schrödinger matter field one has to be content with the old rule that half-integer spin particles are fermions and integer spin particles are bosons. A distinct advantage of the second-quantization procedure, however, is that it incorporates in the commutator algebra all the procedures for a correct and simple account of the symmetry properties of the total wavefunction of a many-particle system.

3.5 Gauge invariance and matter fields

We have already discussed the property of gauge invariance in connection with the electromagnetic field in Chapter 1. There this property was manifest in the description of the field in terms of the four-vector A_μ, since the Lagrangian density of the electromagnetic field is invariant under the addition to A_μ of the four-gradient of a scalar χ which is a function of $\mathbf{x}$ and t, although the equations of motion of A_μ do change under such a transformation. Because the transformation in question is a function of the space-time point, the ensuing gauge invariance is called local. In contrast, a global transformation is one which is independent of $\mathbf{x}$ and t. The question we now wish to consider is if a set of gauge transformations exists also for the various versions of the matter field that we have introduced in the previous sections of this chapter.

First we discuss the possibility of global gauge invariance. A very obvious invariance of the Lagrangian density of any of the complex matter fields, relativistic or not, is with respect to multiplication of the field amplitude by a fixed phase factor $e^{i\Lambda}$ (Λ real). This is so because all free complex matter-field Lagrangian densities contain a pair of field amplitude operators conjugate to each other, which neutralizes the effect of the phase factor. Consider, for example, a complex scalar field for which the gauge transformation can be represented as (Ryder 1986)

$$\phi \to e^{-i\Lambda}\phi \; ; \; \phi^\dagger \to e^{i\Lambda}\phi^\dagger \tag{3.102}$$

Thus the form (3.25) for the Lagrangian density is invariant under this transformation, in the sense that the new Lagrangian density, expressed in terms of the new field amplitudes

$$\phi' = e^{-i\Lambda}\phi \; ; \; \phi^{\dagger'} = e^{i\Lambda}\phi^\dagger \tag{3.103}$$

has exactly the same form (3.25). The real and imaginary components of the field amplitude, however, introduced in (3.36), do change under (3.102). In fact from (3.103)

$$\begin{cases} \phi_1' + i\phi_2' = e^{-i\Lambda}(\phi_1 + i\phi_2) \\ \phi_1' - i\phi_2' = e^{i\Lambda}(\phi_1 - i\phi_2) \end{cases} ;$$
$$\begin{cases} \phi_1' = \phi_1 \cos\Lambda + \phi_2 \sin\Lambda \\ \phi_2' = -\phi_1 \sin\Lambda + \phi_2 \cos\Lambda \end{cases} \tag{3.104}$$

which shows that the transformation may be described as a rotation through an angle Λ in a two-dimensional space where the field amplitude is represented as a vector whose two components along the reference axes are its real and its imaginary parts. Notice that this is a rotation in an internal space and not in real space. It should be expected, however, that a conserved quantity should exist analogous to angular momentum which arises from invariance of the Lagrangian density with respect to rotations in real space. This is actually a consequence of a very general theorem, Noether's theorem, which within the relatively limited scope of this book can be stated in the simplified form: for any continuous symmetry operation which leaves the Lagrangian density of a dynamical system invariant there is a conserved quantity (Doughty 1990). Since angular momentum is known to be the generator of infinitesimal rotations, we are led to study an infinitesimal transformation of the kind (3.104) in order to identify its generator as the conserved quantity. Since it is easy to check that (3.104) leaves the field commutation rules invariant, the transformation must be operated by a unitary operator U, in such a way that

$$\phi_1' = U\phi_1 U^{-1} \; ; \; \phi_2' = U\phi_2 U^{-1} \tag{3.105}$$

Moreover, if the transformation is infinitesimal, with $\delta\Lambda$ as a parameter rather than Λ, from (3.104) one has

$$U\phi_1 U^{-1} = \phi_1 + \delta\Lambda\phi_2 \; ; \; U\phi_2 U^{-1} = \phi_2 - \delta\Lambda\phi_1 \tag{3.106}$$

and from (3.105), since $U = 1 + \delta\Lambda Q$,

$$U\phi_1 U^{-1} = \phi_1 + i\delta\Lambda[Q, \phi_1] \; ; \; U\phi_2 U^{-1} = \phi_2 + i\delta\Lambda[Q, \phi_2] \tag{3.107}$$

Comparing (3.106) and (3.107) one has immediately

$$[Q, \phi_1] = -i\phi_2 \; ; \; [Q, \phi_2] = i\phi_1 \tag{3.108}$$

Using the field amplitude commutation rules, it is easy to convince oneself that on the basis of (3.108) one must have (Henley and Thirring 1962)

$$Q = \frac{1}{\hbar} \int \{\dot{\phi}_1(\mathbf{x})\phi_2(\mathbf{x}) - \dot{\phi}_2(\mathbf{x})\phi_1(\mathbf{x})\} d^3\mathbf{x} \tag{3.109}$$

which is the sought-for constant of motion and which is generally called isospin. Expanding ϕ_1 and ϕ_2 in terms of creation and annihilation operators $a_1, a_1^\dagger, a_2$ and $a_2^\dagger$ introduced in (3.49) it is possible to prove that $Q \mid 0\rangle = 0$, where $\mid 0\rangle$ is the vacuum of the ϕ_1 and ϕ_2 fields. Moreover it is immediate from (3.108) that

$$[Q, \phi] = -\phi \; ; \; [Q, \phi^\dagger] = \phi^\dagger \tag{3.110}$$

Hence

$$Q\phi \mid 0\rangle = -\phi \mid 0\rangle \; ; \; Q\phi^\dagger \mid 0\rangle = \phi^\dagger \mid 0\rangle \tag{3.111}$$

and we see that $\phi \mid 0\rangle$ and $\phi^\dagger \mid 0\rangle$ are eigenvectors of Q with eigenvalues ∓ 1. In view of (3.53)

$$\phi \mid 0\rangle = \frac{1}{V^{1/2}} \sum_{\mathbf{k}} \sqrt{\frac{\hbar}{\omega_k}} \beta_{\mathbf{k}}^\dagger e^{-i\mathbf{k}\cdot\mathbf{x}} \mid 0\rangle \; ;$$

$$\phi^\dagger \mid 0\rangle = \frac{1}{V^{1/2}} \sum_{\mathbf{k}} \sqrt{\frac{\hbar}{\omega_k}} \alpha_{\mathbf{k}}^\dagger e^{-i\mathbf{k}\cdot\mathbf{x}} \mid 0\rangle \tag{3.112}$$

This leads us to identify the particles and antiparticles of the complex Klein-Gordon field as the carriers of isospin ± 1.

The results obtained above demonstrate the value of investigating the matter field from the point of view of gauge transformation. The gauge transformation discussed until now, however, may be considered rather trivial since it is of a global nature. What happens if, as in the case of the e.m. field, we eliminate this constraint by requiring local gauge invariance of the Lagrangian density of the matter field? This can be investigated by generalizing Λ in (3.102) to be a function of $\mathbf{x}$. Consider, for example, the Lagrangian density of the Schrödinger field (3.84), and perform the transformation

$$\psi \to e^{-i\Lambda(\mathbf{x},t)}\psi\,;\ \psi^* \to e^{i\Lambda(\mathbf{x},t)}\psi^* \tag{3.113}$$

This leads to new field amplitudes ψ' and ψ'^* which are related to the old ones by

$$\psi' = e^{-i\Lambda}\psi\,;\ \psi'^* = e^{i\Lambda}\psi^* \tag{3.114}$$

where we have omitted explicit indication of the $\mathbf{x}$ and t dependence of Λ for simplicity of notation. Expressing $\mathcal{L}$ in (3.84) in terms of the new field amplitudes yields

$$\begin{aligned}
&\nabla\psi = e^{i\Lambda}(i\nabla\Lambda\psi' + \nabla\psi');\ \dot{\psi} = e^{i\Lambda}\left(i\dot{\Lambda}\psi' + \dot{\psi}'\right);\\
&\mathcal{L} = -\frac{\hbar^2}{2m}(-i\nabla\Lambda\psi'^* + \nabla\psi'^*)\cdot(i\nabla\Lambda\psi' + \nabla\psi')\\
&\quad -\frac{i\hbar}{2}\left[\left(-i\dot{\Lambda}\psi'^* + \dot{\psi}'^*\right)\psi' - \psi'^*\left(i\dot{\Lambda}\psi' + \dot{\psi}'\right)\right]
\end{aligned} \tag{3.115}$$

The presence of space and time derivatives of Λ in the expression for $\mathcal{L}$ in (3.115) spoils the invariance of the Lagrangian density of the Schrödinger field under local transformation (3.113). If however $\mathcal{L}$ were of the form

$$\begin{aligned}
\mathcal{L}' &= -\frac{\hbar^2}{2m}(-i\mathbf{G}\psi^* + \nabla\psi^*)\cdot(i\mathbf{G}\psi + \nabla\psi)\\
&\quad -\frac{i\hbar}{2}\left[(cG_4\psi^* + \dot{\psi}^*)\psi - \psi^*(-cG_4\psi + \dot{\psi})\right]
\end{aligned} \tag{3.116}$$

rather than of the form (3.84), things might take a different aspect. Suppose in fact that $G_\mu \equiv \{\mathbf{G}, G_4\}$ is a four-vector field, and take G_μ to transform, under the same transformation yielding the matter field amplitude substitutions (3.113), according to

$$G_\mu(\mathbf{x}, t) \to G_\mu(\mathbf{x}, t) + \frac{\partial\Lambda}{\partial x_\mu} \tag{3.117}$$

$$G_\mu = G'_\mu - \frac{\partial \Lambda}{\partial x_\mu} \tag{3.118}$$

Then under the complete (ϕ and G_μ) transformation the Lagrangian density $\mathcal{L}'$ transforms as

$$\begin{aligned}
\mathcal{L}' = &-\frac{\hbar^2}{2m}\left[-i(\mathbf{G}' - \nabla\Lambda)e^{-i\Lambda}\psi'^* + e^{-i\Lambda}(-i\nabla\Lambda\psi'^* + \nabla\psi'^*)\right] \\
&\cdot \left[i(\mathbf{G}' - \nabla\Lambda)e^{i\Lambda}\psi' + e^{i\Lambda}(i\nabla\Lambda\psi' + \nabla\psi')\right] \\
&-\frac{i\hbar}{2}\Big\{\Big[\Big(cG'_4 + i\dot{\Lambda}\Big)e^{-i\Lambda}\psi'^* + e^{-i\Lambda}\Big(-i\dot{\Lambda}\psi'^* + \dot{\psi}'^*\Big)\Big]e^{i\Lambda}\psi' \\
&- e^{-i\Lambda}\psi'^*\Big[-\Big(cG'_4 + i\dot{\Lambda}\Big)e^{i\Lambda}\psi' + e^{i\Lambda}\Big(i\dot{\Lambda}\psi' + \dot{\psi}'\Big)\Big]\Big\} \\
= &-\frac{\hbar^2}{2m}(-i\mathbf{G}'\psi'^* + \nabla\psi'^*)\cdot(i\mathbf{G}'\psi' + \nabla\psi') \\
&-\frac{i\hbar}{2}\left[(cG'_4\psi'^* + \dot{\psi}'^*)\psi' - \psi'^*(-cG'_4\psi' + \dot{\psi}')\right]
\end{aligned} \tag{3.119}$$

and as such it is perfectly invariant under local transformation (3.113) and (3.117). We see that insisting on invariance under local gauge transformation involves a new form (3.116) of the Lagrangian density and the appearance of a new field which is called a gauge field. The problem is that $\mathcal{L}'$ in (3.116) is not likely to be a complete Lagrangian density as it stands. In fact one can write it in the form

$$\begin{aligned}
\mathcal{L}' &= \mathcal{L}_M + \mathcal{L}_i\,; \\
\mathcal{L}_i &= -\frac{\hbar^2}{2m}(\mathbf{G}\psi^*\cdot\mathbf{G}\psi - i\mathbf{G}\psi^*\cdot\nabla\psi + i\nabla\psi^*\cdot\mathbf{G}\psi) \\
&\quad -\frac{i\hbar c}{2}(G_4\psi^*\psi + \psi^* G_4\psi)
\end{aligned} \tag{3.120}$$

Clearly $\mathcal{L}_M$ is the Lagrangian of the Schrödinger field (3.84) which is expressed in terms of the Schrödinger field amplitudes alone, whereas $\mathcal{L}_i$ contains the amplitudes of both the Schrödinger and the gauge field. Consequently $\mathcal{L}_i$ looks very much like an interaction Lagrangian density between the ψ and the G_μ field. What is missing from $\mathcal{L}'$ is the Lagrangian density of the free gauge field, which we call $\mathcal{L}_G$. The question is, does a gauge field exist such that $\mathcal{L}_G$ is invariant under the transformation (3.117), in such a way that the total Lagrangian $\mathcal{L}' + \mathcal{L}_G$ is invariant under the complete gauge transformation (3.113) and (3.117)? The answer is affirmative, since we can put

$$G_\mu = -\frac{q}{\hbar c} A_\mu \; ; \; \Lambda = -\frac{q}{\hbar c}\chi \tag{3.121}$$

where q is a constant and A_μ is the four-potential of the e.m. field and $\chi(\mathbf{x}, t)$ is the scalar function introduced in Section 1.4 in connection with e.m. gauge invariance. In fact, as discussed in Chapter 1, the Lagrangian density of the free e.m. field is invariant under the transformation (1.68), which ensures invariance of $\mathcal{L}_G$ under (3.117), provided G_μ is related to A_μ via (3.121). Therefore we see that a Schrödinger field theory which is also invariant under local gauge transformations requires the presence of the e.m. field and a total Lagrangian density of the form

$$\begin{aligned}
\mathcal{L} &= \mathcal{L}_M + \mathcal{L}_F + \mathcal{L}_i \; ; \\
\mathcal{L}_M &= -\frac{\hbar^2}{2m} \nabla\psi^* \cdot \nabla\psi - \frac{i\hbar}{2}(\dot{\psi}^*\psi - \psi^*\dot{\psi}) \; ; \\
\mathcal{L}_F &= -\frac{1}{16\pi} F_{\mu\nu}F_{\mu\nu} \; ; \\
\mathcal{L}_i &= \frac{1}{2m}\left[\psi^* \frac{q}{c}\mathbf{A} \cdot \left(\frac{\hbar}{i}\nabla\right) - \left(\frac{\hbar}{i}\nabla\psi^*\right) \cdot \frac{q}{c}\mathbf{A}\psi - \psi^* \frac{q^2}{c^2}\mathbf{A}^2\psi\right] \\
&\quad - \psi^* q V \psi
\end{aligned} \tag{3.122}$$

We also see that in this Lagrangian density the interaction term is proportional to the constant q in such a way that for $q = 0$ the matter and the electromagnetic field are decoupled. This constant q is thus called the charge of the Schrödinger field.

We are led to conclude that the matter-field description of the Schrödinger equation, which gives the possibility of introducing the concept of local gauge invariance, is a most fruitful approach and that it leads in a natural way to obtain a matter-electromagnetic field interaction Lagrangian from first principles. It should be noted that the principle of local gauge invariance can be used in connection with other fields different from the e.m. field, such as in connection with electroweak theory and with QCD as discussed in Appendices F and G (Mills 1989). We have selected the Schrödinger field just as an example to show how the principle of local gauge invariance leads to the coupling of this field to the e.m. field; we should remark, however, that the procedure used can be applied along the same lines to the Klein-Gordon field or to the Dirac field. For the latter in particular, as shown in Appendix F, the interaction

Lagrangian density takes the form (Mandl and Shaw 1984)

$$\mathcal{L}_i = q\bar{\phi}\gamma_\mu\phi A_\mu \tag{3.123}$$

Finally, it should be noted that nowhere have we made use of the fact that the e.m. field is quantized or that the matter field is second-quantized. Thus our argumentation applies equally well to classical or first- or second-quantized fields, and it is of very general validity.

References

M. Abramovitz, I.A. Stegun (eds.) (1965). *Handbook of Mathematical Functions* (Dover Publications Inc., New York)

V.B. Berestetskii, E.M. Lifshitz, L.P. Pitaevskii (1982). *Quantum Electrodynamics* (Pergamon Press, London)

E. Corinaldesi, F. Strocchi (1963). *Relativistic Wave Mechanics* (North-Holland Publishing Company, Amsterdam)

N.A. Doughty (1990). *Lagrangian Interaction* (Addison-Wesley Publishing Co., Sydney)

E.M. Henley, W. Thirring (1962). *Elementary Quantum Field Theory* (McGraw-Hill Book Co., New York)

F. Mandl, G. Shaw (1984). *Quantum Field Theory* (John Wiley and Sons, Norwich)

R. Mills (1989). *Am. J. Phys.* **57**, 493

P. Roman (1965). *Advanced Quantum Theory* (Addison-Wesley Publishing Co., Reading, Ma.)

L.H. Ryder (1986). *Quantum Field Theory* (Cambridge University Press, Cambridge, UK)

J.J. Sakurai (1982). *Advanced Quantum Mechanics* (Addison-Wesley Publishing Co., Reading, Ma.)

S.S. Schweber (1964). *An Introduction to Relativistic Quantum Field Theory* (John Weatherhill Inc., Tokyo)

Further reading

A clear introduction to second quantization in nonrelativistic quantum mechanics can be found in

L.D. Landau, E.M. Lifshitz, *Quantum Mechanics* (Pergamon Press, London 1958).

A remarkably complete discusssion of the properties and solutions of relativistic wave equations has recently appeared in the book

W. Greiner, *Relativistic Quantum Mechanics* (Springer-Verlag 1990).

The following book

L.I. Schiff, *Quantum Mechanics* (McGraw-Hill and Kogakusha Company Ltd., Tokyo 1955)

includes an elementary approach to quantization of the Schrödinger field.

A generalized application of local gauge invariance to Yang-Mills fields is discussed in

P. Ramond, *Field Theory: a Modern Primer* (Addison-Wesley Publishing Co., New York 1990).

Internal symmetries of quantum fields are analysed in connection with Noether's theorem in the book

C. Itzykson, J.B. Zuber, *Quantum Field Theory* (McGraw-Hill Book Co., Singapore 1985).

4
Electrodynamics in the presence of sources

Introduction. In the previous chapter we have obtained the form of the coupling between matter and the electromagnetic field. We are thus in a position to treat electrodynamics in the presence of charges and currents. We start by obtaining the Euler-Lagrange equations of the matter-electromagnetic field system. These turn out to be the Maxwell-Lorentz equations, which is a set of coupled equations of motion in which the matter field acts as a source for the electromagnetic field and vice versa. In Section 4.2 we derive the Hamiltonian of the complete system in the Coulomb gauge, and this leads to the so-called minimal coupling Hamiltonian, containing both the electromagnetic potential and the matter-field amplitude. Specializing the nonrelativistic matter field to the case of a neutral atom, consisting of the field of electrons in a static nuclear potential, in Section 4.3 we obtain the atom-photon Hamiltonian in the minimal coupling scheme. Some of the basic processes induced by the atom-photon interaction part of the Hamiltonian are also discussed in this section. This gives us the possibility of introducing at this stage a fundamental simplification of the interaction Hamiltonian, namely the electric dipole approximation. The minimal coupling scheme, however, is not the only possible atom-field coupling. In fact, in Section 4.4 we show that a unitary transformation exists, the so-called Power-Zienau transformation, which leads from the Hamiltonian in the minimal coupling scheme to a new form of the Hamiltonian where the matter-field interaction takes place through the transverse displacement field rather than through the vector potential. This is the Hamiltonian in the multipolar coupling scheme. At the end of this section we derive the electric dipole approximation also for the Hamiltonian in the multipolar coupling scheme. The atom-photon interaction Hamiltonians, both in the minimal coupling and in the multipolar scheme, are second-quantized and expressed in terms of creation and annihilation operators in Section 4.5. In the same section we also introduce the concept of the two-level atom, and we derive model Hamiltonians describing the two-level atom, coupled to the radiation field both in the minimal coupling and in the multipolar scheme. The section is concluded by deriving another simplified form of the atom-field Hamiltonian, the Craig-Power model, in which the coupling takes place through the static electric polarizability tensor of the atom. All these Hamiltonians will be used in future chapters.

4.1 Coupled equations of motion

The fundamental Lagrangian describing the interaction of the Schrödinger field of charge q with the e.m. field (3.122), obtained in the previous chapter, can be cast in the form (Cohen-Tannoudji *et al.* 1989)

$$\mathcal{L} = -\frac{1}{16\pi}F_{\mu\nu}F_{\mu\nu} - \frac{1}{2m}\left(-\frac{\hbar}{i}\nabla - \frac{q}{c}\mathbf{A}\right)\psi^* \cdot \left(\frac{\hbar}{i}\nabla - \frac{q}{c}\mathbf{A}\right)\psi$$
$$- qV\psi^*\psi - \frac{i\hbar}{2}(\dot{\psi}^*\psi - \psi^*\dot{\psi}) \tag{4.1}$$

One can obtain from $\mathcal{L}$ the equation of motion for the matter-field amplitude ψ by means of the Euler-Lagrange equations (3.85). One has

$$\frac{\partial \mathcal{L}}{\partial \dot{\psi}^*} = -\frac{i\hbar}{2}\psi\,; \quad \frac{\partial \mathcal{L}}{\partial \nabla\psi^*} = -\frac{i\hbar}{2m}\left(\frac{\hbar}{i}\nabla - \frac{q}{c}\mathbf{A}\right)\psi\,;$$
$$\frac{\partial \mathcal{L}}{\partial \psi^*} = \frac{1}{2m}\frac{q}{c}\mathbf{A}\cdot\left(\frac{\hbar}{i}\nabla - \frac{q}{c}\mathbf{A}\right)\psi - qV\psi + \frac{i\hbar}{2}\dot{\psi} \tag{4.2}$$

and after some algebra one gets

$$i\hbar\dot{\psi} - \frac{1}{2m}\left(\frac{\hbar}{i}\nabla - \frac{q}{c}\mathbf{A}\right)^2\psi + qV\psi = 0 \tag{4.3}$$

Since the matter field considered is nonrelativistic, in the single-particle case this equation can be interpreted as the Schrödinger equation for a spinless particle of mass m and charge q in an electromagnetic field described by the four-potential A_μ, in which case ψ takes the usual meaning of the wavefunction of the particle. Also the Euler-Lagrange equations for the A_μ field can be obtained from (4.1) and (1.9). One has

$$\frac{\partial \mathcal{L}}{\partial(\partial A_\mu/\partial x_\nu)} = \frac{\partial \mathcal{L}_0}{\partial(\partial A_\mu/\partial x_\nu)} = \frac{1}{4\pi}\left(\frac{\partial A_\nu}{\partial x_\mu} - \frac{\partial A_\mu}{\partial x_\nu}\right) = \frac{1}{4\pi}F_{\mu\nu} \tag{4.4}$$

where $\mathcal{L}_0$ is defined in (1.8), and

$$\frac{\partial \mathcal{L}}{\partial A_\mu} = \frac{1}{c} j_\mu \tag{4.5}$$

Here we have defined the four-vector j_μ as $j_\mu = (\mathbf{j}, ic\rho)$ with

$$\mathbf{j} = \frac{q}{2m}\left\{\psi^*\left(\frac{\hbar}{i}\nabla - \frac{q}{c}\mathbf{A}\right)\psi + \psi\left(-\frac{\hbar}{i}\nabla - \frac{q}{c}\mathbf{A}\right)\psi^*\right\};$$

$$\rho = q\psi^*\psi \tag{4.6}$$

The quantities defined in (4.6) are the electric current density $\mathbf{j}$ and the charge density ρ. Although we are adopting the same symbols $\mathbf{j}$ and ρ as for the probability current density and for the probability density respectively, introduced in the previous chapter, no confusion should arise in what follows, noting the context in which these symbols are being used. Using (4.4) and (4.5) in (1.9) we obtain the equations of motion for the field in any of the three equivalent forms

$$\nabla \cdot \mathbf{E} = 4\pi\rho\,;\ \nabla \times \mathbf{H} - \frac{1}{c}\dot{\mathbf{E}} = \frac{4\pi}{c}\mathbf{j}\,;$$

$$\nabla \cdot \mathbf{H} = 0\,;\ \nabla \times \mathbf{E} + \frac{1}{c}\dot{\mathbf{H}} = 0 \tag{4.7}$$

$$\frac{\partial F_{\mu\nu}}{\partial x_\nu} = \frac{4\pi}{c} j_\mu\,;\ e_{\kappa\lambda\mu\nu}\frac{\partial F_{\mu\nu}}{\partial x_\lambda} = 0 \tag{4.8}$$

$$\frac{\partial^2 A_\mu}{\partial x_\nu^2} - \frac{\partial}{\partial x_\mu}\left(\frac{\partial A_\nu}{\partial x_\nu}\right) = -\frac{4\pi}{c} j_\mu \tag{4.9}$$

The matter field is seen as acting as a source for the classical electromagnetic field through the presence of j_μ in (4.7, 4.8, 4.9). In turn, the matter field is influenced also by the electromagnetic field via the presence of $\mathbf{A}$ in the Schrödinger equation (4.3). Thus the system of equations to solve, analogous to the Maxwell-Lorentz system of equations for a classical particle (see e.g. Heitler 1960) where the Schrödinger equation is substituted by the Lorentz equation, is constituted by any of (4.7, 4.8, 4.9) and by (4.3). This is a system of coupled equations which cannot be solved exactly. The matter field contributes to the total electromagnetic field a part which may be called the radiation reaction field (see e.g. Barut 1980). Calculation of the radiation reaction effects is one of the main tasks of modern electrodynamics, and it forms one of the main subjects of this book.

Specialization of (4.9) to the Lorentz gauge $\partial A_\nu/\partial x_\nu = 0$ yields

$$\nabla^2 \mathbf{A} - \frac{1}{c^2}\ddot{\mathbf{A}} = -\frac{4\pi}{c}\mathbf{j}\,;\ \nabla^2 V - \frac{1}{c^2}\ddot{V} = -4\pi\rho \tag{4.10}$$

whose solutions are time-dependent waves. Because of this particular form, (4.10) can also be regarded as four-dimensional Poisson equations. From this point of view it is easy to convince oneself that

$$\mathbf{A}(\mathbf{x}, t) = \frac{1}{c}\int \frac{\mathbf{j}(\mathbf{x}', t' = t - |\,\mathbf{x} - \mathbf{x}'\,|/c)}{|\,\mathbf{x} - \mathbf{x}'\,|}d^3\mathbf{x}'\,;$$

$$V(\mathbf{x}, t) = \int \frac{\rho(\mathbf{x}', t' = t - |\,\mathbf{x} - \mathbf{x}'\,|/c)}{|\,\mathbf{x} - \mathbf{x}'\,|}d^3\mathbf{x}' \tag{4.11}$$

are particular solutions of (4.10). Naturally the general solutions are found by combining (4.11) with the general solutions of the homogeneous equations (obtained by putting $\mathbf{j} = 0, \rho = 0$) associated with (4.10). Expressions (4.11) are known as retarded potentials. On the other hand, in the Coulomb gauge $\nabla \cdot \mathbf{A} = 0$. Hence $\mathbf{A}$ is a transverse field $\mathbf{A} \equiv \mathbf{A}_\perp$ and from (4.9)

$$\nabla^2 \mathbf{A}_\perp - \frac{1}{c^2}\ddot{\mathbf{A}}_\perp - \frac{1}{c}\nabla\dot{V} = -\frac{4\pi}{c}\mathbf{j}\,;\ \nabla^2 V = -4\pi\rho \tag{4.12}$$

The second equation for V is a three-dimensional Poisson equation, a particular solution of which is

$$V(\mathbf{x}, t) = \int \frac{\rho(\mathbf{x}', t)}{|\,\mathbf{x} - \mathbf{x}'\,|}d^3\mathbf{x}' \tag{4.13}$$

We see that V, in contrast to the case of the Lorentz gauge, propagates instantaneously. This is not contrary to the principle of Lorentz invariance since V and $\mathbf{A}$ are not gauge invariant and as such they are not measurable quantities. The physical entities are $\mathbf{E}$ and $\mathbf{H}$ which propagate with velocity c since they satisfy equation (4.8) which is manifestly Lorentz-invariant. Moreover, the first equation (4.12) can be simplified by introducing longitudinal and transverse current fields as

$$\mathbf{j} = \mathbf{j}_\perp + \mathbf{j}_\parallel\,;\ \nabla \cdot \mathbf{j}_\perp = 0\,;\ \nabla \times \mathbf{j}_\parallel = 0 \tag{4.14}$$

In fact from (4.8) and using $F_{\mu\nu} = -F_{\nu\mu}$, one obtains easily the continuity equation in the following forms

$$\frac{\partial j_\mu}{\partial x_\mu} = 0\,;\ \nabla \cdot \mathbf{j} + \frac{\partial\rho}{\partial t} = 0\,;\ \nabla \cdot \mathbf{j}_\parallel = -\frac{\partial\rho}{\partial t} \tag{4.15}$$

while from the third of (4.14) one can introduce a scalar ϕ such that

$$\mathbf{j}_{\parallel} = \nabla\phi \tag{4.16}$$

Thus

$$\nabla^2\phi = -\frac{\partial\rho}{\partial t} \tag{4.17}$$

which should be compared with the second of (4.12) to yield

$$\phi = \frac{1}{4\pi}\dot{V}\ ;\ \mathbf{j}_{\parallel} = \frac{1}{4\pi}\nabla\dot{V}\ ;\ \mathbf{j} = \mathbf{j}_{\perp} + \frac{1}{4\pi}\nabla\dot{V} \tag{4.18}$$

Substitution of the last expression into the first of (4.12) finally yields

$$\nabla^2\mathbf{A}_{\perp} - \frac{1}{c^2}\ddot{\mathbf{A}}_{\perp} = -\frac{4\pi}{c}\mathbf{j}_{\perp} \tag{4.19}$$

from which V has been eliminated and which is the sought-for simplification.

4.2 Minimal coupling Hamiltonian in the Coulomb gauge

The canonical procedure for deriving the Hamiltonian density from (4.1) is to obtain first the momenta conjugate to the field amplitudes A_μ and ψ and then to form the Hamiltonian density according to the usual prescription which we have followed, e.g. in the case of the Klein-Gordon field. While the problems of redundancy of the generalized coordinates A_μ have been dealt with in Section 1.4 by the gauge-fixing procedure, we have not yet discussed the analogous problems arising in connection with the matter field ψ. That a redundancy problem exists for ψ is obvious since from (4.2) one can see that $(\partial\mathcal{L}/\partial\dot{\psi})$, which should be the momentum conjugate to ψ, is proportional to another generalized coordinate ψ^*. Thus the momenta obtained are not independent of the coordinates, and the canonical procedure is not straightforwardly applicable to part of the model. A technique to reduce the matter field redundancy exists (see e.g. Cohen-Tannoudji *et al.* 1989), but it is slightly involved and it would take us through an unnecessary detour. Thus we shall take a shortcut, and remark that in any case the interaction part $\mathcal{L}_i$ of the Lagrangian density of the coupled system defined in (3.122) does not contain time derivatives of A_μ or of ψ, hence it cannot contribute in any way to the respective conjugate momenta. It follows that the Hamiltonian density of the coupled system consists of the free e.m. field and of the free Schrödinger field Hamiltonian density, minus $\mathcal{L}_i$.

The Hamiltonian density for the free e.m. field is obtained from the corresponding Lagrangian density appearing in (4.1)

$$\mathcal{L}_F = -\frac{1}{16\pi} F_{\mu\nu} F_{\mu\nu} = \frac{1}{8\pi}(\mathbf{E}^2 - \mathbf{H}^2)$$
$$= \frac{1}{8\pi}\left\{\frac{1}{c^2}\dot{\mathbf{A}}^2 + \frac{2}{c}\dot{\mathbf{A}} \cdot \nabla V + (\nabla V)^2 - (\nabla \times \mathbf{A})^2\right\} \tag{4.20}$$

where we have used (1.5) in the form

$$\mathbf{E} = -\nabla V - \frac{1}{c}\dot{\mathbf{A}} \tag{4.21}$$

Since in the Coulomb gauge V is not an independent dynamical variable, as is evident from (4.13), we shall reduce the field variables to the three components of $\mathbf{A}$, and we obtain the canonical momenta of $\mathbf{A}$ as

$$\frac{\partial \mathcal{L}_F}{\partial \dot{\mathbf{A}}} = \frac{1}{4\pi c}\left(\frac{1}{c}\dot{\mathbf{A}} + \nabla V\right) = -\frac{1}{4\pi c}\mathbf{E} \tag{4.22}$$

Then the canonical e.m. field Hamiltonian density is

$$\mathcal{H}_F = \frac{\partial \mathcal{L}_F}{\partial \dot{\mathbf{A}}} \cdot \dot{\mathbf{A}} - \mathcal{L}_F = -\frac{1}{4\pi c}\mathbf{E} \cdot \dot{\mathbf{A}} - \mathcal{L}_F$$
$$= \frac{1}{4\pi}\mathbf{E} \cdot (\mathbf{E} + \nabla V) - \mathcal{L}_F$$
$$= \frac{1}{8\pi}(\mathbf{E}^2 + \mathbf{H}^2) + \frac{1}{4\pi}\mathbf{E} \cdot \nabla V \tag{4.23}$$

In the absence of charges we have $V = 0$, and we obtain again (1.32) which coincides with the energy density. In the presence of charges we cannot set $V = 0$, but in the Coulomb gauge $\dot{\mathbf{A}}$ is a transverse vector, while ∇V is a longitudinal one. Thus from (4.21) we have

$$\mathbf{E} = \mathbf{E}_\perp + \mathbf{E}_\parallel \; ; \; \mathbf{E}_\perp = -\frac{1}{c}\dot{\mathbf{A}} \; ; \; \mathbf{E}_\parallel = -\nabla V \tag{4.24}$$

and $\mathcal{H}_F$ can be written as

$$\mathcal{H}_F = \frac{1}{8\pi}\left(\mathbf{E}_\perp^2 + \mathbf{H}^2 - \mathbf{E}_\parallel^2\right) \tag{4.25}$$

The Hamiltonian density for the free Schrödinger field can be taken to coincide with the energy density which has already been obtained in (3.88) as

$$\mathcal{H}_M = \frac{\hbar^2}{2m}\nabla\psi^* \cdot \nabla\psi \tag{4.26}$$

Thus the total Hamiltonian density in the Coulomb gauge following (3.122) is

$$\mathcal{H} = \mathcal{H}_M + \mathcal{H}_F + \mathcal{H}_i \, ;$$
$$\mathcal{H}_i = -\frac{q}{2mc}\left[\mathbf{A}\cdot\left(\psi^*\frac{\hbar}{i}\nabla\psi - \psi\frac{\hbar}{i}\nabla\psi^*\right) - \frac{q}{c}\psi^*\mathbf{A}^2\psi\right]$$
$$+ q\psi^* V\psi \tag{4.27}$$

The total Hamiltonian is obtained by integration over the whole space

$$H = \frac{1}{8\pi}\int (\mathbf{E}_\perp^2 + \mathbf{H}^2)d^3\mathbf{x} - \frac{1}{8\pi}\int \mathbf{E}_\parallel^2 d^3\mathbf{x} + \frac{\hbar^2}{2m}\int \nabla\psi^*\cdot\nabla\psi d^3\mathbf{x}$$
$$-\frac{q}{2mc}\int\left(\psi^*\frac{\hbar}{i}\nabla\psi - \psi\frac{\hbar}{i}\nabla\psi^*\right)\cdot\mathbf{A}_\perp d^3\mathbf{x}$$
$$+\frac{q^2}{2mc^2}\int \psi^*\mathbf{A}_\perp^2\psi d^3\mathbf{x} + q\int \psi^* V\psi d^3\mathbf{x} \tag{4.28}$$

Assuming that $V\nabla V$ vanishes on the surface at infinity, one has

$$\frac{1}{8\pi}\int \mathbf{E}_\parallel^2 d^3\mathbf{x} = \frac{1}{8\pi}\int \nabla V\cdot\nabla V d^3\mathbf{x}$$
$$= \frac{1}{8\pi}\int \nabla\cdot(V\nabla V)d^3\mathbf{x} - \frac{1}{8\pi}\int V\nabla^2 V d^3\mathbf{x}$$
$$= -\frac{1}{8\pi}\int V\nabla^2 V d^3\mathbf{x} = \frac{1}{2}\int V\rho d^3\mathbf{x}$$
$$= \frac{1}{2}q\int \psi^* V\psi d^3\mathbf{x} \tag{4.29}$$

where we have made use of (4.12). Moreover, exploiting the assumption that the field ψ and its gradient also vanish sufficiently fast at infinity, one can convert the integrals over the Schrödinger field coordinates appearing in (4.28) as follows

$$\int \nabla\psi^*\cdot\nabla\psi d^3\mathbf{x} = \int \nabla\cdot(\psi^*\nabla\psi)d^3\mathbf{x} - \int \psi^*\nabla^2\psi d^3\mathbf{x}$$
$$= -\int \psi^*\nabla^2\psi d^3\mathbf{x} \, ;$$
$$\int \psi\nabla\psi^* d^3\mathbf{x} = \int \nabla(\psi\psi^*)d^3\mathbf{x} - \int \psi^*\nabla\psi d^3\mathbf{x}$$
$$= -\int \psi^*\nabla\psi d^3\mathbf{x} \tag{4.30}$$

Thus H takes the form

$$H = \frac{1}{8\pi}\int (\mathbf{E}_\perp^2 + \mathbf{H}^2)d^3\mathbf{x} - \frac{\hbar^2}{2m}\int \psi^*\nabla^2\psi d^3\mathbf{x}$$
$$- \frac{q}{mc}\int \psi^*\mathbf{A}_\perp \cdot \frac{\hbar}{i}\nabla\psi d^3\mathbf{x} + \frac{q^2}{2mc^2}\int \psi^*\mathbf{A}_\perp^2\psi d^3\mathbf{x}$$
$$+ \frac{1}{2}q\int \psi^* V\psi d^3\mathbf{x} \tag{4.31}$$

This Hamiltonian forms the basis for the canonical treatment of a Schrödinger matter field interacting with the e.m. field in the Coulomb gauge. The form of the coupling between matter and field appearing in the last three terms of (4.31) is generally called minimal coupling.

4.3 Minimal coupling for a neutral atom

We now specialize the Hamiltonian (4.31) to the case of a neutral atom coupled to the radiation field. Such an atom is usually modelled as a positive charge Ze concentrated in a point-like nucleus, whose dynamics is treated classically in view of its relatively large mass, surrounded by Z spinless electrons each of charge $q = -e$. In our model the nucleus is taken as fixed at point $\mathbf{R}$ in space. In view of the nature of the model, it is convenient to separate off V the part V_Z due to the nucleus. This part satisfies a Poisson equation like (4.12) with $\rho = Ze\delta(\mathbf{x} - \mathbf{R})$. Hence

$$V_Z = \frac{Ze}{|\mathbf{R} - \mathbf{x}|} \tag{4.32}$$

The remaining part V_e of V, neglecting the interaction of the nucleus with itself which is infinite but obviously does not play any role in the dynamics of the model, is due to the electron-electron interaction. For the other part we substitute $\rho = -e\psi^*\psi$ into (4.13) thereby obtaining

$$V_e = -e\int \frac{\psi^*(\mathbf{x}')\psi(\mathbf{x}')}{|\mathbf{x} - \mathbf{x}'|}d^3\mathbf{x}' \tag{4.33}$$

Consequently the atom-field Hamiltonian (4.31) becomes

$$H = H_A + H_F + H_{AF}\ ;$$
$$H_A = -\frac{\hbar^2}{2m}\int \psi^*(\mathbf{x})\nabla^2\psi(\mathbf{x})d^3\mathbf{x} - Ze^2\int \psi^*(\mathbf{x})\frac{1}{|\mathbf{R} - \mathbf{x}|}\psi(\mathbf{x})d^3\mathbf{x}$$
$$+ \frac{1}{2}e^2\int\int \psi^*(\mathbf{x})\psi^*(\mathbf{x}')\frac{1}{|\mathbf{x} - \mathbf{x}'|}\psi(\mathbf{x}')\psi(\mathbf{x})d^3\mathbf{x}d^3\mathbf{x}'\ ;$$

$$H_F = \frac{1}{8\pi}\int \left\{\frac{1}{c^2}\dot{\mathbf{A}}_\perp^2(\mathbf{x}) + [\nabla \times \mathbf{A}_\perp(\mathbf{x})]^2\right\} d^3\mathbf{x} \, ;$$

$$H_{AF} = \frac{e}{mc}\int \psi^*(\mathbf{x})\mathbf{A}_\perp(\mathbf{x}) \cdot \frac{\hbar}{i}\nabla\psi(\mathbf{x})d^3\mathbf{x} + \frac{e^2}{2mc^2}\int \psi^*(\mathbf{x})\mathbf{A}_\perp^2(\mathbf{x})\psi(\mathbf{x})d^3\mathbf{x} \tag{4.34}$$

In this expression H_A contains only the generalized coordinates of the electron field around the nucleus, H_F contains only the generalized coordinates of the transverse e.m. field, and H_{AF} contains both kinds of generalized coordinates and represents an interaction between the two fields.

Up to now both the electron field and the e.m. field have been treated classically. We now proceed to second quantization by assigning to the field amplitudes the status of operators

$$H = H_A^{min} + H_F^{min} + H_{AF}^{min} \, ;$$

$$H_A^{min} = -\frac{\hbar^2}{2m}\int \psi^\dagger(\mathbf{x})\nabla^2\psi(\mathbf{x})d^3\mathbf{x} - Ze^2\int \psi^\dagger(\mathbf{x})\frac{1}{|\mathbf{R}-\mathbf{x}|}\psi(\mathbf{x})d^3\mathbf{x} + \frac{1}{2}e^2\int\int \psi^\dagger(\mathbf{x})\psi^\dagger(\mathbf{x}')\frac{1}{|\mathbf{x}-\mathbf{x}'|}\psi(\mathbf{x}')\psi(\mathbf{x})d^3\mathbf{x}d^3\mathbf{x}' \, ;$$

$$H_F^{min} = \frac{1}{8\pi}\int \left\{\frac{1}{c^2}\dot{\mathbf{A}}_\perp^2(\mathbf{x}) + [\nabla \times \mathbf{A}_\perp(\mathbf{x})]^2\right\} d^3\mathbf{x} \, ;$$

$$H_{AF}^{min} = \frac{e}{mc}\int \psi^\dagger(\mathbf{x})\mathbf{A}_\perp(\mathbf{x}) \cdot \frac{\hbar}{i}\nabla\psi(\mathbf{x})d^3\mathbf{x} + \frac{e^2}{2mc^2}\int \psi^\dagger(\mathbf{x})\mathbf{A}_\perp^2(\mathbf{x})\psi(\mathbf{x})d^3\mathbf{x} \tag{4.35}$$

where the superscript *min* has been added to emphasize that the atom-field interaction is described in the minimal coupling representation. In order to introduce the appropriate creation and annihilation operators, we first consider the electron field. The eigenstates and eigenvalues of the single-particle Hamiltonian

$$H_P = -\frac{\hbar^2}{2m}\nabla^2 - \frac{Ze^2}{|\mathbf{R}-\mathbf{x}|} \tag{4.36}$$

are the well-known hydrogenic eigensolutions. The eigenvalue spectrum has a discrete and a continuum part. The eigenvalues of the discrete part are given by

$$E_N = -\frac{mZ^2e^4}{2\hbar^2 N^2} \quad (N = 1, 2, \ldots) \tag{4.37}$$

where N is the principal quantum number. $E = 0$ is clearly an accumulation point for the discrete spectrum. The continuous spectrum is in the range of all positive E. The eigenfunctions u_{NLM} for the discrete spectrum corresponding to E_N can be labelled by three integer quantum numbers N, L, M, the second of which for each N ranges from 0 to $N - 1$, while M for each L ranges from $-L$ to L. L is the quantum number of the angular momentum of the particle in state u_{NLM}, and M is the quantum number of its component along the z-axis of a reference system. The quantum average value of the distance of the particle from the centre of attraction at $\mathbf{R}$ in an eigenstate u_{NLM} is finite for negative E and increases as N^2 (see e.g. Landau and Lifshitz 1958); hence the motion of the particle is bound to the centre of attraction although in principle it may attain very large radii for large N. The large N states are called Rydberg states. For positive E the discrete positive number N is substituted by a continous index q and the eigenfunctions $u_{qLM}(\mathbf{x})$ are labelled also by the integers L and M. In this range L can take all discrete values from 0 to infinity, while for each L, M ranges from $-L$ to $+L$ as in the discrete part of the spectrum. The average distance of the particles from the centre of attraction diverges and the motion is unbound for any value of q. In the language of atomic physics the states with $E > 0$ are ionized states. In order to avoid the non-essential algebraic difficulties connected with the continuous part of the eigenvalue spectrum of (4.35) we shall imagine that the system is contained in a very large box, centred on the nucleus. This will have negligible effects on most of the discrete part of the energy spectrum and will transform the continuous part into a discrete spectrum of closely spaced levels. This trick enables us to develop the field amplitude ψ as in (3.94)

$$\psi(\mathbf{x}, t) = \sum_k a_k(t) u_k(\mathbf{x}) \tag{4.38}$$

where $u_k(\mathbf{x})$ are the hydrogenic wavefunctions discussed above, approximate eigenfunctions of the Hamiltonian (4.36) according to

$$H_P u_k(\mathbf{x}) = E_k u_k(\mathbf{x}) \tag{4.39}$$

and where the index k represents an appropriate set of quantum numbers. Since electrons are fermions in spite of the fact that our considerations have not involved the spin degrees of freedom, we must transform the amplitudes a_k into operators using the anticommutation rule prescription (3.100). Substitution of (4.38) into H_A^{min} and use of (4.39) yields

$$H_A^{min} = \sum_k E_k a_k^\dagger a_k + \frac{1}{2}e^2 \sum_{kk'k''k'''} D_{kk'}^{k''k'''}(|\mathbf{x}-\mathbf{x}'|) a_k^\dagger a_{k'}^\dagger a_{k''} a_{k'''} ;$$

$$D_{kk'}^{k''k'''}(|\mathbf{x}-\mathbf{x}'|) = \int\int u_k^*(\mathbf{x}) u_{k'}^*(\mathbf{x}') \frac{1}{|\mathbf{x}-\mathbf{x}'|} u_{k''}(\mathbf{x}') u_{k'''}(\mathbf{x}) d^3\mathbf{x} d^3\mathbf{x}' \qquad (4.40)$$

The effect of the electrostatic repulsion among electrons is contained in the e^2 term in (4.40). In the case of the hydrogen atom $Z = 1$, and the electrostatic repulsion vanishes as it should, because in the one-electron subspace the matrix elements of two consecutive annihilation operators vanish. It is the task of atomic physics to find approximate solutions of H_A^{min} in the form (4.40) for various Z. These solutions are obtained by Hartree or Hartree-Fock techniques which fall outside the scope of this book. Consequently we shall adopt a simplified point of view. Namely we shall assume that it is possible to find one-electron states $u_n(\mathbf{x})$ and corresponding energy eigenvalues which, differently from those defined by (4.39), take account of the electrostatic interaction. These lead to new creation and annihilation one-electron operators $c_k^\dagger$ and c_k, and we assume that the electron field can be expanded in terms of these as

$$\psi(\mathbf{x}, t) = \sum_n c_n u_n(\mathbf{x}) \qquad (4.41)$$

such that, when (4.41) is substituted into H_A^{min}, the latter takes the simple diagonal form

$$H_A^{min} = \sum_n E_n c_n^\dagger c_n \qquad (4.42)$$

where E_n also takes into account in an approximate way the electrostatic repulsion among electrons.

As for second quantization of the field, we remark that it has already been performed in Chapter 2, so that we may use directly result (2.10) and write

$$H_F^{min} = \sum_{\mathbf{k}j} \hbar\omega_k \left(a_{\mathbf{k}j}^\dagger a_{\mathbf{k}j} + \frac{1}{2} \right) \qquad (4.43)$$

Moreover we can also use the first of (2.9) to express H_{AF}^{min} in the form

$$H_{AF}^{min} = \frac{e}{m}\sqrt{\frac{2\pi\hbar}{V}}\sum_{nn'}\sum_{\mathbf{k}j}\frac{1}{\sqrt{\omega_k}}$$

$$\times\left(\mathbf{e}_{\mathbf{k}j}\cdot\int e^{i\mathbf{k}\cdot\mathbf{x}}u_n^*(\mathbf{x})\frac{\hbar}{i}\nabla u_{n'}(\mathbf{x})\right)a_{\mathbf{k}j}c_n^\dagger c_{n'}$$

$$+\frac{e^2}{2m}\frac{2\pi\hbar}{V}\sum_{nn'}\sum_{\mathbf{k}\mathbf{k}'jj'}\frac{1}{\sqrt{\omega_k\omega_{k'}}}\left\{\left(\mathbf{e}_{\mathbf{k}j}\cdot\mathbf{e}_{\mathbf{k}'j'}\right)\int d^3\mathbf{x}e^{i(\mathbf{k}+\mathbf{k}')\cdot\mathbf{x}}\right.$$

$$\times u_n^*(\mathbf{x})u_{n'}(\mathbf{x})a_{\mathbf{k}j}a_{\mathbf{k}'j'}c_n^\dagger c_{n'}+\left(\mathbf{e}_{\mathbf{k}j}\cdot\mathbf{e}_{\mathbf{k}'j'}^*\right)\int d^3\mathbf{x}e^{i(\mathbf{k}-\mathbf{k}')\cdot\mathbf{x}}$$

$$\left.\times u_n^*(\mathbf{x})u_{n'}(\mathbf{x})a_{\mathbf{k}j}a_{\mathbf{k}'j'}^\dagger c_n^\dagger c_{n'}\right\}+\text{h.c.} \tag{4.44}$$

Thus the atom-field interaction, in the minimal coupling form, consists of two parts (see e.g. Sakurai 1982). The first, denoted by $H_{AF}^{(1)}$, is of order e and describes processes by which one of the atomic electrons absorbs one photon $\mathbf{k}j$ and jumps from state n' to state n. Also, the Hermitian conjugate process is possible, by which the electron jumps from n to n' emitting one photon $\mathbf{k}j$. The second part, denoted by $H_{AF}^{(2)}$, is of order e^2 and describes processes which involve two photons $\mathbf{k}j$ and $\mathbf{k}'j'$. The basic processes due to $H_{AF}^{(1)}$ are pictorially represented by the Feynman diagrams in Figure 4.1, and those due to $H_{AF}^{(2)}$ are represented by the Feynman

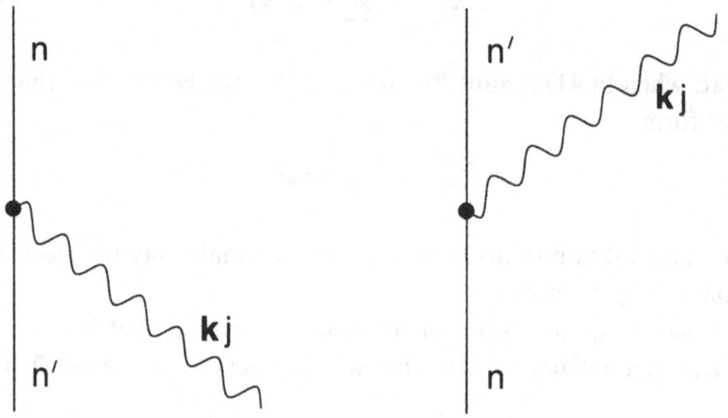

Fig. 4.1 The diagram on the left describes processes by which an atomic electron in state n' absorbs a photon $\mathbf{k}j$ and jumps to the state n. The opposite process is described by the diagram on the right and leads to the emission of a photon $\mathbf{k}j$.

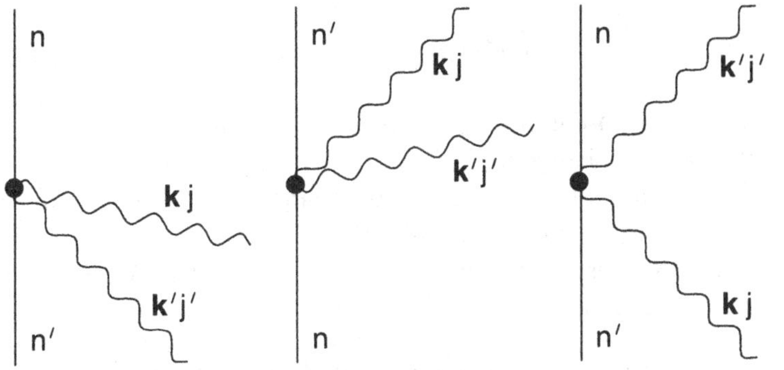

Fig. 4.2 The diagram on the left describes processes by which an atomic transition from state n' to state n is accompanied by absorption of two photons $\mathbf{k}j$ and $\mathbf{k}'j'$. The opposite process is described by the central diagram. The diagram on the right describes an atomic transition from state n' to state n accompanied by absorption of one photon $\mathbf{k}j$ and by emission of one photon $\mathbf{k}'j'$.

diagrams of Figure 4.2. The transition amplitudes of the processes due to $H_{AF}^{(1)}$ are proportional to the integral

$$\int e^{i\mathbf{k}\cdot\mathbf{x}} u_n^*(\mathbf{x}) \frac{\hbar}{i} \nabla u_{n'}(\mathbf{x}) d^3\mathbf{x} \tag{4.45}$$

If the wavelengths of the photons involved in the processes are much larger than atomic dimensions, this integral can be evaluated in the dipole approximation, which consists of approximating

$$\begin{aligned} &e^{i\mathbf{k}\cdot\mathbf{x}} = 1 - i\mathbf{k}\cdot\mathbf{x} - \frac{1}{2}(\mathbf{k}\cdot\mathbf{x})^2 + \ldots \sim 1 \quad (\mathbf{R} = 0)\ ; \\ &e^{i\mathbf{k}\cdot\mathbf{x}} \simeq e^{i\mathbf{k}\cdot\mathbf{R}} \quad (\mathbf{R} \neq 0) \end{aligned} \tag{4.46}$$

This is a good approximation for optical photons ($\lambda \sim 10^3$Å) which are of special interest for our treatment and which are expected to interact strongly with the atomic electrons, since the differences between low-lying atomic states tend to fall within the optical range. Thus (4.45) in the dipole approximation can be approximated as ($\mathbf{R} = 0$)

$$\int u_n^*(\mathbf{x}) \frac{\hbar}{i} \nabla u_{n'}(\mathbf{x}) d^3\mathbf{x} \tag{4.47}$$

which is at the origin of the selection rules familiar to spectroscopists (see e.g. Cohen-Tannoudji *et al*. 1977). Within the same dipole approximation the analogous integrals in $H_{AF}^{(2)}$ are proportional to

$$\int u_n^*(\mathbf{x})u_{n'}(\mathbf{x})d^3\mathbf{x} = \delta_{nn'} \tag{4.48}$$

which shows that in the dipole approximation the interaction due to the $\mathbf{A}_\perp^2$ terms does not change the atomic internal state.

Finally, we remark that the dipole approximation (4.46) can be directly performed on the interaction Hamiltonian in (4.35) rather than on its matrix elements. This is obtained by substituting $\mathbf{A}_\perp(\mathbf{x})$ by $\mathbf{A}_\perp(\mathbf{R})$ in H_{AF}^{min}, which thereby takes the form

$$\begin{aligned} H_{AF}^{min} &= \frac{e}{mc}\int \psi^\dagger(\mathbf{x})\mathbf{A}_\perp(\mathbf{R})\cdot\frac{\hbar}{i}\nabla\psi(\mathbf{x})d^3\mathbf{x} \\ &+ \frac{e^2}{2mc^2}\int \psi^\dagger(\mathbf{x})\mathbf{A}_\perp^2(\mathbf{R})\psi(\mathbf{x})d^3\mathbf{x} \end{aligned} \tag{4.49}$$

and which clearly shows that the dipole approximation is obtained by assuming that the vector potential is constant over the atomic volume.

4.4 Multipolar coupling for a neutral atom

We will now discuss another description of the atom-photon interaction, which was originally introduced by Power and Zienau (1959). This new description is unitarily related, and as such fully equivalent, to the minimal coupling scheme obtained in the previous section. It has the interesting feature, however, that the coupling between charges and radiation is expressed in terms of the electric and magnetic fields rather than in terms of the vector potential. We return to the atom-photon Hamiltonian (4.35) which can be expressed in second quantization by substituting the Schrödinger field amplitude c-functions ψ, ψ^* with operators $\psi, \psi^\dagger$ subject to quantization rules (3.101). We rearrange the terms to cast (4.35) in the form

$$\begin{aligned} H &= -\frac{1}{2m}\int \psi^\dagger(\mathbf{x})\left[\hbar\nabla + i\frac{e}{c}\mathbf{A}_\perp(\mathbf{x})\right]^2\psi(\mathbf{x})d^3\mathbf{x} \\ &- Ze^2\int \psi^\dagger(\mathbf{x})\frac{1}{|\mathbf{R}-\mathbf{x}|}\psi(\mathbf{x})d^3\mathbf{x} \\ &+ \frac{1}{2}e^2\int\int \psi^\dagger(\mathbf{x})\psi^\dagger(\mathbf{x}')\frac{1}{|\mathbf{x}-\mathbf{x}'|}\psi(\mathbf{x}')\psi(\mathbf{x})d^3\mathbf{x}d^3\mathbf{x}' \\ &+ \frac{1}{8\pi}\int\left\{\frac{1}{c^2}\dot{\mathbf{A}}_\perp^2(\mathbf{x}) + [\nabla\times\mathbf{A}_\perp(\mathbf{x})]^2\right\}d^3\mathbf{x} \end{aligned} \tag{4.50}$$

Consider now the unitary operator e^{is} where

$$s = \int \psi^\dagger(\mathbf{x}) S(\mathbf{x}) \psi(\mathbf{x}) d^3\mathbf{x} \; ;$$
$$S(\mathbf{x}) = \frac{1}{\hbar c} \int \mathbf{p}(\mathbf{r}, \mathbf{x}) \cdot \mathbf{A}_\perp(\mathbf{r}) d^3\mathbf{r} \tag{4.51}$$

where $\mathbf{p}(\mathbf{r}, \mathbf{x})$ is the polarization field defined in (B.9). We subject all operators appearing in (4.50) to the unitary transformation induced by e^{is}, thereby obtaining transformed operators which we indicate by a tilde. Thus (Power and Thirunamachandran 1983)

$$\tilde{\psi}(\mathbf{x}) = e^{is}\psi(\mathbf{x})e^{-is} = \psi(\mathbf{x}) + [is, \psi(\mathbf{x})] + \frac{1}{2!}[is, [is, \psi(\mathbf{x})]] + \dots \tag{4.52}$$

Using the commutation relation (3.101) it is easy to obtain

$$[s, \psi(\mathbf{x})] = \int [\psi^\dagger(\mathbf{x}') S(\mathbf{x}') \psi(\mathbf{x}'), \psi(\mathbf{x})] d^3\mathbf{x}' = -S(\mathbf{x})\psi(\mathbf{x}) \; ;$$
$$[s, [s, \psi(\mathbf{x})]] = -\,[s, S(\mathbf{x})\psi(\mathbf{x})] = S^2(\mathbf{x})\psi(\mathbf{x}) \tag{4.53}$$

Substituting the infinite chain of commutators (4.53) into (4.52) yields for the electron field

$$\tilde{\psi}(\mathbf{x}) = e^{-iS(\mathbf{x})}\psi(\mathbf{x}) \tag{4.54}$$

On the other hand, since $\mathbf{A}_\perp(\mathbf{x})$ and $\mathbf{A}_\perp(\mathbf{r})$ commute for any $\mathbf{x}$ and $\mathbf{r}$ and since s is not a function of $\mathbf{x}$, one immediately has

$$\tilde{\mathbf{A}}_\perp(\mathbf{x}) = e^{is}\mathbf{A}_\perp(\mathbf{x})e^{-is} = \mathbf{A}_\perp(\mathbf{x}) \; ; \; \tilde{\nabla} = \nabla \tag{4.55}$$

The same is not true for $\dot{\mathbf{A}}_\perp(\mathbf{x})$, which because of the commutation relation (2.5) does not always commute with $\mathbf{A}_\perp(\mathbf{r})$. In fact

$$\tilde{\dot{\mathbf{A}}}_\perp(\mathbf{x}) = e^{is}\dot{\mathbf{A}}_\perp(\mathbf{x})e^{-is} = \dot{\mathbf{A}}_\perp(\mathbf{x}) + \left[is, \dot{\mathbf{A}}_\perp(\mathbf{x})\right] + \frac{1}{2!}\left[is, \left[is, \dot{\mathbf{A}}_\perp(\mathbf{x})\right]\right] + \dots \tag{4.56}$$

and

$$\begin{aligned}
\left[s, \dot{\mathbf{A}}_\perp(\mathbf{x})\right] &= \int \left[\psi^\dagger(\mathbf{x}') S(\mathbf{x}') \psi(\mathbf{x}'), \dot{\mathbf{A}}_\perp(\mathbf{x})\right] d^3\mathbf{x}' \\
&= \frac{1}{\hbar c} \int\int \psi^\dagger(\mathbf{x}') \left[p_j(\mathbf{r}, \mathbf{x}') A_{\perp j}(\mathbf{r}), \dot{\mathbf{A}}_\perp(\mathbf{x})\right] \psi(\mathbf{x}') d^3\mathbf{r} d^3\mathbf{x}' \\
&= \frac{1}{\hbar} \int \psi^\dagger(\mathbf{x}') \left[\mathbf{E}_\perp(\mathbf{x}), p_j(\mathbf{r}, \mathbf{x}') A_{\perp j}(\mathbf{r})\right] \psi(\mathbf{x}') d^3\mathbf{r} d^3\mathbf{x}'
\end{aligned} \tag{4.57}$$

Consequently, using (2.5),

$$\begin{aligned}
[s, \dot{A}_{\perp i}(\mathbf{x})] &= \frac{1}{\hbar}\int \psi^\dagger(\mathbf{x}')\left[E_{\perp i}(\mathbf{x}), p_j(\mathbf{r},\mathbf{x}')A_{\perp j}(\mathbf{r})\right]\psi(\mathbf{x}')d^3\mathbf{r}d^3\mathbf{x}' \\
&= i4\pi c \int \psi^\dagger(\mathbf{x}')p_j(\mathbf{r},\mathbf{x}')\psi(\mathbf{x}')d^3\mathbf{x}'\delta_{\perp ij}(\mathbf{x}-\mathbf{r})d^3\mathbf{r} \\
&= i4\pi c \int p_j(\mathbf{r})\delta_{\perp ij}(\mathbf{x}-\mathbf{r})d^3\mathbf{r} = i4\pi c p_{\perp i}(\mathbf{x}) \qquad (4.58)
\end{aligned}$$

where according to (1.62) $\mathbf{p}_\perp$ is the transverse part of $\mathbf{p}$. We conclude that

$$\left[s, \dot{\mathbf{A}}_\perp(\mathbf{x})\right] = i4\pi c\mathbf{p}_\perp(\mathbf{x}) \qquad (4.59)$$

Since

$$\left[s, \left[s, \dot{\mathbf{A}}_\perp(\mathbf{x})\right]\right] = i4\pi c[s, \mathbf{p}_\perp(\mathbf{x})] = 0 \qquad (4.60)$$

the final result is

$$\tilde{\dot{\mathbf{A}}}_\perp(\mathbf{x}) = \dot{\mathbf{A}}_\perp(\mathbf{x}) - 4\pi c\mathbf{p}_\perp(\mathbf{x}) \qquad (4.61)$$

Using a normal mode expansion as in (2.9), it is also easy to convince ourselves that $[\nabla\times\mathbf{A}(\mathbf{x})]_i$ commutes with $A_{\perp j}(\mathbf{r})$ for any $\mathbf{x}$ and $\mathbf{r}$, hence

$$\widetilde{\nabla\times\mathbf{A}_\perp} = \nabla\times\mathbf{A}_\perp(\mathbf{x}) \qquad (4.62)$$

Results (4.54), (4.55), (4.61) and (4.62) can be used to express all operators appearing in (4.50) in terms of the new tilde operators. In particular, the Hamiltonian H takes the form

$$\begin{aligned}
H = &-\frac{1}{2m}\int \tilde{\psi}^\dagger(\mathbf{x})e^{-iS(\mathbf{x})}\left[\hbar\nabla + i\frac{e}{c}\tilde{\mathbf{A}}_\perp(\mathbf{x})\right]^2 e^{iS(\mathbf{x})}\tilde{\psi}(\mathbf{x})d^3\mathbf{x} \\
&- Ze^2\int \tilde{\psi}^\dagger(\mathbf{x})\frac{1}{|\mathbf{R}-\mathbf{x}|}\tilde{\psi}(\mathbf{x})d^3\mathbf{x} \\
&+\frac{1}{2}e^2\int\int \tilde{\psi}^\dagger(\mathbf{x})\tilde{\psi}^\dagger(\mathbf{x}')\frac{1}{|\mathbf{x}-\mathbf{x}'|}\tilde{\psi}(\mathbf{x}')\tilde{\psi}(\mathbf{x})d^3\mathbf{x}d^3\mathbf{x}' \\
&+\frac{1}{8\pi}\int\left\{\frac{1}{c^2}\left[\tilde{\dot{\mathbf{A}}}_\perp(\mathbf{x}) + 4\pi c\mathbf{p}_\perp(\mathbf{x})\right]^2\right. \\
&+\left.\left[\widetilde{\nabla\times\mathbf{A}_\perp(\mathbf{x})}\right]^2\right\}d^3\mathbf{x} \qquad (4.63)
\end{aligned}$$

In this expression we can work out explicitly

$$\begin{aligned}
&e^{-iS(\mathbf{x})}\left[\hbar\nabla + i\frac{e}{c}\mathbf{A}_\perp(\mathbf{x})\right]^2 e^{iS(\mathbf{x})} \\
&= \left\{e^{-iS(\mathbf{x})}\left[\hbar\nabla + i\frac{e}{c}\mathbf{A}_\perp(\mathbf{x})\right]e^{iS(\mathbf{x})}\right\}^2 \\
&= \left\{\hbar\nabla + i\left[\hbar\nabla S(\mathbf{x}) + \frac{e}{c}\mathbf{A}_\perp(\mathbf{x})\right]\right\}^2 \qquad (4.64)
\end{aligned}$$

and one can also set, using (4.61)

$$-\frac{1}{c}\dot{\tilde{\mathbf{A}}}_\perp(\mathbf{x}) = \mathbf{D}_\perp(\mathbf{x}) \tag{4.65}$$

where $\mathbf{D}(\mathbf{x}) = \mathbf{E}(\mathbf{x}) + 4\pi\mathbf{p}(\mathbf{x})$ is the displacement field. Thus (4.63) takes the form

$$\begin{aligned} H = &-\frac{1}{2m}\int \tilde{\psi}^\dagger(\mathbf{x})\left\{\hbar\nabla + i\left[\hbar\nabla S(\mathbf{x}) + \frac{e}{c}\mathbf{A}_\perp(\mathbf{x})\right]\right\}^2 \tilde{\psi}(\mathbf{x})d^3\mathbf{x} \\ &- Ze^2\int \tilde{\psi}^\dagger(\mathbf{x})\frac{1}{|\mathbf{R}-\mathbf{x}|}\tilde{\psi}(\mathbf{x})d^3\mathbf{x} \\ &+\frac{1}{2}e^2\int\int \tilde{\psi}^\dagger(\mathbf{x})\tilde{\psi}^\dagger(\mathbf{x}')\frac{1}{|\mathbf{x}-\mathbf{x}'|}\tilde{\psi}(\mathbf{x}')\tilde{\psi}(\mathbf{x})d^3\mathbf{x}d^3\mathbf{x}' \\ &+\frac{1}{8\pi}\int\left\{[\mathbf{D}_\perp(\mathbf{x}) - 4\pi\mathbf{p}_\perp(\mathbf{x})]^2 + \mathbf{H}^2(\mathbf{x})\right\}d^3\mathbf{x} \end{aligned} \tag{4.66}$$

Consider now the vector

$$\begin{aligned} \mathbf{n}(\mathbf{x}',\mathbf{x}) \times \mathbf{H}(\mathbf{x}') = &-e(\mathbf{x}-\mathbf{R}) \times [\nabla' \times \mathbf{A}_\perp(\mathbf{x}')] \\ &\times \int_0^1 \lambda\delta[\mathbf{x}' - \mathbf{R} - \lambda(\mathbf{x}-\mathbf{R})]d\lambda \end{aligned} \tag{4.67}$$

where $\mathbf{n}$ is defined in (B.27) or (B.33) and $\nabla' = \partial/\partial\mathbf{x}'$. Using $[\mathbf{a}\times(\mathbf{b}\times\mathbf{c})]_i = a_j b_i c_j - a_j b_j c_i$, the components of this vector can be written as

$$\begin{aligned} [\mathbf{n}(\mathbf{x}',\mathbf{x}) \times \mathbf{H}(\mathbf{x}')]_i = &\, e\Big\{(\mathbf{x}-\mathbf{R})_j\left[\nabla'_j A_{\perp i}(\mathbf{x}')\right] \\ &-(\mathbf{x}-\mathbf{R})_j\left[\nabla'_i A_{\perp j}(\mathbf{x}')\right]\Big\}\int_0^1 \lambda\delta[\mathbf{x}'-\mathbf{R}-\lambda(\mathbf{x}-\mathbf{R})]d\lambda \\ = &\, e\Big\{-(\mathbf{x}-\mathbf{R})_j A_{\perp i}(\mathbf{x}')\nabla'_j + (\mathbf{x}-\mathbf{R})_j A_{\perp j}(\mathbf{x}')\nabla'_i\Big\} \\ &\times \int_0^1 \lambda\delta[\mathbf{x}'-\mathbf{R}-\lambda(\mathbf{x}-\mathbf{R})]d\lambda \\ &+ e(\mathbf{x}-\mathbf{R})_j\Big\{\nabla'_j A_{\perp i}(\mathbf{x}')\int_0^1 \lambda\delta[\mathbf{x}'-\mathbf{R}-\lambda(\mathbf{x}-\mathbf{R})]d\lambda \\ &-\nabla'_i A_{\perp j}(\mathbf{x}')\int_0^1 \lambda\delta[\mathbf{x}'-\mathbf{R}-\lambda(\mathbf{x}-\mathbf{R})]d\lambda\Big\} \end{aligned} \tag{4.68}$$

Therefore, integrating over a large volume bounded by a surface on which $A_{\perp i}$ can be taken to vanish,

$$\int [\mathbf{n}(\mathbf{x}',\mathbf{x}) \times \mathbf{H}(\mathbf{x}')]_i d^3\mathbf{x}' = e \int \Big\{ -(\mathbf{x}-\mathbf{R})_j A_{\perp i}(\mathbf{x}')\nabla'_j$$
$$+(\mathbf{x}-\mathbf{R})_j A_{\perp j}\nabla'_i \Big\} \int_0^1 \lambda\delta[\mathbf{x}'-\mathbf{R}-\lambda(\mathbf{x}-\mathbf{R})]d\lambda d^3\mathbf{x}'$$
$$= -e\Big\{ \int A_{\perp i}(\mathbf{x}')(\mathbf{x}-\mathbf{R})\cdot\nabla' \int_0^1 \lambda\delta[\mathbf{x}'-\mathbf{R}-\lambda(\mathbf{x}-\mathbf{R})]d\lambda d^3\mathbf{x}'$$
$$+ \int (\mathbf{x}-\mathbf{R})\cdot\mathbf{A}_\perp(\mathbf{x}')\nabla_i \int \delta[\mathbf{x}'-\mathbf{R}-\lambda(\mathbf{x}-\mathbf{R})]d\lambda d^3\mathbf{x}' \Big\} \tag{4.69}$$

where in the last step $\partial f(x-y)/\partial y = -\partial f(x-y)/\partial x$ has been used, and consequently the last of the ∇_i acts on $\mathbf{x}$ rather than on $\mathbf{x}'$. Moreover, use of (B.35) yields

$$\int [\mathbf{n}(\mathbf{x}',\mathbf{x}) \times \mathbf{H}(\mathbf{x}')]_i d^3\mathbf{x}'$$
$$= e\Big\{ \int A_{\perp i}(\mathbf{x}') \int_0^1 \lambda \frac{d}{d\lambda}\delta[\mathbf{x}'-\mathbf{R}-\lambda(\mathbf{x}-\mathbf{R})]d\lambda d^3\mathbf{x}'$$
$$- \int (\mathbf{x}-\mathbf{R})\cdot\mathbf{A}_\perp(\mathbf{x}')\nabla_i \int_0^1 \delta[\mathbf{x}'-\mathbf{R}-\lambda(\mathbf{x}-\mathbf{R})]d\lambda d^3\mathbf{x}' \Big\}$$
$$= e\Big\{ \int A_{\perp i}(\mathbf{x}') \int_0^1 \left(1+\lambda\frac{d}{d\lambda}\right)\delta[\mathbf{x}'-\mathbf{R}-\lambda(\mathbf{x}-\mathbf{R})]d\lambda d^3\mathbf{x}'$$
$$- \nabla_i \int (\mathbf{x}-\mathbf{R})\cdot\mathbf{A}_\perp(\mathbf{x}') \int_0^1 \delta[\mathbf{x}'-\mathbf{R}-\lambda(\mathbf{x}-\mathbf{R})]d\lambda d^3\mathbf{x}' \Big\}$$
$$= e\Big\{ \int A_{\perp i}(\mathbf{x}') \int_0^1 \frac{d}{d\lambda}(\lambda\delta[\mathbf{x}'-\mathbf{R}-\lambda(\mathbf{x}-\mathbf{R})])d\lambda d^3\mathbf{x}'$$
$$-\nabla_i(\mathbf{x}-\mathbf{R})\cdot \int \mathbf{A}_\perp(\mathbf{x}') \int_0^1 \delta[\mathbf{x}'-\mathbf{R}-\lambda(\mathbf{x}-\mathbf{R})]d\lambda d^3\mathbf{x}' \Big\}$$
$$= eA_{\perp i}(\mathbf{x}) + \nabla_i \int \mathbf{p}(\mathbf{x}',\mathbf{x})\cdot\mathbf{A}_\perp(\mathbf{x}')d^3\mathbf{x}'$$
$$= eA_{\perp i}(\mathbf{x}) + \hbar c\nabla_i S(\mathbf{x}) \tag{4.70}$$

Hence

$$\hbar\nabla S(\mathbf{x}) + \frac{e}{c}\mathbf{A}(\mathbf{x}) = \frac{1}{c}\int \mathbf{n}(\mathbf{x}',\mathbf{x}) \times \mathbf{H}(\mathbf{x}')d^3\mathbf{x}' \tag{4.71}$$

and (4.66) takes the form

$$H = -\frac{1}{2m}\int \tilde{\psi}^\dagger(\mathbf{x})\left[\hbar\nabla + \frac{i}{c}\int \mathbf{n}(\mathbf{x}',\mathbf{x}) \times \mathbf{H}(\mathbf{x}')d^3\mathbf{x}'\right]^2 \tilde{\psi}(\mathbf{x})d^3\mathbf{x}$$
$$- Ze^2\int \tilde{\psi}^\dagger(\mathbf{x})\frac{1}{|\mathbf{R}-\mathbf{x}|}\tilde{\psi}(\mathbf{x})d^3\mathbf{x}$$
$$+\frac{1}{2}e^2\int\int \tilde{\psi}^\dagger(\mathbf{x})\tilde{\psi}^\dagger(\mathbf{x}')\frac{1}{|\mathbf{x}-\mathbf{x}'|}\tilde{\psi}(\mathbf{x}')\tilde{\psi}(\mathbf{x})d^3\mathbf{x}d^3\mathbf{x}'$$
$$+\frac{1}{8\pi}\int\left\{[\mathbf{D}_\perp(\mathbf{x}) - 4\pi\mathbf{p}_\perp(\mathbf{x})]^2 + \mathbf{H}^2(\mathbf{x})\right\}d^3\mathbf{x} \tag{4.72}$$

Let us now transform the velocity of the Schrödinger field (B.15) to the multipolar form. Using the rules defined earlier on in this section we have

$$\dot{\mathbf{X}} = -\frac{i}{2m}\int\left\{\tilde{\psi}^\dagger(\mathbf{x})e^{-iS(\mathbf{x})}\left[\hbar\nabla + i\frac{e}{c}\mathbf{A}_\perp(\mathbf{x})\right]e^{iS(\mathbf{x})}\tilde{\psi}(\mathbf{x})\right.$$
$$\left. + \tilde{\psi}(\mathbf{x})e^{iS(\mathbf{x})}\left[-\hbar\nabla + i\frac{e}{c}\mathbf{A}_\perp(\mathbf{x})\right]e^{-iS(\mathbf{x})}\tilde{\psi}^\dagger(\mathbf{x})\right\}d^3\mathbf{x}$$
$$= -\frac{i}{2m}\int\left\{\tilde{\psi}^\dagger(\mathbf{x})\left[\hbar\nabla + \frac{i}{c}\int \mathbf{n}(\mathbf{x}',\mathbf{x}) \times \mathbf{H}(\mathbf{x}')d^3\mathbf{x}'\right]\tilde{\psi}(\mathbf{x})\right.$$
$$\left. + \tilde{\psi}(\mathbf{x})\left[-\hbar\nabla + \frac{i}{c}\int \mathbf{n}(\mathbf{x}',\mathbf{x}) \times \mathbf{H}(\mathbf{x}')d^3\mathbf{x}'\right]\tilde{\psi}^\dagger(\mathbf{x})\right\}d^3\mathbf{x} \tag{4.73}$$

Thus in the multipolar representation the velocity field operator, analogous to (B.16), takes the form

$$\dot{\mathbf{x}} = \frac{1}{m}\left[\frac{\hbar}{i}\nabla + \frac{1}{c}\int \mathbf{n}(\mathbf{x}',\mathbf{x}) \times \mathbf{H}(\mathbf{x}')d^3\mathbf{x}\right];$$
$$\hbar\nabla = im\dot{\mathbf{x}} - \frac{i}{c}\int \mathbf{n}(\mathbf{x}',\mathbf{x}) \times \mathbf{H}(\mathbf{x}')d^3\mathbf{x}' \tag{4.74}$$

Consequently

$$\left[\hbar\nabla + \frac{i}{c}\int \mathbf{n}(\mathbf{x}',\mathbf{x}) \times \mathbf{H}(\mathbf{x}')d^3\mathbf{x}'\right]^2$$
$$= \hbar^2\nabla^2 - \frac{m}{c}\left\{\dot{\mathbf{x}}\cdot\int \mathbf{n}(\mathbf{x}',\mathbf{x}) \times \mathbf{H}(\mathbf{x}')d^3\mathbf{x}'\right.$$
$$\left. + \int \mathbf{n}(\mathbf{x}',\mathbf{x}) \times \mathbf{H}(\mathbf{x}')d^3\mathbf{x}'\cdot\dot{\mathbf{x}}\right\} + \frac{1}{c^2}\left[\int \mathbf{n}(\mathbf{x}',\mathbf{x}) \times \mathbf{H}(\mathbf{x}')d^3\mathbf{x}'\right]^2$$
$$= \hbar^2\nabla^2 + \frac{m}{c}\left\{\int \mathbf{n}(\mathbf{x}',\mathbf{x}) \times \dot{\mathbf{x}}\cdot\mathbf{H}(\mathbf{x}')d^3\mathbf{x}'\right.$$
$$\left. - \int \dot{\mathbf{x}} \times \mathbf{n}(\mathbf{x}',\mathbf{x})\cdot\mathbf{H}(\mathbf{x}')d^3\mathbf{x}'\right\} + \frac{1}{c^2}\left[\int \mathbf{n}(\mathbf{x}',\mathbf{x}) \times \mathbf{H}(\mathbf{x}')d^3\mathbf{x}'\right]^2$$
$$= \hbar^2\nabla^2 + 2m\int m(\mathbf{x}',\mathbf{x})\cdot\mathbf{H}(\mathbf{x}')d^3\mathbf{x}'$$
$$+\frac{1}{c^2}\left[\int \mathbf{n}(\mathbf{x}',\mathbf{x}) \times \mathbf{H}(\mathbf{x}')d^3\mathbf{x}'\right]^2 \tag{4.75}$$

where some obvious rules of vector algebra have been used together with (B.29). Substituting (4.75) into (4.72) yields

$$\begin{aligned}
H = &-\frac{\hbar^2}{2m}\int \tilde{\psi}^\dagger(\mathbf{x})\nabla^2\tilde{\psi}(\mathbf{x})d^3\mathbf{x} \\
&-\int\int \tilde{\psi}^\dagger(\mathbf{x})\mathbf{m}(\mathbf{x}',\mathbf{x})\cdot\mathbf{H}(\mathbf{x}')\tilde{\psi}(\mathbf{x})d^3\mathbf{x}d^3\mathbf{x}' \\
&-\frac{1}{2mc^2}\int \tilde{\psi}^\dagger(\mathbf{x})\left[\int \mathbf{n}(\mathbf{x}',\mathbf{x})\times\mathbf{H}(\mathbf{x}')d^3\mathbf{x}'\right]^2\tilde{\psi}(\mathbf{x})d^3\mathbf{x} \\
&-Ze^2\int \tilde{\psi}^\dagger(\mathbf{x})\frac{1}{|\mathbf{R}-\mathbf{x}|}\tilde{\psi}(\mathbf{x})d^3\mathbf{x} \\
&+\frac{1}{2}e^2\int\int \tilde{\psi}^\dagger(\mathbf{x})\tilde{\psi}^\dagger(\mathbf{x}')\frac{1}{|\mathbf{x}-\mathbf{x}'|}\tilde{\psi}(\mathbf{x}')\tilde{\psi}(\mathbf{x})d^3\mathbf{x}d^3\mathbf{x}' \\
&+\frac{1}{8\pi}\int\left\{[\mathbf{D}_\perp(\mathbf{x})-4\pi\mathbf{p}_\perp(\mathbf{x})]^2+\mathbf{H}^2(\mathbf{x})\right\}d^3\mathbf{x}
\end{aligned} \tag{4.76}$$

This Hamiltonian is exactly the same as (4.50), although it is expressed in terms of the new multipolar operators. We will partition it as

$$\begin{aligned}
H &= H_A^{mul} + H_F^{mul} + H_{AF}^{mul}\ ; \\
H_A^{mul} &= -\frac{\hbar^2}{2m}\int \tilde{\psi}^\dagger(\mathbf{x})\nabla^2\tilde{\psi}(\mathbf{x})d^3\mathbf{x} - Ze^2\int \tilde{\psi}^\dagger(\mathbf{x})\frac{1}{|\mathbf{R}-\mathbf{x}|}\tilde{\psi}(\mathbf{x})d^3\mathbf{x} \\
&\quad+\frac{1}{2}e^2\int\int \tilde{\psi}^\dagger(\mathbf{x})\tilde{\psi}^\dagger(\mathbf{x}')\frac{1}{|\mathbf{x}-\mathbf{x}'|}\tilde{\psi}(\mathbf{x}')\tilde{\psi}(\mathbf{x})d^3\mathbf{x}d^3\mathbf{x}' \\
&\quad+2\pi\int\left\{\int \tilde{\psi}^\dagger(\mathbf{x})\mathbf{p}_\perp(\mathbf{x}',\mathbf{x})\tilde{\psi}(\mathbf{x})d^3\mathbf{x}\right\}^2 d^3\mathbf{x}'\ ; \\
H_F^{mul} &= \frac{1}{8\pi}\int\{\mathbf{D}_\perp^2(\mathbf{x})+\mathbf{H}^2(\mathbf{x})\}d^3\mathbf{x}\ ; \\
H_{AF}^{mul} &= -\int\int \tilde{\psi}^\dagger(\mathbf{x})\mathbf{p}_\perp(\mathbf{x}',\mathbf{x})\cdot\mathbf{D}_\perp(\mathbf{x}')\tilde{\psi}(\mathbf{x})d^3\mathbf{x}d^3\mathbf{x}' \\
&\quad-\int\int \tilde{\psi}^\dagger(\mathbf{x})\mathbf{m}(\mathbf{x}',\mathbf{x})\cdot\mathbf{H}(\mathbf{x}')\tilde{\psi}(\mathbf{x})d^3\mathbf{x}d^3\mathbf{x}' \\
&\quad-\frac{1}{2mc^2}\int \tilde{\psi}^\dagger(\mathbf{x})\left[\int \mathbf{n}(\mathbf{x}',\mathbf{x})\times\mathbf{H}(\mathbf{x}')d^3\mathbf{x}'\right]^2\tilde{\psi}(\mathbf{x})d^3\mathbf{x}
\end{aligned} \tag{4.77}$$

This expression is the atom-photon interaction Hamiltonian in multipolar form (Power 1964, Craig and Thirunamachandran 1984). It should be noted that although, as already remarked, it is exactly the same as in the minimal coupling form, the partitioning is not the same as in (4.35). In fact it is easy to realize that $H_A^{min} \neq H_A^{mul}$, $H_F^{min} \neq H_F^{mul}$, $H_{AF}^{min} \neq H_{AF}^{mul}$

(Ackerhalt and Milonni 1984, Power and Thirunamachandran 1985). This is not the only difference, however. In fact the electron field operator $\tilde{\psi}$ in the multipolar representation is different from the corresponding minimal coupling ψ. Since the two fields are related by a unitary transformation, one might use (4.52) and (4.38) to obtain

$$\tilde{\psi}(\mathbf{x}) = e^{is}\psi(\mathbf{x})e^{-is} = \sum_k e^{is}a_k e^{-is}u_k(\mathbf{x}) = \sum_k \tilde{a}_k u_k(\mathbf{x}) \ ;$$

$$\left\{\tilde{a}_k, \tilde{a}^\dagger_{k'}\right\} = \delta_{kk'} \ ; \ \{\tilde{a}_k, \tilde{a}_{k'}\} = 0 \tag{4.78}$$

The disadvantage of this procedure, however, is that the u_k are approximate eigenfunctions of the single-particle hydrogenic Hamiltonian H_P in (4.36), where no account is taken of the electrostatic interaction nor of the last $\mathbf{p}_\perp^2$ term appearing in H_A^{mul}. Thus expansion (4.78) is not likely to be very useful in practice, and in analogy with the minimal coupling case one is led to introduce new fermion creation and annihilation operators $c_n'^\dagger$ and c_n' and to define a new set of single-electron states $u_n'(\mathbf{x})$ which take into account all the effects coming from the non-hydrogenic terms in H_A^{mul}. Thus one assumes that

$$\tilde{\psi}(\mathbf{x}) = \sum_n c_n' u_n'(\mathbf{x}) \tag{4.79}$$

is such that, when substituted in (4.77), H_A^{mul} takes the form

$$H_A^{mul} = \sum_n E_n' c_n'^\dagger c_n' \tag{4.80}$$

We remark that the difference $E_n' - E_n$ is due entirely to the $\mathbf{p}_\perp^2$ term in H_A^{mul}. This term can be shown to be generally small and to contribute to the Lamb shift in the multipolar scheme (see Appendix J).

As for H_F^{mul}, we remark that using (4.65) and (2.9)

$$\begin{aligned}\mathbf{D}_\perp(\mathbf{x}) &= -\frac{1}{c}\dot{\tilde{\mathbf{A}}}_\perp(\mathbf{x}) = \tilde{\mathbf{E}}_\perp(\mathbf{x}) \\ &= i\sum_{\mathbf{k}j}\sqrt{\frac{2\pi\hbar\omega_k}{V}}\left\{\mathbf{e}_{\mathbf{k}j}\tilde{a}_{\mathbf{k}j}e^{i\mathbf{k}\cdot\mathbf{x}} - \mathbf{e}^*_{\mathbf{k}j}\tilde{a}^\dagger_{\mathbf{k}j}e^{-i\mathbf{k}\cdot\mathbf{x}}\right\}\end{aligned} \tag{4.81}$$

Since the $\tilde{a}_{\mathbf{k}j}$ operators are unitarily related to the $a_{\mathbf{k}j}$ ones, they satisfy Bose commutation relations. Nevertheless they are different from the $a_{\mathbf{k}j}$; consequently the photons in the multipolar scheme, being elementary excitations of field $\mathbf{D}_\perp$, are different from photons in the minimal coupling scheme, which are elementary excitations of field $\mathbf{E}_\perp$. When (4.81) is substituted into (4.77) one has

$$H_F^{mul} = \sum_{\mathbf{k}j} \hbar\omega_k \left(\tilde{a}_{\mathbf{k}j}^\dagger \tilde{a}_{\mathbf{k}j} + 1/2\right) \tag{4.82}$$

Expansion (4.79) for $\tilde{\psi}$, (4.81) for $\mathbf{D}_\perp$ and (2.9) for $\mathbf{H}$ can also be substituted in H_{AF}^{mul} to obtain the atom-photon interaction in terms of creation and annihilation operators. Rather than proceeding in this way, it is more instructive to introduce some approximations at this stage. Assuming the vector potential constant over the atomic volume according to (4.46) leads to vanishing of the integrals

$$\int \mathbf{m}(\mathbf{x}', \mathbf{x}) \cdot \mathbf{H}(\mathbf{x}') d^3\mathbf{x}' \,;$$
$$\int \mathbf{n}(\mathbf{x}', \mathbf{x}) \times \mathbf{H}(\mathbf{x}') d^3\mathbf{x}' \tag{4.83}$$

because $\mathbf{n}(\mathbf{x}', \mathbf{x})$, and consequently $\mathbf{m}(\mathbf{x}', \mathbf{x})$, as functions of $\mathbf{x}'$ are strongly localized inside the atom, as is evident from (B.27), where $\mathbf{H} = \nabla \times \mathbf{A}$ vanishes. Therefore the same assumption that leads to the dipole approximation (4.49) leads also to the disappearance of the second and third terms in the interaction Hamiltonian in (4.77). Also, (4.46) involves $\mathbf{D}_\perp$ constant within the atom in view of expression (4.81). Since $\mathbf{p}(\mathbf{x}')$ is strongly localized within the atom, we should expect

$$\begin{aligned}
&\int\int \tilde{\psi}^\dagger(\mathbf{x}) \mathbf{p}_\perp(\mathbf{x}', \mathbf{x}) \cdot \mathbf{D}_\perp(\mathbf{x}') \tilde{\psi}(\mathbf{x}) d^3\mathbf{x} d^3\mathbf{x}' \\
&= \int \mathbf{p}_\perp(\mathbf{x}') \cdot \mathbf{D}_\perp(\mathbf{x}') d^3\mathbf{x}' \\
&= \int \mathbf{p}(\mathbf{x}') \cdot \mathbf{D}_\perp(\mathbf{x}') d^3\mathbf{x}' \simeq \int \mathbf{p}(\mathbf{x}') \cdot \mathbf{D}_\perp(\mathbf{R}) d^3\mathbf{x}' \\
&= \int\int \tilde{\psi}^\dagger(\mathbf{x}) \mathbf{p}(\mathbf{x}', \mathbf{x}) \cdot \mathbf{D}_\perp(\mathbf{R}) \tilde{\psi}(\mathbf{x}) d^3\mathbf{x} d^3\mathbf{x}' \\
&= \int\int \tilde{\psi}^\dagger(\mathbf{x}) \boldsymbol{\mu}(\mathbf{x}) \delta(\mathbf{x}' - \mathbf{R}) \cdot \mathbf{D}_\perp(\mathbf{R}) \tilde{\psi}(\mathbf{x}) d^3\mathbf{x} d^3\mathbf{x}' \\
&= \int \tilde{\psi}^\dagger(\mathbf{x}) \boldsymbol{\mu}(\mathbf{x}) \tilde{\psi}(\mathbf{x}) d^3\mathbf{x} \cdot \mathbf{D}_\perp(\mathbf{R})
\end{aligned} \tag{4.84}$$

Higher-order multipole contributions to $\mathbf{p}(\mathbf{x}', \mathbf{x})$ do not appear in view of the vanishing derivatives of $\mathbf{D}_\perp$ within the atomic volume. Thus constant $\mathbf{A}_\perp$ within this volume implies that the only surviving contribution to H_{AF}^{mul} is

$$H_{AF}^{mul} = -\boldsymbol{\mu} \cdot \mathbf{D}_{\perp}(\mathbf{R}) \; ;$$
$$\boldsymbol{\mu} = \int \tilde{\psi}^{\dagger}(\mathbf{x}) \boldsymbol{\mu}(\mathbf{x}) \tilde{\psi}(\mathbf{x}) d^3\mathbf{x} \tag{4.85}$$

This expression is the counterpart of H_{AF}^{mul} in (4.49), and it is called the electric-dipole approximation to the multipolar interaction (Power 1964, Craig and Thirunamachandran 1984). Account of nonvanishing derivatives of $\mathbf{A}_{\perp}$ within the atomic volume yields higher order multipolar terms, both electric and magnetic, arising from all of the three contributions to H_{AF}^{mul} in (4.77).

4.5 The interaction Hamiltonian in dipole approximation

We shall now express the interaction Hamiltonians in dipole approximation, obtained in the previous section, in terms of creation and annihilation operators of the various fields.

We start with the minimal coupling form (4.49) and use expansions (4.41) and (2.9), to obtain

$$\begin{aligned} H_{AF}^{min} = & \frac{e}{m} \sum_{nn'} \sum_{\mathbf{k}j} \sqrt{\frac{2\pi\hbar}{V\omega_k}} \left(\mathbf{p}_{nn'} \cdot \mathbf{e}_{\mathbf{k}j} c_n^{\dagger} c_{n'} a_{\mathbf{k}j} e^{i\mathbf{k}\cdot\mathbf{R}} + \text{h.c.} \right) \\ & + \frac{\pi\hbar e^2}{mV} \sum_n c_n^{\dagger} c_n \sum_{\mathbf{k}\mathbf{k}'jj'} \frac{1}{\sqrt{\omega_k \omega_{k'}}} \\ & \times \left\{ \left(\mathbf{e}_{\mathbf{k}j} \cdot \mathbf{e}_{\mathbf{k}'j'} \right) a_{\mathbf{k}j} a_{\mathbf{k}'j'} e^{i(\mathbf{k}+\mathbf{k}')\cdot\mathbf{R}} \right. \\ & + \left(\mathbf{e}_{\mathbf{k}j} \cdot \mathbf{e}_{\mathbf{k}'j'}^{*} \right) a_{\mathbf{k}j} a_{\mathbf{k}'j'}^{\dagger} e^{i(\mathbf{k}-\mathbf{k}')\cdot\mathbf{R}} \\ & + \left(\mathbf{e}_{\mathbf{k}j}^{*} \cdot \mathbf{e}_{\mathbf{k}'j'} \right) a_{\mathbf{k}j}^{\dagger} a_{\mathbf{k}'j'} e^{-i(\mathbf{k}-\mathbf{k}')\cdot\mathbf{R}} \\ & \left. + \left(\mathbf{e}_{\mathbf{k}j}^{*} \cdot \mathbf{e}_{\mathbf{k}'j'}^{*} \right) a_{\mathbf{k}j}^{\dagger} a_{\mathbf{k}'j'}^{\dagger} e^{-i(\mathbf{k}+\mathbf{k}')\cdot\mathbf{R}} \right\} \end{aligned} \tag{4.86}$$

where

$$\mathbf{p}_{nn'} = \int u_n^{*}(\mathbf{x}) \frac{\hbar}{i} \nabla u_{n'}(\mathbf{x}) d^3\mathbf{x} \tag{4.87}$$

and where the linear term describes absorption and emission of one photon with internal atomic state changes, whereas the quadratic term describes two-photon processes which are not accompanied by changes of the internal atomic state. In fact the second summation can give an energy shift common to all atomic states. Consider now the commutator

$$[(\mathbf{x}-\mathbf{R}), H_A^{min}] = -\frac{\hbar^2}{2m}\int \psi^\dagger(\mathbf{x})\,[\mathbf{x}, \nabla^2]\,\psi(\mathbf{x})d^3\mathbf{x}$$
$$= \frac{\hbar^2}{m}\int \psi^\dagger(\mathbf{x})\nabla\psi(\mathbf{x})d^3\mathbf{x} \tag{4.88}$$

Hence, using (4.41),

$$\sum_{nn'}\left(\int u_n^*(\mathbf{x})\frac{\hbar}{i}\nabla u_{n'}(\mathbf{x})d^3\mathbf{x}\right)c_n^\dagger c_{n'} = -i\frac{m}{\hbar}[(\mathbf{x}-\mathbf{R}), H_A^{min}] \tag{4.89}$$

Take two eigenstates $|+n_1\rangle$, $|-n_2\rangle$ of H_A^{min} which are identical except for one electron being displaced from the single-particle state $u_{n_2}(\mathbf{x})$ to the single-particle state $u_{n_1}(\mathbf{x})$. Explicitly, in terms of occupation numbers, $|+n_1\rangle \equiv |n_1 = 1, n_2 = 0, n_3, n_4, \ldots\rangle$, $|-n_2\rangle \equiv |n_1 = 0, n_2 = 1, n_3, n_4, \ldots\rangle$. Let E_{n_1} and E_{n_2} be the eigenvalues of H_A^{min} corresponding to $|+n_1\rangle$ and to $|-n_2\rangle$ respectively. Then the matrix element of the LHS of (4.89) between these two states is

$$\langle -n_2 | \sum_{nn'}\int u_n^*(\mathbf{x})\frac{\hbar}{i}\nabla u_{n'}(\mathbf{x})d^3\mathbf{x}c_n^\dagger c_{n'}\,|+n_1\rangle$$
$$= \int u_{n_2}^*(\mathbf{x})\frac{\hbar}{i}\nabla u_{n_1}(\mathbf{x})d^3\mathbf{x} \equiv \mathbf{p}_{n_2 n_1} \tag{4.90}$$

The matrix element of the RHS of (4.89) between the same states is

$$-i\frac{m}{\hbar}\langle -n_2 | (\mathbf{x}-\mathbf{R}) | +n_1\rangle(E_{n_1} - E_{n_2})$$
$$= -i\frac{m}{\hbar}\int u_{n_2}^*(\mathbf{x})(\mathbf{x}-\mathbf{R})u_{n_1}(\mathbf{x})d^3\mathbf{x}(E_{n_1} - E_{n_2})$$
$$\equiv -i\frac{m}{\hbar}(E_{n_1} - E_{n_2})(\mathbf{x}-\mathbf{R})_{n_2 n_1} \tag{4.91}$$

Consequently

$$\mathbf{p}_{nn'} = i\frac{m}{e\hbar}(E_{n'} - E_n)\int u_n^*(\mathbf{x})\boldsymbol{\mu}(\mathbf{x})u_{n'}(\mathbf{x})d^3\mathbf{x}$$
$$= i\frac{m}{e\hbar}(E_{n'} - E_n)\boldsymbol{\mu}_{nn'} = i\frac{m}{e}\omega_{n'n}\boldsymbol{\mu}_{nn'} \tag{4.92}$$

where $\omega_{n'n} = (E_{n'} - E_n)/\hbar$ and

$$\boldsymbol{\mu}(\mathbf{x}) = -e(\mathbf{x}-\mathbf{R}) \tag{4.93}$$

is the dipole moment operator and $\boldsymbol{\mu}_{nn'}$ is its matrix element connecting single-particle states u_n and $u_{n'}$. Thus the atom-field interaction in the minimal coupling scheme and in the dipole approximation takes the form

$$
\begin{aligned}
H_{AF}^{min} &= i\sum_{nn'}\sum_{\mathbf{k}j}\sqrt{\frac{2\pi\hbar}{V\omega_k}}\left(\omega_{n'n}\mu_{nn'}\cdot\mathbf{e}_{\mathbf{k}j}e^{i\mathbf{k}\cdot\mathbf{R}}c_n^\dagger c_{n'}a_{\mathbf{k}j} - \text{h.c.}\right) \\
&+ \frac{\pi\hbar e^2}{mV}\sum_n c_n^\dagger c_n \sum_{\mathbf{k}\mathbf{k}'jj'}\frac{1}{\sqrt{\omega_k\omega_{k'}}} \\
&\times\left\{\left(\mathbf{e}_{\mathbf{k}j}\cdot\mathbf{e}_{\mathbf{k}'j'}\right)a_{\mathbf{k}j}a_{\mathbf{k}'j'}e^{i(\mathbf{k}+\mathbf{k}')\cdot\mathbf{R}}\right. \\
&+ \left(\mathbf{e}_{\mathbf{k}j}\cdot\mathbf{e}^*_{\mathbf{k}'j'}\right)a_{\mathbf{k}j}a^\dagger_{\mathbf{k}'j'}e^{i(\mathbf{k}-\mathbf{k}')\cdot\mathbf{R}} \\
&+ \left(\mathbf{e}^*_{\mathbf{k}j}\cdot\mathbf{e}_{\mathbf{k}'j'}\right)a^\dagger_{\mathbf{k}j}a_{\mathbf{k}'j'}e^{-i(\mathbf{k}-\mathbf{k}')\cdot\mathbf{R}} \\
&+ \left.\left(\mathbf{e}^*_{\mathbf{k}j}\cdot\mathbf{e}^*_{\mathbf{k}'j'}\right)a^\dagger_{\mathbf{k}j}a^\dagger_{\mathbf{k}'j'}e^{-i(\mathbf{k}+\mathbf{k}')\cdot\mathbf{R}}\right\}
\end{aligned}
\tag{4.94}
$$

Turning to the dipole-approximated multipolar coupling form of the atom-photon interaction (4.85), and using expansions (4.79) and (4.81), one immediately obtains

$$
\begin{aligned}
H_{AF}^{mul} &= -i\sum_{nn'}\sum_{\mathbf{k}j}\sqrt{\frac{2\pi\hbar\omega_k}{V}}\left(\mu'_{nn'}\cdot\mathbf{e}_{\mathbf{k}j}e^{i\mathbf{k}\cdot\mathbf{R}}c_n'^\dagger c'_{n'}\tilde{a}_{\mathbf{k}j} - \text{h.c.}\right); \\
\mu'_{nn'} &= \int u_n'^*(\mathbf{x})\mu(\mathbf{x})u'_{n'}(\mathbf{x})d^3\mathbf{x}
\end{aligned}
\tag{4.95}
$$

Comparison of (4.94) and (4.95) shows the evident simplicity of the latter in comparison with the former. Some remarks concerning the two forms of interaction are in order.

In both forms (4.94) and (4.95) the operator for the total number of electrons commutes with the total Hamiltonian, and it is thus a constant of motion. Thus in (4.94) $\sum_n c_n^\dagger c_n$, where the sum includes also summation over the continuum of the ionized states, is equivalent to a constant which is Z for a neutral atom. Consequently the $\mathbf{A}_\perp^2$ term in (4.94) does not contain the electron coordinates, and it can for many purposes be included in H_F^{min}. Moreover, H_F^{min} plus this $\mathbf{A}_\perp^2$ term is a quadratic form in the creation and annihilation operators of the e.m. field, which can be diagonalized exactly, at least in principle, yielding new creation and annihilation operators $a'_{\mathbf{k}j}$ and $a'^\dagger_{\mathbf{k}j}$ and a new diagonal field Hamiltonian in the form

$$
H_F'^{min} = \sum_{\mathbf{k}j}\hbar\omega'_k\left(a'^\dagger_{\mathbf{k}j}a'_{\mathbf{k}j} + 1/2\right) \tag{4.96}
$$

Hence the total Hamiltonian in the minimal coupling scheme and within dipole approximation takes the form

$$H = \sum_n E_n c_n^\dagger c_n + \sum_{\mathbf{k}j} \hbar\omega'_k \left(a'^\dagger_{\mathbf{k}j} a'_{\mathbf{k}j} + 1/2 \right)$$

$$+ i \sum_{nn'} \sum_{\mathbf{k}j} \sqrt{\frac{2\pi\hbar}{V\omega'_k}} \left(\omega_{n'n} \boldsymbol{\mu}_{nn'} \cdot \mathbf{e}'_{\mathbf{k}j} e^{i\mathbf{k}\cdot\mathbf{R}} c_n^\dagger c_{n'} a'_{\mathbf{k}j} - \text{h.c.} \right) \tag{4.97}$$

This form should be contrasted with the multipolar form within the same approximation

$$H = \sum_n E'_n c'^\dagger_n c'_n + \sum_{\mathbf{k}j} \hbar\omega_k \left(\tilde{a}^\dagger_{\mathbf{k}j} \tilde{a}_{\mathbf{k}j} + 1/2 \right)$$

$$- i \sum_{nn'} \sum_{\mathbf{k}j} \sqrt{\frac{2\pi\hbar\omega_k}{V}} \left(\boldsymbol{\mu}'_{nn'} \cdot \mathbf{e}_{\mathbf{k}j} e^{i\mathbf{k}\cdot\mathbf{R}} c'^\dagger_n c'_{n'} \tilde{a}_{\mathbf{k}j} - \text{h.c.} \right) \tag{4.98}$$

A second point worth emphasizing is that within the dipole approximation it is inconsistent to extend the summation over the photon modes $\mathbf{k}j$ to arbitrarily large k, since when k^{-1} becomes smaller than atomic dimensions the assumption of uniformity of $\mathbf{A}_\perp$ over the atomic volume breaks down. Fortunately the interaction of the short-wavelength modes with the atom is likely to be small because the atom-field interaction tends to average out upon volume integration. Thus this problem is usually solved by introducing a cut-off frequency in the $\sum_{\mathbf{k}j}$ summation and a corresponding cut-off k_M such that

$$k_M = \frac{2\pi}{a} \tag{4.99}$$

where a is of the order of the Bohr radius. It should be noticed that the cut-off wavelength $\lambda_M \sim a$ falls within the γ-ray region, well beyond the range of frequencies of interest in quantum optics.

Further, we remark that, for some problems in quantum optics when only two of the infinitely many atomic levels play a role in the atom-photon dynamics, the physics of the system can be described by introducing the so-called two-level atom. This is a fictitious atom with only two nondegenerate energy eigenvalues E_1 and E_2 ($E_2 > E_1$), corresponding to free atomic states $|\,1\rangle$ and $|\,2\rangle$, eigenstates of the two-level atomic Hamiltonian H_A. The following relations of the electron field operators acting in this two-dimensional subspace are easily obtainable from the fermion anticommutation rules (3.100)

$$\begin{aligned}\left[c_1^\dagger c_2, c_2^\dagger c_1\right] &= c_1^\dagger c_1 - c_2^\dagger c_2 \ ;\\ \left[c_1^\dagger c_2, c_1^\dagger c_1 - c_2^\dagger c_2\right] &= -2c_1^\dagger c_2 \ ;\\ \left[c_2^\dagger c_1, c_1^\dagger c_1 - c_2^\dagger c_2\right] &= 2c_2^\dagger c_1 \ ;\\ \left(c_1^\dagger c_2\right)^2 &= \left(c_2^\dagger c_1\right)^2 = 0\end{aligned} \tag{4.100}$$

These are identical with the $S = 1/2$ angular momentum commutation relations, provided we put

$$\begin{aligned}&S_+ = c_2^\dagger c_1 \ ; \ S_- = c_1^\dagger c_2 \ ;\\ &S_z = \frac{1}{2}\left(c_2^\dagger c_2 - c_1^\dagger c_1\right) \ ; \ c_1^\dagger c_1 + c_2^\dagger c_2 = 1\end{aligned} \tag{4.101}$$

since in this case (4.100) are equivalent to

$$\begin{aligned}&[S_+, S_-] = 2S_z \ ; \ [S_+, S_z] = -S_+ \ ;\\ &[S_-, S_z] = S_- \ ; \ S_+^2 = S_-^2 = 0\end{aligned} \tag{4.102}$$

Thus one has

$$c_1^\dagger c_1 = \frac{1}{2} - S_z \ ; \ c_2^\dagger c_2 = \frac{1}{2} + S_z \tag{4.103}$$

and

$$\begin{aligned}H_A = E_1 c_1^\dagger c_1 + E_2 c_2^\dagger c_2 &= \frac{1}{2}(E_1 + E_2)\\ + (E_2 - E_1)S_z &= \frac{1}{2}(E_1 + E_2) + \hbar\omega_0 S_z\end{aligned} \tag{4.104}$$

where $(E_1 + E_2)/2$ amounts to an unobservable common shift of the atomic energy levels which can be completely eliminated and where $\omega_o = \omega_{21}$. Moreover, in order to simplify notation, we shall drop all dashes and tildes appearing in (4.97) and (4.98), which should not cause any ambiguity as long as one does not change from the minimal to the multipolar scheme or vice versa in the course of calculations. In any case, in order to keep track of the scheme one is using, we shall indicate the two forms of the atom-field Hamiltonian appearing in (4.97) and (4.98) by H^{min} and H^{mul} respectively. Moreover, when dealing with two-level atoms, here and in the future we will assume real polarization vectors $\mathbf{e}_{\mathbf{k}j} = \mathbf{e}_{\mathbf{k}j}^*$. Thus in terms of the pseudospin operators defined above, and for the two-level atom approximation, the two forms of the atom-field

Hamiltonian can be modelled as

$$
\begin{aligned}
H^{min} &= \hbar\omega_o S_z + \sum_{\mathbf{k}j} \hbar\omega_k \left(a^\dagger_{\mathbf{k}j} a_{\mathbf{k}j} + 1/2\right) \\
&+ \sum_{\mathbf{k}j} \left(\epsilon_{\mathbf{k}j} S_+ + \epsilon^*_{\mathbf{k}j} S_-\right)\left(e^{i\mathbf{k}\cdot\mathbf{R}} a_{\mathbf{k}j} + e^{-i\mathbf{k}\cdot\mathbf{R}} a^\dagger_{\mathbf{k}j}\right); \\
\epsilon_{\mathbf{k}j} &= -i\sqrt{\frac{2\pi\hbar\omega_0^2}{V\omega_k}}\mu_{21}\cdot\mathbf{e}_{\mathbf{k}j} \\
H^{mul} &= \hbar\omega_0 S_z + \sum_{\mathbf{k}j} \hbar\omega_k \left(a^\dagger_{\mathbf{k}j} a_{\mathbf{k}j} + 1/2\right) \\
&+ \sum_{\mathbf{k}j} \left(\epsilon_{\mathbf{k}j} S_+ - \epsilon^*_{\mathbf{k}j} S_-\right)\left(e^{i\mathbf{k}\cdot\mathbf{R}} a_{\mathbf{k}j} - e^{-i\mathbf{k}\cdot\mathbf{R}} a^\dagger_{\mathbf{k}j}\right); \\
\epsilon_{\mathbf{k}j} &= -i\sqrt{\frac{2\pi\hbar\omega_k}{V}}\mu_{21}\cdot\mathbf{e}_{\mathbf{k}j}
\end{aligned}
\tag{4.105}
$$

The similarity of the two forms (4.105) shows that within the two-level approximation the same mathematical methods can be applied to investigate the eigenvalue problem of H^{min} and H^{mul}. It is appropriate to emphasize, however, that this similarity is only formal, since "atom" and "photon" refer to two quite different objects in the two forms (4.105), as is indicated also by the different dependence of the coupling constant $\epsilon_{\mathbf{k}j}$ from the frequency ω_k as well as from ω_0. The two forms in (4.105) are often called Dicke models for the atom-photon interaction (see e.g. Leonardi *et al.* 1986).

It should be noted that the e.m. field in this section has been expanded in plane waves. Another useful expansion is in spherical waves. Using the continuum expansion (2.22) for the vector potential, the term linear in $\mathbf{A}_\perp$ in H^{min}_{AF}, as expressed in (4.35), takes the form

$$
\begin{aligned}
&\frac{e}{mc}\int \psi^\dagger(\mathbf{x})\mathbf{A}_\perp(\mathbf{x})\cdot\mathbf{p}\psi(\mathbf{x})d^3\mathbf{x} \\
&= \frac{2e}{m}\sqrt{\frac{\hbar}{c}}\sum_{\ell m}\int_0^\infty \Big\{ a(k,\mathcal{M},\ell,m) \\
&\times \int \psi^\dagger(\mathbf{x}) j_\ell(kx)\mathbf{Y}_{\ell m0}(\theta,\phi)\cdot\mathbf{p}\psi(\mathbf{x})d^3\mathbf{x} + a(k,\mathcal{E},\ell,m)\frac{1}{\sqrt{2\ell+1}} \\
&\times \Big[\sqrt{\ell}\int \psi^\dagger(\mathbf{x}) j_{\ell+1}(kx)\mathbf{Y}_{\ell m+}(\theta,\phi)\cdot\mathbf{p}\psi(\mathbf{x})d^3\mathbf{x} \\
&-\sqrt{\ell+1}\int \psi^\dagger(\mathbf{x}) j_{\ell-1}(kx)\mathbf{Y}_{\ell m-}(\theta,\phi)\cdot\mathbf{p}\psi(\mathbf{x})d^3\mathbf{x}\Big]\Big\} \\
&\times k^{1/2}dk + \text{h.c.}
\end{aligned}
\tag{4.106}
$$

As we have already remarked, the $\ell = 0$ term is missing from the sum in (4.106). Thus the lowest order spherical Bessel function appearing is $j_0(kx)$ coming from the $\ell = 1$ term and appearing in the third integral on the RHS of (4.106). This is also the only surviving term if one assumes $kx \sim 0$, that is, uniformity of $\mathbf{A}_\perp$ over the volume of the atom (assumed to be centred at $R = 0$ for simplicity), since $j_\ell(kx) \sim 0$ for $kx \sim 0$ and $\ell \geq 1$. Thus in the dipole approximation (4.106) reduces to

$$\begin{aligned}
&\frac{e}{mc}\int \psi^\dagger(\mathbf{x})\mathbf{A}_\perp(0)\cdot\mathbf{p}\psi(\mathbf{x})d^3\mathbf{x} \\
&= -\frac{2e}{m}\sqrt{\frac{\hbar}{c}}\sqrt{\frac{2}{3}}\sum_m\int_0^{k_M} a(k,\mathcal{E},1,m) \\
&\times \int \psi^\dagger(\mathbf{x})\mathbf{Y}_{1m-}(\theta,\phi)\cdot\mathbf{p}\psi(\mathbf{x})d^3\mathbf{x}k^{1/2}dk + \text{h.c.} \\
&= -2\sqrt{\frac{2}{3}}\frac{e}{m}\sqrt{\frac{\hbar}{c}}\sum_m\sum_{nn'}\int u_n^*(\mathbf{x})\mathbf{Y}_{1m-}(\theta,\phi)\cdot\mathbf{p}u_{n'}(\mathbf{x})d^3\mathbf{x} \\
&\times \int_0^{k_M} c_n^\dagger c_{n'} a(k,\mathcal{E},1,m)k^{1/2}dk + \text{h.c.}
\end{aligned} \tag{4.107}$$

In the same approximation, the electric and magnetic field expansions (2.22) reduce to

$$\mathbf{E}(\mathbf{x}) = -i2\sqrt{\frac{2}{3}}\sqrt{\hbar c}\sum_m \mathbf{Y}_{1m-}(\theta,\phi)\int_0^{k_M} a(k,\mathcal{E},1,m)k^{3/2}dk + \text{h.c.}$$

$$\begin{aligned}
\mathbf{H}(\mathbf{x}) = -i2\sqrt{\frac{2}{3}}\sqrt{\hbar c}\sum_m \mathbf{Y}_{1m-}(\theta,\phi)\int_0^{k_M} a(k,\mathcal{M},1,m)k^{3/2}dk \\
+ \text{h.c.}
\end{aligned} \tag{4.108}$$

Thus, as is clear from the form of (4.107), the atom-photon coupling in the dipole approximation takes place through the electric field alone.

Finally we shall discuss yet another form of the atom-radiation coupling which is really an approximated effective Hamiltonian, but which has been used several times in applications, e.g. in connection with van der Waals forces. The main advantage of this new coupling is that the atom-radiation interaction is expressed in terms of the ground-state polarizability of the atom, and that the internal atomic dynamics do not play any explicit role. We shall derive this approximate interaction within the electric dipole approximation of the multipolar Hamiltonian

$$
\begin{aligned}
H &= H_A^{mul} + H_F^{mul} + H_{AF}^{mul} \; ; \; H_A^{mul} = \sum_n E'_n c'^{\dagger}_n c'_n \; ; \\
H_F^{mul} &= \sum_{\mathbf{k}j} \hbar\omega_k \left(\tilde{a}^{\dagger}_{\mathbf{k}j} \tilde{a}_{\mathbf{k}j} + 1/2 \right) ; \\
H_{AF}^{mul} &= -\boldsymbol{\mu} \cdot \mathbf{D}_{\perp}(\mathbf{R})
\end{aligned}
\tag{4.109}
$$

Preliminarily, we introduce the following notation for the eigenvalues and eigenfunctions of the unperturbed Hamiltonian $H_F^{mul} + H_A^{mul}$

$$
\begin{aligned}
& H_A^{mul} \mid \{n_i\}\rangle = \left(\sum_i E'_i \right) \mid \{n_i\}\rangle \; ; \\
& H_F^{mul} \mid \{n_{\mathbf{k}j}\}\rangle = \left(\sum_{\mathbf{k}j} \hbar\omega_k (n_{\mathbf{k}j} + 1/2) \right) \mid \{n_{\mathbf{k}j}\}\rangle \; ; \\
& E_{ik} = \sum_i E'_i + \sum_{\mathbf{k}j} \hbar\omega_k (n_{\mathbf{k}j} + 1/2) \; ; \\
& \mid i, k\rangle = \mid \{n_i\}\rangle \otimes \mid \{n_{\mathbf{k}j}\}\rangle
\end{aligned}
\tag{4.110}
$$

Clearly $\{n_i\}$ represents a distribution of electrons among the eigenstates $u'_i(\mathbf{x})$ of the single-particle atomic Hamiltonian, and $\{n_{\mathbf{k}j}\}$ represents a distribution of photons within the various $\mathbf{k}j$ modes of H_F^{mul}. A Hamiltonian H_N equivalent to H can be obtained by performing a unitary transformation

$$
\begin{aligned}
H_N = e^{-N} H e^{N} &= H + [H, N] + \frac{1}{2!}[[H, N], N] + \dots \\
&= H_A^{mul} + H_F^{mul} + H_{AF}^{mul} + \left[H_A^{mul} + H_F^{mul}, N \right] \\
&+ \left[H_{AF}^{mul}, N \right] + \frac{1}{2!} \left[\left[H_A^{mul} + H_F^{mul}, N \right], N \right] \\
&+ \frac{1}{2!} \left[\left[H_{AF}^{mul}, N \right], N \right] + \dots
\end{aligned}
\tag{4.111}
$$

One can formally eliminate the terms linear in H_{AF}^{mul} from (4.111) by choosing N in such a way that

$$
\begin{aligned}
& H_{AF}^{mul} + \left[H_A^{mul} + H_F^{mul}, N \right] = 0 \\
& \text{or } \langle ik \mid N \mid i'k'\rangle = \frac{\langle ik \mid H_{AF}^{mul} \mid i'k'\rangle}{E_{i'k'} - E_{ik}}
\end{aligned}
\tag{4.112}
$$

where $\mid i'k'\rangle = \mid \{n_i\}'\rangle \otimes \mid \{n'_{\mathbf{k}j}\}\rangle$ corresponds to an eigenstate of $H_A^{mul} + H_F^{mul}$ having primed population distributions in general different from the unprimed ones. In fact with choice (4.112) for N, (4.111) reduces to

$$H_N = H_A^{mul} + H_F^{mul} + \frac{1}{2}\left[H_{AF}^{mul}, N\right] + \dots \tag{4.113}$$

We remark that the terms omitted in (4.113) are $O(H_{AF}^3)$ in view of (4.112), which means $O(e^3)$ in view of the form of the atom-photon coupling. Because of this, one is entitled to neglect them altogether in a perturbation approach spirit. Thus terms of order e are missing from (4.113), and one can concentrate on the form of the commutator which is obviously of order e^2. Rather than working directly on this approximated version of H_N, we shall derive from (4.113) an effective Hamiltonian under the assumption that the electron distribution $\{n_i\}$ within the atom is rigid and does not change even in presence of the interaction with the e.m. field. Thus one is entitled to introduce, instead of H_N in (4.113), the effective photon Hamiltonian

$$\begin{aligned}\langle\{n_i\} \mid H_N \mid \{n_i\}\rangle &= \langle\{n_i\} \mid H_A^{mul} \mid \{n_i\}\rangle + H_F^{mul} \\ &+ \frac{1}{2}\sum_{kk'} \mid k\rangle\langle ik \mid \left[H_{AF}^{mul}, N\right] \mid ik'\rangle\langle k' \mid \end{aligned} \tag{4.114}$$

where $\mid k\rangle$ and $\mid k'\rangle$ refer to two different photon distributions. On the other hand, on the basis of (4.112),

$$\begin{aligned}\langle ik \mid \left[H_{AF}^{mul}, N\right] \mid ik'\rangle &= \sum_{i''k''}\big(\langle ik \mid H_{AF}^{mul} \mid i''k''\rangle\langle i''k'' \mid N \mid ik'\rangle \\ &- \langle ik \mid N \mid i''k''\rangle\langle i''k'' \mid H_{AF}^{mul} \mid ik'\rangle\big) \\ &= \sum_{i''k''}\left(\frac{\langle ik \mid H_{AF}^{mul} \mid i''k''\rangle\langle i''k'' \mid H_{AF}^{mul} \mid ik'\rangle}{E_{ik'} - E_{i''k''}}\right. \\ &+ \left.\frac{\langle ik \mid H_{AF}^{mul} \mid i''k''\rangle\langle i''k'' \mid H_{AF}^{mul} \mid ik'\rangle}{E_{ik} - E_{i''k''}}\right)\end{aligned} \tag{4.115}$$

Moreover, from (4.85)

$$\begin{aligned}&\langle ik \mid H_{AF}^{mul} \mid i''k''\rangle \\ &= -\langle\{n_i\} \mid \boldsymbol{\mu} \mid \{n_i\}''\rangle \cdot \langle\{n_{\mathbf{k}j}\} \mid \mathbf{D}_\perp(\mathbf{R}) \mid \{n_{\mathbf{k}j}\}''\rangle \\ &= -\sum_{nn'}(\mu'_{nn'})_\ell\langle\{n_i\} \mid c_n'^{\dagger} c'_{n'} \mid \{n_i''\}\rangle \\ &\times \langle\{n_{\mathbf{k}j}\} \mid D_{\perp\ell}(\mathbf{R}) \mid \{n_{\mathbf{k}j}\}''\rangle\end{aligned} \tag{4.116}$$

It is then clear that the two electron distributions $\{n_i\}$ and $\{n_i\}''$ connected by H_{AF}^{mul} differ by one electron displaced from the single-particle state $u'_{n'}$

to u'_n. Moreover, since $D_{\perp\ell}(\mathbf{R})$ is a one-photon operator, the photon distributions $\{n_{\mathbf{k}j}\}$ and $\{n_{\mathbf{k}j}\}''$ (or $\{n_{\mathbf{k}j}\}'$ and $\{n_{\mathbf{k}j}\}''$) appearing in (4.115) differ by one photon of energy $\hbar\omega_k$. Thus both energy denominators on the RHS of (4.115) are of the form $E'_n - E'_{n'} \pm \hbar\omega_k$. Restricting our considerations to low energy photons, for which $\hbar\omega_k \ll E'_n - E'_{n'}$ for any pair of atomic single particle states $u'_{n'}$ and u'_n, the energy denominators can be approximated as $E'_n - E'_{n'}$, and (4.115) takes the form

$$\begin{aligned}
&\sum_{i''k''}\sum_{nn'n''n'''} \frac{(\mu'_{nn'})_\ell (\mu'_{n''n'''})_m}{E'_n - E'_{n'}} \langle\{n_i\} \mid c'^{\dagger}_n c'_{n'} \mid \{n_i\}''\rangle \\
&\times \langle\{n_i\}'' \mid c'^{\dagger}_{n''} c'_{n'''} \mid \{n_i\}\rangle\langle\{n_{\mathbf{k}j}\} \mid D_{\perp\ell}(\mathbf{R}) \mid \{n_{\mathbf{k}j}\}''\rangle \\
&\times \langle\{n_{\mathbf{k}j}\}'' \mid D_{\perp m}(\mathbf{R}) \mid \{n_{\mathbf{k}j}\}'\rangle \\
&= 2\sum_{nn'n''n'''} \frac{(\mu'_{nn'})_\ell (\mu'_{n''n'''})_m}{E'_n - E'_{n'}} \langle\{n_i\} \mid c'^{\dagger}_n c'_{n'} c'^{\dagger}_{n''} c'_{n'''} \mid \{n_i\}\rangle \\
&\times \langle\{n_{\mathbf{k}j}\} \mid D_{\perp\ell}(\mathbf{R}) D_{\perp m}(\mathbf{R}) \mid \{n_{\mathbf{k}j}\}'\rangle \\
&= 2\sum_{nn'} (\mu'_{nn'})_\ell (\mu'_{n'n})_m \frac{1}{E'_n - E'_{n'}} \langle\{n_i\} \mid c'^{\dagger}_n c'_n (1 - c'^{\dagger}_{n'} c'_{n'}) \mid \{n_i\}\rangle \\
&\times \langle\{n_{\mathbf{k}j}\} \mid D_{\perp\ell}(\mathbf{R}) D_{\perp m}(\mathbf{R}) \mid \{n_{\mathbf{k}j}\}'\rangle
\end{aligned} \tag{4.117}$$

where we have used the fermion commutation rules and we have assumed definite parity for $u'_n(\mathbf{x})$, so that $\mu'_{nn} = 0$. Defining the static electric polarizability tensor for the atom in the state characterized by the electronic distribution $\{n_i\}$ as

$$\alpha_{\ell m}(\{n_i\}) = -2\sum_{nn'} \frac{(\mu'_{nn'})_\ell (\mu'_{n'n})_m}{E'_n - E'_{n'}} \langle\{n_i\} \mid c'^{\dagger}_n c'_n \left(1 - c'^{\dagger}_{n'} c'_{n'}\right) \mid \{n_i\}\rangle \tag{4.118}$$

expression (4.115) can be cast in the form

$$\langle ik \mid \left[H^{mul}_{AF}, N\right] \mid ik'\rangle = -\alpha_{\ell m}(\{n_i\})\langle\{n_{\mathbf{k}j}\} \mid D_{\perp\ell}(\mathbf{R}) D_{\perp m}(\mathbf{R}) \mid \{n_{\mathbf{k}j}\}'\rangle \tag{4.119}$$

When (4.119) is substituted into (4.114), one obtains (Craig and Power 1969)

$$\langle\{n_i\} \mid H_N \mid \{n_i\}\rangle = E_A(\{n_i\}) + H^{mul}_F - \frac{1}{2}\alpha_{\ell m}(\{n_i\}) D_{\perp\ell}(\mathbf{R}) D_{\perp m}(\mathbf{R}) \tag{4.120}$$

where $E_A(\{n_i\})$ is the atomic energy in the electronic configuration $\{n_i\}$. Thus (4.120) plays the role of an effective Hamiltonian which accounts for the effect of the atom-photon coupling on the low-frequency modes of the e.m. field, that is on those modes of frequency ω_k much smaller than the typical atomic energy difference.

References

J.R. Ackerhalt, P.W. Milonni (1984). *J. Opt. Soc. Am. B* **1,** 116

A.O. Barut (1980). *Electrodynamics and Classical Theory of Fields and Particles* (Dover Publications Inc., New York)

C. Cohen-Tannoudji, B. Diu, F. Laloë (1977). *Quantum Mechanics* (John Wiley and Sons, Paris)

C. Cohen-Tannoudji, J. Dupont-Roc, G. Grynberg (1989). *Photons and Atoms* (John Wiley and Sons, New York)

D.P. Craig, E.A. Power (1969). *Int. J. Quantum Chem.* **3,** 903

D.P. Craig, T. Thirunamachandran (1984). *Molecular Quantum Electrodynamics* (Academic Press Inc., London)

W. Heitler (1960). *The Quantum Theory of Radiation* (Oxford University Press, London)

L.D. Landau, E.M. Lifshitz (1958). *Quantum Mechanics* (Pergamon Press, London)

C. Leonardi, F. Persico, G. Vetri (1986). *Rivista Nuovo Cimento* **9,** 1

E.A. Power (1964). *Introductory Quantum Electrodynamics* (Longmans, Green and Co. Ltd., London)

E.A. Power, T. Thirunamachandran (1983). *Phys. Rev. A* **28,** 2649

E.A. Power, T. Thirunamachandran (1985). *J. Opt. Soc. Am. B* **2,** 1100

E.A. Power, S. Zienau (1959). *Phil. Trans. Roy. Soc. A* **251,** 427

J.J. Sakurai (1982). *Advanced Quantum Mechanics* (Addison-Wesley Publishing Co., Reading, Ma.)

Further reading

An interesting discussion on the potentialities of Rydberg atoms has been given in

L. Moi, P. Goy, M. Gross, J.M. Raimond, C. Fabre, S. Haroche, *Phys. Rev. A* **27**, 2043 (1983)

P. Goy, L. Moi, M. Gross, J.M. Raimond, C. Fabre, S. Haroche, *Phys. Rev. A* **27**, 2065 (1983).

The various elementary processes in atom-photon interaction have been extensively discussed in

C. Cohen-Tannoudji, J. Dupont-Roc, G. Grynberg, S. Haroche, *Atom-Photon Interactions* (John Wiley and Sons Inc., New York 1992).

The theory of unitary transformations relating the minimal coupling and the multipolar forms of the atom-photon Hamiltonian has also been discussed in

C. Cohen-Tannoudji, J. Dupont-Roc, G. Grynberg, *Photons and Atoms* (John Wiley and Sons Inc., New York 1989).

5
Atoms dressed by a real e.m. field

Introduction. The second part of the book is dedicated to the dressed atom, and it begins with this chapter, which deals mainly with the quantum-dynamical description of an atom dressed by a real electromagnetic field. Here the emphasis is on the adjective 'real', by which we mean that the field is in an excited state populated by real photons and not just by the zero-point photon background. Due to the coupling discussed in the first part of the book, atom-photon correlations are established which admix, shift and split the levels of the system atom plus radiation field. The admixed and correlated states are called dressed-atom states. In Section 5.2 we obtain the Hamiltonian for an atom in a cavity with perfectly reflecting walls. The cavity selects a discrete set of field modes, and this leads us naturally to consider the simplest possible nontrivial atom-field system: the Jaynes-Cummings model describing a two-level atom coupled to a single-mode system. In Section 5.3 we develop the theory, based on a unitary transformation, to dress a two-level atom by a mode of the cavity populated by real photons. A necessary preliminary for dealing with more complicated atom-field models is the theory of spontaneous emission in free space, which is discussed in Section 5.4 in a Wigner-Weisskopf framework. This serves as an introduction to the theory of resonance fluorescence by a two-level atom, which is presented in Section 5.5 as a phenomenon of spontaneous decay by a dressed atom: one of the modes of the electromagnetic field is strongly populated and dresses the atom. This dressed atom emits photons in the other modes of the field (the fluorescent modes); the spectrum of the emitted radiation is obtained and its features are discussed in detail. Section 5.6 is dedicated to a simplified discussion of the forces exerted on a neutral two-level atom by a real electromagnetic field. The external degrees of freedom of the atom are introduced in the Hamiltonian of a two-level atom coupled to the radiation field. This Hamiltonian is diagonalized by the dressing transformation of Section 5.3. Expressions are obtained for the dissipative force and for the reactive force acting on the atom.

5.1 Qualitative introduction to dressed atoms

As we have remarked in Section 4.1, the requirement of local gauge invariance of the Lagrangian density leads necessarily to the coupling of a charged matter field with the electromagnetic field. In fact the matter field acts as a source for the electromagnetic field through the presence of current and charge density appearing in the equations of motion for the electromagnetic field amplitudes, whereas the amplitude of the matter field is influenced also by the presence of the electromagnetic field amplitude which appears in the Schrödinger equation, interpreted as the equation of motion of the matter field amplitude. Consequently the concept of a pure electromagnetic field, which does not take account of the matter present, as well as the concept of a "bare" matter field whose behaviour is calculated as if the interaction with the electromagnetic field were absent, are ill-defined in principle. This is not to say, however, that the concepts of bare atom and bare field are without use; one can indeed use them as limiting concepts of the same nature as those of point mass and point charge in classical physics, having clear in mind that in real cases the matter field and the electromagnetic fields are inextricably related to each other. This line of thought leads to introducing the concept of elementary excitations of the coupled matter-electromagnetic field, very much in the same way as it is normally done with various kinds of fields in condensed matter physics such as magnons, polaritons and similar (see e.g. Anderson 1963, Kittel 1963, Haken 1988). These elementary excitations are not pure quanta of matter or pure photons, but they share features of both fields. In the cases of interest for this book, the matter field is usually that of the electrons in the potential generated by a fixed nucleus, and this leads us to consider atomic-photon excitations and eventually to the concept of dressed atom.

It should be realized from the very beginning that the expression "dressed atom" in nonrelativistic QED is used with reference to two quite different physical situations as far as the number occupancy is concerned. In the first situation one has a ground-state atom interacting with the vacuum electromagnetic field. Taking the total atom-field system to be in the lowest possible energy state, the zero-point quantum fluctuations of the field discussed in Chapter 2 induce virtual absorption and re-emission processes of photons by the atom which take place continuously, thereby creating a cloud of virtual photons around the bare atom. The complex object, bare atom plus cloud of virtual photons, is what one calls a dressed atom. In this sense the concept described by this expression is closely related to that of "dressed source" introduced long ago in the theory of quantum fields (Van Hove 1955, 1956). The problem of the structure of the virtual field around the atom and of its relationship to the physical properties of the atom, as modified by the presence of the cloud, will form the subject of the next chapters of this book.

In this chapter we shall concentrate on a second physical situation which is also normally described by the expression "dressed atom". In this second context the atom is in the presence of real photons. Thus the atom-field system is not in its ground state, an external electromagnetic field being provided e.g. by a laser. This laser field, much stronger than that provided by the zero-point vacuum fluctuations within the laser bandwidth, admixes, shifts and splits the levels of the system atom plus laser field. The resulting atom-field energy levels display correlations between bare atomic occupation numbers and photon numbers. Also, these admixed and correlated states are called dressed atomic states, although their nature is quite different from that of the dressed atoms discussed above. This second kind of dressed state has been widely used in the past in connection with noteworthy phenomena such as resonance fluorescence, atomic dynamics in a cavity, atomic motion in laser beams and similar (see e.g. Cohen-Tannoudji *et al.* 1992). In this chapter we shall discuss some of these phenomena from the point of view of dressed atoms.

5.2 Atom in a cavity

In order to discuss in a transparent way the features of an atom dressed by a field of real photons we shall adopt a simplified model which is based on the dynamical properties of an atom inside a perfect cavity. A perfect cavity is defined as a portion of space bounded by a perfect electrically conducting surface. As discussed in Section 1.11, the constraints imposed on the surface of this cavity are that the components of the electric field parallel to the surface, as well as the components of the magnetic field normal to the surface, should vanish. We have also remarked that these boundary conditions on the e.m. field select a set of solutions of Maxwell's equations corresponding to a discrete, albeit infinite, set of numbers which characterize the form of the field inside the cavity. Each solution, which is called a normal mode of the field, corresponds to a frequency of vibration of the field belonging to that particular solution. This frequency is called a normal frequency of the field.

The normal modes of a cavity can be obtained by developing solutions of equation (1.75) for the vector potential in the form (see e.g. Louisell 1990)

$$\mathbf{A}_\perp(\mathbf{x}, t) = \sum_\ell A_\ell(t)\mathbf{f}_\ell(\mathbf{x}) \tag{5.1}$$

where ℓ represents a set of discrete numbers. This set identifies the normal mode $\mathbf{f}_\ell(\mathbf{x})$, while $A_\ell(t)$ are expansion coefficients. Substitution of (5.1) into (1.75) and separation of time and space variables yields

$$\nabla^2\mathbf{f}_\ell(\mathbf{x}) + \frac{\omega_\ell^2}{c^2}\mathbf{f}_\ell(\mathbf{x}) = 0 \tag{5.2}$$

where the separation constant ω_ℓ is the frequency of the normal mode. The boundary and transversality conditions on $\mathbf{A}_\perp$ imply

$$\begin{aligned}&[\mathbf{f}_\ell(\mathbf{x})]_{\mathrm{tan}} = 0, \ [\nabla \times \mathbf{f}_\ell(\mathbf{x})]_{\mathrm{nor}} = 0 \ \text{ on the cavity surface;}\\ &\nabla \cdot \mathbf{f}_\ell(\mathbf{x}) = 0 \ \text{ inside the cavity}\end{aligned} \tag{5.3}$$

The solutions of the infinite set of equations (5.2) can be chosen to be orthogonal and normalized in the sense that

$$\int \mathbf{f}_\ell^*(\mathbf{x}) \cdot \mathbf{f}_{\ell'}(\mathbf{x}) d^3\mathbf{x} = V\delta_{\ell\ell'} \tag{5.4}$$

where V is the cavity volume. The infinite set $\mathbf{f}_\ell$ is complete in the sense that any field configuration inside the cavity can be expressed as a superposition of $\mathbf{f}_\ell$. It is also apparent that the solutions $\mathbf{f}_\ell(\mathbf{x})$ depend on the form of the cavity. For example, in the parallelepiped cavity represented in Figure 5.1, the components of normal modes are

$$\begin{aligned}(\mathbf{f}_\ell)_1 &= \sqrt{8}e_1 \cos k_1x_1 \sin k_2x_2 \sin k_3x_3\\ (\mathbf{f}_\ell)_2 &= \sqrt{8}e_2 \sin k_1x_1 \cos k_2x_2 \sin k_3x_3\\ (\mathbf{f}_\ell)_3 &= \sqrt{8}e_3 \sin k_1x_1 \sin k_2x_2 \cos k_3x_3\end{aligned} \tag{5.5}$$

where

$$k_i = \pi\ell_i L_i^{-1} \ (\ell_i = \text{positive integer}) \tag{5.6}$$

in order to satisfy the boundary conditions on the surface of the cavity in (5.3), and where the unity vector $\mathbf{e}$ is perpendicular to $\mathbf{k}$ in order to satisfy the transversality condition. Substitution of (5.5) in (5.2) yields the normal

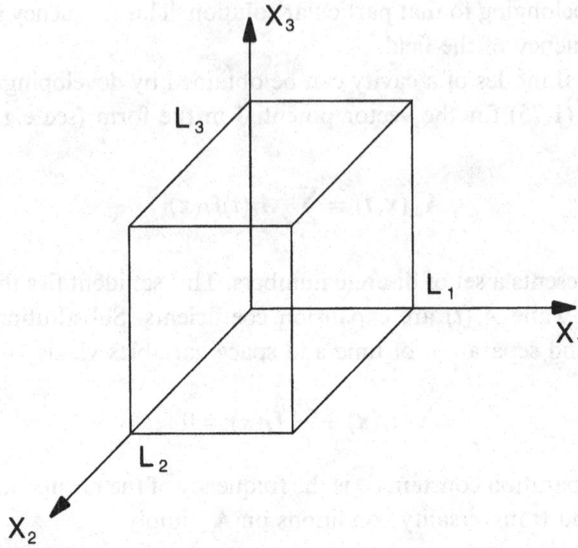

Fig. 5.1 The surface of a parallelepiped cavity. The edges of the cavity are of length L_1, L_2 and L_3.

frequencies

$$\omega_\ell^2 \equiv \omega^2\{\ell_i\} = c^2k^2 \quad (k^2 = k_1^2 + k_2^2 + k_3^2) \tag{5.7}$$

The electric and magnetic fields inside the cavity can be expressed in terms of normal modes as

$$\mathbf{E}_\perp(\mathbf{x}) = -\frac{1}{c}\dot{\mathbf{A}}_\perp(\mathbf{x}) = -\frac{1}{c}\sum_\ell \dot{A}_\ell(t)\mathbf{f}_\ell(\mathbf{x}) \; ;$$
$$\mathbf{H}(\mathbf{x}) = \nabla \times \mathbf{A}_\perp(\mathbf{x}) = \sum_\ell A_\ell(t)\nabla \times \mathbf{f}(\mathbf{x}) \tag{5.8}$$

The field inside the cavity can be quantized by promoting the A_ℓ to the status of operators as follows

$$A_\ell(t) = \sqrt{\frac{2\pi\hbar c^2}{V\omega_\ell}}\left(a_\ell e^{-i\omega_\ell t} + a_\ell^\dagger e^{i\omega_\ell t}\right) \; ;$$
$$\dot{A}_\ell(t) = -i\sqrt{\frac{2\pi\hbar c^2\omega_\ell}{V}}\left(a_\ell e^{-i\omega_\ell t} - a_\ell^\dagger e^{i\omega_\ell t}\right) \; ;$$
$$\left[a_\ell, a_{\ell'}^\dagger\right] = \delta_{\ell\ell'} \; ; \; [a_\ell, a_{\ell'}] = 0 \tag{5.9}$$

Substitution of (5.9) into (5.8), with the assumption that $\mathbf{f}_\ell(\mathbf{x})$ is real, yields

$$\mathbf{E}_\perp(\mathbf{x}) = i\sum_\ell \sqrt{\frac{2\pi\hbar\omega_\ell}{V}}\left(a_\ell - a_\ell^\dagger\right)\mathbf{f}_\ell(\mathbf{x}) \; ;$$

$$\mathbf{H}(\mathbf{x}) = \sum_\ell \sqrt{\frac{2\pi\hbar c^2}{V\omega_\ell}}\left(a_\ell + a_\ell^\dagger\right)\nabla \times \mathbf{f}_\ell(\mathbf{x}) \qquad (5.10)$$

The multipolar scheme has been shown to be much more convenient than the minimal coupling in order to describe the atom-field interaction in a cavity. The former, in fact, automatically takes into account the effect of the image charges of the atom on the cavity walls (Power and Thirunamachandran 1982). This is in fact a more general feature of the transformation from the minimal to the multipolar scheme. This transformation, when applied to a multi-atom system, formally eliminates from the Hamiltonian all interatomic Coulomb interactions, leaving only the intra-atomic electrostatic binding energies (Power and Zieanu 1959, Woolley 1971, Babiker *et al.* 1974). The interatomic forces in the multipolar scheme are mediated by the transverse electromagnetic field. The same feature holds for the interaction between an atom in a cavity and its image charges induced on the walls; the transformation from minimal to multipolar formally eliminates these interactions from the Hamiltonian of the system. Thus the atom-photon Hamiltonian in the multipolar scheme and in the electric dipole approximation for our model is

$$H = \sum_n E'_n c_n^{\dagger\prime} c'_n + \frac{1}{8\pi}\int \left\{\mathbf{D}_\perp^2(\mathbf{x}) + \mathbf{H}^2(\mathbf{x})\right\} d^3\mathbf{x} - \boldsymbol{\mu}\cdot\mathbf{D}_\perp(\mathbf{R}) \qquad (5.11)$$

where $\mathbf{R}$ is the position of the atom inside the cavity. Performing the unitary transformation to the multipolar scheme for the e.m. field, we obtain from (5.10)

$$\mathbf{D}_\perp(\mathbf{x}) = \tilde{\mathbf{E}}_\perp(\mathbf{x}) = i\sum_\ell \sqrt{\frac{2\pi\hbar\omega_\ell}{V}}\left(\tilde{a}_\ell - \tilde{a}_\ell^\dagger\right)\mathbf{f}_\ell(\mathbf{x}) \; ;$$

$$\mathbf{H}(\mathbf{x}) = \tilde{\mathbf{H}}(\mathbf{x}) = \sum_\ell \sqrt{\frac{2\pi\hbar c^2}{V\omega_\ell}}\left(\tilde{a}_\ell + \tilde{a}_\ell^\dagger\right)\nabla \times \mathbf{f}_\ell(\mathbf{x}) \qquad (5.12)$$

From these expressions we have

$$\int \mathbf{D}_\perp^2(\mathbf{x})d^3\mathbf{x} = -\frac{2\pi\hbar}{V}\sum_{\ell\ell'}\sqrt{\omega_\ell\omega_{\ell'}}\left(\tilde{a}_\ell\tilde{a}_{\ell'} - \tilde{a}_\ell\tilde{a}_{\ell'}^\dagger - \tilde{a}_\ell^\dagger\tilde{a}_{\ell'} + \tilde{a}_\ell^\dagger\tilde{a}_{\ell'}^\dagger\right)$$
$$\times \int \mathbf{f}_\ell(\mathbf{x})\cdot\mathbf{f}_{\ell'}(\mathbf{x})d^3\mathbf{x}$$
$$\int \mathbf{H}^2(\mathbf{x})d^3\mathbf{x} = \frac{2\pi\hbar c^2}{V}\sum_{\ell\ell'}\frac{1}{\sqrt{\omega_\ell\omega_{\ell'}}}\left(\tilde{a}_\ell\tilde{a}_{\ell'} + \tilde{a}_\ell\tilde{a}_{\ell'}^\dagger + \tilde{a}_\ell^\dagger\tilde{a}_{\ell'} + \tilde{a}_\ell^\dagger\tilde{a}_{\ell'}^\dagger\right)$$
$$\times \int (\nabla\times\mathbf{f}_\ell(\mathbf{x}))\cdot(\nabla\times\mathbf{f}_{\ell'}(\mathbf{x}))d^3\mathbf{x} \tag{5.13}$$

The well-known vector identity

$$\nabla\cdot(\mathbf{A}\times\mathbf{B}) = \mathbf{B}\cdot(\nabla\times\mathbf{A}) - \mathbf{A}\cdot(\nabla\times\mathbf{B}) \tag{5.14}$$

yields

$$\nabla\cdot[\mathbf{f}_\ell\times(\nabla\times\mathbf{f}_{\ell'})] = (\nabla\times\mathbf{f}_\ell)\cdot(\nabla\times\mathbf{f}_{\ell'}) - \mathbf{f}_\ell\cdot(\nabla\times\nabla\times\mathbf{f}_{\ell'}) \tag{5.15}$$

Consequently

$$\int (\nabla\times\mathbf{f}_\ell)\cdot(\nabla\times\mathbf{f}_{\ell'})d^3\mathbf{x} = \int \nabla\cdot[\mathbf{f}_\ell\times(\nabla\times\mathbf{f}_{\ell'})]d^3\mathbf{x}$$
$$+\int \mathbf{f}_\ell\cdot(\nabla\times\nabla\times\mathbf{f}_{\ell'})d^3\mathbf{x} \tag{5.16}$$

The first integral on the RHS is transformed, using Gauss' theorem, into the surface integral

$$\int_S [\mathbf{f}_\ell\times(\nabla\times\mathbf{f}_{\ell'})]\cdot\mathbf{n}dS \tag{5.17}$$

where S is the cavity surface. In view of the boundary conditions (5.3) the surface integral vanishes since $\mathbf{f}_\ell\times(\nabla\times\mathbf{f}_{\ell'})$ is parallel to the cavity walls. Thus

$$\int (\nabla\times\mathbf{f}_\ell)\cdot(\nabla\times\mathbf{f}_{\ell'})d^3\mathbf{x} = \int \mathbf{f}_\ell\cdot(\nabla\times\nabla\times\mathbf{f}_{\ell'})d^3\mathbf{x}$$
$$= \int \mathbf{f}_\ell\cdot\nabla(\nabla\cdot\mathbf{f}_{\ell'})d^3\mathbf{x} - \int \mathbf{f}_\ell\nabla^2\mathbf{f}_{\ell'}d^3\mathbf{x} \tag{5.18}$$

The first integral on the RHS vanishes in view of the transversality of $\mathbf{f}_\ell$, and the second is immediately transformed using (5.2). Thus

$$\int (\nabla\times\mathbf{f}_\ell)\cdot(\nabla\times\mathbf{f}_{\ell'})d^3\mathbf{x} = \frac{\omega_{\ell'}^2}{c^2}\int \mathbf{f}_\ell\cdot\mathbf{f}_{\ell'}d^3\mathbf{x} = \frac{\omega_\ell^2}{c^2}V\delta_{\ell\ell'} \tag{5.19}$$

Substituting (5.19) into (5.13) and using (5.4) again gives

$$\int \{\mathbf{D}_\perp^2(\mathbf{x}) + \mathbf{H}^2(\mathbf{x})\} d^3\mathbf{x} = 4\pi \sum_\ell \hbar\omega_\ell \left(\tilde{a}_\ell \tilde{a}_\ell^\dagger + \tilde{a}_\ell^\dagger \tilde{a}_\ell \right) \tag{5.20}$$

Finally, substitution of (5.12) into (5.11) and use of (4.85) and of (4.79) in the interaction term yields the following form for the atom-photon Hamiltonian in a cavity, in the multipolar scheme and in the dipole approximation

$$\begin{aligned} H = & \sum_n E_n' c_n'^\dagger c_n' + \sum_\ell \hbar\omega_\ell \left(\tilde{a}_\ell^\dagger \tilde{a}_\ell + \frac{1}{2} \right) \\ & - i \sum_{nn'} \sum_\ell \sqrt{\frac{2\pi\hbar\omega_\ell}{V}} \left(\mu_{nn'} \cdot \mathbf{f}_\ell(\mathbf{R}) c_n'^\dagger c_{n'}' \tilde{a}_\ell - \text{h.c.} \right) \end{aligned} \tag{5.21}$$

which may be compared with (4.98), valid when periodic boundary conditions are imposed within the same approximation.

Suppose now that the normal modes of the cavity are well separated in frequency; as is clear from (5.7) this can always be obtained, at least in principle, by choosing linear dimensions L_i small enough. Further, suppose that only one of the modes, of frequency $\omega_\ell \equiv \omega$, is strongly populated, the others being completely devoid of photons. Also, assume that this strongly populated mode can influence the dynamics of only a pair of atomic levels with $n = 1, 2$, the dynamics of the others being scarcely modified by the presence of the mode. This situation can be obtained in practice both at microwave frequency, when the atomic levels 1 and 2 are Rydberg levels defined in Section 4.3 (Rempe *et al.* 1987), and at optical frequency (Thompson *et al.* 1992). With these assumptions and using the pseudospin formalism leading to (4.104), (5.21) can be approximated as

$$\begin{aligned} & H = \hbar\omega_0 S_z + \hbar\omega \tilde{a}^\dagger \tilde{a} + (\epsilon S_+ - \epsilon^* S_-)(\tilde{a} - \tilde{a}^\dagger) \; ; \\ & \epsilon = -i \sqrt{\frac{2\pi\hbar\omega}{V}} \mu_{21} \cdot \mathbf{f}(\mathbf{R}) \end{aligned} \tag{5.22}$$

The model Hamiltonian (5.22) represents a two-level atom interacting with a single mode, and it is the multipolar version of the Jaynes-Cummings (JC) Hamiltonian (Jaynes and Cummings 1963; for a simple discussion see e.g. Allen and Eberly 1987). An even more simple version of the two-level atom, single mode radiation Hamiltonian can be obtained by observing that if the effects of the interaction term in (5.22) can be neglected, the Heisenberg equations of motion for $\tilde{a}$ and S_- give

$$\tilde{a}(t) = \tilde{a}(0)e^{-i\omega t} \; ; S_-(t) = S_-(0)e^{-i\omega_0 t} \tag{5.23}$$

Thus near resonance ($\omega_0 \sim \omega$) and when the number of photons is not so large that the time-dependences (5.23) are not even approximately true, the terms in $\tilde{a}S_-$ and $\tilde{a}^\dagger S_+$ change relatively fast in comparison with those in $\tilde{a}S_+$ and $\tilde{a}^\dagger S_-$. Therefore the effect of the fast terms on the dynamics of the system can be expected to average out in comparison with that of the slow terms, and (5.22) can be approximated as

$$H = \hbar\omega_0 S_z + \hbar\omega\tilde{a}^\dagger\tilde{a} + \epsilon\tilde{a}S_+ + \epsilon^*\tilde{a}^\dagger S_- \tag{5.24}$$

The latter Hamiltonian is the JC Hamiltonian in the Rotating Wave Approximation (RWA). The two versions (5.22) and (5.24) of the JC Hamiltonian form the basis for a large part of Quantum Optics (see e.g. Shore and Knight 1993).

The theory developed so far in the present section has been concerned only with "perfect" cavities. A real cavity, however, is not perfect in the sense that it must have at least a port for communications with the outside world and its walls present a finite electrical resistance. Both these facts lead to damping of the internal fields as well as to losses which can be characterized by the introduction of the so-called quality factor Q of the cavity. For a cavity whose fundamental frequency is ω, this factor is defined as

$$Q = \omega\tau/2\pi \tag{5.25}$$

where τ is the damping time of the fields. Another measure of the quality of a cavity, which is useful at optical frequencies for one-dimensional cavities, is the finesse F defined as

$$F = \frac{\lambda}{2L}Q = \frac{c}{2L}\tau \tag{5.26}$$

where L is the length of the cavity. Since $c/2L$ is the spacing in frequency between neighbouring modes and since the damping τ^{-1} is a measure of the width of the mode frequencies, a high-finesse cavity is one in which the normal modes are well separated in frequency in spite of their broadening due to losses. Thus an atom in a high-Q or high-F cavity can interact with a single mode, approaching the ideal situation described earlier in this section. A single-atom, single-mode microwave cavity is often called a micromaser, exploiting superconducting walls in which values of $Q \sim 10^{11}$ are not uncommon. Micromaser experiments must obviously be performed at very low temperatures (Meschede 1992). Optical cavities with values of $F \sim 8 \cdot 10^4$ have also been used (Thompson *et al.* 1992).

5.3 Atom dressed by a populated cavity mode

We are now ready to introduce on a more quantitative basis the concept of dressed atoms in a field of real photons. We shall assume that the conditions are such that one can describe the atom-field system by the JC Hamiltonian in the RWA (5.24). Thus we put

$$H = H_0 + V \; ; \; H_0 = \hbar\omega_0 S_z + \hbar\omega a^\dagger a \; ; \; V = \epsilon a S_+ + \epsilon^* a^\dagger S_- \qquad (5.27)$$

where we have dropped the tildes for simplicity of notation. The eigenstates and eigenvalues of H_0 can be written in any of the following equivalent forms

$$\begin{aligned} &| \, n, \pm 1/2\rangle \; ; \; | \, n, \uparrow (\downarrow)\rangle \; ; \; | \, n, \pm\rangle \; ; \\ &E^0_{n\pm} = \hbar(n\omega \pm \omega_0/2) \end{aligned} \qquad (5.28)$$

Except for the ground state $| \, 0, \downarrow\rangle$, any other eigenstate of H_0 belongs to one of an infinite set of two-dimensional subspaces $(| \, n, \uparrow\rangle, | \, n+1, \downarrow\rangle)$. It is immediate to see that each two-dimensional subspace is characterized by an eigenvalue of the operator

$$\mathcal{N} = a^\dagger a + S_z + 1/2 \qquad (5.29)$$

The ground state of H_0 is also an eigenstate of $\mathcal{N}$, corresponding to the eigenvalue 0, but it constitutes a one-dimensional space by itself. Clearly $[V, \mathcal{N}] = 0$, and consequently V can connect only states with the same eigenvalue of $\mathcal{N}$. Thus, except for the ground state of H_0 which is not influenced by V, the infinite matrix representing H on the basis of the eigenstates of H_0 splits into an infinite set of 2×2 matrices

$$\begin{vmatrix} \hbar\omega n - \hbar\omega_0/2 & \epsilon^*\sqrt{n} \\ \epsilon\sqrt{n} & \hbar\omega(n-1) + \hbar\omega_0/2 \end{vmatrix} \qquad (5.30)$$

yielding the following eigenstates and eigenvalues

$$\begin{aligned} | \, u_n^{(+)}\rangle &= \cos\frac{\theta}{2} \, | \, n, \downarrow\rangle - \frac{\epsilon}{|\,\epsilon\,|} \sin\frac{\theta}{2} \, | \, n-1, \uparrow\rangle \; ; \\ | \, u_n^{(-)}\rangle &= \cos\frac{\theta}{2} \, | \, n-1, \uparrow\rangle + \frac{\epsilon^*}{|\,\epsilon\,|} \sin\frac{\theta}{2} \, | \, n, \downarrow\rangle \; ; \\ E_n^{(\pm)} &= \left(n - \frac{1}{2}\right)\hbar\omega \pm \Delta/2 \end{aligned} \qquad (5.31)$$

where

$$\sin\frac{\theta}{2} = \frac{1}{\sqrt{2}}[1+\delta/\Delta]^{1/2}\ ;\ \cos\frac{\theta}{2} = -\frac{1}{\sqrt{2}}[1-\delta/\Delta]^{1/2}\ ;$$

$$\delta = \hbar(\omega_0 - \omega)\ ;\ \Delta = \left[\delta^2 + 4 \mid \epsilon \mid^2 n\right]^{1/2} \tag{5.32}$$

The states $| u_n^{(\pm)} \rangle$ defined in (5.31) are known as "dressed states" (Cohen-Tannoudji 1968, Cohen-Tannoudji and Haroche 1969). They are characterized by entanglement between atomic and field states, which for $\delta^2 \ll 4 \mid \epsilon \mid^2 n$ is quite strong. In the same limit, the eigenvalue spectrum of H is characterized by a set of doublets split by the energy $\Delta \sim 2 \mid \epsilon \mid \sqrt{n}$, as shown in Figure 5.2 for the lowest part of the spectrum. The distance between a doublet and the next is $\hbar\omega \sim \hbar\omega_0$, and for the time being we shall assume that the cavity field is not so strong as to violate the constraint

$$\hbar\omega \sim \hbar\omega_0 \gg 2 \mid \epsilon \mid \sqrt{n} \tag{5.33}$$

It should be remarked that if (5.33) does not hold true, the assumptions leading to the RWA are not guaranteed and the whole model considered breaks down. Moreover, since for a coherent cavity field $2 \mid \epsilon \mid \sqrt{n}$ is proportional to the electric field inside the cavity, while $\hbar\omega_0$ is of the same order as the atomic Coulomb field, (5.33) simply means that the cavity field is taken to be smaller than the Coulomb field of the nucleus acting on the atomic electron.

The simple description presented above does not immediately display the justification for calling dressed states those defined in (5.31). We shall now take up a slightly more formal approach to the problem which will enable us to identify more clearly the concept of a dressed atom in the present context. Consider in fact the unitary operator (Compagno and Persico 1984)

$$\begin{aligned} T &= \exp\left\{-\theta\left[4 \mid \epsilon \mid^2 \mathcal{N}\right]^{-1/2}\left(\epsilon a S_+ - \epsilon^* a^\dagger S_-\right)\right\} \\ &= \cos\frac{\hat{\theta}}{2} - \left[\mid \epsilon \mid^2 \mathcal{N}\right]^{-1/2} \sin\frac{\hat{\theta}}{2}\left(\epsilon a S_+ - \epsilon^* a^\dagger S_-\right) \end{aligned} \tag{5.34}$$

where $\hat{\theta}$ is an operator defined as

$$\begin{aligned} &\sin\hat{\theta} = -\left[4 \mid \epsilon \mid^2 \mathcal{N}\right]^{1/2} \hat{\Delta}^{-1}\ ; \\ &\cos\hat{\theta} = -\delta\hat{\Delta}^{-1}\ ;\ \hat{\Delta} = \left[\delta^2 + 4 \mid \epsilon \mid^2 \mathcal{N}\right]^{1/2}\ ; \\ &\sin\frac{\hat{\theta}}{2} = \frac{1}{\sqrt{2}}\left[1 + \delta\hat{\Delta}^{-1}\right]^{1/2}\ ; \\ &\cos\frac{\hat{\theta}}{2} = -\frac{1}{\sqrt{2}}\left[1 - \delta\hat{\Delta}^{-1}\right]^{1/2} \end{aligned} \tag{5.35}$$

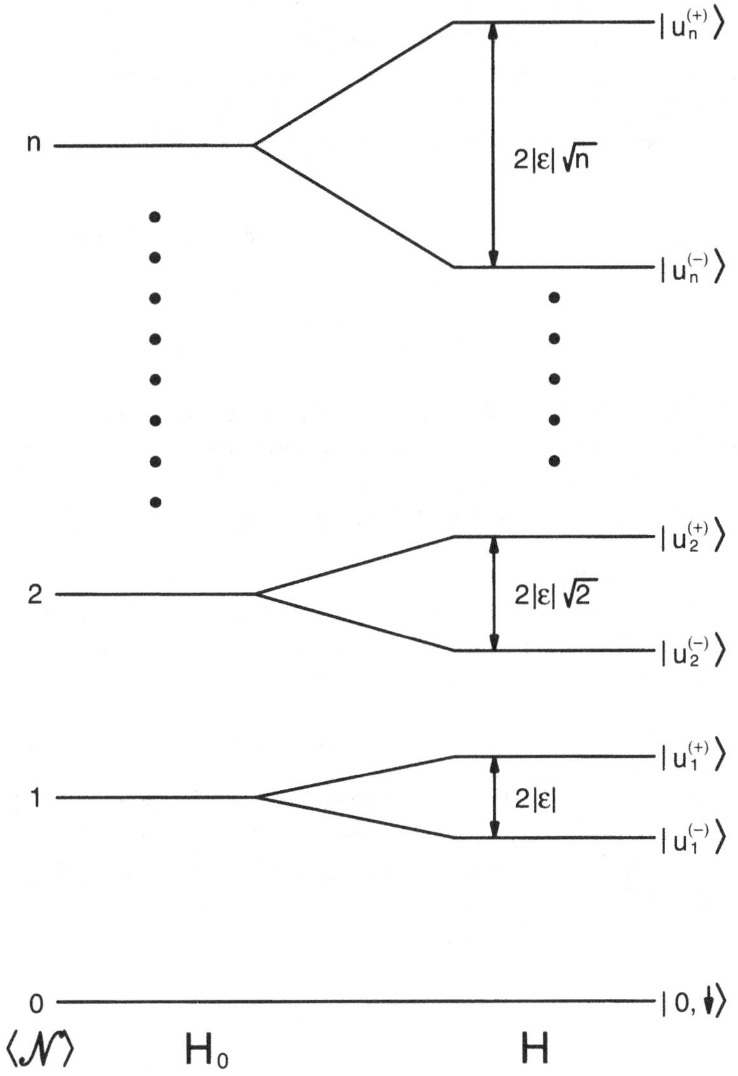

Fig. 5.2 The left side of the figure represents the low-lying eigenvalues of the Jaynes-Cummings Hamiltonian with $\epsilon = 0$ and $\omega_0 = \omega$. In these conditions each eigenvalue is doubly degenerate except the ground state. The degeneracy is lifted for $\epsilon \neq 0$ as shown by the right side of the figure. The new eigenstates $| u_n^{(\pm)}\rangle$ are dressed states.

We note that the eigenvalues of operators like $\sin\hat{\theta}/2$, $\cos\hat{\theta}/2$, $\hat{\Delta}$ ... within a two-dimensional subspace of the Hilbert space of H, characterized by an eigenvalue of $\mathcal{N}$, coincide with $\sin\theta/2$, $\cos\theta/2$, Δ ... as given by (5.32).

Define next the new dressed operators $\bar{A} = TAT^{-1}$. Using (5.34) one obtains

$$\bar{a}^\dagger\bar{a} = a^\dagger a + S_z\left(1 - \cos\hat{\theta}\right) - (\epsilon a S_+ + \epsilon^* S_-)[4\,|\,\epsilon\,|\,\mathcal{N}]^{-1/2}\sin\hat{\theta}$$

$$\bar{S}_z = S_z\cos\hat{\theta} + \left(\epsilon a S_+ + \epsilon^* a^\dagger S_-\right)\left[4\,|\,\epsilon\,|^2\,\mathcal{N}\right]^{-1/2}\sin\hat{\theta}$$

$$\epsilon\bar{a}\bar{S}_+ + \epsilon^*\bar{a}^\dagger\bar{S}_- = \left(\epsilon a S_+ + \epsilon^* a^\dagger S_-\right)\cos\hat{\theta} - S_z[4\,|\,\epsilon\,|\,\mathcal{N}]^{1/2}\sin\hat{\theta} \qquad (5.36)$$

Expressions (5.36) are easily inverted to obtain the bare operators in terms of the dressed ones. The result is then substituted in (5.27), which gives

$$H = \hbar\omega\bar{a}^\dagger\bar{a} + \left(\hbar\omega - \hat{\Delta}\right)\bar{S}_z \qquad (5.37)$$

H here is the same Hamiltonian as the original (5.27), except that it is expressed in terms of dressed operators. Since dressed photon operators commute with dressed atom operators, in view of the unitarity of T, Hamiltonian (5.37) can be partitioned into two commuting parts

$$H = \bar{H}_F + \bar{H}_A\ ;\ \bar{H}_F = \hbar\omega\bar{a}^\dagger\bar{a}\ ;\ \bar{H}_A = \left(\hbar\omega - \hat{\Delta}\right)\bar{S}_z \qquad (5.38)$$

Thus the eigenstates of H can be written as $|\,\bar{n}, \bar{\uparrow}(\bar{\downarrow})\rangle$ where $\bar{n} = 0, 1, \ldots$ are the dressed photon occupation numbers and $\bar{\uparrow}(\bar{\downarrow})$ are the two states of the dressed atom corresponding to the two eigenvalues $\pm 1/2$ of $\bar{S}_z$. Naturally the dressed eigenstates $|\,\bar{n}, \bar{\uparrow}(\bar{\downarrow})\rangle$ are related to the bare ones $|\,n, \uparrow(\downarrow)\rangle$ by the unitary relations

$$|\,\bar{n}, \bar{\downarrow}\rangle = T\,|\,n, \downarrow\rangle = \cos\frac{\hat{\theta}}{2}\,|\,n, \downarrow\rangle - \frac{\epsilon}{|\,\epsilon\,|}\sin\frac{\hat{\theta}}{2}\,|\,n-1, \uparrow\rangle$$

$$|\,\bar{n}-1, \bar{\uparrow}\rangle = T\,|\,n-1, \uparrow\rangle = \cos\frac{\hat{\theta}}{2}\,|\,n-1, \uparrow\rangle + \frac{\epsilon^*}{|\,\epsilon\,|}\sin\frac{\hat{\theta}}{2}\,|\,n, \downarrow\rangle \qquad (5.39)$$

The dressed eigenstates (5.39) coincide with $|\,u_n^{(\pm)}\rangle$ obtained by the method used earlier in this section (apart from unimportant phase factors), and also the dressed eigenvalues, which are immediately obtained from (5.38), coincide with $E_n^{(\pm)}$ in (5.31). Using the dressing operator T, however, has enabled us to give a clear definition of the dressed atom, which is now defined as a two-level system whose Hamiltonian is $\hat{H}_A$.

The following remarks are in order.

In the Heisenberg representation the equation of motion of S_z is

$$\dot{S}_z = -\frac{i}{\hbar}[S_z, H] = -\frac{i}{\hbar}(\epsilon a S_+ - \epsilon^* a^\dagger S_-)$$
$$= -\frac{i}{\hbar}\left(S_z - \frac{1}{4}\delta K^{-1}\right)2K \tag{5.40}$$

where

$$K = \delta S_z + \epsilon a S_+ + \epsilon^* a^\dagger S_- \tag{5.41}$$

is a constant of motion, since $[K, H] = 0$. It is then easy to check that the solution of (5.40) in terms of constant operators is

$$S_z(t) = \left[S_z(0) - \frac{1}{4}\delta K^{-1}\right]e^{-(i/\hbar)2Kt} + \frac{1}{4}\delta K^{-1} \tag{5.42}$$

From (5.36) and squaring (5.41) we also find

$$K = -\hat{\Delta}\bar{S}_z \; ; \; K^2 = \frac{1}{4}\hat{\Delta}^2 \tag{5.43}$$

In view of (5.38), the first of these expressions tells us that the eigenstates of K are the dressed states (5.39) or (5.31). The second of (5.43) tells us that the eigenvalues of K are $\pm\Delta/2$. Hence

$$K \mid u_n^{(\pm)}\rangle = \pm\frac{1}{2}\Delta \mid u_n^{(\pm)}\rangle \tag{5.44}$$

We are now in the position, starting from the solution (5.42), to answer the question: suppose at $t = 0$ the atom is in the bare excited state and the field has $n - 1$ photons, what will be the atomic state at $t > 0$? In fact, since inverting (5.31) yields

$$\mid n-1, \uparrow\rangle = -\frac{\epsilon^*}{\mid\epsilon\mid}\sin\frac{\theta}{2} \mid u_n^{(+)}\rangle + \cos\frac{\theta}{2} \mid u_n^{(-)}\rangle \tag{5.45}$$

we have, using (5.44) and after some algebra

$$\langle n-1, \uparrow \mid S_z(t) \mid n-1, \uparrow\rangle = \frac{1}{2}\left(\sin^2\theta\cos\frac{\Delta}{\hbar}t + \cos^2\theta\right) \tag{5.46}$$

Thus the atom oscillates without any damping between the bare excited state and a state where the average value of S_z is $\frac{1}{2}\cos 2\theta$ at a frequency $\Delta/\hbar$. These oscillations are called Rabi oscillations and $\Delta/\hbar$ is called the

Rabi frequency (see e.g. Knight and Milonni 1980). When the atom is resonant with the cavity mode ($\omega_0 = \omega$), one has $\delta = 0$, $\Delta = 2 \mid \epsilon \mid \sqrt{n}$, $\cos\theta = 0$ and $\sin\theta = -1$. Consequently on resonance the atom oscillates between the bare ground and the bare excited state. Since $\mathcal{N}$ is constant in time, one cavity photon is continuously emitted and reabsorbed during the atomic oscillations. In contrast, no dynamics take place if the atom is placed at $t = 0$ in its dressed excited state, because $\bar{S}_z$ is a constant of motion. However the essential element of the bare atomic dynamics, namely the Rabi frequency, is incorporated in the renormalized splitting of the dressed atomic levels appearing in (5.37) which is $\hbar\omega - \Delta$ rather than $\hbar\omega_0$. The Rabi oscillations are also present for $n = 1$, when the initial state is $\mid 0, \uparrow\rangle$ and there is no photon in the cavity at $t = 0$. In this case the Rabi frequency is $2 \mid \epsilon \mid$, and one photon is periodically absorbed and re-emitted by the atom. Rabi oscillations have been experimentally observed in a micromaser device (Rempe *et al.* 1987, Bernardot *et al.* 1992). Rabi oscillations were first experimentally observed at radio frequency in the form of sinusoidal variations of atomic populations (Rabi 1937). In this case the field can be considered as fully classical. If the quantum features of the field cannot be neglected, in spite of the fact that it is in a coherent state of the kind described in Section 2.3, novel phenomena occur. In fact in this case the Rabi oscillations do not persist indefinitely, but display an infinite series of collapses and revivals due to the quantum nature of the field. Here we cannot discuss adequately the physics of these collapses and revivals, and we simply refer the interested reader to the recent review by Shore and Knight (1993).

Further, it is important to note that the ground state of $\mid 0, \downarrow\rangle$ of H_0 is not perturbed by the presence of V, since $V \mid 0, \downarrow\rangle = 0$. This, like all the results obtained in the present section hitherto, is an exact result for the RWA Hamiltonian (5.27). This Hamiltonian, however, is only an approximation of the more complete Hamiltonian (5.22), which we write in the form

$$H = H_0 + V_1 + V_2 \ ; \ H_0 = \hbar\omega_0 S_z + \hbar\omega a^\dagger a \ ;$$
$$V_1 = \epsilon a S_+ + \epsilon^* a^\dagger S_- \ ; \ V_2 = -\epsilon a^\dagger S_+ - \epsilon^* a S_- \qquad (5.47)$$

The rotating terms are represented by V_1 and the counterrotating ones by V_2 as a perturbation on the dressed states, which are the eigenstates of $H_0 + V_1$. One immediately realizes that V_2 does not connect states belonging to the same eigenvalue of $\mathcal{N}$, but rather states belonging to doublets whose eigenvalues of $\mathcal{N}$ differ by ± 2. If the splitting caused by V_1

within the same doublets, which is $O(\epsilon\sqrt{n})$, is much smaller than the average energy difference $\hbar\omega$ (for $\omega \sim \omega_0$) between contiguous doublets, the dressed states connected by V_2 will have energies differing by $\sim 2\hbar\omega$. Thus the energy shifts induced by V_2 in the lowest possible order of perturbation theory are of the order of $| V_2 |^2 /2\hbar\omega \sim | \epsilon |^2 n/2\hbar\omega \ll \epsilon\sqrt{n}$. This substantiates the argument in favour of validity of RWA given at the end of Section 5.2, under the assumption of near resonance ($\omega_0 \sim \omega$) and n not too large. The same argument is not valid, however, for the ground state $| 0, \downarrow\rangle$ of H_0, which as we have just discussed is also the ground state of $H_0 + V_1$ and which is not connected to any other state by V_1. Using the results of Appendix C, this state in second-order perturbation theory under the action of V_2 is changed into

$$\begin{aligned}
| 0, \downarrow\rangle' = | 0, \downarrow\rangle + \Bigg\{ & (1 - P_0)\frac{1}{E_0 - H_0} V_2 \\
& + \frac{1}{E_0 - H_0}(1 - P_0) V_1 \frac{1}{E_0 - H_0} V_2 \\
& - \frac{1}{2}\langle 0, \downarrow | \, V_2 (1 - P_0)\frac{1}{E_0 - H_0} \\
& \times (1 - P_0)\frac{1}{E_0 - H_0} V_2 \, | 0, \downarrow\rangle \Bigg\} | 0, \downarrow\rangle
\end{aligned} \tag{5.48}$$

where P_0 is the projection operator onto $| 0, \downarrow\rangle$ and $E_0 = -\hbar\omega_0/2$. The state (5.48) is normalized up to terms of order ϵ^2, and it is explicitly given by

$$\begin{aligned}
| 0, \downarrow\rangle' = & \left\{ 1 - \frac{1}{2} \frac{| \epsilon |^2}{\hbar^2(\omega_0 + \omega)^2} \right\} | 0, \downarrow\rangle + \frac{\epsilon}{\hbar(\omega_0 + \omega)} | 1, \uparrow\rangle \\
& - \frac{| \epsilon |^2 /\sqrt{2}}{\hbar^2 \omega(\omega_0 + \omega)} | 2, \downarrow\rangle
\end{aligned} \tag{5.49}$$

Thus there is some admixture of $| 1, \uparrow\rangle$ and of $| 2, \downarrow\rangle$ into $| 0, \downarrow\rangle$, which means that in the ground state of the atom-photon system, due to the action of V_2, the atom spends part of the time in the excited state and there are one or two photons in the field. We shall be very interested in the future in this sort of effect, since it is obviously representative of the kind of zero-point dressing which we have touched upon in Section 5.1. At the moment we wish only to point out that the correction to the state caused by V_2 is associated with a shift in the energy of $| 0, \downarrow\rangle$ which in second-order perturbation theory is given by (C.10) as

$$E_0' - E_0 = -\frac{|\epsilon|^2}{\hbar(\omega_0 + \omega)} \tag{5.50}$$

This shift is due to the virtual radiative process and it is a prototype of a radiative shift.

Finally, we remark that the dressed photons, defined in terms of the occupation number of the dressed field given by the eigenvalues of $\bar{a}^\dagger \bar{a}$, have a physical nature which is different from the bare ones. We shall not dwell on this problem here and we refer the interested reader to the specialized literature on this subject.

5.4 Spontaneous decay of an excited atom in free space

In the last section we have studied the concept of an atom dressed by a single-mode field. Such a field can in practice be obtained inside a cavity which selects a discrete numbers of modes, one of them being assumed to interact with the atom more efficiently than the others. In contrast, when the atom is placed in unbounded space, it is in contact with a continuum of modes of the field, or at least with a quasi-continuum if the field is subjected to periodic boundary conditions on the surface of large volume. Consequently, in the latter conditions there is no reason to assume that one of the quasi-continuum modes interacts with the atom more strongly than any of the many others which are in the same range of frequency. Therefore the problem becomes mathematically much more involved than the single-mode one discussed in the previous section even if one is prepared to work within a two-level atom model. Even in this case in fact the form of either Hamiltonian (4.105) does not lend itself to a simple mathematical analysis of the sort given in the previous section, in view of the role played by the large number modes interacting with the atom.

In order to show how the concept of the dressed atom can be applied to some extent also in the presence of a continuum of modes, it is useful to discuss preliminarily the dynamical evolution of an atom initially in an excited state with no photons present in the free-space field. This is the analogue of the problem discussed in the previous section for the single mode field case in the RWA, where we have seen that the atom undergoes periodic Rabi oscillations between the excited and the ground state, one real photon being emitted and reabsorbed synchronously with the atomic oscillations. We shall adopt the following Hamiltonian

$$H = \hbar\omega_0 S_z + \sum_{\mathbf{k}j} \hbar\omega_k \left(a^\dagger_{\mathbf{k}j} a_{\mathbf{k}j} + 1/2\right)$$
$$+ \sum_{\mathbf{k}j} \left(\epsilon_{\mathbf{k}j} a_{\mathbf{k}j} S_+ + \epsilon^*_{\mathbf{k}j} a^\dagger_{\mathbf{k}j} S_-\right)$$
$$+ \lambda \sum_{\mathbf{k}j} \left(\epsilon_{\mathbf{k}j} a^\dagger_{\mathbf{k}j} S_+ + \epsilon^*_{\mathbf{k}j} a_{\mathbf{k}j} S_-\right) \qquad (5.51)$$

which is equivalent to either Hamiltonian (4.105) for $\mathbf{R} = 0$ and $\lambda = \pm 1$. On the other hand putting $\lambda = 0$ yields the RWA for either forms (4.105). We shall work in the Heisenberg representation and obtain

$$\dot{S}_z = -\frac{i}{\hbar}[S_z, H] = -\frac{i}{\hbar}\left\{\sum_{\mathbf{k}j}\left(\epsilon_{\mathbf{k}j} a_{\mathbf{k}j} S_+ - \epsilon^*_{\mathbf{k}j} a^\dagger_{\mathbf{k}j} S_-\right)\right.$$
$$\left. + \lambda \sum_{\mathbf{k}j}\left(\epsilon_{\mathbf{k}j} a^\dagger_{\mathbf{k}j} S_+ - \epsilon^*_{\mathbf{k}j} a_{\mathbf{k}j} S_-\right)\right\} \qquad (5.52)$$

Due to the presence of the λ terms, as well as to the presence of many modes, (5.52) is much more involved than its single-mode counterpart (5.40), and it is not possible to introduce constants of motion at this stage. Thus we are led to investigate the time-development of the operators on the RHS of (5.52), such as

$$(a_{\mathbf{k}j}\dot{S}_+) = -\frac{i}{\hbar}\left[a_{\mathbf{k}j} S_+, H\right] = -\frac{i}{\hbar}\left\{\hbar(\omega_k - \omega_0) a_{\mathbf{k}j} S_+\right.$$
$$\left. + \epsilon^*_{\mathbf{k}j}(S_z + 1/2) + 2\sum_{\mathbf{k}'j'} \epsilon^*_{\mathbf{k}'j'}\left(a^\dagger_{\mathbf{k}'j'} + \lambda a_{\mathbf{k}'j'}\right) a_{\mathbf{k}j} S_z\right\} \qquad (5.53)$$

Also this equation is not exactly soluble because of the appearance of the two-photon terms on its RHS. Further, the equations of motion for the latter terms would involve three-photon terms, and we would be confronted with an infinite hierarchy of coupled differential equations, which would lead us nowhere. For this reason at this point one usually performs the first Born approximation, which consists of neglecting terms which would yield in the final results contributions $O(\epsilon^3)$. The two-photon terms in (5.53) fall into this category, since their quantum average on the initial state $|\{0_{\mathbf{k}j}\}\rangle$ vanishes; consequently the first terms with nonvanishing quantum average in their equations of motions must be at least $O(\epsilon)$, which would contribute terms $O(\epsilon^3)$ in the quantum average of (5.52). Hence (5.53) in the first Born approximation takes the form

$$(a_{\mathbf{k}j}\dot{S}_+) = -\frac{i}{\hbar}\left\{\hbar(\omega_k - \omega_0)a_{\mathbf{k}j}S_+ + \epsilon^*_{\mathbf{k}j}(S_z + 1/2)\right\} \tag{5.54}$$

An analogous procedure for $a_{\mathbf{k}j}S_-$ yields in the first Born approximation

$$(a_{\mathbf{k}j}\dot{S}_-) = -\frac{i}{\hbar}\{\hbar(\omega_k + \omega_0)a_{\mathbf{k}j}S_- - \lambda\epsilon_{\mathbf{k}j}(S_z - 1/2)\} \tag{5.55}$$

Equations (5.54) and (5.55) can be easily integrated to give

$$\begin{aligned}(a_{\mathbf{k}j}S_+)_t &= (a_{\mathbf{k}j}S_+)_0 e^{-i(\omega_k-\omega_0)t} \\ &\quad - \frac{i}{\hbar}\epsilon^*_{\mathbf{k}j}e^{-i(\omega_k-\omega_0)t}\int_0^t e^{i(\omega_k-\omega_0)t'}[S_z(t') + 1/2]dt' \\ (a_{\mathbf{k}j}S_-)_t &= (a_{\mathbf{k}j}S_-)_0 e^{-i(\omega_k+\omega_0)t} \\ &\quad + \frac{i}{\hbar}\lambda\epsilon_{\mathbf{k}j}e^{-i(\omega_k+\omega_0)t}\int_0^t e^{-i(\omega_k+\omega_0)t'}[S_z(t') - 1/2]dt'\end{aligned} \tag{5.56}$$

Substitution of (5.56) into (5.52) and taking the quantum mechanical average on the initial state leads to disappearance of all $t = 0$ terms such as $\langle a_{\mathbf{k}j}S_+\rangle_0$ which vanish exactly. The result is (Leonardi *et al.* 1986)

$$\begin{aligned}\langle\dot{S}_z(t)\rangle &= -\frac{2}{\hbar^2}\sum_{\mathbf{k}j}|\epsilon_{\mathbf{k}j}|^2\left\{\int_0^t \cos[(\omega_k - \omega_0)(t - t')]\right. \\ &\times (\langle S_z(t')\rangle + 1/2)dt' \\ &\left. + \lambda^2\int_0^t \cos[(\omega_k + \omega_0)(t - t')](\langle S_z(t')\rangle - 1/2)dt'\right\}\end{aligned} \tag{5.57}$$

An equation of the form (5.57) is called non-Markoffian since the rate of change of S_z at time t depends on its previous history. A Markoffian equation, however, can be easily obtained if the decay time of the atomic system is assumed to be much longer than ω_0^{-1}. In this case in fact one can take the slow variable $\langle S_z(t')\rangle$ out of the integrals in (5.57) and, for $t \gg \omega_0^{-1}$, one can also let the upper limit of the integrations go to infinity.

Equation (5.57) then becomes, with $\tau = t - t'$,

$$\langle \dot{S}_z(t)\rangle = -\frac{2}{\hbar^2}\sum_{\mathbf{k}j} |\epsilon_{\mathbf{k}j}|^2 \left\{ \left(\langle S_z(t)\rangle + \frac{1}{2}\right) \int_0^{t\to\infty} \cos(\omega_k - \omega_0)\tau d\tau \right.$$

$$\left. + \lambda^2(\langle S_z(t)\rangle - 1/2) \int_0^{t\to\infty} \cos(\omega_k + \omega_0)\tau d\tau \right\}$$

$$= -(\langle S_z(t)\rangle + 1/2)\frac{2\pi}{\hbar^2}\sum_{\mathbf{k}j} |\epsilon_{\mathbf{k}j}|^2\, \delta(\omega_k - \omega_0)$$

$$- \lambda^2(\langle S_z(t)\rangle - 1/2)\frac{2\pi}{\hbar^2}\sum_{\mathbf{k}j} |\epsilon_{\mathbf{k}j}|^2\, \delta(\omega_k + \omega_0) \tag{5.58}$$

where we have used Heitler's (1960) definition

$$\delta(x) = \frac{1}{\pi}\int_0^{t\to\infty} \cos \tau x d\tau \tag{5.59}$$

Since the second term on the RHS of (5.58) vanishes, the counterrotating terms whose effect is proportional to λ^2 are seen not to contribute in this approximation to the time development of $\langle S_z(t)\rangle$. Moreover, in view of transversality of the field, and taking $\mathbf{e}_{\mathbf{k}j}$ and μ_{21} as real

$$\sum_j \left(\mu_{21}\cdot \mathbf{e}_{\mathbf{k}j}\right)^2 = |\mu_{21}|^2 - \left(\mu_{21}\cdot\hat{\mathbf{k}}\right)^2 \tag{5.60}$$

where $\hat{\mathbf{k}}$ is the unit vector of $\mathbf{k}$. Therefore, using the minimal coupling value for $\epsilon_{\mathbf{k}j}$ given in (4.105), one has for any $g(\omega_k)$

$$\sum_{\mathbf{k}j} |\epsilon_{\mathbf{k}j}|^2\, g(\omega_k) = \frac{2\pi\hbar}{V}\omega_0^2 \sum_{\mathbf{k}j}\omega_k^{-1}\left(\mathbf{e}_{\mathbf{k}j}\cdot\mu_{21}\right)^2 g(\omega_k)$$

$$= \frac{2\pi\hbar}{V}\omega_0^2 |\mu_{21}|^2 \sum_{\mathbf{k}}\omega_k^{-1}\left[1 - (\hat{\mathbf{k}}\cdot\hat{\mu}_{21})^2\right] g(\omega_k)$$

$$= \frac{4\pi^2\hbar |\mu_{21}|^2 \omega_0^2}{(2\pi)^3 c}\int_0^{k_M} k g(\omega_k)\int_0^{\pi}(1 - \cos^2\theta)\sin\theta d\theta dk$$

$$= \frac{\gamma\hbar^2}{2\pi\omega_0}\int_0^{\omega_M}\omega_k g(\omega_k) d\omega_k \; ; \; \gamma = \frac{4 |\mu_{21}|^2 \omega_0^3}{3\hbar c^3} \tag{5.61}$$

where k_M is the cut-off introduced in (4.99) and ω_M the corresponding cut-off frequency (see e.g. Loudon 1981). It should be noted that γ coincides with the Einstein A coefficient. Using instead the multipolar expression in (4.105) for $\epsilon_{\mathbf{k}j}$ one has

$$\sum_{\mathbf{k}j} |\epsilon_{\mathbf{k}j}|^2 g(\omega_k) = \frac{2\pi\hbar}{V} \sum_{\mathbf{k}j} \omega_k (\mathbf{e}_{\mathbf{k}j} \cdot \mu_{21})^2 g(\omega_k)$$
$$= \frac{4\pi^2\hbar \,|\, \mu_{21} \,|^2\, c}{(2\pi)^3} \int_0^{k_M} k^3 g(\omega_k) \int_0^{\pi} (1 - \cos^2\theta) \sin\theta d\theta dk$$
$$= \frac{\gamma\hbar^2}{2\pi\omega_0^3} \int_0^{\omega_M} \omega_k^3 g(\omega_k) d\omega_k \tag{5.62}$$

It is easy to convince oneself that results (5.61) and (5.62) are valid for real $\mathbf{e}_{\mathbf{k}j}$ and for real or imaginary μ_{21}. Moreover it is clear that for $g(\omega_k) = \delta(\omega_k - \omega_0)$ the two sums in (5.61) and (5.62) are equal. Substitution into (5.58) yields in the minimal as well as in the multipolar case

$$\langle \dot{S}_z(t) \rangle = -\gamma(\langle S_z(t) \rangle + 1/2) \tag{5.63}$$

Thus our approximations, which retain only the contribution of energy-conserving terms to the time evolution of $\langle S_z \rangle$, as is evident in view of the appearance of the δ function in (5.58), yield exactly the same equation for $\langle S_z(t) \rangle$. This might have easily been predicted, since the two expressions (4.105) for $\epsilon_{\mathbf{k}j}$ coincide for $\omega_k = \omega_0$. For the same reason in our approximation the counterrotating terms, which are evidently non-energy conserving, do not contribute to the dynamics of $\langle S_z \rangle$.

The solution of (5.63) is immediate and it yields the well-known Wigner-Weisskopf exponential decay law which is sketched in Figure 5.3 for $\langle S_z(0) \rangle = 1/2$ and which is more generally given by

$$\langle S_z(t) \rangle = -\frac{1}{2} + (\langle S_z(0) \rangle + 1/2) e^{-\gamma t} \tag{5.64}$$

Thus γ is called spontaneous decay rate in free space. It is appropriate to compare this exponential and irreversible behaviour of the excited atom in free space with the periodic and reversible behaviour of the atom in a single-mode cavity obtained in the previous section.

The simultaneous use of Born and Markoff approximations is not always equivalent to the RWA. This can be seen by considering the time development of the transverse component of $\langle \mathbf{S} \rangle$. The Heisenberg equation of motion for S_+ is

$$\dot{S}_+ = i\omega_0 S_+ - \frac{2i}{\hbar} \sum_{\mathbf{k}j} \epsilon_{\mathbf{k}j}^* \left(a_{\mathbf{k}j}^\dagger + \lambda a_{\mathbf{k}j} \right) S_z \tag{5.65}$$

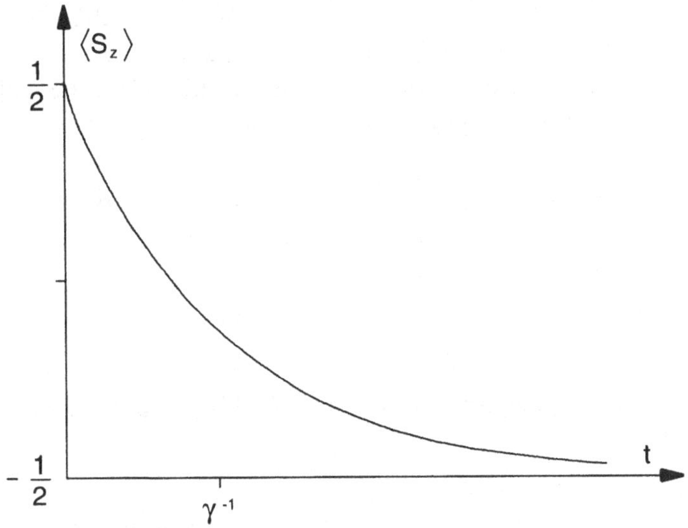

Fig. 5.3 The Wigner-Weisskopf exponential decay law for a two-level atom and $\langle S_z(0)\rangle = 1/2$.

and the Heisenberg equation for $a_{\mathbf{k}j}S_z$, in the Born approximation, is

$$\left(a_{\mathbf{k}j}\dot{S}_z\right) = -i\omega_k a_{\mathbf{k}j}S_z + \frac{i}{2\hbar}\left(\epsilon^*_{\mathbf{k}j}S_- - \lambda\epsilon_{\mathbf{k}j}S_+\right) \tag{5.66}$$

The solution of (5.66) is

$$\begin{aligned}(a_{\mathbf{k}j}S_z)_t &= (a_{\mathbf{k}j}S_z)_0 e^{-i\omega_k t} \\ &+ \frac{i}{2\hbar}e^{-i\omega_k t}\int_0^t e^{i\omega_k t'}\left[\epsilon^*_{\mathbf{k}j}S_-(t') - \lambda\epsilon_{\mathbf{k}j}S_+(t')\right]dt'\end{aligned} \tag{5.67}$$

A Markoffian equation is obtained by using the free evolution of the atomic system as

$$S_-(t') \simeq S_-(t)e^{-i\omega_0(t'-t)} \tag{5.68}$$

and by replacing, for $t \gg \omega_0^{-1}$, the upper limit of integration by $+\infty$. Thus (5.67) is approximated as

$$\begin{aligned}\left(a_{\mathbf{k}j}S_z\right)_t &= \left(a_{\mathbf{k}j}S_z\right)_0 e^{-i\omega_k t} + \frac{i}{2\hbar}\left\{\epsilon^*_{\mathbf{k}j}S_-(t)\int_0^\infty e^{-i(\omega_k-\omega_0-i\eta)\tau}d\tau\right. \\ &\left. - \lambda\epsilon_{\mathbf{k}j}S_+(t)\int_0^\infty e^{-i(\omega_k+\omega_0-i\eta)\tau}d\tau\right\} \\ &= \left(a_{\mathbf{k}j}S_z\right)_0 e^{-i\omega_k t} + \frac{1}{2\hbar}\left\{\epsilon^*_{\mathbf{k}j}S_-(t)\frac{1}{\omega_k-\omega_0-i\eta}\right. \\ &\left. -\lambda\epsilon_{\mathbf{k}j}S_+(t)\frac{1}{\omega_k+\omega_0-i\eta}\right\}\end{aligned} \tag{5.69}$$

where $\tau = t - t'$ and η is a positive and real infinitesimal which has been introduced in order to regularize the integrals. Substituting (5.69) and its h.c. into (5.65), taking the quantum average on the initial state with the atom excited and no photons in the field, taking μ_{21} as real and letting $\eta \to 0$ leads to

$$\langle \dot{S}_+ \rangle = i\omega_0 \langle S_+ \rangle + \langle S_+ \rangle [-i(\Omega_- - \lambda^2 \Omega_+) - \gamma/2] + \lambda \langle S_- \rangle [i(\Omega_- - \Omega_+) - \gamma/2] \tag{5.70}$$

where

$$\Omega_\pm = \frac{1}{\hbar^2} \sum_{\mathbf{k}j} | \epsilon_{\mathbf{k}j} |^2 P \frac{1}{\omega_k \pm \omega_0} \tag{5.71}$$

Neglecting the coupling term in $\langle S_- \rangle$ in (5.70) this equation can be solved immediately as

$$\langle S_+(t) \rangle = \langle S_+(0) \rangle e^{i(\omega_0 - \Omega_- + \lambda^2 \Omega_+)t} e^{-\gamma t/2} \tag{5.72}$$

Thus the transverse components of $\langle \mathbf{S} \rangle$ are seen to be damped in a time $\sim 2/\gamma$, while oscillating at a frequency which differs from the bare-atom value by the amount $\Omega_- - \lambda^2 \Omega_+$. Thus the two-level atom behaves as if in the presence of a renormalization of the bare atom resonance frequency ω_0. The corresponding energy shift is the Lamb shift that we shall discuss in more detail in the next chapter, and it is a generalization of the prototype concept introduced for the atom in a single-mode cavity. This shift is seen to arise both from the rotating (Ω_-) and from the counterrotating terms (Ω_+). The limits of this "dressing" concept will however become evident in Section 6.5.

On theoretical grounds, deviations from a simple exponential decay law should be expected in view of the approximations adopted to obtain (5.64) and (5.72). Both short- and long-time deviations have been discussed. It has been argued that the time derivative of $\langle S_z(t) \rangle$ should vanish at $t = 0$, which is incompatible with exponential decay at very short times (Fonda *et al.* 1978, Gaemers and Visser 1988 and references therein). Long-time deviations proportional to t^{-2} have also been predicted (see e.g. Knight and Milonni 1976, Nussenzweig 1984, Davidovich 1975 and references therein). The smallness of these effects prevents their experimental detection at present. Another interesting effect is the so-called Zeno effect, which is related to the the vanishing of the time derivative of $\langle S_z(t) \rangle$ at $t = 0$. This is the inhibition of spontaneous decay by frequent measurement (Misra and Sudarshan 1977) which may have been observed recently (Hitano *et al.* 1990), although the interpretation of the experimental results is still a matter of debate (Petrosky *et al.* 1991).

Finally, we turn to consider the spectrum of the radiation emitted by the atom during the spontaneous decay process. To this aim, we first solve the equation of motion for S_+, which is obtained, as already mentioned, by substituting (5.69) and its h.c. into (5.65) as

$$\dot{S}_+ = i\omega_0 S_+ - \frac{i}{\hbar^2}\sum_{\mathbf{k}j} \mid \epsilon_{\mathbf{k}j} \mid^2$$
$$\times \left\{ S_+ \left(\frac{1}{\omega_k - \omega_0 + i\eta} - \lambda^2 \frac{1}{\omega_k + \omega_0 - i\eta} \right) \right.$$
$$\left. -\lambda S_- \left(\frac{1}{\omega_k - \omega_0 - i\eta} - \frac{1}{\omega_k + \omega_0 + i\eta} \right) \right\}$$
$$-\frac{2i}{\hbar}\sum_{\mathbf{k}j} \epsilon^*_{\mathbf{k}j} \left\{ \left(a^\dagger_{\mathbf{k}j} S_z \right)_0 e^{i\omega_k t} + \lambda \left(a_{\mathbf{k}j} S_z \right)_0 e^{-i\omega_k t} \right\} \tag{5.73}$$

This equation is approximated by neglecting all terms in S_- and $a_{\mathbf{k}j}$. Occasionally this is called a rotating wave approximation at second level (Ackerhalt *et al.* 1973, Ackerhalt and Eberly 1974). This yields, letting $\eta \to 0$,

$$\dot{S}_+ = i\omega_0 S_+ - i(\Omega_- - \lambda^2 \Omega_+ - i\gamma/2) S_+ - \frac{2i}{\hbar}\sum_{\mathbf{k}j} \epsilon^*_{\mathbf{k}j} \left(a^\dagger_{\mathbf{k}j} S_z \right)_0 e^{i\omega_k t} \tag{5.74}$$

Equation (5.74) can be solved exactly, obtaining

$$S_+(t) = S_+(0) e^{i(\omega_0 - \beta)t} - \frac{2i}{\hbar}\sum_{\mathbf{k}j} \epsilon^*_{\mathbf{k}j} \left(a^\dagger_{\mathbf{k}j} S_z \right)_0 F_k(t) \tag{5.75}$$

where

$$\beta = (\Omega_- - \lambda^2 \Omega_+) - i\gamma/2 \, ;$$
$$F_k(t) = e^{i(\omega_0 - \beta)t} \int_0^t e^{-i(\omega_0 - \beta)t'} e^{i\omega_k t'} dt'$$

Moreover, from

$$\dot{a}_{\mathbf{k}j} = -\frac{i}{\hbar}\left[a_{\mathbf{k}j}, H \right] = -i\omega_k a_{\mathbf{k}j} - \frac{i}{\hbar}\left(\epsilon^*_{\mathbf{k}j} S_- + \lambda \epsilon_{\mathbf{k}j} S_+ \right) \tag{5.77}$$

one immediately obtains the formal solution

$$a_{\mathbf{k}j}(t) = a_{\mathbf{k}j}(0) e^{-i\omega_k t} - \frac{i}{\hbar}\left\{ \epsilon^*_{\mathbf{k}j} e^{-i\omega_k t} \int_0^t e^{i\omega_k t'} S_-(t') dt' \right.$$
$$\left. + \lambda \epsilon_{\mathbf{k}j} e^{-i\omega_k t} \int_0^t e^{i\omega_k t'} S_+(t') dt' \right\} \tag{5.78}$$

From (5.78) and its h.c. one obtains a rather long expression for $(a^\dagger_{\mathbf{k}j}a_{\mathbf{k}j})_t$, which can be averaged in the initial state devoid of photons, yielding

$$\langle a^\dagger_{\mathbf{k}j}a_{\mathbf{k}j}\rangle_t = \frac{1}{\hbar^2} \mid \epsilon_{\mathbf{k}j} \mid^2 \int_0^t \int_0^t e^{-i\omega_k(t'-t'')} \{\langle S_+(t')S_-(t'')\rangle$$
$$- \lambda\langle S_+(t')S_+(t'')\rangle - \lambda\langle S_-(t')S_-(t'')\rangle$$
$$+\lambda^2\langle S_-(t')S_+(t'')\rangle\} dt'dt'' \qquad (5.79)$$

Noting that (5.79) is already $O(\epsilon^2)$, one can substitute in it the zero-order solutions $S_+(t) = S_+(0)e^{i(\omega_0-\beta)t}$, which is within the limits of accuracy of our procedure. Thus the contribution of terms in λ to (5.79) vanishes while the contribution of the term in λ^2 can be evaluated roughly by neglecting β as

$$\lambda^2 \frac{1}{\hbar^2} \mid \epsilon_{\mathbf{k}j} \mid^2 \langle S_-(0)S_+(0)\rangle \int_0^t \int_0^t e^{-i\omega_k(t'-t'')} e^{-i\omega_0(t'-t'')} dt'dt''$$
$$= \lambda^2 \frac{1}{\hbar^2} \mid \epsilon_{\mathbf{k}j} \mid^2 \langle S_-(0)S_+(0)\rangle \left| \int_0^t e^{-i(\omega_k+\omega_0)t'} dt' \right|^2$$
$$= \lambda^2 \frac{1}{\hbar^2} \mid \epsilon_{\mathbf{k}j} \mid^2 \langle S_-(0)S_+(0)\rangle \frac{\left|e^{-i(\omega_k+\omega_0)t} - 1\right|^2}{(\omega_k+\omega_0)^2} \qquad (5.80)$$

Thus this contribution contains a large nonvanishing denominator, and it is presumably small with respect to the first term in (5.79). This is hardly surprising, since the two λ factors indicate that this contribution comes from the counterrotating terms, which do not normally yield real photons since they do not conserve energy. In conclusion, one is left with (Cohen-Tannoudji 1977)

$$\langle a^\dagger_{\mathbf{k}j}a_{\mathbf{k}j}\rangle_t = \frac{1}{\hbar^2} \mid \epsilon_{\mathbf{k}j} \mid^2 \int_0^t \int_0^t e^{-i\omega_k(t'-t'')} \langle S_+(t')S_-(t'')\rangle dt'dt''$$
$$= \frac{1}{\hbar^2} \mid \epsilon_{\mathbf{k}j} \mid^2 \int_0^t \int_0^t e^{-i\omega_k(t'-t'')} \langle S_+(t')S_-(t'')\rangle$$
$$\times [\theta(t'-t'') + \theta(t''-t')] dt'dt''$$
$$= \frac{1}{\hbar^2} \mid \epsilon_{\mathbf{k}j} \mid^2 \left\{ \int_0^t \int_0^{t''} e^{-i\omega_k(t'-t'')} \langle S_+(t')S_-(t'')\rangle dt'dt'' \right.$$
$$\left. + \int_0^t \int_0^{t'} e^{-i\omega_k(t'-t'')} \langle S_+(t')S_-(t'')\rangle dt'dt'' \right\}$$
$$= \frac{2}{\hbar^2} \mid \epsilon_{\mathbf{k}j} \mid^2 \mathrm{Re}\left\{ \int_0^t \int_0^{t'} e^{-i\omega_k(t'-t'')} \langle S_+(t')S_-(t'')\rangle dt''dt' \right\} \qquad (5.81)$$

where Re(z) stands for the real part of z. Using the same substitution as before one has, after some algebra and for $t \gg \gamma^{-1}$,

$$\langle a^{\dagger}_{\mathbf{k}j} a_{\mathbf{k}j} \rangle_t = \frac{1}{\hbar^2} \mid \epsilon_{\mathbf{k}j} \mid^2 \langle S_+(0) S_-(0) \rangle \frac{1}{(\omega_k - \omega_0 + \mathrm{Re}\beta)^2 + \gamma^2/4} \; ; \quad (t \gg \gamma^{-1}) \tag{5.82}$$

where $\mathrm{Re}\beta = \Omega_- - \lambda^2 \Omega_+$ is proportional to the Lamb shift (see Section 6.5 and Appendix J). Thus the radiation emitted by the spontaneously decaying atom is not monochromatic at the bare atomic frequency ω_0, but it has a Lorentz-shaped spectrum peaked at the renormalized frequency $\omega_0 - \Omega_- + \lambda^2 \Omega_+$. The contribution Ω_+ is related to the Bloch-Siegert shift (Bloch and Siegert 1940). The width of the Lorentzian is γ as shown in Figure 5.4. The height of the peak is proportional to $\langle S_+(0) S_-(0) \rangle = \langle S_z(0) \rangle + 1/2$, which is the probability of finding the atom excited at $t = 0$. The energy of the excited atomic level may then thought of as being shifted by $\mathrm{Re}\beta$ with respect to the ground state and spread over a range of energy $\hbar\gamma$ because of the interaction with the vacuum fluctuations. The energy spread may also be interpreted as a consequence of the finite lifetime γ^{-1} of the atom in the excited state and

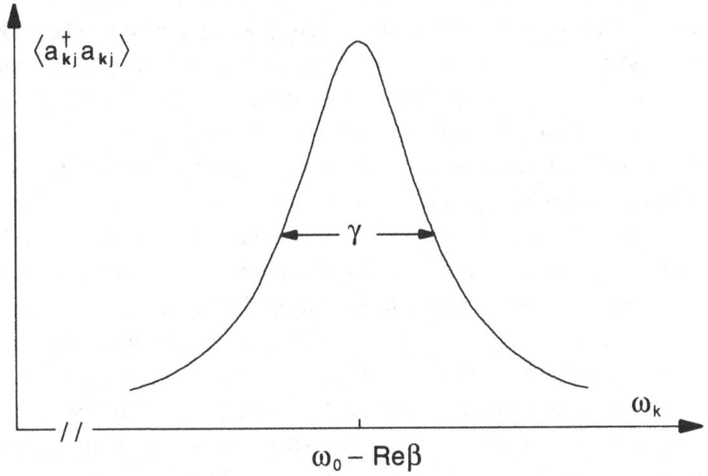

Fig. 5.4 The Lorentzian shape of the spectrum of the light emitted during spontaneous decay of a two-level atom. The width is γ and the position of the peak is shifted with respect to the bare atomic resonance frequency ω_0. The height of the peak is in arbitrary units.

of the Heisenberg time-energy uncertainty relation, since

$$\Delta E \sim \hbar/\Delta t = \hbar\gamma \tag{5.83}$$

It should be stressed that, as is evident from the form of (5.81), the RWA would yield the same form for the spectrum of the spontaneously emitted line, but that $\mathrm{Re}\beta$ would be simply Ω_-, the Lamb shift in the RWA being contributed only by the energy-conserving terms. Since this is in practice a tiny effect, and since in any case the Lamb shift may be incorporated in a renormalization of the atomic resonant frequency ω_0, in the future we shall work within the RWA when our main interest lies in the spectrum of the spontaneously emitted radiation. It should be remarked at this point that interesting problems remain in connection with the use of different kinds of atom-field couplings related by unitary transformations (Power 1993, Milonni *et al.* 1989).

5.5 Resonance fluorescence and dressed atoms

Suppose that the two-level atom of the previous section is immersed in an external monochromatic radiation field of frequency ω in the neighbourhood of ω_0. Here ω is the frequency of one of the $\mathbf{k}j$ normal modes, which we assume to be properly excited, for example by a laser source, the other modes being assumed to be initially devoid of real photons. This situation is realized in an experiment schematically represented in Figure 5.5, where a beam of effective two-level atoms crosses a laser beam. The light is absorbed and re-emitted or scattered into other field modes by the atoms while they are in the intersection region of the two beams. This light is called the fluorescent light and all modes different from that populated by the laser are called fluorescent modes. For normal densities of the atomic beam, the atoms can be regarded as behaving independently of each other. Thus we may appropriately investigate how the dynamics of a two-level atom and its simple spectrum of radiation emitted in vacuum, which we have discussed in Section 5.4, are modified by the presence of the external field. This phenomenon, which in the weak field regime was investigated theoretically some time ago by Heitler (1960) and called resonance fluorescence, will be discussed here in the strong laser field regime. We shall qualify later on what we mean by weak and strong fields, after we have shown that one can treat the laser field and the two-level atom within the dressed atom formalism presented in Section 5.3, and that the phenomenon of resonance fluorescence can be discussed in terms of spontaneous relaxation of the dressed atom.

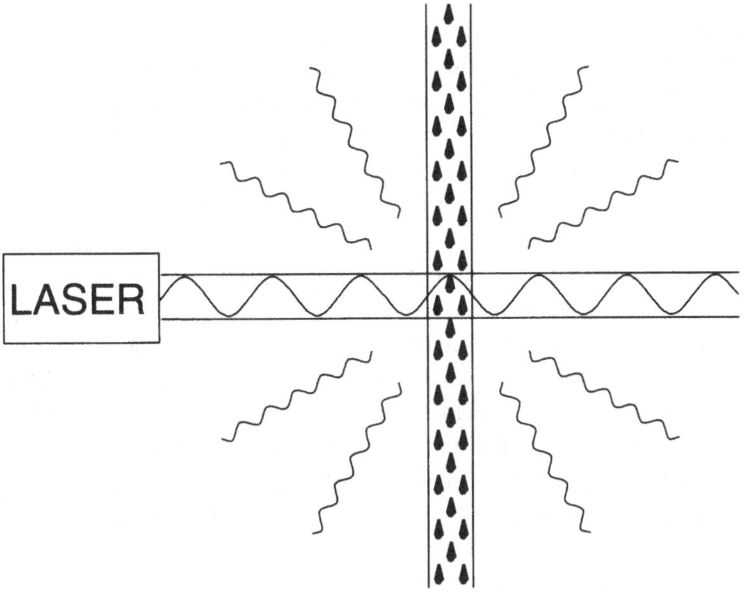

Fig. 5.5 Schematic representation of an experiment of resonance fluorescence. A beam of atoms moving upwards crosses a laser beam at a right angle. Fluorescent radiation is scattered by the atoms in different directions.

It is clear, however, that at low enough laser power one should expect the atom to absorb a single laser photon at a time, which is successively re-emitted spontaneously into the fluorescent modes. Thus energy conservation in this weak-field case ensures that the photon emitted should have the same frequency as the one absorbed, thereby yielding a single-peaked fluorescence spectrum at frequency ω, although it should be mentioned that several subtle problems exist even in the low-intensity regime for which we refer to Heitler's treatment. Thus in the limit of weak laser field it is likely that one can treat both the laser-atom and the atom-fluorescent modes interaction by a perturbative scheme. The same is not true, however, for intensities of the laser field such that the rate of absorption of laser photons becomes comparable with or larger than γ^{-1}, the rate of spontaneous emission *in vacuo*, since then the perturbation induced by the laser on the atomic energy spectrum is nonnegligible with respect to the corresponding spread γ induced by the vacuum fluctuations. In these conditions all we know is that a certain number L of laser photons of frequency ω disappear and that L fluorescent photons appear. Energy conservation tells us only that the total energy of the L fluorescent photons must equal $L\hbar\omega$, but it gives no information as to their individual energy. This means that for the strong laser case it is not correct to treat on the same footing the effects on the two-level atom coming from the laser field and those coming from the fluorescent modes. This in turn suggests that one should treat in an accurate, non-perturbative fashion the laser-photon interaction, and possibly limit perturbation theory to a treatment of the fluorescent modes-atom interaction.

Here the point of view will be adopted of first dressing the atom by the laser field, and successively evaluating the effect of the interaction with the vacuum modes (see e.g. Cohen-Tannoudji 1977), although it should be mentioned that some of the early theories, also capable of giving very accurate results, did not use the concept of dressed atom (see e.g. Mollow 1969 and, more recently, Knight and Milonni 1980). In fact, the main advantage of the dressed atom approach to resonance fluorescence, which is pursued here, is its transparency in physical and mathematical terms. Within this approach it is possible to describe the system by the two-level Hamiltonian which is obtained from (5.51)

$$\begin{aligned} H &= \hbar\omega_0 S_z + \hbar a^\dagger a + \epsilon a S_+ + \epsilon^* a^\dagger S_- \\ &\quad + \sum_{\mathbf{k}j} \hbar\omega_k a^\dagger_{\mathbf{k}j} a_{\mathbf{k}j} + \sum_{\mathbf{k}j} \left(\epsilon_{\mathbf{k}j} a_{\mathbf{k}j} S_+ + \epsilon^*_{\mathbf{k}j} a^\dagger_{\mathbf{k}j} S_- \right) \end{aligned} \tag{5.84}$$

in the RWA and by separating the laser mode of frequency ω (a and $a^\dagger$ operators) from the rest of the $\mathbf{k}j$ vacuum modes. Similarly to the dressed atom framework introduced in Section 5.3 and with the considerations developed earlier in the present section, (5.84) is partitioned as follows

$$\begin{aligned} H &= H_0 + V + \sum_{\mathbf{k}j} \left(\epsilon_{\mathbf{k}j} a_{\mathbf{k}j} S_+ + \epsilon^*_{\mathbf{k}j} a^\dagger_{\mathbf{k}j} S_- \right) ; \\ H_0 &= \hbar\omega a^\dagger a + \hbar\omega_0 S_z + \sum_{\mathbf{k}j} \hbar\omega_k a^\dagger_{\mathbf{k}j} a_{\mathbf{k}j} ; \\ V &= \epsilon a S_+ + \epsilon^* a^\dagger S_- \end{aligned} \tag{5.85}$$

and the eigenstates of H_0 are defined as $| n, \{n_{\mathbf{k}j}\}, \uparrow\downarrow\rangle$, where n is the number of dressing field photons and $\{n_{\mathbf{k}j}\}$ is the photon distribution within the fluorescent modes.

Expressing bare operators in terms of dressed ones as described in (5.36) leads to

$$H_0 + V = \hbar\omega\bar{a}^\dagger\bar{a} + \left(\hbar\omega - \hat{\Delta}\right)\bar{S}_z + \sum_{\mathbf{k}j} \hbar\omega_k a^\dagger_{\mathbf{k}j} a_{\mathbf{k}j} \tag{5.86}$$

as from (5.38). The low-lying eigenvalues of (5.86) with $\{n_{\mathbf{k}j}\} \equiv \{0_{\mathbf{k}j}\}$ are represented in Figure 5.2. One has also to transform the fluorescent part of H in (5.85). Clearly the dressing operator T as defined in (5.34) does not act on the fluorescent mode operators $a_{\mathbf{k}j}$, but it does change atomic operators. Unfortunately the change of S_+ and S_- cannot be obtained exactly in closed form, but it is possible to show that if one neglects $O(\mathcal{N}^{-1/2})$ with respect to $O(1)$, which is a good approximation for a strong laser field, one can approximate the bare operator S_+ in terms of dressed operators as

$$\begin{aligned} S_+ = T^{-1}\bar{S}_+ T = \bar{S}_+ \frac{1}{2}\left(1 + \cos\hat{\theta}\right) + \epsilon^* \bar{a}^\dagger \bar{S}_z \left[\mid \epsilon \mid^2 \mathcal{N}\right]^{-1/2} \\ - \epsilon^{*2}\bar{a}^{\dagger 2}\bar{S}_- \mid \epsilon \mid^2 \mathcal{N} \frac{1}{2}\left(1 - \cos\hat{\theta}\right) \end{aligned} \tag{5.87}$$

Thus (5.85), including the interaction of the dressed atom with the fluorescent field, takes the form

$$\begin{aligned} H &= \hbar\omega\bar{a}^\dagger\bar{a} + \left(\hbar\omega - \hat{\Delta}\right)\bar{S}_z + \sum_{\mathbf{k}j} \hbar\omega_k a^\dagger_{\mathbf{k}j} a_{\mathbf{k}j} + \sum_{i=1}^{3} V_i\ ; \\ V_1 &= \sum_{\mathbf{k}j} \epsilon_{\mathbf{k}j} a_{\mathbf{k}j} \bar{S}_+ \frac{1}{2}\left(1 + \cos\hat{\theta}\right) + \text{h.c.}\ ; \\ V_2 &= \sum_{\mathbf{k}j} \epsilon_{\mathbf{k}j}\epsilon^* a_{\mathbf{k}j} \bar{a}^\dagger \bar{S}_z \left[\mid \epsilon \mid^2 \mathcal{N}\right]^{-1/2} \sin\hat{\theta} + \text{h.c.}\ ; \\ V_3 &= -\sum_{\mathbf{k}j} \epsilon_{\mathbf{k}j}\epsilon^{*2} a_{\mathbf{k}j} \bar{a}^{\dagger 2} \bar{S}_- \mid \epsilon \mid^2 \mathcal{N} \frac{1}{2}\left(1 - \cos\hat{\theta}\right) + \text{h.c.} \end{aligned} \tag{5.88}$$

The increased complication of the atom-fluorescent field interaction in the form (5.88) of H is the price which has to be paid in order to eliminate formally the laser-atom interaction which was explicitly present in (5.84). Such a price, however, is worth paying because the effects of V_i on the dressed atom are amenable to a simple qualitative interpretation. Consider in fact, in a first approximation, the effect of each V_i independently of those of the other two. Clearly V_1 connects pairs of states of the dressed atom with $\Delta\mathcal{N} = \pm 1$, $\Delta\bar{S}_z = \pm 1$; V_2 connects pairs of states with $\Delta\mathcal{N} = \pm 1$, $\Delta\bar{S}_z = 0$; V_3 connects pairs of states with $\Delta\mathcal{N} = \pm 1$, $\Delta\bar{S}_z = \mp 1$ as shown in Figure 5.6 for a pair of neighbouring doublets in the eigenvalue spectrum of the dressed atom and for $\langle\mathcal{N}\rangle \gg 1$. Thus, in the spirit of the first-order time-dependent perturbation approach and in view of energy conservation for the dressed atom-fluorescent field system, V_1 should induce spontaneous decay from $|\,\bar{n}, \bar{\uparrow}\rangle$ to $|\,\bar{n}, \bar{\downarrow}\rangle$ and a fluorescent peak near frequency $\omega_k \sim \omega - \Delta/\hbar$; V_2 should induce spontaneous decay from $|\,\bar{n}+1, \bar{\downarrow}\rangle$ to $|\,\bar{n}, \bar{\downarrow}\rangle$ and from $|\,\bar{n}, \bar{\uparrow}\rangle$ to $|\,\bar{n}-1, \bar{\uparrow}\rangle$, and two coincident fluorescence peaks near frequency $\omega_k \sim \omega$; V_3 should induce decay from $|\,\bar{n}+1, \bar{\downarrow}\rangle$ to $|\,\bar{n}-1, \bar{\uparrow}\rangle$ and a fluorescent peak at frequency $\omega_k \sim \omega + \Delta/\hbar$. In conclusion, on the basis of the naive perturbative treatment proposed above, which is valid near resonance $\omega \sim \omega_0$ and in which the action of each V_i is treated independently of the action of the other two, one should expect the

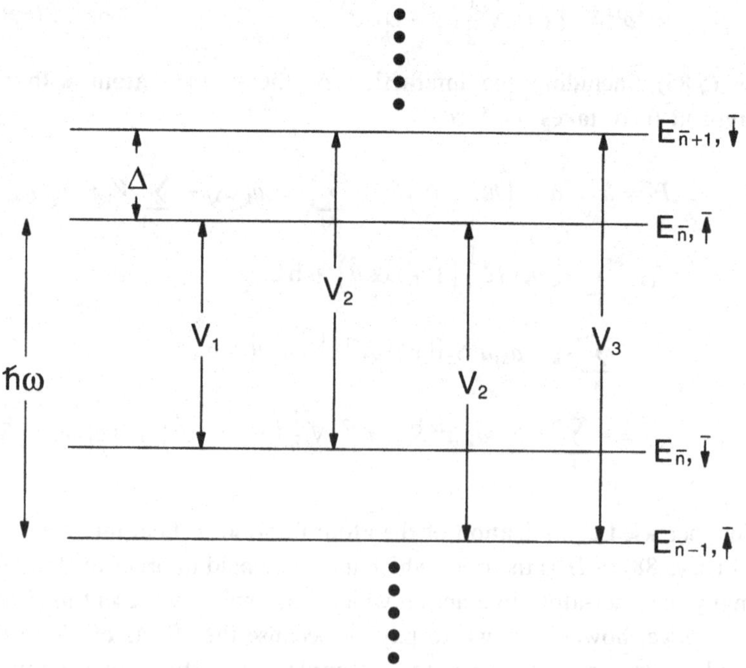

Fig. 5.6 Two neighbouring doublets in the eigenvalue spectrum of the Jaynes-Cummings Hamiltonian, for large numbers of photons. The states belonging to the doublets are connected by different parts V_1, V_2, V_3 of the interaction describing the coupling of the dressed atom to the fluorescent modes, as shown in the figure.

fluorescent radiation spectrum, emitted spontaneously by the dressed atom, to consist of three peaks symmetrically placed about the central frequency $\omega_k = \omega_0$. This is indeed borne out by resonance fluorescence experiments (see e.g. Wu *et al.* 1975, Schuda *et al.* 1974). The width of each peak should be $\sim \gamma$ according to this naive theory and the intensity of the central peak should be twice that of the lateral peaks. These two predictions are at variance with experiment, however, since the experimental width of each of the two lateral peaks is 3/2 that of the central peak and their intensity is only 1/3 the intensity of the central component, as qualitatively shown in Figure 5.7, although the integrated intensity is in the ratio 2:1. The reason for this discrepancy is that each of the V_i has been assumed to act independently of the other two. In fact, it has been shown explicitly (Swain 1975, Cohen-Tannoudji and Reynaud 1977) that all emission acts leading to the same final state during the emission of fluorescent photons interfere, and that taking into account this interference leads to the correct result for the width and intensity of the fluorescent spectrum.

This can be shown quantitatively starting from (5.79) in the form

$$\langle a_{\mathbf{k}j}^{\dagger} a_{\mathbf{k}j} \rangle_t = \frac{2}{\hbar^2} \mid \epsilon_{\mathbf{k}j} \mid^2 \mathrm{Re} \sum_{\ell m} \int_0^t \int_0^{t'} e^{-i\omega_k(t'-t'')} \langle s_\ell^{\dagger}(t') s_m(t'') \rangle dt'' dt' \quad (5.89)$$

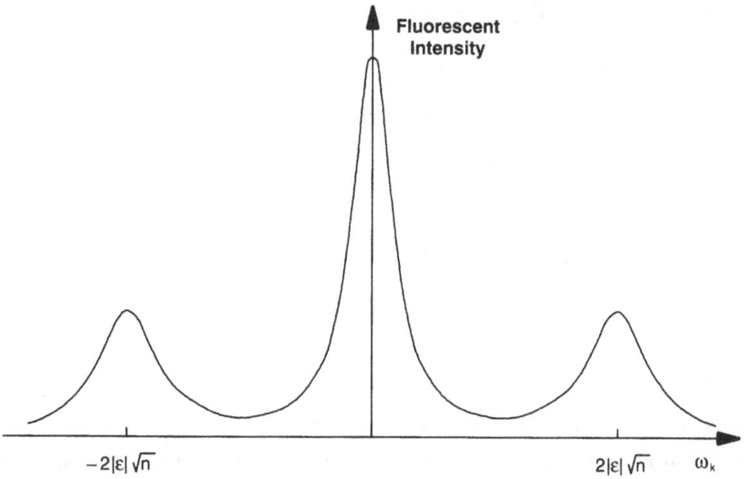

Fig. 5.7 Qualitative representation of the fluorescence spectrum of a two-level atom in arbitrary units. The intensity and width of the two side peaks are 1/3 and 3/2 respectively that of the central line.

where (5.87) has been used and where the dressed operators

$$s_1 = \bar{S}_- \frac{1}{2}\left(1+\cos\hat{\theta}\right) \; ; \; s_2 = \epsilon\bar{a}\bar{S}_z\left[|\,\epsilon\,|^2 \mathcal{N}\right]^{-1/2}\sin\hat{\theta} \; ;$$
$$s_3 = -\epsilon^2\bar{a}^2\bar{S}_+\,|\,\epsilon\,|^2 \mathcal{N}\frac{1}{2}\left(1-\cos\hat{\theta}\right)$$

have been introduced for convenience of notation. The procedure follows rather closely that in Section 5.4 for the spectrum of spontaneous emission, except that here we have three operators s_i instead of a single one S_-. The approximate equation of motion for s_i is obtained as (Leonardi *et al*. 1986)

$$\dot{s}_i = -\frac{i}{\hbar}[s_i, H_0 + V] - \frac{1}{2}\gamma\sum_\ell\left(\left[s_i, s_\ell^\dagger\right]s_\ell - s_\ell^\dagger[s_i, s_\ell]\right) + F(s_i) \qquad (5.90)$$

where the Lamb shifts have been incorporated in the atomic natural frequency ω_0, where γ is the same as in the spontaneous case, and where $F(s_i)$ is an operator whose matrix elements vanish when evaluated between states in which the fluorescent field is devoid of photons. Equation (5.90) shows clearly that the motion of each s_i is influenced by that of the other two. Thus the approach we are following takes into account the interference effects between different transitions in Figure 5.6. In fact, after some algebra needed to express $s_\ell(\ell \neq i)$ in terms of s_i, and taking diagonal matrix elements of (5.90) on eigenstates of $H_0 + V$ in which the fluorescent field is empty, one obtains

$$\langle\dot{s}_1\rangle = -(i\omega_1 + \Gamma_1)\langle s_1\rangle \; ;$$
$$\langle\dot{s}_2\rangle = -(i\omega_2 + \Gamma_2)\langle s_2\rangle - \gamma\sin\theta\cos\theta\langle\mathcal{N}^{-1/2}a(0)\rangle e^{-i\omega t} \; ;$$
$$\langle\dot{s}_3\rangle = -(i\omega_3 + \Gamma_3)\langle s_3\rangle \qquad (5.91)$$

where

$$\omega_1 = \omega - \Delta/\hbar \; ; \; \omega_2 = \omega \; ; \; \omega_3 = \omega + \Delta/\hbar \; ;$$
$$\Gamma_1 = \Gamma_3 = \frac{1}{4}\gamma(3 - \cos^2\theta) \; ; \; \Gamma_2 = \frac{1}{2}\gamma(1 + \cos^2\theta) \qquad (5.92)$$

Thus the effective damping of the three transition operators is not the same as guessed in the naive perturbation approach discussed qualitatively earlier. They also depend on the detuning between the laser and the atomic frequency through the parameter $\cos^2\theta$. This is due to the interference effect between the different transitions. On resonance $\cos\theta = 0$ and $\Gamma_1 = \Gamma_3 = 3\gamma/4$, $\Gamma_2 = \gamma/2$, in agreement with experiments.

Also the correlation functions $\langle s_\ell^\dagger(t')s_m(t'')\rangle$ can be evaluated and substituted in (5.89), with the result

$$\langle a_{\mathbf{k}j}^\dagger a_{\mathbf{k}j}\rangle = \frac{1}{4\hbar^2}\gamma \mid \epsilon_{\mathbf{k}j} \mid^2 \frac{\sin^4\theta}{1+\cos^2\theta}\left\{\frac{1}{4}\frac{3-\cos^2\theta}{(\omega_k-\omega_1)^2+\Gamma_1^2}\right.$$
$$\left.+\frac{1}{4}\frac{3-\cos^2\theta}{(\omega_k-\omega_3)^2+\Gamma_3^2}+\frac{\sin^2\theta}{(\omega_k-\omega_2)^2+\Gamma_2^2}\right\}$$
$$+\frac{2\pi}{\hbar^2}\mid \epsilon_{\mathbf{k}j}\mid^2 \frac{\sin^2\theta\cos^2\theta}{(1+\cos^2\theta)^2}\delta(\omega_k-\omega) \tag{5.93}$$

The part of (5.93) within curly brackets is called the inelastic spectrum. It has the Mollow (1969) three-peaked structure as anticipated and it is in agreement with the experiment as far as intensities and widths of the peaks are concerned. The last term in (5.93), whose presence was not anticipated on the basis of the naive perturbation approach, is due to the laser light scattered elastically by the atom into the vacuum modes which have the same frequency as the laser field. For this reason it is called the elastic part of the spectrum. Its meaning can be better appreciated by solving explicitly (5.91) for $t \gg \gamma^{-1}$, which gives $\langle s_1(t)\rangle = 0$, $\langle s_3(t)\rangle = 0$ and

$$\langle s_2(t)\rangle = -\frac{\sin\theta\cos\theta}{1+\cos^2\theta}\langle \mathcal{N}^{-1/2}a(0)\rangle e^{-i\omega t} \tag{5.94}$$

Thus for large t, $\langle s_1\rangle$ and $\langle s_3\rangle$ are damped to zero, but $\langle s_2\rangle$ keeps oscillating with a finite, time-independent amplitude. Substituting this result into the complex conjugate of the quantum average of (5.87) one has, for $t \gg \gamma^{-1}$, that the average bare atomic dipole reaches a steady state with an amplitude

$$\mid \langle S_-\rangle \mid = \frac{\sin\theta\cos\theta}{1+\cos^2\theta} \tag{5.95}$$

Since the intensity of the elastic component in (5.93) is proportional to $\mid \langle S_-\rangle \mid^2$, it is the amplitude (5.95) which is responsible for the elastically scattered radiation, very much in the same way as a driven harmonic oscillator, consisting of an electric charge bound by elastic forces, acts as a source of scattering for a driving e.m. wave (see e.g. Heitler 1960). Actually the elastic component cannot be infinitely sharp as suggested by (5.93) because of the finite transit time of the two-level atom within the laser beam. It is so sharp, however, that it cannot be detected without severe instrumental distortions, possibly due to the finite response time of the detector of the fluorescent light and to the effects of the driving laser bandwidth.

It should be noted that expression (5.95) coincides with the steady-state value $|\langle S_-\rangle_{st}|$ obtained by solving the Bloch equations (see e.g. Cohen-Tannoudji *et al.* 1992)

$$|\langle S_-\rangle_{st}| = 2\epsilon\langle\mathcal{N}^{1/2}\rangle \frac{|\delta - i\gamma/2|}{\delta^2 + \Delta^2 + \gamma^2/2} \tag{5.96}$$

provided the terms of the order γ/Δ are neglected. This discrepancy is a consequence of the simplified approach used in deriving (5.91) and which we have not discussed here (Compagno and Persico 1980). It should also be noted that equation (5.90) for s_i has been obtained by neglecting terms which oscillate at frequencies $\omega_\ell \neq \omega_i$. This approximation is reminiscent of the rotating wave approximation at the second level we have already used in connection with spontaneous decay in the absence of any external real driving field. Consequently, when the width of each component of the inelastic spectrum (5.93) is large enough to give a substantial overlap between neighbouring peaks, the development in terms of dressed atoms presented in this section breaks down. This provides a scale for driving field strengths, and shows that what we have called here strong field means $\Delta \sim \epsilon\sqrt{n} > \hbar\gamma$. It should be clear now that the validity of the dressed atom picture for resonance fluorescence is limited to fields such that $\epsilon\sqrt{n} > \hbar\gamma$. Modifications of the triplet spectrum (5.93) have been predicted for an atom inside a cavity (Lewenstein *et al.* 1987) and more recently observed (Lange and Walther 1993) and explained theoretically (Agarwal *et al.* 1993).

We can hardly close this session without mentioning briefly recent experimental observations of the vacuum Rabi splitting both in an optical (Thompson *et al.* 1992) and in a microwave cavity (Bernardot *et al.* 1992). This splitting amounts to $2|\epsilon|$ and corresponds to the absorption doublet for single-photon transitions leading from the ground state $|0,\downarrow\rangle$ to the first two excited states $|u_1^{(\pm)}\rangle$ of Figure 5.2.

5.6 Radiative forces on atoms

As a further example of application of the theory of atoms dressed by a real e.m. field, we shall discuss here the forces which are applied by an external e.m. field onto a neutral atom (see e.g. Kazantsev *et al.* 1985). Up to now we have considered an atom as fixed at a defined position **R** in space, that is, we have suppressed the external degrees of freedom of the atoms by which they can change their position in space. In this section we shall restore these external degrees of freedom, endowing the atom with a mass M and with a momentum $\mathbf{P} = -i\hbar\partial/\partial\mathbf{R}$. For such an atom interacting with an external field unbounded in space and in the electric dipole approximation, the Hamiltonian can be obtained by the following simple generalization of the multipolar form (4.98)

$$H = \frac{1}{2M}\mathbf{P}^2 + \sum_n E_n c_n^\dagger c_n + \sum_{\mathbf{k}j} \hbar\omega_k \left(a_{\mathbf{k}j}^\dagger a_{\mathbf{k}j} + 1/2\right)$$
$$-i\sum_{nn'}\sum_{\mathbf{k}j}\sqrt{\frac{2\pi\hbar\omega_k}{V}}(\boldsymbol{\mu}_{nn'}\cdot\mathbf{e}_{\mathbf{k}j}e^{i\mathbf{k}\cdot\mathbf{R}}c_n^\dagger c_{n'}a_{\mathbf{k}j} - \text{h.c.}) \qquad (5.97)$$

and, for an atom-field system enclosed in a perfect cavity, by generalizing (5.21) as

$$H = \frac{1}{2M}\mathbf{P}^2 + \sum_n E_n c_n^\dagger c_n + \sum_\ell \hbar\omega_\ell \left(a_\ell^\dagger a_\ell + 1/2\right)$$
$$-i\sum_{nn'}\sum_\ell\sqrt{\frac{2\pi\hbar\omega_\ell}{V}}(\boldsymbol{\mu}_{nn'}\cdot\mathbf{f}_\ell(\mathbf{R})c_n^\dagger c_{n'}a_\ell - \text{h.c.}) \qquad (5.98)$$

All the superscripts indicating use of the multipolar scheme have been suppressed here for simplicity of notation, and the Röntgen force (Baxter *et al.* 1993) has been neglected, so that the presence of the translational degrees of freedom is indicated only by the atomic kinetic energy term. Reduction of (5.97, 5.98) to the two-level atom model and introduction of the RWA yields, for real $\mathbf{e}_{\mathbf{k}j}$ and $\mathbf{f}_\ell(\mathbf{R})$,

$$H = \frac{1}{2M}\mathbf{P}^2 + \hbar\omega_0 S_z + \sum_{\mathbf{k}j}\hbar\omega_k\left(a_{\mathbf{k}j}^\dagger a_{\mathbf{k}j} + 1/2\right)$$
$$+\sum_{\mathbf{k}j}(\epsilon_{\mathbf{k}j}e^{i\mathbf{k}\cdot\mathbf{R}}a_{\mathbf{k}j}S_+ + \text{h.c.})\ ;$$
$$\epsilon_{\mathbf{k}j} = -i\sqrt{\frac{2\pi\hbar\omega_k}{V}}\boldsymbol{\mu}_{21}\cdot\mathbf{e}_{\mathbf{k}j} \qquad (5.99)$$

$$H = \frac{1}{2M}\mathbf{P}^2 + \hbar\omega_0 S_z + \sum_\ell \hbar\omega_\ell\left(a_\ell^\dagger a_\ell + 1/2\right)$$
$$+\sum_\ell(\epsilon_\ell(\mathbf{R})a_\ell S_+ + \text{h.c.})\ ;$$
$$\epsilon_\ell(\mathbf{R}) = -i\sqrt{\frac{2\pi\hbar\omega_\ell}{V}}\boldsymbol{\mu}_{21}\cdot\mathbf{f}_\ell(\mathbf{R}) \qquad (5.100)$$

The influence of the e.m. field on the external atomic degrees of freedom is easily seen in the Heisenberg representation, since the operator for the atomic velocity

$$\dot{\mathbf{R}} = -\frac{i}{\hbar}[\mathbf{R}, H] = -\frac{i}{\hbar}\left[\mathbf{R}, \frac{1}{2M}\mathbf{P}^2\right] = \frac{1}{M}\mathbf{P} \tag{5.101}$$

does not commute with H. The operator for atomic acceleration is

$$\ddot{\mathbf{R}} = \frac{1}{M}\dot{\mathbf{P}} = -\frac{i}{\hbar}\left[\frac{1}{M}\mathbf{P}, H\right] = -\frac{1}{M}\nabla H_{AF} \tag{5.102}$$

which does not vanish since H_{AF} in (5.99, 5.100), represented by the terms proportional to ϵ, is a function of $\mathbf{R}$. Consequently, one has a force acting on the external degrees of freedom of the atom, corresponding to a force operator which can be defined as (see e.g. Cohen-Tannoudji 1991)

$$\mathbf{F} = -\nabla H_{AF} \tag{5.103}$$

Assuming only one of the infinitely many modes of the e.m. field to be populated by real photons, one can try to simplify the discussion by neglecting all the other modes in (5.99, 5.100). This is permissible only if the effect of the other modes on the atomic dynamics can be considered small in comparison with the corresponding effects coming from the interaction with the populated mode. Hence we shall assume that spontaneous emission by the atom towards the vacuum modes can be completely neglected, which is true for large populations of the only retained mode and for times of interaction of the atom with the field which are not too long. In this way many of the interesting subtleties of the theory of the radiative forces on atoms (Cohen-Tannoudji 1991) will unfortunately be out of reach of the present treatment, the advantage being that the main ideas will be presented here in an easily accessible form.

Using the above ideas leads to considering, for the case of an atom interacting with an unbounded field, the following single-mode Hamiltonian, which is obtained from (5.99)

$$\begin{aligned} H = {} & \frac{1}{2M}\mathbf{P}^2 + \hbar\omega_0 S_z + \hbar\omega_k a^\dagger_{\mathbf{k}j} a_{\mathbf{k}j} \\ & + \epsilon_{\mathbf{k}j} e^{i\mathbf{k}\cdot\mathbf{R}} a_{\mathbf{k}j} S_+ + \epsilon^*_{\mathbf{k}j} e^{-i\mathbf{k}\cdot\mathbf{R}} a^\dagger_{\mathbf{k}j} S_- \end{aligned} \tag{5.104}$$

This is the Hamiltonian of a two-level atom of mass M interacting with a plane wave mode of the field populated with photons, each of which carries a momentum $\hbar\mathbf{k}$. The force operator acting on the atom is obtained from (5.103) as

$$\mathbf{F}(\mathbf{R}) = -i\mathbf{k}\left(\epsilon_{\mathbf{k}j}e^{i\mathbf{k}\cdot\mathbf{R}}a_{\mathbf{k}j}S_{+} - \epsilon^{*}_{\mathbf{k}j}e^{-i\mathbf{k}\cdot\mathbf{R}}a^{\dagger}_{\mathbf{k}j}S_{-}\right) \tag{5.105}$$

On the other hand, the rate of change of the photon number in the mode corresponds to the operator

$$\begin{aligned}\dot{\left(a^{\dagger}_{\mathbf{k}j}a_{\mathbf{k}j}\right)} &= -\frac{i}{\hbar}\left[a^{\dagger}_{\mathbf{k}j}a_{\mathbf{k}j}, H\right] \\ &= \frac{i}{\hbar}\left(\epsilon_{\mathbf{k}j}e^{i\mathbf{k}\cdot\mathbf{R}}a_{\mathbf{k}j}S_{+} - \epsilon^{*}_{\mathbf{k}j}e^{-i\mathbf{k}\cdot\mathbf{R}}a^{\dagger}_{\mathbf{k}j}S_{-}\right)\end{aligned} \tag{5.106}$$

Then the following operator relation, obtained from (5.105) and (5.106),

$$\mathbf{F} = -\hbar\mathbf{k}\dot{n}_{\mathbf{k}j} \tag{5.107}$$

expresses the simple fact that in the interaction process with the strongly populated mode, the atom absorbs from the photons translational momentum as well as energy. This explains in a very simple way the origin of the force acting on the atom due to the interaction with a resonant plane wave of momentum $\mathbf{k}$. This force is called the dissipative force because the energy absorbed by the atom from the strongly populated mode for $\omega_k \sim \omega_0$ is partly channelled into the empty modes of the field by processes of spontaneous emission. Since the latter is not directional because the spontaneous photons can be emitted in all directions, on the average the contribution of spontaneous emission does not change the atomic momentum and the atom can be accelerated under the action of $\langle\mathbf{F}\rangle = -\hbar\mathbf{k}\langle\dot{n}_{\mathbf{k}j}\rangle$, where angular brackets indicate an appropriate quantum average of the bracketed operators. It should be noted that if the atom reaches an internal dynamical equilibrium, then its rate of absorption of resonant $\mathbf{k}j$ photons as well as its rate of spontaneously emitted photons into the empty modes reach a steady state. Since the latter rate of emission cannot exceed γ, as discussed in Section 5.4, and since in steady state conditions the two rates must be equal, the dissipative force acting on the atom is limited by the value $\langle F\rangle \sim \hbar k\gamma$, yielding a maximum atomic acceleration $\langle\ddot{R}\rangle \sim \frac{\hbar k}{M}\gamma$. Since $\hbar k/M$ is the recoil velocity associated with a single act of spontaneous emission or absorption, the expression obtained for $\langle\ddot{R}\rangle$ represents the change in atomic velocity during a fluorescent act lasting a time γ^{-1}. This is the physical explanation of the experimental observation of Frisch

(1933) of deflection of atoms by a light beam. Although the recoil velocity of atoms at optical frequencies is only of the order of $\sim 1\ \mathrm{cm\,s^{-1}}$, the spontaneous emission time is very short, being of the order of 10^{-8}s. As a result the acceleration of suitably prepared atoms due to the dissipative force can be of the order of 10^6 m/s^2, which is quite large. Indeed the experimental techniques for stopping atomic beams using resonant laser radiation are based on this principle (Cohen-Tannoudji 1991).

Consider now an atom interacting with the e.m. field inside a perfect cavity, in which one of the modes is strongly populated. If the interaction with the other modes can be neglected, from (5.100) one has

$$H = \frac{1}{2M}\mathbf{P}^2 + \hbar\omega_0 S_z + \hbar\omega_\ell a_\ell^\dagger a_\ell + \epsilon_\ell(\mathbf{R})a_\ell S_+ + \epsilon_\ell^*(\mathbf{R})a_\ell^\dagger S_- \qquad (5.108)$$

Proceeding by analogy with the plane wave case, one has

$$\mathbf{F}(\mathbf{R}) = -\nabla\epsilon_\ell(\mathbf{R})a_\ell S_+ - \nabla\epsilon_\ell^*(\mathbf{R})a_\ell^\dagger S_- \,;$$

$$\dot{n}_\ell = \frac{i}{\hbar}\left\{\epsilon_\ell(\mathbf{R})a_\ell S_+ - \epsilon_\ell^*(\mathbf{R})a_\ell^\dagger S_-\right\} \qquad (5.109)$$

It is clear that, contrary to the plane wave case, in the cavity mode case there is no obvious relation between $\mathbf{F}$ and $\dot{n}_\ell$. The physical reason is that in the plane wave case absorption of a photon from the strongly populated mode involves changing the atomic momentum in view of overall momentum conservation. On the other hand, the photons in the cavity have no net momentum, and in fact we shall see that the nature of the forces exerted by a standing wave is quite different from that of the forces exerted by a travelling wave. To investigate this new situation we first specialize to a rectangular cavity. The normal modes of this cavity are given in (5.5). Thus from (5.100) one has

$$\begin{aligned}\epsilon_\ell(\mathbf{R}) = -i4\sqrt{\frac{\pi\hbar\omega_\ell}{V}}\{&(\mu_{21})_1 e_1 \cos k_1R_1 \sin k_2R_2 \sin k_3R_3 \\ &+ (\mu_{21})_2 e_2 \sin k_1R_1 \cos k_2R_2 \sin k_3R_3 \\ &+ (\mu_{21})_3 e_3 \sin k_1R_1 \sin k_2R_2 \cos k_3R_3\}\end{aligned} \qquad (5.110)$$

Further, in order to simplify the treatment, we assume that the atom is constrained to move along a straight line with $R_2 = \pi/2k_2$, $R_3 = \pi/2k_3$ and that μ_{21} is pure imaginary, which can always be obtained by an appropriate choice of the phases of the atomic states. With these conditions ϵ_ℓ is real and given by

$$\epsilon_\ell(R_1) = 4\sqrt{\frac{\pi\hbar\omega_\ell}{V}}\mathrm{Im}(\mu_{21})_1 e_1 \cos k_1 R_1 \tag{5.111}$$

where $\mathrm{Im}(z)$ stands for the imaginary part of z. Dropping the superflous subscripts, for the one-dimensional version obtained from (5.108) it is

$$H = \frac{1}{2M}P^2 + \hbar\omega_0 S_z + \hbar\omega a^\dagger a + \epsilon(x)\big(aS_+ + a^\dagger S_-\big) ;$$
$$\epsilon(x) = \epsilon_0 \cos kx ; \; P = \frac{\hbar}{i}\frac{\partial}{\partial x} \tag{5.112}$$

where $R_1 \equiv x$. Hamiltonian (5.112) can be partitioned as

$$H = H_0 + \frac{1}{2M}P^2 ; \; H_0 = \hbar\omega_0 S_z + \hbar\omega a^\dagger a + \epsilon(x)\big(aS_+ + a^\dagger S_-\big) \tag{5.113}$$

and the dressing technique of Section 5.3 can be applied to H_0. In fact through the unitary operator

$$T(x) = \exp\Big\{-\theta(x) 2\mathcal{N}^{1/2}\big(aS_+ - a^\dagger S_-\big)\Big\} \tag{5.114}$$

H_0 can be expressed in terms of the dressed operators defined in (5.36) as

$$H_0 = \hbar\omega(\mathcal{N} - 1/2) + \Big[\delta\cos\hat{\theta}(x) + 2\epsilon(x)\mathcal{N}^{1/2}\sin\hat{\theta}(x)\Big]\bar{S}_z \tag{5.115}$$

with $\delta = \hbar(\omega_0 - \omega)$ provided

$$\hat{\theta}(x) = \arctan\frac{2\epsilon(x)\mathcal{N}^{1/2}}{\delta} \tag{5.116}$$

It is important to remark that, if it can be shown that

$$T^{-1}(x)\frac{1}{2M}P^2 T(x) \sim \frac{1}{2M}P^2 \tag{5.117}$$

then the complete Hamiltonian (5.112) can be put in the form

$$H = \hbar\omega(\mathcal{N} - 1/2) + \frac{1}{2M}P^2 + \hat{V}(x)\bar{S}_z ;$$
$$\hat{V}(x) = \delta\cos\hat{\theta}(x) + 2\epsilon(x)\mathcal{N}^{1/2}\sin\hat{\theta}(x) \tag{5.118}$$

The advantage of form (5.118) for H over form (5.112) is evident, since the internal atomic degrees of freedom in (5.118) are diagonal and can be easily treated, as we shall see in a moment. The point is that (5.117) can be shown to hold true to a good approximation provided we make the choice

$$\sin\hat{\theta}(x) = -\text{sgn}[\epsilon(x)]2\epsilon(x)\mathcal{N}^{1/2}\hat{\Delta}^{-1}\,;$$
$$\cos\hat{\theta}(x) = -\text{sgn}[\epsilon(x)]\delta\hat{\Delta}^{-1}\,\left(|\,\delta\,| \ll 2\epsilon_0\langle\mathcal{N}\rangle^{1/2}\right) \quad (5.119)$$
$$\sin\hat{\theta}(x) = -\text{sgn}(\delta)2\epsilon(x)\mathcal{N}^{1/2}\hat{\Delta}^{-1}\,;$$
$$\cos\hat{\theta}(x) = |\,\delta\,|\,\hat{\Delta}^{-1}\,\left(|\,\delta\,| > 2\epsilon_0\langle\mathcal{N}\rangle^{1/2}\right) \quad (5.120)$$

where $\hat{\Delta} = [\delta^2 + 4\epsilon^2(x)\mathcal{N}]^{1/2}$. Avoiding the mathematical complications arising in the transition region $|\,\delta\,| \sim 2\epsilon_0\mathcal{N}^{1/2}$(Compagno *et al*. 1982), we shall concentrate on the two regions of the parameters δ and $\epsilon_0\mathcal{N}^{1/2}$ defined by (5.120). Substituting (5.119, 5.120) into (5.118) one has

$$\hat{V}(x) = -\text{sgn}(\cos kx)\hat{\Delta}\,\left(|\,\delta\,| \ll 2\epsilon_0\langle\mathcal{N}\rangle^{1/2}\right)\,;$$
$$\hat{V}(x) = \text{sgn}(\delta)\hat{\Delta}\,\left(|\,\delta\,| > 2\epsilon_0\langle\mathcal{N}\rangle^{1/2}\right) \quad (5.121)$$

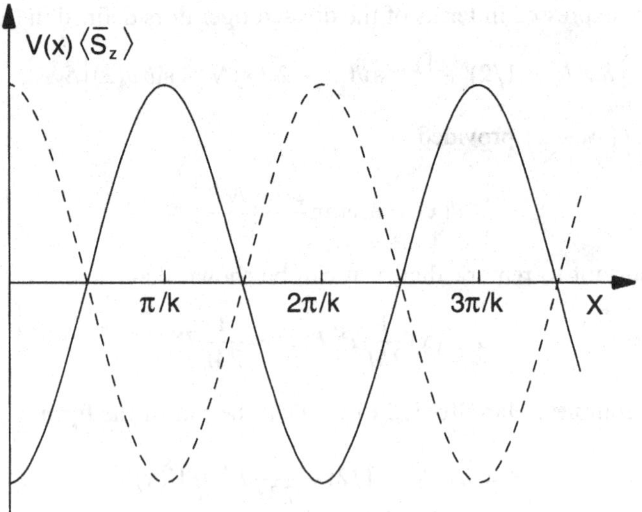

Fig. 5.8 The effective potential acting on a two-level atom in the field of a standing wave of wavelength $\lambda = 2\pi/k$ and for $\delta = 0$ (resonant case). The phase of this periodic potential depends on the internal state of the dressed atom. The continuous and broken lines correspond to $\langle\bar{S}_z\rangle = +1/2$ and $-1/2$ respectively.

The quantum average of H on an eigenstate of $\mathcal{N}$ is, neglecting additive constants,

$$\langle H \rangle \equiv H_{eff} = \frac{1}{2m} P^2 + V(x)\bar{S}_z \; ; \; V(x) = \langle \hat{V}(x) \rangle \tag{5.122}$$

This is an effective Hamiltonian describing the one-dimensional motion of a dressed atom of mass M in a potential which, because of the presence of $\bar{S}_z$, depends on the internal state of the dressed atom in a very simple way. In fact in each of the regions defined by $|\,\delta\,| \ll 2\epsilon_0 \langle \mathcal{N}^{1/2} \rangle$ and by $|\,\delta\,| > 2\epsilon_0 \langle \mathcal{N}^{1/2} \rangle$ the potential seen by the atom is two-valued according to $\langle \bar{S}_z \rangle = \pm 1/2$. The situation is graphically represented in Figures 5.8 and 5.9. The effective potential seen by the atom is periodic, leading to a band structure of the atomic motion. Near resonance, the atom is pushed towards the points $x = 2n\pi/k$ if $\langle \bar{S}_z \rangle = +1/2$ and towards $x = (2n+1)\pi/k$ if $\langle \bar{S}_z \rangle = -1/2$. Off resonance, the period of the

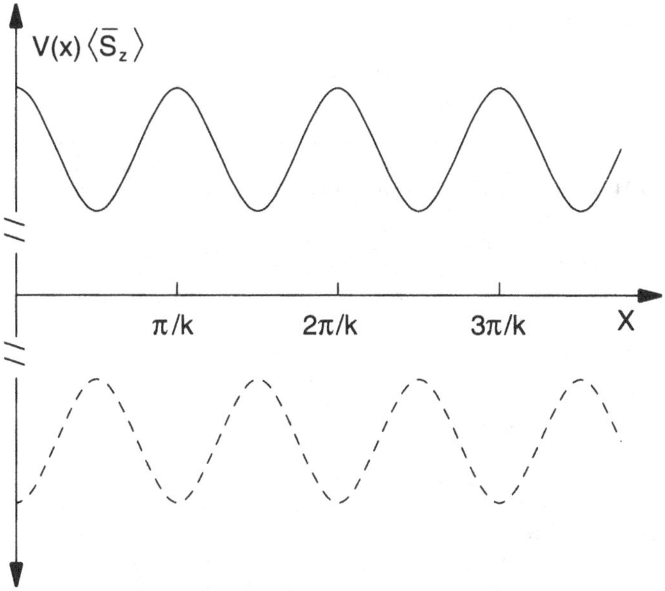

Fig. 5.9 The effective potential acting on a two-level atom in the field of a standing wave of wavelength $\lambda = 2\pi/k$ and $\delta > 2\epsilon_0 \langle \mathcal{N}^{1/2} \rangle$. Both the phase and average value of this periodic potential depend on the internal state of the dressed atom. Continuous and broken lines correspond to $\langle \bar{S}_z \rangle = +1/2$ and $-1/2$ respectively. Note that the period of the effective potential is here half that on resonance.

potential is smaller by a factor 2, and the atom is pushed towards $x = (2n+1)\pi/k$ for $\langle \bar{S}_z \rangle = +1/2$ and towards $x = n\pi/k$ for $\langle \bar{S}_z \rangle = -1/2$. The force acting on the atom is called the reactive force (dipole force). Contrary to the dissipative force, no limit exists for large field intensity, since the slope of the potential increases with $\langle \mathcal{N} \rangle$.

We wish to conclude this section by mentioning some of the early observations of the force by a travelling wave on a atomic beam (see Schieder *et al.* 1972). Moreover important experimental work has more recently been performed on the forces by a standing-wave field on atoms. These forces, due to the space periodicity of the standing wave laser field, have led to the observation of diffractive effects on a beam of neutral Na atoms (Arimondo *et al.* 1979, Moscowitz *et al.* 1983). Atomic beam cooling has also been achieved, exploiting slowing down of spontaneously emitting Cs atoms in a strong standing-wave field by the so-called Sisyphus cooling (Aspect *et al.* 1986). A very thorough review of the semiclassical theory of laser cooling has been published by Stenholm (1986).

References

J.R. Ackerhalt, J.H. Eberly (1974). *Phys. Rev. D* **10**, 3350

J.R. Ackerhalt, P.L. Knight, J.H. Eberly (1973). *Phys. Rev. Lett.* **30**, 456

G.S. Agarwal, W. Lange, H. Walther (1993). *Phys. Rev. A* **48**, 4555

L. Allen, J.H. Eberly (1987). *Optical Resonance and Two-Level Atoms* (Dover Publications Inc., New York)

P.W. Anderson (1963). *Concepts of Solids* (W.A. Benjamin Inc., New York)

E. Arimondo, H. Lew, T. Oka (1979). *Phys. Rev. Lett.* **43**, 753

A. Aspect, J. Dalibard, A. Heidman, C. Salomon, C. Cohen-Tannoudji (1986). *Phys. Rev. Lett.* **57**, 1688

M.A. Babiker, E.A. Power, T. Thirunamachandran (1974). *Proc. Roy. Soc. A* **338**, 235

C. Baxter, M. Babiker, R. Loudon (1993). *Phys. Rev. A* **47**, 1278

F. Bernardot, P. Nussenzweig, R. Brune, J.M. Raimond, S. Haroche (1992). *Europhys. Lett.* **17**, 33

F. Bloch, A.J.F. Siegert (1940). *Phys. Rev.* **57**, 522

C. Cohen-Tannoudji (1968). In *Cargese Lectures in Physics* vol. 2, M. Levy (ed.) (Gordon and Breach, New York)

C. Cohen-Tannoudji (1977). In *Frontiers in Laser Spectroscopy*, R. Balian, S. Haroche, S. Liberman (eds.) (North-Holland Publishing Company, Amsterdam), p. 5

C. Cohen-Tannoudji (1991). In *Fundamental Systems in Quantum Optics,* Les Houches 1990, J. Dalibard, J.M. Raimond, J. Zinn-Justin (eds.), (Elsevier Science Publishers B.V.)

C. Cohen-Tannoudji, J. Dupont-Roc, G. Grynberg (1992). *Atom-Photon Interactions* (John Wiley and Sons Inc., New York)

C. Cohen-Tannoudji, S. Haroche (1969). *J. Physique* **30**, 125, 153
C. Cohen-Tannoudji, S Reynaud (1977). *J. Phys. B* **10**, 345
G. Compagno, J.S. Peng, F. Persico (1982). *Phys. Rev. A* **26**, 2065
G. Compagno, F. Persico (1980). *Phys. Rev. A* **22**, 2108
G. Compagno, F. Persico (1984). *Acta Phys. Austr.* **56**, 21
L. Davidovich (1975). PhD Thesis, Rochester
L. Fonda, G.C. Ghirardi, A. Rimini (1978). *Rep. Progr. Phys.* **41**, 587
O.R. Frisch (1933). *Z. Phys.* **86**, 42
K.J.F. Gaemers, T.D. Visser (1988). *Physica A* **153**, 234
H. Haken (1988). *Quantum Field Theory of Solids* (North-Holland Publishing Company, Amsterdam)
W. Heitler (1960). *The Quantum Theory of Radiation* (Oxford University Press, London)
W.M. Hitano, D.J. Heinzen, J.J. Bollinger, D.J. Wineland (1990). *Phys. Rev. A* **41**, 2295
E.T. Jaynes, F.W. Cummings (1963). *Proc. IEEE* **51**, 89
A.P. Kazantsev, G.A. Ryakenko, G.I. Surdutovich, V.P. Yakovlev (1985). *Phys. Rep.* **129**, 75
A.P. Kazantsev, G.I. Surdutovich, V.P. Yakovlev (1990). *Mechanical Action of Light on Atoms* (World Scientific, Singapore)
C. Kittel (1963). *Quantum Theory of Solids* (John Wiley and Sons Inc., New York)
P.L. Knight, P.W. Milonni (1976). *Phys. Lett. A* **56**, 275
P.L. Knight, P.W. Milonni (1980). *Phys. Rep.* **66**, 22
W. Lange, H. Walther (1993). *Phys. Rev. A* **48**, 4551
C. Leonardi, F. Persico, G. Vetri (1986). *Rivista Nuovo Cimento* **9**, 1
M. Lewenstein, T.W. Mossberg, R.J. Glauber (1987). *Phys. Rev. Lett.* **59**, 775
R. Loudon (1981). *The Quantum Theory of Light* (Oxford University Press, London)
W.H. Louisell (1990). *Quantum Statistical Properties of Radiation* (John Wiley and Sons Inc., New York)
D. Meschede (1992) *Phys. Rep.* **211**, 201
P.W. Milonni, R.J. Cook, J.R. Ackerhalt (1989). *Phys. Rev. A* **40**, 3764
B. Misra, E.C.G. Sudarshan (1977). *J. Math. Phys.* **18**, 756
B.R. Mollow (1969). *Phys. Rev.* **188**, 1969
P.E. Moscowitz, P.L. Gould, S.R. Atlas, D.E. Pritchard (1983). *Phys. Rev. Lett.* **51**, 370
H.M. Nussenzweig (1984). In *QED and Quantum Optics*, A.O. Barut (ed.) (Plenum Press, New York), p. 341
T. Petrosky, S. Tasaki, I. Prigogine (1991). *Physica A* **170**, 306
E.A. Power (1993) In *Physics and Probability*, W.T. Grandy, Jr. and P.W. Milonni (eds.) (Cambridge University Press), p. 101
E.A. Power, T. Thirunamachandran (1982). *Phys. Rev. A* **25**, 2473
E.A. Power, S. Zieanu (1959). *Phil. Trans. Roy. Soc. A* **251**, 427
I.I. Rabi (1937). *Phys. Rev.* **51**, 652
G. Rempe, H. Walther, N. Klein (1987). *Phys. Rev. Lett.* **58**, 353
R. Schieder, H. Walther, L. Wöste (1972). *Opt. Commun.* **5**, 337
F. Schuda, C.R. Stroud, M. Hercher (1974). *J. Phys. B* **7**, L198
B.W. Shore, P.L. Knight (1993). *J. Mod. Opt.* **40**, 1195
S. Swain (1975). *J. Phys. B* **8**, L437
R.J. Thompson, G. Rempe, H.J. Kimble (1992). *Phys. Rev. Lett.* **68**, 1132
L. Van Hove (1955). *Physica* **21**, 901

L. Van Hove (1956). *Physica* **22**, 343
R.G. Woolley (1971). *Proc. Roy. Soc. A* **321**, 557
F.Y. Wu, R.E. Grove, E. Ezekiel (1975). *Phys. Rev. Lett.* **35**, 1426

Further reading

An account of resonant processes with particular attention to Rabi oscillations is given by
P. Stehle, *Phys. Rep.* **156**, 67 (1987).
A review of experimental results of cavity QED has recently been given by
H. Walther, *Phys. Rep.* **219**, 263 (1992).
Theoretical analysis and experimental results on the micromaser can be found in the following papers
P. Filipowicz, J. Javanainen, P. Meystre, *Opt. Commun.* **58**, 327 (1986)
J. Krause, M.O. Scully, H. Walther, *Phys. Rev. A* **34**, 2032 (1986)
G. Rempe, H. Walther, *Phys. Rev A* **42**, 1650 (1990)
G. Rempe, F. Schmidt-Kaler, H. Walther, *Phys. Rev. Lett.* **64**, 2783 (1990).
Squeezing in the JC model has recently been reviewed by
F.L. Kien, A.S. Shumovsky, *Int. J. Mod. Phys. B* **5**, 2287 (1991).
A detailed account of the theory of spontaneous emission can be found in
G.S. Agarwal, *Quantum Statistical Theories of Spontaneous Emission and Their Relation To Other Approaches* (Springer-Verlag, Berlin 1974).
A recent discussion at an elementary level of nonexponential features in decaying systems has been given by
D.S. Onley, A. Kumar, *Am. J. Phys.* **60**, 432 (1992).
Dressed state lasers and masers have been discussed by
G.S. Agarwal, *Phys. Rev. A* **42**, 686 (1990).
The concept of dressed state has been applied to the theory of multiphoton ionization by
P.E. Coleman, P.L. Knight, *Phys. Lett. A* **81**, 379 (1981); *J. Phys. B* **14**, 2139 (1981); *J. Phys. B* **15**, L235 (1982).
An early discussion of the shift of atomic levels due to interaction with a real radiation field has been given by
C. Cohen-Tannoudji, A. Kastler, in *Progress in Optics* vol. 5, E. Wolf (ed.) (North-Holland Publishing Company, Amsterdam 1966) p. 1.
A discussion of the theory of dressed photons is given by
L. Lo Cascio, F. Persico, *J. Mod. Opt.* **39**, 87 (1992).
Time-dependent resonance fluorescence has been investigated by, among others,
J.H. Eberly, C.V. Kunasz, K. Wodkiewicz, *J. Phys. B* **13**, 217 (1980).
A thorough review of the semiclassical theory of laser cooling is by
S. Stenholm, *Rev. Mod. Phys.* **58**, 699 (1986).
A qualitative discussion of laser cooling has been given by
C. Cohen-Tannoudji, W.D. Phillips, *Physics Today*, Oct. 1990, p. 33.
A useful introduction to the theory of non-Markoffian processes is given by
L. Takacs, *Stochastic Processes* (Methuen and Co. Ltd., London 1960).
For a review of new directions in the field of mechanical effects of light on atoms, see
D.F. Walls, M.J. Collett, E.P. Storey, S.M. Tan, *Aust. J. Phys.* **46**, 61 (1993).

6
Dressing by zero-point fluctuations

Introduction. The purpose of Chapter 6 is to discuss from a general point of view the dressing of a source by the vacuum fluctuations of the field coupled to the source. In Section 6.1 we show that in quantum optics, as well as in different branches of physics, virtual quanta of the field are present in the ground state of the source-field system. Three examples are considered: a two-level atom coupled to the vacuum electromagnetic field, a static model of a nucleon coupled to the vacuum meson field and an electron coupled to the optical phonon modes of a semiconductor (Fröhlich polaron). Section 6.2 is dedicated to a qualitative discussion of the physical nature of these virtual quanta and of their spatial distribution around the source. The dressed source is then defined as the bare source together with the virtual quanta surrounding it. This virtual cloud is shown is Section 6.3 to lead to a change of the energy levels of a nonrelativistic free electron interacting with the vacuum electromagnetic field. This kind of self-energy effect can be represented by a mass renormalization of the free electron. Self-energy effects due to the virtual cloud are discussed in Section 6.4 for each of the three examples of dressed sources considered in Section 6.1. In particular, the shift of the ground-state two-level atom (related to the Lamb shift) is shown to be the same for the minimal coupling and for the multipolar coupling scheme. The problem of the virtual cloud is taken up in Section 6.5 for an excited state of the source, and the difficulties connected with the Lee-Friedrichs model are discussed. Section 6.6 is dedicated to a simplified presentation of the Van Hove theory of the dressed states, which is successively applied to describe various possible configurations of a two-level atom coupled to a radiation field.

6.1 Number of virtual quanta in the ground state

Consider a two-level atom in free space, whose interaction with the electromagnetic field is described by Hamiltonian (5.51). First we wish to obtain by perturbation theory the new ground state of the system which takes into account the presence of the atom-field interaction

$$\begin{aligned} H_{AF} &= \sum_{\mathbf{k}j}\left(\epsilon_{\mathbf{k}j}a_{\mathbf{k}j}S_+ + \epsilon^*_{\mathbf{k}j}a^\dagger_{\mathbf{k}j}S_-\right) \\ &\quad + \lambda\sum_{\mathbf{k}j}\left(\epsilon_{\mathbf{k}j}a^\dagger_{\mathbf{k}j}S_+ + \epsilon^*_{\mathbf{k}j}a_{\mathbf{k}j}S_-\right) \\ &= V_1 + \lambda V_2 \end{aligned} \tag{6.1}$$

The eigenstates of the unperturbed Hamiltonian

$$H_0 = H_A + H_F = \hbar\omega_0 S_z + \sum_{\mathbf{k}j}\hbar\omega_k\left(a^\dagger_{\mathbf{k}j}a_{\mathbf{k}j} + 1/2\right) \tag{6.2}$$

are $\mid \{n_{\mathbf{k}j}\}, \uparrow\downarrow\rangle$ where $\{n_{\mathbf{k}j}\}$ is the photon distribution among the $\mathbf{k}j$ modes of the field. In the RWA, $\lambda = 0$ and the ground state $\mid \{0_{\mathbf{k}j}\}, \downarrow\rangle$ of H_0 is not perturbed by H_{AF}, generalizing the single-mode case discussed in Section 5.3. For $\lambda \neq 0$, the new normalized ground state is, up to terms of order ϵ^2 and using (C.13),

$$\begin{aligned} \mid \{0_{\mathbf{k}j}\}, \downarrow\rangle' &= \Bigg\{1 + \lambda\frac{1}{E_0 - H_0}(1 - P_0)V_2 \\ &\quad + \lambda\frac{1}{E_0 - H_0}(1 - P_0)V_1\frac{1}{E_0 - H_0}(1 - P_0)V_2 \\ &\quad - \frac{\lambda^2}{2}\langle\{0_{\mathbf{k}j}\}, \downarrow\mid V_2\frac{1}{(E_0 - H_0)^2}(1 - P_0)V_2 \mid \{0_{\mathbf{k}j}\}, \downarrow\rangle\Bigg\} \\ &\quad \times \mid 0_{\mathbf{k}j}, \downarrow\rangle \end{aligned} \tag{6.3}$$

where $E_0 = -\hbar\omega_0/2$. Using this expression, it is easy to see that the average number of $\mathbf{k}j$ photons in the ground state of the total Hamiltonian, accurate to terms of order ϵ^2, is

$$\begin{aligned} \langle a^\dagger_{\mathbf{k}j}a_{\mathbf{k}j}\rangle &\equiv {}'\langle\{0_{\mathbf{k}j}\}, \downarrow\mid a^\dagger_{\mathbf{k}j}a_{\mathbf{k}j} \mid \{0_{\mathbf{k}j}\}, \downarrow\rangle' \\ &= \lambda^2\langle\{0_{\mathbf{k}j}\}, \downarrow\mid V_2\frac{1}{E_0 - H_0}a^\dagger_{\mathbf{k}j}a_{\mathbf{k}j}\frac{1}{E_0 - H_0}V_2 \mid \{0_{\mathbf{k}j}\}, \downarrow\rangle \\ &= \frac{\lambda^2 \mid \epsilon_{\mathbf{k}j}\mid^2}{\hbar^2(\omega_0 + \omega_k)^2} \end{aligned} \tag{6.4}$$

Thus the total number of photons in the ground state of $H = H_0 + H_{AF}$, using (5.61) in the minimal coupling scheme, is

$$\langle n \rangle = \frac{\lambda^2}{\hbar^2} \sum_{\mathbf{k}j} \frac{|\epsilon_{\mathbf{k}j}|^2}{(\omega_0 + \omega_k)^2} = \lambda^2 \frac{\gamma}{2\pi\omega_0} \int_0^{\omega_M} \frac{\omega_k}{(\omega_0 + \omega_k)^2} d\omega_k$$
$$= \lambda^2 \frac{\gamma}{2\pi\omega_0^3} \left[\ln \frac{\omega_M + \omega_0}{\omega_0} - \frac{\omega_M}{\omega_M + \omega_0} \right] \tag{6.5}$$

and using (5.62) in the multipolar coupling scheme, is

$$\langle n \rangle = \lambda^2 \frac{\gamma}{2\pi\omega_0^3} \int_0^{\omega_M} \frac{\omega_k^3}{(\omega_0 + \omega_k)^2} d\omega_k$$
$$= \lambda^2 \frac{\gamma}{2\pi\omega_0} \left[3 \ln \frac{\omega_M + \omega_0}{\omega_0} - \frac{\omega_M}{\omega_M + \omega_0} - \frac{2\omega_M}{\omega_0} + \frac{\omega_M^2}{2\omega_0^2} \right] \tag{6.6}$$

The two expressions (6.5, 6.6), although both divergent with ω_M, are different, which is not surprising due to the different physical nature of the photons considered in each of the two schemes as discussed in Section 4.4 (Drummond 1986, Compagno *et al.* 1990). What we want to emphasize here is the presence of photons in the ground state of the system. These photons are entirely due to the presence of the counterrotating terms in (6.1) and disappear in the RWA with $\lambda = 0$. Thus they originate from processes which do not conserve the bare energy, and for this reason they are called virtual quanta.

The existence of virtual quanta in the ground state of a source-field system is by no means limited to atomic physics. For example, the simplest possible model for a nucleon interacting with the neutral meson field is that of the static source linearly coupled to a scalar field (Henley and Thirring 1962, Bolsterli 1991). In this model, which can be solved exactly, the source is fixed at the origin of the reference frame and it is rigid, in the sense that it is described by a source density $\rho(\mathbf{r})$ which is unchanged by the interaction with the meson field. The latter consists of a real Klein-Gordon field, which in second quantization takes the form

$$H_F = \sum_{\mathbf{k}} \hbar\omega_k \left(a_{\mathbf{k}}^\dagger a_{\mathbf{k}} + 1/2 \right) \; ; \; \omega_k = \left[c^2k^2 + m^2c^4/\hbar^2 \right]^{1/2} \; ;$$
$$\left[a_{\mathbf{k}}, a_{\mathbf{k}'}^\dagger \right] = \delta_{\mathbf{k}\mathbf{k}'} \; ; \; [a_{\mathbf{k}}, a_{\mathbf{k}'}] = 0 \tag{6.7}$$

where m is the rest mass of the mesons as in (3.43). The source-field interaction is taken to be of the form

$$H_{SF} = -g \int \rho(\mathbf{x})\phi(\mathbf{x})d^3\mathbf{x} = -g\sum_{\mathbf{k}} \sqrt{\frac{\hbar}{2V\omega_k}}\left(\rho_{\mathbf{k}}^* a_{\mathbf{k}} + \rho_{\mathbf{k}} a_{\mathbf{k}}^\dagger\right) ;$$

$$\rho_{\mathbf{k}} = \int \rho(\mathbf{x})e^{i\mathbf{k}\cdot\mathbf{x}}d^3\mathbf{x} \tag{6.8}$$

where g is the nucleon-meson coupling constant (strong force) and where the field amplitude has been quantized with periodic boundary conditions on the surface of a cubic box of volume V according to

$$\phi(\mathbf{x}) = \sum_{\mathbf{k}} \sqrt{\frac{\hbar}{2V\omega_k}}\left(a_{\mathbf{k}}e^{i\mathbf{k}\cdot\mathbf{x}} + a_{\mathbf{k}}^\dagger e^{-i\mathbf{k}\cdot\mathbf{x}}\right) \tag{6.9}$$

as in (3.45). The total Hamiltonian is $H = H_F + H_{SF}$, since the source has no internal degrees of freedom, and in second-quantized form

$$H = \sum_{\mathbf{k}} \hbar\omega_k\left(a_{\mathbf{k}}^\dagger a_{\mathbf{k}} + 1/2\right) - g\sum_{\mathbf{k}} \sqrt{\frac{\hbar}{2V\omega_k}}\left(\rho_{\mathbf{k}}^* a_{\mathbf{k}} + \rho_{\mathbf{k}} a_{\mathbf{k}}^\dagger\right) \tag{6.10}$$

Thus it can be partitioned into a sum of Hamiltonians, one for each $\mathbf{k}$, which commute with each other. Each of these sub-Hamiltonians corresponds to that of a shifted harmonic oscillator, which can be diagonalized exactly by a unitary shift or coherent displacement operator, of the same kind as that introduced in section 2.3 for the coherent states,

$$D_{\mathbf{k}} = \exp\left\{-g[2\hbar V\omega_k^3]^{-1/2}\left(\rho_{\mathbf{k}}^* a_{\mathbf{k}} - \rho_{\mathbf{k}} a_{\mathbf{k}}^\dagger\right)\right\} \tag{6.11}$$

Thus the ground state $|\,\tilde{0}\rangle$ of H can be obtained from the ground state $|\,\{0_{\mathbf{k}}\}\rangle$ of H_0, where no mesons are present in any of the field modes, as

$$|\,\tilde{0}\rangle = \prod D_{\mathbf{k}}\,|\,\{0_{\mathbf{k}}\}\rangle \tag{6.12}$$

One may now ask how many mesons are present in the new ground state $|\,\tilde{0}\rangle$. The number of mesons in the $\mathbf{k}$ mode is obtained as

$$\langle a_{\mathbf{k}}^\dagger a_{\mathbf{k}}\rangle \equiv \langle\tilde{0}\,|\,a_{\mathbf{k}}^\dagger a_{\mathbf{k}}\,|\,\tilde{0}\rangle = \langle 0\,|\prod_{\mathbf{k}'} D_{\mathbf{k}'}^{-1} a_{\mathbf{k}}^\dagger a_{\mathbf{k}} \prod_{\mathbf{k}''} D_{\mathbf{k}''}\,|\,0\rangle$$

$$= \langle 0\,|\,D_{\mathbf{k}}^{-1}a_{\mathbf{k}}^\dagger D_{\mathbf{k}} D_{\mathbf{k}}^{-1} a_{\mathbf{k}} D_{\mathbf{k}}\,|\,0\rangle \tag{6.13}$$

It is now easy to check that

$$D_{\mathbf{k}}^{-1} a_{\mathbf{k}} D_{\mathbf{k}} = a_{\mathbf{k}} + g\left[2\hbar V \omega_k^3\right]^{-1/2} \rho_{\mathbf{k}} \tag{6.14}$$

Substituting this result into (6.13) one immediately obtains

$$\langle a_{\mathbf{k}}^\dagger a_{\mathbf{k}} \rangle = \frac{g^2 \mid \rho_{\mathbf{k}} \mid^2}{2\hbar V \omega_k^3} \tag{6.15}$$

Thus the total number of mesons in the ground state of H is

$$\langle n \rangle = \sum_{\mathbf{k}} \langle a_{\mathbf{k}}^\dagger a_{\mathbf{k}} \rangle = \frac{g^2}{2\hbar V} \sum_{\mathbf{k}} \frac{\mid \rho_{\mathbf{k}} \mid^2}{\omega_k^3} = \frac{g^2}{2\hbar} \frac{1}{(2\pi)^3} \int \frac{\mid \rho_{\mathbf{k}} \mid^2}{\omega_k^3} d^3\mathbf{k} \tag{6.16}$$

where we have taken the continuum limit according to the usual recipe

$$\frac{1}{V} \sum_{\mathbf{k}} = \frac{1}{(2\pi)^3} \int d^3\mathbf{k} \tag{6.17}$$

In order to proceed further an explicit model for the shape $\rho(\mathbf{r})$ of the source is needed. Assuming for simplicity a uniform spherical source of radius a, centred at the origin

$$\rho(\mathbf{r}) = \frac{3}{4\pi a^3} \theta(a - r) \tag{6.18}$$

one has from (6.8), after elementary integrations,

$$\rho_{\mathbf{k}} = \frac{3}{a^3 k} \int_0^a r \sin kr dk = \frac{3}{a^3 k^3} (\sin ka - ka \cos ka) \tag{6.19}$$

Thus $\rho_{\mathbf{k}}$ can be approximated as ~ 1 for $ka < 1$ and as 0 for $ka > 1$. Consequently in (6.16) with $k_M = a^{-1}$

$$\int \frac{\mid \rho_{\mathbf{k}} \mid^2}{\omega_k^3} d^3\mathbf{k} \sim 4\pi \int_0^{k_M} \frac{k^2}{\omega_k^3} dk = \frac{4\pi}{c^3} \int_0^{k_M} \frac{k^2}{[k^2 + m^2c^2/\hbar^2]^{3/2}} dk \tag{6.20}$$

Moreover, if the source is to represent a nucleon, its core radius a is much smaller than the meson Compton wavelength $\hbar/mc$ and one can further approximate

$$\int_0^{k_M} \frac{k^2}{[k^2 + m^2c^2/\hbar^2]^{3/2}} dk \sim \left(\frac{\hbar}{mc}\right)^3 \int_0^{mc/\hbar} k^2 dk + \int_{mc/\hbar}^{k_M} \frac{1}{k} dk$$
$$= \frac{1}{3} + \ln \frac{\hbar k_M}{mc} \tag{6.21}$$

and finally from (6.16)

$$\langle n \rangle \sim \frac{g^2}{4\pi^2 \hbar c^3}\left(\frac{1}{3} + \ln \frac{\hbar k_M}{mc}\right) \tag{6.22}$$

which is the average number of virtual mesons in the ground state of the model.

Another entirely different system displaying virtual quanta in the ground state is the Fröhlich polaron. In a semiconductor crystal it is possible to excite an electron from the valence to the conduction band. If the crystal possesses a sufficiently ionic character, the electron would be expected to interact mainly with the longitudinal optical phonon modes of the crystal, which are capable of generating large electric polarization fields. Moreover, if the electron created is not too energetic, it is appropriate to approximate the rigid lattice band as parabolic. In these conditions the system can be modelled as (Fröhlich 1963).

$$\begin{aligned} H &= H_0 + V ; \\ H_0 &= \frac{1}{2m}\sum_{\mathbf{p}} p^2 c_{\mathbf{p}}^{\dagger} c_{\mathbf{p}} + \hbar\omega \sum_{\mathbf{k}} a_{\mathbf{k}}^{\dagger} a_{\mathbf{k}} ; \\ V &= \sum_{\mathbf{pk}} \left(V_k a_{\mathbf{k}} c_{\mathbf{p}+\hbar\mathbf{k}}^{\dagger} c_{\mathbf{p}} + V_k^* a_{\mathbf{k}}^{\dagger} c_{\mathbf{p}}^{\dagger} c_{\mathbf{p}+\hbar\mathbf{k}} \right) \end{aligned} \tag{6.23}$$

where m is the effective mass of the slow electron in the rigid lattice, that is in the absence of phonons. The sum over $\mathbf{p}$ runs over the electronic band states of wavevector $\mathbf{p}/\hbar$, $c_{\mathbf{p}}^{\dagger}$ and $c_{\mathbf{p}}$ (Fermi operators) create and annihilate respectively a spinless electron in band state $\mathbf{p}$. ω is the frequency of the longitudinal optical modes (taken equal for all modes for simplicity), $a_{\mathbf{k}}^{\dagger}$ and $a_{\mathbf{k}}$ (Bose operator) create and annihilate respectively an optical phonon with wavevector $\mathbf{k}$. The electron-phonon coupling constant is (Pines 1963)

$$\begin{aligned} V_k &= i\frac{\hbar\omega}{k}\left(\frac{\hbar}{2m\omega}\right)^{1/4}\left(\frac{4\pi\alpha_p}{V}\right)^{1/2} ; \\ \alpha_p &= \frac{e^2}{\hbar\bar{\epsilon}}\left(\frac{m}{2\hbar\omega}\right)^{1/2} ; \ \bar{\epsilon} = \left(\frac{\epsilon(0)\epsilon(\infty)}{\epsilon(0) - \epsilon(\infty)}\right) \end{aligned} \tag{6.24}$$

In the above expression $\epsilon(0)$ and $\epsilon(\infty)$ are respectively the static and the high-frequency dielectric constants of the semiconductor, so that $\bar{\epsilon}$ is its effective dielectric constant, while the physical meaning of α_p will become more clear in the following. Since H in (6.23) obviously conserves the overall wavevector of the system, the operator

$$\mathbf{K} = \frac{1}{\hbar}\sum_{\mathbf{p}} \mathbf{p} c_{\mathbf{p}}^{\dagger} c_{\mathbf{p}} + \sum_{\mathbf{k}} \mathbf{k} a_{\mathbf{k}}^{\dagger} a_{\mathbf{k}} \tag{6.25}$$

commutes with H, and $\mathbf{K}$ is a constant of motion. Thus when using perturbation theory to get approximate expressions of the eigenstates of H, we may limit our consideration to a subspace of H characterized by an eigenvalue $\langle\mathbf{K}\rangle$ of $\mathbf{K}$. Consider for example the state $\mid \mathbf{p} = \hbar\langle\mathbf{K}\rangle, \{0_{\mathbf{k}}\}\rangle$ which is an eigenstate of H_0 in which the electron has a momentum $\hbar\langle\mathbf{K}\rangle$ and there are no phonons. This is not exactly the ground state of the unperturbed Hamiltonian, except for $\langle\mathbf{K}\rangle = 0$, but we may consider it as near to the ground state if $\langle\mathbf{K}\rangle$ is not too large. This state is connected by the electron-phonon coupling to the set of eigenstates of H_0 $\mid \mathbf{p}' = \hbar(\langle\mathbf{K}\rangle - \mathbf{k}), 1_{\mathbf{k}}\rangle$, where there is only one phonon with wavevector $\mathbf{k}$. The energy difference between the states of this set and $\mid \mathbf{p} = \hbar\langle\mathbf{K}\rangle, \{0_{\mathbf{k}}\}\rangle$ is

$$\Delta_{\mathbf{k}} = \frac{1}{2m}p^2 - \frac{1}{2m}p'^2 - \hbar\omega = -\frac{\hbar k^2}{2m} + \frac{\hbar^2}{m}\langle K\rangle k\cos\theta - \hbar\omega \tag{6.26}$$

where θ is the angle between $\langle\mathbf{K}\rangle$ and $\mathbf{k}$. We make sure that (6.26) does not vanish for any k by requiring that the electron is so slow that

$$\frac{1}{2m}p^2 < \hbar\omega \tag{6.27}$$

Thus we are entitled to use perturbation theory at least if $V_{\mathbf{k}}$ is small enough, which we will take to be the case ("small coupling"). Use of the first of (C.13) yields

$$\begin{aligned}
\mid \mathbf{p} = \hbar\langle\mathbf{K}\rangle, \{0_{\mathbf{k}}\}\rangle^{(1)} &= \left\{1 + \frac{1}{\frac{\hbar^2\langle K\rangle^2}{2m} - H_0}(1 - P_0)V\right\} \\
&\quad \times \mid \mathbf{p} = \hbar\langle\mathbf{K}\rangle, \{0_{\mathbf{k}}\}\rangle \\
&= \mid \mathbf{p} = \hbar\langle\mathbf{K}\rangle, \{0_{\mathbf{k}}\}\rangle + \frac{1}{\frac{\hbar^2\langle K\rangle^2}{2m} - H_0} \\
&\quad \times \sum_{\mathbf{k}} V_{\mathbf{k}}^* \mid \mathbf{p}' = \hbar(\langle\mathbf{K}\rangle - \mathbf{k}), 1_{\mathbf{k}}\rangle \\
&= \mid \mathbf{p} = \hbar\langle\mathbf{K}\rangle, \{0_{\mathbf{k}}\}\rangle \\
&\quad + \sum_{\mathbf{k}} \frac{V_k^*}{\Delta_{\mathbf{k}}} \mid \mathbf{p}' = \hbar(\langle\mathbf{K}\rangle - \mathbf{k}), 1_{\mathbf{k}}\rangle
\end{aligned} \tag{6.28}$$

The quantum average of $a_{\mathbf{k}}^{\dagger}a_{\mathbf{k}}$ on (6.28) yields the average number of virtual phonons in mode **k** as

$$\langle a_{\mathbf{k}}^{\dagger}a_{\mathbf{k}}\rangle = \frac{|V_k|^2}{\Delta_{\mathbf{k}}^2} \tag{6.29}$$

and the total number of phonons is, assuming $\langle \mathbf{K}\rangle \sim 0$ and using (6.24) and (6.26),

$$\begin{aligned}\langle n\rangle &= \sum_{\mathbf{k}}\langle a_{\mathbf{k}}^{\dagger}a_{\mathbf{k}}\rangle = \sum_{\mathbf{k}}\frac{|V_k|^2}{\Delta_{\mathbf{k}}^2} = \frac{1}{V}\sum_{\mathbf{k}}\left(\frac{\hbar\omega}{k}\right)^2\frac{4\pi\alpha_p A^{1/2}}{(\hbar\omega+\hbar^2k^2/2m)^2}\\ &= 4\pi\alpha_p A^{1/2}\frac{1}{V}\sum_{\mathbf{k}}\frac{1}{k^2(1+Ak^2)^2}\\ &\sim \frac{2}{\pi}\alpha_p A^{1/2}\int_0^{\infty}\frac{dk}{(1+Ak^2)^2} = \frac{\alpha_p}{2}\end{aligned} \tag{6.30}$$

where we have approximated the integral by extending the upper limit to infinity and where we have put $A = \hbar/2m\omega$. Thus we are led to conclude that phonons exist also in, or near, the ground state of an electron coupled to the phonons of a crystal. We also understand why α_p has been factored off expression (6.24) for V_k, since the total number of phonons obtained using perturbation theory at the lowest possible order is simply related to α_p.

6.2 The physical nature of the virtual cloud

From the examples developed in the previous section, we are led to conclude that the occurrence of quanta of the field in the ground state of source-radiation systems is quite common in physics. This seems to run somehow contrary to intuition, in the sense that one would expect that the lowest possible state of a source coupled to a field should be "dark", or devoid of quanta of radiation. Thus we shall devote this section to qualitative considerations in order to understand the physical nature of the ground state where we have shown quanta of radiation to exist.

It is appropriate to begin by remarking that, contrary to what happens in classical physics, a quantum mechanical source coupled to a field is subjected to quantum fluctuations and that this can have an influence on the field generated by the source. Consider, for example, a set of charges in classical physics constrained within a finite volume V, whose surface forms the boundary between the system and the vacuum. Whatever the initial state of the system, except for particular "nonradiating" dynamical states in which the motion of different parts of the charge leads to destructive interference (Kim and Wolf 1986, Meyer-Vernet 1989) and which are not of interest in the present context, the system will radiate away all of its available energy by electromagnetic processes and it will freeze down in a configuration of zero kinetic energy and of minimum potential energy, if such a minimum exists. In this final configuration, which can be appropriately called the zero-temperature configuration, the field generated in space by the system coincides with the longitudinal field

$$\mathbf{E}_{\parallel}(\mathbf{x}) = -\nabla V(\mathbf{x}) \tag{6.31}$$

where $V(\mathbf{x})$ is the electric potential of the set of charges in the Coulomb gauge. The form of the scalar $V(\mathbf{x})$ is strictly related to the geometrical features of the zero-temperature configuration attained by the charges. Consider now the same set of charges in quantum physics. An electrostatic description of the field inside as well as outside volume V, of the kind provided by (6.31), is obviously incomplete at zero temperature because of the residual zero-point motion of the charges constrained within volume V. In fact this zero-point motion leads to the possibility for the system of absorbing and emitting quanta of transverse electromagnetic radiation, even in its ground state. Naturally these photons cannot behave as normal photons, since energy conservation does not allow them to escape to infinity, where they would carry away energy which is not available because the system is in the ground state. Thus they must remain somehow localized near volume V.

This example can be generalized to different kinds of source-field systems other than electromagnetic, and the localization, in the ground state of the system, of the quanta in a region near to the source or to the system of sources can be understood on the basis of energy considerations. In fact at zero temperature the quanta of the field which appear in the course of a quantum fluctuation lead to an increase of the energy of the source-field system, if the source-field interaction energy is not taken into account. In other words, the bare energy E of the source-field system is not necessarily conserved during such a fluctuation, and states having a bare energy $E' \neq E$ can be excited. The magnitude of the energy imbalance $\delta E = \mid E' - E \mid$, however, is constrained by the Heisenberg uncertainty principle

$$\delta E \sim \hbar / \tau \tag{6.32}$$

where τ is the duration of the fluctuation, in which time the energy balances again and the extra quanta of the field are reabsorbed. It is important to emphasize that we are discussing here fluctuations of a purely quantum nature, which take place also at zero temperature. Since these fluctuations take place continuously, the source can be described as surrounded by a steady-state cloud of virtual quanta continuously emitted into, and reabsorbed from, the field. In contrast with the behaviour of a real quantum, which is emitted during an energy conserving process with $E' = E$, and which in the absence of boundaries can abandon the source forever, a virtual quantum can only attain a finite distance from the source roughly given by

$$r \simeq c\tau \simeq \hbar c/\delta E \tag{6.33}$$

where c is a velocity scale for the quantum. Consequently one should expect the linear dimensions of the virtual cloud surrounding the source to coincide approximately with r given by (6.33) for virtual transitions characterized by an energy imbalance δE.

In the case of the two-level atom discussed in the previous section, the energy imbalance due to the fluctuations induced by the counterrotating terms, which lead the atom from the ground to the excited state and which contribute to the number of virtual photons of mode $\mathbf{k}j$ given by (6.4), is $\delta E = \hbar(\omega_0 + \omega_k)$ and c coincides with the velocity of light. If these are substituted in (6.33), and long-wavelength photons are considered with $\omega_k \ll \omega_0$, then $r \sim c/\omega_0 = \lambda_0/2\pi$, where λ_0 is the wavelength of the atomic frequency involved in the process. Thus, for this kind of fluctuation, and neglecting the possibility of two-photon processes, one obtains typical linear dimensions of the dressed atom which are of the order of 10^{-4} to 10^{-5} cm. On the other hand, in the elementary model for the nucleon which we have considered, the source is rigid; thus the minimum energy required to create a meson from the vacuum is equal to mc^2, while the velocity scale for the bare field is chosen to be again the velocity of light. Introduction of these values into (6.33) yields $r \sim \hbar/mc$ which is the meson Compton wavelength. For π mesons $m \sim 140$ MeV, and one finds $r \sim 1.4 \cdot 10^{-13}$ cm, which is of the right order of magnitude for the experimental mean square radius of the proton and of the neutron (see e.g. Kenyon 1987). Finally in the case of the Fröhlich polaron, which is occasionally called the optical polaron, the electron is much lighter than an atom or a nucleon, and in practice the only mobile particle in view of the small group velocity of the optical phonons. Consequently its recoil cannot be neglected, unlike the previous two cases. Thus the appropriate value for the energy fluctuation to be substituted in (6.33) is $\delta E = \hbar\omega + p^2/2m$. Moreover, one should expect that the strongest interaction is with phonons of wavelength λ such that in a time $2\pi/\omega$ the electron travels over a distance corresponding to λ. In fact this permits the electron to feel a phonon electric field which is always of the same sign. This yields $2\pi p/m\omega \simeq \lambda$ or $p \simeq m\omega/k$. Moreover, assuming the velocity acquired by the electron during a fluctuation to originate only from overall wavevector conservation in the process of emission of the phonon in a one-phonon process (small coupling assumption), one has $p = \hbar k$, which together with $p \simeq m\omega/k$ gives $p^2/2m \simeq \hbar\omega/2$. This satisfies condition (6.27) and in addition yields $\delta E \sim 3\hbar\omega/2$ for the amplitude of the energy fluctuation. Moreover, in view of the already mentioned small velocity of the optical phonons, one should choose for c the velocity of the electron $p/m \simeq \sqrt{\hbar\omega/m}$. Substitution in (6.33) of the values obtained for δE and c gives $r \simeq \frac{2}{3}\sqrt{\hbar/\omega m}$. Using the free electron mass for m and $\omega \sim 10^{14}$ Hz one thus obtains $r \sim 10^{-7}$ cm or ~ 10Å, in good order of magnitude agreement with the so-called polaron radius evaluated from experimental data (Kartheuser 1972).

It should be noted, however, that this qualitative representation of the virtual cloud which dresses a source has been obtained under the tacit assumption that the quanta are emitted and reabsorbed one at a time, since each fluctuation has been considered individually. To check if this is indeed the case, one can use the results of the previous section concerning the average number of quanta. In fact, if this number is much smaller than one, it is likely that the picture based on independent acts of absorption or re-emission is valid, but if $\langle n\rangle \gg 1$ the qualitative picture of the virtual cloud we have given in this section is likely to break down. Thus, turning to (6.6) under the assumption of multipolar coupling for an atom of radius a, one can approximate, since $\omega_M = 2\pi c/a \gg \omega_0$,

$$\langle n\rangle \simeq \frac{\gamma\omega_M^2}{4\pi\omega_0^3} = \frac{|\mu_{21}|^2\,\omega_M^2}{3\pi\hbar c^3} = \frac{4\pi}{3}\frac{e^2}{\hbar c} \sim 3\cdot 10^{-2} \tag{6.34}$$

where expression (5.61) for γ has been used together with $|\mu_{21}| \sim ea$. One can see that in view of the smallness of the fine structure constant $e^2/\hbar c$ the cloud of photons in the atomic case is dilute enough to make plausible the independent fluctuation assumption. In the simple nucleon model with spherical source case, which leads to (6.22), we adopt the strong coupling constant to be $g\sqrt{\frac{\hbar}{mc}} \sim 0.2$ GeV $\sim 3\cdot 10^{-4}$ erg (see e.g. Griffiths 1987), hence $g \sim 10^3$ in Gaussian units. Moreover, taking $\hbar k_M/mc$ to be of the order of five yields the number of virtual mesons $\langle n \rangle \sim 1$. This indicates that the strong force case discussed is in an intermediate region where the assumption of independent fluctuations is of uncertain validity. In fact also, lowest-order perturbation theory is likely to break down in this case, and the exact approach of the previous section is to be preferred. Finally, for the optical polaron case, (6.30) indicates a strong correlation between the number of virtual phonons in the cloud dressing the electron and the constant α_p defined in (6.24). The experimentally measured values of α_p (Kartheuser 1972) range from small values of the order of 10^{-2} in the case of semiconductors of the III-IV group like GaAs and InSb, for which the independent fluctuations model, and consequently the perturbative approach described in the previous section, are adequate, to large values of three to five in the case of the alkali halides like KBr and NaCl for which perturbation theory is likely to break down.

The discussion in the present section, together with the discussion on quark-gluon dressed states in Appendix G, substantiates our statement in Section 6.1 that the concept of the dressed atom applies also in the absence of an external driving field, since the atom can get dressed also by the zero-point fluctuations of the e.m. field. Furthermore, we have shown that the origin of this kind of dressing is of a very general nature and that it is valid for quite a large number of systems, well outside the domain of the atom-radiation interaction. In what follows we shall consider some of the observable effects induced by this kind of dressing.

6.3 Self-energy effects and the free electron

In this section we shall consider the effects of the zero-point fluctuations on the dynamics of a very simple system, which is a free nonrelativistic and spinless electron in free space. A free electron possesses an electric charge, and as such it must be considered a source of the e.m. field. Thus one should expect that it is surrounded by a cloud of virtual photons. We shall discuss the shape of this cloud in some detail in the future. Here we shall ask if its presence induces observable effects on the dynamics of the electron.

Although one could obviously use a multipolar scheme (Craig and Thirunamachandran 1984), since a free electron is not a neutral object and since the problem does not possess an obvious centre of symmetry, it is appropriate to work in the minimal coupling scheme and to use the second-quantized Hamiltonian (4.31)

$$H = -\frac{\hbar^2}{2m}\int \psi^\dagger \nabla^2 \psi d^3\mathbf{x} + \frac{1}{8\pi}\int (\mathbf{E}_\perp^2 + \mathbf{H}^2) d^3\mathbf{x}$$
$$+ \frac{e}{mc}\int \psi^\dagger \mathbf{A}_\perp \cdot \frac{\hbar}{i}\nabla\psi d^3\mathbf{x} + \frac{e^2}{2mc^2}\int \psi^\dagger \mathbf{A}_\perp^2 \psi d^3\mathbf{x} \qquad (6.35)$$

where m and e are the mass and charge of the electron and where, as already remarked in the case of the hydrogen atom briefly discussed after (4.40), the Coulomb term does not play any role in the one-electron subspace due to the ordering of creation and annihilation operators. This ordering ensures that the longitudinal Coulomb energy

$$\frac{1}{2}q\int \psi^* V \psi d^3\mathbf{x} = \frac{1}{8\pi}\int \mathbf{E}_\parallel^2 d^3\mathbf{x} \qquad (6.36)$$

present in (4.31), does not appear in (6.35). This longitudinal energy term is sometimes called Coulomb self energy and can be thought of as contributing to the mass m of the electron in (6.35) (see e.g. Cohen-Tannoudji *et al*. 1989).

We partition (6.35) as

$$H = H_0 + V_1 + V_2 ;$$
$$H_0 = -\frac{\hbar^2}{2m}\int \psi^\dagger \nabla^2 \psi d^3\mathbf{x} + \frac{1}{8\pi}\int (\mathbf{E}_\perp^2 + \mathbf{H}^2) d^3\mathbf{x} ;$$
$$V_1 = \frac{e}{mc}\int \psi^\dagger \mathbf{A}_\perp \cdot \frac{\hbar}{i}\nabla\psi d^3\mathbf{x} ;$$
$$V_2 = \frac{e^2}{2mc^2}\int \psi^\dagger \mathbf{A}_\perp^2 \psi d^3\mathbf{x} \qquad (6.37)$$

and consider H_0 as the unperturbed Hamiltonian. In this scheme V_1 and V_2 are considered perturbations yielding corrections to the eigenvectors and eigenvalues of H_0. It should be noted that V_1 is of order e whereas V_2 is of order e^2. We express H in terms of creation and annihilation operators using (2.9) and (4.38) in the form

$$\psi(\mathbf{x}) = \frac{1}{\sqrt{V}}\sum_{\mathbf{p}} c_{\mathbf{p}} e^{(i/\hbar)\mathbf{p}\cdot\mathbf{x}} ;$$
$$\mathbf{A}_\perp(\mathbf{x}) = \sum_{\mathbf{k}j}\sqrt{\frac{2\pi\hbar c^2}{V\omega_k}}\left(\mathbf{e}_{\mathbf{k}j} a_{\mathbf{k}j} e^{i\mathbf{k}\cdot\mathbf{x}} + \mathbf{e}_{\mathbf{k}j}^* a_{\mathbf{k}j}^\dagger e^{-i\mathbf{k}\cdot\mathbf{x}}\right) \qquad (6.38)$$

where periodic boundary conditions have been imposed on both fields at the surface of a large cubic box of volume V. Clearly, $\mathbf{p}$ are the values of the electron momentum compatible with these boundary conditions and $c_{\mathbf{p}}$ is the annihilation operator for an electron of momentum $\mathbf{p}$. Substituting (6.38) into (6.37) one finds after some algebra

$$H_0 = \sum_{\mathbf{p}} \frac{1}{2m} p^2 c_{\mathbf{p}}^\dagger c_{\mathbf{p}} + \sum_{\mathbf{k}j} \hbar\omega_k \left(a_{\mathbf{k}j}^\dagger a_{\mathbf{k}j} + 1/2 \right) ;$$

$$V_1 = \frac{e}{m} \sum_{\mathbf{p}\mathbf{k}j} \sqrt{\frac{2\pi\hbar}{V\omega_k}} \left(\mathbf{e}_{\mathbf{k}j} \cdot \mathbf{p} c_{\mathbf{p}+\hbar\mathbf{k}}^\dagger c_{\mathbf{p}} a_{\mathbf{k}j} + \text{h.c.} \right) ;$$

$$V_2 = \frac{\pi\hbar e^2}{mV} \sum_{\mathbf{p}\mathbf{p}'} \sum_{\mathbf{k}\mathbf{k}'jj'} \frac{1}{\sqrt{\omega_k \omega_{k'}}}$$
$$\times \left(\mathbf{e}_{\mathbf{k}j} \cdot \mathbf{e}_{\mathbf{k}'j'} c_{\mathbf{p}}^\dagger c_{\mathbf{p}'} a_{\mathbf{k}j} a_{\mathbf{k}'j'} \delta_{\mathbf{p}-\mathbf{p}',\hbar(\mathbf{k}+\mathbf{k}')} \right.$$
$$+ \mathbf{e}_{\mathbf{k}j} \cdot \mathbf{e}_{\mathbf{k}'j'}^* c_{\mathbf{p}}^\dagger c_{\mathbf{p}'} a_{\mathbf{k}j} a_{\mathbf{k}'j'}^\dagger \delta_{\mathbf{p}-\mathbf{p}',\hbar(\mathbf{k}-\mathbf{k}')}$$
$$\left. - \frac{1}{2} c_{\mathbf{p}}^\dagger c_{\mathbf{p}} \delta_{\mathbf{p}\mathbf{p}'} \delta_{\mathbf{k}\mathbf{k}'} \delta_{jj'} + \text{h.c.} \right) \tag{6.39}$$

In the one-electron subspace, the eigenstates and eigenvalues of H_0 are

$$| \mathbf{p}, \{n_{\mathbf{k}j}\} \rangle ;$$
$$E_0(\mathbf{p}, \{n_{\mathbf{k}j}\}) = \frac{1}{2m} p^2 + \sum_{\mathbf{k}j} \hbar\omega_k (n_{\mathbf{k}j} + 1/2) \tag{6.40}$$

We note that both V_1 and V_2 give rise to momentum-conserving processes. Restricting our attention to states where no real photons are present, $\{n_{\mathbf{k}j}\} = \{0_{\mathbf{k}j}\}$, we shall apply perturbation theory in order to obtain the changes induced by V_1 and V_2 on the eigenstates (6.40). A perturbative approach of the kind discussed in Appendix C is valid for sufficiently small V_1 and V_2 at any order for states $| \mathbf{p}, \{0_{\mathbf{k}j}\} \rangle$ in spite of infinite degeneracy of the latter. In fact, momentum conservation prevents the existence of any path which can connect $| \mathbf{p}, \{0_{\mathbf{k}j}\} \rangle$ and $| \mathbf{p}', \{0_{\mathbf{k}j}\} \rangle$ ($p = p'$ but $\mathbf{p} \neq \mathbf{p}'$) at any order. Use of (C.12) yields the state, correct to order e^2,

$$| \mathbf{p}, \{0_{\mathbf{k}j}\} \rangle' = \left\{ 1 + \frac{1}{E_0 - H_0}(1 - P_0) V_1 + \frac{1}{E_0 - H_0}(1 - P_0) V_2 \right.$$
$$+ \frac{1}{E_0 - H_0}(1 - P_0) V_1 \frac{1}{E_0 - H_0}(1 - P_0) V_1$$
$$- \langle \phi | V_1 | \phi \rangle \frac{1}{(E_0 - H_0)^2}(1 - P_0) V_1$$
$$\left. - \frac{1}{2} \langle \phi | V_1 \frac{1}{(E_0 - H_0)^2}(1 - P_0) V_1 | \phi \rangle \right\} | \phi \rangle \tag{6.41}$$

where $| \phi\rangle \equiv | \mathbf{p}, \{0_{\mathbf{k}j}\}\rangle$. The quantum average of the photon number for the state (6.41) to order e^2 is

$$\begin{aligned}
\langle a^\dagger_{\mathbf{k}j} a_{\mathbf{k}j}\rangle &= \langle\phi \mid V_1(1-P_0)\frac{1}{E_0-H_0} a^\dagger_{\mathbf{k}j} a_{\mathbf{k}j} \frac{1}{E_0-H_0}(1-P_0)V_1 \mid \phi\rangle \\
&= \frac{e}{m}\sqrt{\frac{2\pi\hbar}{V\omega_k}}\mathbf{e}^*_{\mathbf{k}j}\cdot\mathbf{p}\langle\phi\mid V_1\frac{1}{(E_0-H_0)^2}\mid \mathbf{p}-\hbar\mathbf{k}, 1_{\mathbf{k}j}\rangle \\
&= \frac{e}{m}\sqrt{\frac{2\pi\hbar}{V\omega_k}}\mathbf{e}^*_{\mathbf{k}j}\cdot\mathbf{p}\frac{1}{\left[\frac{1}{2m}p^2-\frac{1}{2m}(\mathbf{p}-\hbar\mathbf{k})^2-\hbar\omega_k\right]^2} \\
&\quad\times\langle\phi\mid V_1\mid\mathbf{p}-\hbar\mathbf{k}, 1_{\mathbf{k}j}\rangle
\end{aligned} \tag{6.42}$$

where $1_{\mathbf{k}j}$ indicates a distribution of photons constituted by only one photon in mode $\mathbf{k}j$. We now point out that for a fully nonrelativistic calculation one has both $p \ll mc$ and $\hbar k \ll mc$ for electrons and photons respectively. From the former inequality we have $\frac{\hbar}{m}\mathbf{k}\cdot\mathbf{p} \ll \hbar ck$ and from the latter $\frac{\hbar^2k^2}{2m} \ll \hbar ck$. Consequently, the energy in the denominator of (6.42) can be approximated as

$$\frac{1}{2m}p^2 - \frac{1}{2m}(\mathbf{p}-\hbar\mathbf{k})^2 - \hbar\omega_k = \frac{\hbar}{m}\mathbf{k}\cdot\mathbf{p} - \frac{\hbar^2k^2}{2m} - \hbar\omega_k \sim -\hbar\omega_k \tag{6.43}$$

Substituting (6.43) into (6.42) and completing the calculation, we find

$$\langle a^\dagger_{\mathbf{k}j} a_{\mathbf{k}j}\rangle = \frac{e^2}{\hbar m^2}\frac{2\pi}{V}\frac{1}{\omega_k^3}\mid \mathbf{e}_{\mathbf{k}j}\cdot\mathbf{p}\mid^2 \tag{6.44}$$

The total number of photons in the cloud surrounding the electron is thus given by

$$\begin{aligned}
\langle n\rangle &= \frac{e^2}{\hbar m^2c^3}\frac{2\pi}{V}\sum_{\mathbf{k}j}\frac{1}{k^3}\mid \mathbf{e}_{\mathbf{k}j}\cdot\mathbf{p}\mid^2 \\
&= \frac{e^2}{\hbar m^2c^3}\frac{2\pi}{V}\sum_{\mathbf{k}j}\frac{1}{k^3}p_mp_n\left(\mathbf{e}^*_{\mathbf{k}j}\right)_m\left(\mathbf{e}_{\mathbf{k}j}\right)_n \\
&= \frac{e^2}{\hbar m^2c^3}\frac{2\pi}{V}\sum_{\mathbf{k}}\frac{1}{k^3}p_mp_n\left(\delta_{mn}-\hat{k}_m\hat{k}_n\right) \\
&= \frac{e^2}{\hbar m^2c^3}\frac{2\pi}{V}\sum_{\mathbf{k}j}\frac{1}{k^3}\left[p^2-(\mathbf{p}\cdot\mathbf{k})^2\right] \\
&= \frac{e^2}{\hbar m^2c^3}p^2\frac{1}{(2\pi)^2}\int\frac{1}{k^3}\sin^2\theta d^3\mathbf{k} \\
&= \frac{e^2}{\hbar m^2c^3}\frac{4}{3\pi}p^2\int_0^{mc/\hbar}\frac{1}{k}dk
\end{aligned} \tag{6.45}$$

where (2.6) has been used, $\hat{k}_m$ being the m component of the unit vector of $\mathbf{k}$ and where the integration over k has been cut off at the electron Compton wavevector, consistently with the nonrelativistic nature of the calculation. Thus we find that the free nonrelativistic electron is surrounded by a cloud of photons. In fact their number is infinite, since the last integral in (6.45) diverges at the lower limit of integration. This divergence is in fact a manifestation of the so-called infrared divergence which makes its appearance in the emission of radiation by an accelerated charge. In that circumstance one is not much worried by this divergence since it is irrelevant to count the number of emitted long-wavelength photons, in a situation where the energy is the only physically measurable quantity (Itzykson and Zuber 1985). In fact we shall show immediately that in the present case the energy shift associated with the cloud of photons does not suffer from this infrared catastrophe.

In order to evaluate the energy shift of $|\phi\rangle \equiv |\mathbf{p}, \{0_{\mathbf{k}j}\}\rangle$ due to the perturbation, it is appropriate to remark that V_1 has no diagonal matrix element on the basis (6.40). Hence the lowest possible correction is of order e^2, and use of (C.10) shows that this is

$$E^{(2)} = E_0 + \langle\phi \mid V_2 \mid \phi\rangle + \langle\phi \mid V_1 \frac{1}{E_0 - H_0}(1 - P_0)V_1 \mid \phi\rangle \tag{6.46}$$

where, indicating by ZP the zero-point contribution, we have set

$$E_0 \equiv E_0\big(\mathbf{p}, \{0_{\mathbf{k}j}\}\big) = \frac{1}{2m}p^2 + ZP \tag{6.47}$$

Using (6.39) we obtain

$$\langle\phi \mid V_2 \mid \phi\rangle = \frac{\hbar e^2}{m}\frac{\pi}{V}\sum_{\mathbf{k}j}\frac{1}{\omega_k} = \frac{e^2}{\hbar c}\frac{\hbar^2}{m}\frac{1}{\pi}\int_0^{mc/\hbar} kdk \tag{6.48}$$

In the absence of the cutoff $mc/\hbar$, this correction would obviously diverge, but it might be regarded as a limit of a shift which, however large, is common to all the states of the form $|\mathbf{p}, \{0_{\mathbf{k}j}\}\rangle$ and independent of $\mathbf{p}$. Hence it cannot influence the dynamics of the electron and it can be neglected. The elementary processes giving rise to this shift may be represented by the Feynman diagram in Figure 6.1, where the same criteria as for the diagrams in Figure 4.2 have been used. The third term in (6.46) can instead be represented by the diagram in Figure 6.2. Its

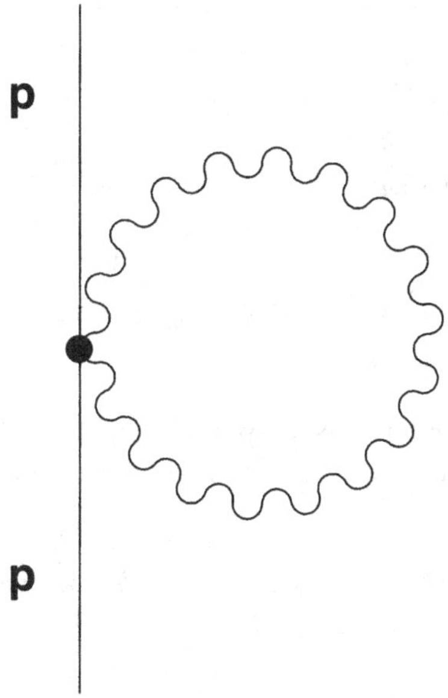

Fig. 6.1 Diagram representing the p-independent contribution to the self-energy of a free electron. This contribution can be discarded because it does not influence the dynamics of the electron.

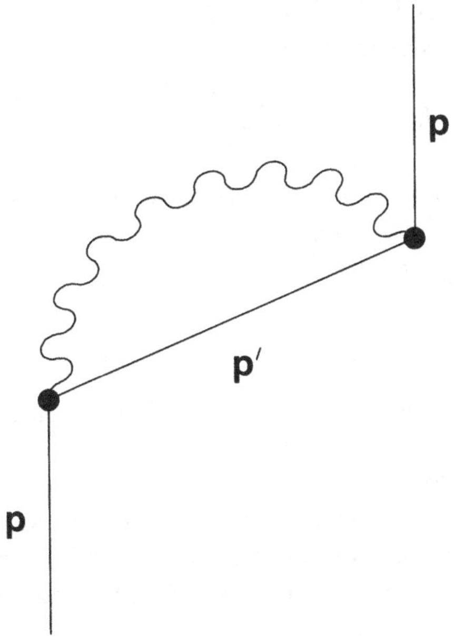

Fig. 6.2 Diagram representing the p-dependent contribution to the self-energy of the free electron. This contribution yields mass renormalization.

evaluation leads to

$$\begin{aligned}
&\langle \phi \mid V_1 \frac{1}{E_0 - H_0}(1 - P_0)V_1 \mid \phi\rangle \\
&= \frac{e}{m}\sum_{\mathbf{k}j}\sqrt{\frac{2\pi\hbar}{V\omega_k}}\,\mathbf{e}^*_{\mathbf{k}j}\cdot\mathbf{p}\langle\phi \mid V_1 \frac{1}{E_0 - H_0} \mid \mathbf{p} - \hbar\mathbf{k}, 1_{\mathbf{k}j}\rangle \\
&= \frac{e}{m}\sum_{\mathbf{k}j}\sqrt{\frac{2\pi\hbar}{V\omega_k}}\,\mathbf{e}^*_{\mathbf{k}j}\cdot\mathbf{p}\frac{1}{\frac{1}{2m}p^2 - \frac{1}{2m}(\mathbf{p} - \hbar\mathbf{k})^2 - \hbar\omega_k} \\
&\times \langle\phi \mid V_1 \mid \mathbf{p} - \hbar\mathbf{k}, 1_{\mathbf{k}j}\rangle
\end{aligned} \tag{6.49}$$

Adopting again the nonrelativistic approximation (6.43) and completing the calculation according to the method which led to (6.45) yields

$$\begin{aligned}
\langle \phi \mid V_1 \frac{1}{E_0 - H_0}(1 - P_0)V_1 \mid \phi\rangle &= -\frac{e^2}{\hbar c^2}\frac{4\hbar}{3\pi}\frac{1}{2m^2}p^2\int_0^{mc/\hbar} dk \\
&= -\frac{e^2}{\hbar c}\frac{4}{3\pi}\frac{1}{2m}p^2
\end{aligned} \tag{6.50}$$

It should be noted that the linear "ultraviolet" divergence (6.50) has been tamed by the cut-off in the upper limit of integration as in (6.45). Substituting these results in (6.46), introducing the fine structure constant $\alpha = e^2/\hbar c$ and eliminating the infinities which do not influence the dynamics of the system, gives

$$E^{(2)} = \left(1 - \frac{4\alpha}{3\pi}\right)\frac{1}{2m}p^2 = \frac{1}{2m^*}p^2 \tag{6.51}$$

Noting that overall accuracy of the calculation is up to terms of order e^2, in the last expression (6.51) we have introduced the effective, or dressed, mass of the free electron

$$m^* = \frac{m}{1 - 4\alpha/3\pi} \sim m\left(1 + \frac{4\alpha}{3\pi}\right) \tag{6.52}$$

This is an interesting result because it shows that the self-interaction of the electron, due to the dressing by the photons of the transverse electromagnetic field which are emitted and reabsorbed during the interaction with the vacuum fluctuations, results in an increase of the electron mass by an amount of the order of α. Thus the energy shift for a free electron can be written in terms of a mass increase δm

$$E^{(2)} = \frac{1}{2m}p^2\left(1 - \frac{\delta m}{m}\right)\ ;\ \delta m = \frac{4\alpha}{3\pi}m \tag{6.53}$$

which expresses the difference between the dressed mass and the bare mass. As argued earlier, the infrared divergence in the number of photons in the cloud does not cause any difficulty with the energy shift, because the energy of each of the infinitely many infrared photons is vanishingly small. Since the interaction with the transverse e.m. field should be considered inherent for any charged particle it is evident that the electron mass which can be observed and measured under ordinary circumstances is the dressed mass $m^* = m + \delta m$ and not the bare mass in terms of which (6.35) has been written. Thus it is natural to re-express the Hamiltonian in terms of m^*. We have from (6.35)

$$\begin{aligned} H = & -\frac{\hbar^2}{2(m^* - \delta m)} \int \psi^\dagger \nabla^2 \psi d^3\mathbf{x} + \frac{1}{8\pi} \int (\mathbf{E}_\perp^2 + \mathbf{H}^2) d^3\mathbf{x} \\ & + \frac{e/c}{m^* - \delta m} \int \psi^\dagger \mathbf{A}_\perp \cdot \frac{\hbar}{i} \nabla \psi d^3\mathbf{x} + \frac{e^2/2c^2}{m^* - \delta m} \int \psi^\dagger \mathbf{A}_\perp^2 \psi d^3\mathbf{x} \\ = & -\frac{\hbar^2}{2m^*} \int \psi^\dagger \nabla^2 \psi d^3\mathbf{x} + \frac{1}{8\pi} \int (\mathbf{E}_\perp^2 + \mathbf{H}^2) d^3\mathbf{x} \\ & + \frac{e}{m^* c} \int \psi^\dagger \mathbf{A}_\perp \cdot \frac{\hbar}{i} \nabla \psi d^3\mathbf{x} \\ & + \frac{e^2}{2m^* c^2} \int \psi^\dagger \mathbf{A}_\perp^2 \psi d^3\mathbf{x} - \frac{\delta m}{m^*} \frac{\hbar^2}{2m^*} \int \psi^\dagger \nabla^2 \psi d^3\mathbf{x} \\ & + O(e^3) \end{aligned} \tag{6.54}$$

This procedure is an example of a more general procedure which is called mass renormalization. It is clear that, neglecting $O(e^3)$ and evaluating the energy shifts of the dressed unperturbed Hamiltonian

$$H_0^* = -\frac{\hbar^2}{2m^*} \int \psi^\dagger \nabla^2 \psi d^3\mathbf{x} + \frac{1}{8\pi} \int (\mathbf{E}_\perp^2 + \mathbf{H}^2) d^3\mathbf{x} \tag{6.55}$$

due to the perturbation

$$\begin{aligned} & V_1^* + V_2^* - \frac{\delta m}{m^*} \frac{\hbar^2}{2m^*} \int \psi^\dagger \nabla^2 \psi d^3\mathbf{x} \,; \\ & V_1^* = \frac{e}{m^* c} \int \psi^\dagger \mathbf{A}_\perp \cdot \frac{\hbar}{i} \nabla \psi d^3\mathbf{x} \,; \\ & V_2^* = \frac{e^2}{2m^* c^2} \int \psi^\dagger \mathbf{A}_\perp^2 \psi d^3\mathbf{x} \end{aligned} \tag{6.56}$$

in second-order perturbation theory as previously, yields zero due to exact cancellation at this order between the contribution of $V_1^* + V_2^*$ and that of the term in δm. Thus mass renormalization as we have introduced it accounts for the self-energy effects on a free electron up to terms of order e^2.

6.4 Self-energy effects and energy shifts

As shown in Section 6.3, the energy of a free electron is shifted due to the interaction with the transverse e.m. vacuum fluctuations, and this gives rise to mass renormalization. The question arises whether similar self-energy shifts occur in other cases in which the source-field interaction gives rise to a cloud of virtual quanta. We shall take up again the three examples considered in Section 6.1, where we have shown that the sources are surrounded by clouds of virtual quanta of the field with which each of them is assumed to interact, and we shall discuss the shifts due to self-energy effects for each example.

Starting with the Fröhlich polaron, assuming validity of perturbation expansion and using (C.10) on Hamiltonian (6.23) one has the energy correction at second order in V as

$$E^{(2)} = E_0 + \langle \phi \mid V \mid \phi \rangle + \langle \phi \mid V \frac{1}{E_0 - H_0} (1 - P_0) V \mid \phi \rangle \tag{6.57}$$

where we choose

$$\mid \phi \rangle \equiv \mid \mathbf{p} = \hbar \langle \mathbf{K} \rangle, \{0_\mathbf{k}\} \rangle \; ; \; E_0 = \frac{\hbar^2 \langle K \rangle^2}{2m} \tag{6.58}$$

with the same notation as in Section 6.1. It is easily seen that $\langle \phi \mid V \mid \phi \rangle = 0$, and a simple calculation yields

$$\begin{aligned}
&\langle \phi \mid V \frac{1}{E_0 - H_0} (1 - P_0) V \mid \phi \rangle \\
&= \sum_\mathbf{K} \frac{\mid V_k \mid^2}{\frac{-\hbar^2 k^2}{2m} + \frac{\hbar^2}{m} \langle K \rangle k \cos\theta - \hbar\omega} \\
&= \frac{1}{2\pi^2} \hbar\omega \left(\frac{\hbar}{2m\omega} \right)^{1/2} \\
&\times \alpha_p \int \frac{k^{-2}}{1 + \frac{\hbar}{2m\omega} k^2 - \frac{\hbar}{2m\omega} 2 \langle K \rangle k \cos\theta} d^3\mathbf{k}
\end{aligned} \tag{6.59}$$

where (6.24) has been used. If $\langle K \rangle$ is sufficiently small it is admissible to approximate

$$\frac{1}{1+\frac{\hbar}{2m\omega}k^2 - \frac{\hbar}{2m\omega}2\langle K \rangle k\cos\theta}$$
$$\equiv \frac{1}{1+\frac{\hbar}{2m\omega}k^2}\left\{1+\left(1+\frac{\hbar k^2}{2m\omega}\right)^{-1}\frac{\hbar}{2m\omega}2\langle K \rangle k\cos\theta\right.$$
$$\left.+\left(1+\frac{\hbar k^2}{2m\omega}\right)^{-2}\left(\frac{\hbar}{2m\omega}\right)^2 4\langle K \rangle^2 k^2\cos^2\theta\right\} \tag{6.60}$$

When (6.60) is used in (6.59) and after some amount of elementary integrations, we have the result

$$\langle \phi \mid V\frac{1}{E_0 - H_0}(1-P_0)V \mid \phi \rangle = -\hbar\omega\alpha_p - \frac{1}{6}\alpha_p\frac{\hbar^2\langle K \rangle^2}{2m} \tag{6.61}$$

Substituting in (6.57) this result yields

$$E^{(2)} = -\hbar\omega\alpha_p + \left(1-\frac{\alpha_p}{6}\right)\frac{\hbar^2\langle K \rangle^2}{2m}$$
$$= -\hbar\omega\alpha_p + \frac{\hbar^2\langle K \rangle^2}{2m^*}\ ;\ m^* = \frac{m}{1-\alpha_p/6} \tag{6.62}$$

Thus, as in the case of the electron in the vacuum e.m. field, the self-energy effects of the polaron can be represented by a mass renormalization yielding a dressed mass m^*, at least in the weak-coupling limit $\alpha_p < 1$ where perturbation theory can be used. In fact we see that the dressed electron mass in (6.62) appears to diverge for $\alpha_p \sim 6$, which indicates complete breakdown of the perturbation expansion (see e.g. Pines 1963). The increase of the dressed mass with respect to the bare one can be understood in the case of the optical polaron by observing that the negative charge of the electron will attract the positive ions and repel the negative ones, thereby inducing a distortion in the lattice which is the substance of the virtual phonon cloud. Any displacement of the electron through the lattice involves a rearrangement of the distortion, which is expected to follow the electron adiabatically, at least for small velocities of the latter. Thus a force acting on the electron does not see the rigid-ion electronic mass alone, but also a kind of local lattice mass, that depends on the mass of the ions involved in the distortion.

Next we consider the static source model of the nucleon introduced in Section 6.1. It is convenient to define

$$D = \prod_{\mathbf{k}} D_{\mathbf{k}} \tag{6.63}$$

where $D_{\mathbf{k}}$ is the unitary operator defined in (6.11). Thus also D is unitary. Applying D to Hamiltonian (6.10), one obtains a new Hamiltonian

$$\tilde{H} = DHD^{-1} \tag{6.64}$$

which one can appropriately call the dressed Hamiltonian. In view of the unitarity of D, $\tilde{H}$ has the same eigenvalue spectrum as H. Using (6.14)

$$\begin{aligned}
D_{\mathbf{k}}^{-1} a_{\mathbf{k}}^{\dagger} a_{\mathbf{k}} D_{\mathbf{k}} &= D_{\mathbf{k}}^{-1} a_{\mathbf{k}}^{\dagger} D_{\mathbf{k}} D_{\mathbf{k}}^{-1} a_{\mathbf{k}} D_{\mathbf{k}} \\
&= \left(a_{\mathbf{k}}^{\dagger} + g\left[2\hbar V\omega_k^3\right]^{-1/2} \rho_{\mathbf{k}}^{*}\right)\left(a_{\mathbf{k}} + g\left[2\hbar V\omega_k^3\right]^{-1/2} \rho_{\mathbf{k}}\right) \\
&= a_{\mathbf{k}}^{\dagger} a_{\mathbf{k}} + g\left[2\hbar V\omega_k^3\right]^{-1/2} \rho_{\mathbf{k}} a_{\mathbf{k}}^{\dagger} + g\left[2\hbar V\omega_k^3\right]^{-1/2} \rho_{\mathbf{k}}^{*} a_{\mathbf{k}} \\
&\quad + \frac{1}{2}\frac{1}{\hbar V\omega_k^3} g^2 \mid \rho_{\mathbf{k}} \mid^2 ; \\
D_{\mathbf{k}}^{-1}&\left(\rho_{\mathbf{k}}^{*} a_{\mathbf{k}} + \rho_{\mathbf{k}} a_{\mathbf{k}}^{\dagger}\right) D_{\mathbf{k}} \\
&= \rho_{\mathbf{k}}^{*} a_{\mathbf{k}} + \rho_{\mathbf{k}} a_{\mathbf{k}}^{\dagger} + g\sqrt{\frac{2}{\hbar V\omega_k^3}} \mid \rho_{\mathbf{k}} \mid^2
\end{aligned} \tag{6.65}$$

Hence

$$\begin{aligned}
D_{\mathbf{k}}^{-1} \hbar\omega_k a_{\mathbf{k}}^{\dagger} a_{\mathbf{k}} D_{\mathbf{k}} &= \hbar\omega_k a_{\mathbf{k}}^{\dagger} a_{\mathbf{k}} + g\sqrt{\frac{\hbar}{2V\omega_k}}\left(\rho_{\mathbf{k}} a_{\mathbf{k}}^{\dagger} + \rho_{\mathbf{k}}^{*} a_{\mathbf{k}}\right) \\
&\quad + \frac{1}{2V\omega_k{}^2} g^2 \mid \rho_{\mathbf{k}} \mid^2 ; \\
D_{\mathbf{k}}^{-1} g\sqrt{\frac{\hbar}{2V\omega_k}}\left(\rho_{\mathbf{k}} a_{\mathbf{k}}^{\dagger} + \rho_{\mathbf{k}}^{*} a_{\mathbf{k}}\right) D_{\mathbf{k}} &= g\sqrt{\frac{\hbar}{2V\omega_k}}\left(\rho_{\mathbf{k}} a_{\mathbf{k}}^{\dagger} + \rho_{\mathbf{k}}^{*} a_{\mathbf{k}}\right) \\
&\quad + \frac{1}{V\omega_k^2} g^2 \mid \rho_{\mathbf{k}} \mid^2
\end{aligned} \tag{6.66}$$

Consequently

$$\tilde{H} = DHD^{-1} = \sum_{\mathbf{k}} \hbar\omega_k \left(a_{\mathbf{k}}^{\dagger} a_{\mathbf{k}} + 1/2\right) - \frac{1}{2} g^2 \frac{1}{V} \sum_{\mathbf{k}} \frac{1}{\omega_k^2} \mid \rho_{\mathbf{k}} \mid^2 \tag{6.67}$$

which shows that the eigenvalues of H, which essentially coincide with the eigenvalues of $\tilde{H}$, are simply the eigenvalues of the free meson field in the absence of sources, shifted by a common amount (Henley and Thirring 1962)

$$\Delta = -\frac{1}{2}g^2\frac{1}{V}\sum_{\mathbf{k}}\frac{1}{\omega_k^2}\mid\rho_{\mathbf{k}}\mid^2 = -\frac{1}{2}g^2\frac{1}{(2\pi)^3}\int\frac{1}{\omega_k^2}\mid\rho_{\mathbf{k}}\mid^2 d^3\mathbf{k} \tag{6.68}$$

This shift can be regarded as a renormalization of the energy if, instead of starting from Hamiltonian H, one starts from $H+\Delta$. In this case the exact treatment of H we have given yields a correction which cancels Δ, leaving the free renormalized Hamiltonian.

Turning finally to the model for the ground-state two-level atom in free space used in Section 6.1, the ground-state energy corresponding to state (6.3) obtained in second order perturbation theory is given by (C.10) in the form

$$\begin{aligned} E^{(2)} &= E_0 + \lambda\langle\{0_{\mathbf{k}j}\},\downarrow\mid V_2\mid\{0_{\mathbf{k}j}\},\downarrow\rangle \\ &\quad + \lambda^2\langle\{0_{\mathbf{k}j}\},\downarrow\mid V_2\frac{1}{E_0-H_0}(1-P_0)V_2\mid\{0_{\mathbf{k}j}\},\downarrow\rangle \end{aligned} \tag{6.69}$$

since $\mid\{0_{\mathbf{k}j}\},\downarrow\rangle$ is not perturbed by V_1, as discussed at the beginning of Section 6.1. Since it is also $\langle\{0_{\mathbf{k}j}\},\downarrow\mid V_2\mid\{0_{\mathbf{k}j}\},\downarrow\rangle = 0$, only the third term contributes to the RHS of (6.69), yielding a shift

$$\begin{aligned} \Delta_L &= \langle\{0_{\mathbf{k}j}\},\downarrow\mid V_2\frac{1}{E_0-H_0}(1-P_0)V_2\mid\{0_{\mathbf{k}j}\},\downarrow\rangle \\ &= -\sum_{\mathbf{k}}\frac{\mid\epsilon_{\mathbf{k}j}\mid^2}{\hbar(\omega_0+\omega_k)} \end{aligned} \tag{6.70}$$

This shift is related to the Lamb shift of the atomic ground state. It is due to the virtual processes which induce fluctuations of the atom+field bare energy. These are the same processes leading to the formation of the virtual photon cloud surrounding the ground-state atom. Comparison with (5.71) shows that $\Delta_L = \hbar\Omega_+$, and this throws light on the physical origin of the frequency shift which had been found in the dynamics of the transverse component of $\langle\mathbf{S}\rangle$ during spontaneous decay of a two-level atom in Section 5.4. For an estimate of the Lamb shift in a more realistic model, see Appendix J.

The result (6.70) may look rather strange, because it depends on the form of $\epsilon_{\mathbf{k}j}$. Since the latter is different for the minimal coupling scheme and for the multipolar one, the shift of the ground state is apparently dependent on the scheme of coupling one is using. In fact using (5.61) one has from (6.70) in the minimal coupling scheme

$$\begin{aligned}\Delta_L &= -\frac{\gamma\hbar}{2\pi\omega_0}\int_0^{\omega_M}\frac{\omega_k}{\omega_0+\omega_k}d\omega_k\\ &= \frac{\gamma\hbar}{2\pi}\left(-\frac{\omega_M}{\omega_0}+\ln\frac{\omega_M}{\omega_0}\right)\\ &= \frac{2\mid\mu_{21}\mid^2}{3\pi c^3}\left(-\omega_M\omega_0^2+\omega_0^3\ln\frac{\omega_M}{\omega_0}\right)\end{aligned} \tag{6.71}$$

whereas, using (5.62) in the multipolar scheme,

$$\begin{aligned}\Delta_L &= \frac{\gamma\hbar}{2\pi\omega_0^3}\int_0^{\omega_M}\frac{\omega_k^3}{\omega_0+\omega_k}d\omega_k\\ &= \frac{\gamma\hbar}{2\pi}\left(-\frac{1}{3}\frac{\omega_M^3}{\omega_0^3}+\frac{1}{2}\frac{\omega_M^2}{\omega_0^2}-\frac{\omega_M}{\omega_0}+\ln\frac{\omega_M}{\omega_0}\right)\\ &= \frac{2\mid\mu_{21}\mid^2}{3\pi c^3}\left(-\frac{1}{3}\omega_M^3+\frac{1}{2}\omega_M^2\omega_0-\omega_M\omega_0^2+\omega_0^3\ln\frac{\omega_M}{\omega_0}\right)\end{aligned} \tag{6.72}$$

which is clearly different from (6.71). The apparent contradiction is resolved if we remind ourselves of the warning issued after introducing the two Hamiltonian forms (4.105), which are not unitarily related and consequently really refer to two different objects. If we go back to the forms (4.97) and (4.98) the origin of the difficulty is evident. The prime superscript in (4.97) comes from incorporating the $\mathbf{A}_\perp^2$ term in the free-photon field Hamiltonian, while the prime superscript in (4.98) originates from incorporation of the $\mathbf{p}^2$ term in the free-atom Hamiltonian. Clearly this spoils the unitary relationship. In order to be able to compare ground-state shifts we must start from the two Hamiltonian forms for multi-electron atoms in the electric dipole approximation

$$\begin{aligned}H &= \sum_n E_n c_n^\dagger c_n + \sum_{\mathbf{k}j}\hbar\omega_k\left(a_{\mathbf{k}j}^\dagger a_{\mathbf{k}j}+1/2\right)\\ &+ i\sum_{nn'}\sum_{\mathbf{k}j}\sqrt{\frac{2\pi\hbar}{V\omega_k}}(\omega_{n'n}\boldsymbol{\mu}_{nn'}\cdot\mathbf{e}_{\mathbf{k}j}c_n^\dagger c_{n'}a_{\mathbf{k}j}-\text{h.c.})\\ &+\frac{e^2}{2mc^2}\int\psi^\dagger(\mathbf{x})\mathbf{A}_\perp^2(0)\psi(\mathbf{x})d^3\mathbf{x}\end{aligned} \tag{6.73}$$

$$H = \sum_n E_n \tilde{c}_n^\dagger \tilde{c}_n + 2\pi \int \left\{ \int \tilde{\psi}^\dagger(\mathbf{x}) \mathbf{p}_\perp(\mathbf{x}', \mathbf{x}) \tilde{\psi}(\mathbf{x}) d^3\mathbf{x} \right\}^2 d^3\mathbf{x}'$$
$$+ \sum_{\mathbf{k}j} \hbar\omega_k \left(\tilde{a}_{\mathbf{k}j}^\dagger \tilde{a}_{\mathbf{k}j} + 1/2 \right)$$
$$- i \sum_{nn'} \sum_{\mathbf{k}j} \sqrt{\frac{2\pi\hbar\omega_k}{V}} (\boldsymbol{\mu}_{nn'} \cdot \mathbf{e}_{\mathbf{k}j} \tilde{c}_n^\dagger \tilde{c}_{n'} \tilde{a}_{\mathbf{k}j} - \text{h.c.}) \tag{6.74}$$

which can be obtained according to the discussion of Sections 4.3-4.5 and which are unitarily related. In these expressions E_n are single-electron atomic eigenvalues obtained by taking approximately into account Coulomb repulsion, corresponding to single-electron approximate eigenfunctions $u_n(\mathbf{x})$, $a_{\mathbf{k}j}$ are the annihilation operators coming from expansion (2.9),

$$\psi(\mathbf{x}) = \sum_n c_n u_n(\mathbf{x}) \tag{6.75}$$

is the electron field operator (4.41) in second quantized form and for any operator O

$$\tilde{O} = e^{is} O e^{-is} \tag{6.76}$$

with s given by (4.51). As discussed in Section 4.4 the two forms (6.73) and (6.74) are (within the electric dipole approximation) related by unitary transformation (6.76), and they should give equal ground-state shifts. In order to show that this is the case, we consider the expression for the polarization field

$$p_{\perp\ell}(\mathbf{x}', \mathbf{x}) = \mu_m(\mathbf{x}) \delta_{\perp\ell m}(\mathbf{x}') \tag{6.77}$$

which can be obtained from (B.7) by simple algebraic manipulations involving the transverse δ-function defined in (1.63). Therefore in (6.74)

$$\left\{ \int \tilde{\psi}^\dagger(\mathbf{x}) \mathbf{p}_\perp(\mathbf{x}', \mathbf{x}) \tilde{\psi}(\mathbf{x}) d^3\mathbf{x} \right\}^2$$
$$= \int \tilde{\psi}^\dagger(\mathbf{x}) p_{\perp\ell}(\mathbf{x}', \mathbf{x}) \tilde{\psi}(\mathbf{x}) d^3\mathbf{x} \int \tilde{\psi}^\dagger(\mathbf{x}) p_{\perp\ell}(\mathbf{x}', \mathbf{x}) \tilde{\psi}(\mathbf{x}) d^3\mathbf{x}$$
$$= \int \tilde{\psi}^\dagger(\mathbf{x}) p_{\perp\ell}(\mathbf{x}', \mathbf{x}) \delta_{\perp\ell m}(\mathbf{x}') \tilde{\psi}(\mathbf{x}) d^3\mathbf{x}$$
$$\times \int \tilde{\psi}^\dagger(\mathbf{x}) \mu_m(\mathbf{x}) \tilde{\psi}(\mathbf{x}) d^3\mathbf{x} \tag{6.78}$$

and

$$\int\left\{\int \tilde{\psi}^{\dagger}(\mathbf{x})\mathbf{p}_{\perp}(\mathbf{x}',\mathbf{x})\tilde{\psi}(\mathbf{x})d^3\mathbf{x}\right\}^2 d^3\mathbf{x}'$$
$$= \int \tilde{\psi}^{\dagger}(\mathbf{x})\left\{\int p_{\perp\ell}(\mathbf{x}',\mathbf{x})\delta_{\perp\ell m}(\mathbf{x}')d^3\mathbf{x}'\right\}\tilde{\psi}(\mathbf{x})d^3\mathbf{x}$$
$$\times \int \tilde{\psi}^{\dagger}(\mathbf{x})\mu_m(\mathbf{x})\tilde{\psi}(\mathbf{x})d^3\mathbf{x}$$
$$= \int \tilde{\psi}^{\dagger}(\mathbf{x})p_{\perp m}(0,\mathbf{x})\tilde{\psi}(\mathbf{x})d^3\mathbf{x}\int \tilde{\psi}^{\dagger}(\mathbf{x})\mu_m(\mathbf{x})\tilde{\psi}(\mathbf{x})d^3\mathbf{x}$$
$$= \delta_{\perp m\ell}(0)\int \tilde{\psi}^{\dagger}(\mathbf{x})\mu_\ell(\mathbf{x})\tilde{\psi}(\mathbf{x})d^3\mathbf{x}$$
$$\times \int \tilde{\psi}^{\dagger}(\mathbf{x})\mu_m(\mathbf{x})\tilde{\psi}(\mathbf{x})d^3\mathbf{x} \tag{6.79}$$

where (1.62) and (6.77) have been used. This expression would diverge, were it not for the fact that all our treatment is within dipole approximation. Coherently with this assumption we should substitute for

$$\delta_{\perp m\ell}(0) = \lim_{\mathbf{x}\to 0}\frac{1}{(2\pi)^3}\int\left(\delta_{m\ell} - \hat{k}_m\hat{k}_\ell\right)e^{i\mathbf{k}\cdot\mathbf{x}}d^3\mathbf{k} \tag{6.80}$$

the quantity (Craig and Thirunamachandran 1984) obtained by introducing a cutoff at k_M,

$$\delta_{\perp m\ell}(0) \sim \lim_{x\to 0}\frac{4\pi}{(2\pi)^3}\int_0^{k_M}\left\{(\delta_{m\ell} - \hat{x}_m\hat{x}_\ell)\frac{\sin kx}{kx}\right.$$
$$\left. + (\delta_{m\ell} - 3\hat{x}_m\hat{x}_\ell)\left(\frac{\cos kx}{k^2x^2} - \frac{\sin kx}{k^3x^3}\right)\right\}k^2dk$$
$$= \frac{4\pi}{(2\pi)^3}\int_0^{k_M}\left\{(\delta_{m\ell} - \hat{x}_m\hat{x}_\ell) - \frac{1}{6}(\delta_{m\ell} - 3\hat{x}_m\hat{x}_\ell)\right\}k^2dk$$
$$= \frac{1}{9\pi^2}k_M^3\delta_{m\ell} \tag{6.81}$$

Sustituting (6.81) in (6.79) and using (6.75) yields

$$\int\left\{\int \tilde{\psi}^{\dagger}(\mathbf{x})\mathbf{p}_{\perp}(\mathbf{x}',\mathbf{x})\tilde{\psi}(\mathbf{x})d^3\mathbf{x}\right\}^2 d^3\mathbf{x}'$$
$$= \sum_{nn'n''n'''}\frac{k_M^3}{9\pi^2}(\mu_m)_{nn'}(\mu_m)_{n''n'''}\tilde{c}_n^{\dagger}\tilde{c}_{n'}\tilde{c}_{n''}^{\dagger}\tilde{c}_{n'''} \tag{6.82}$$

Moreover in (6.73) use of (6.75) yields

$$\int \psi^\dagger(\mathbf{x})\mathbf{A}_\perp^2(0)\psi(\mathbf{x})d^3\mathbf{x} = \mathbf{A}_\perp^2(0)\sum_n c_n^\dagger c_n \tag{6.83}$$

Since both (6.82) and (6.83) contribute terms of order e^2 in H, it is sufficient to evaluate their quantum average on the unperturbed atom-field ground state. If the atomic one-electron states $u_n(\mathbf{x})$ have definite parity, which is usually the case, the diagonal matrix elements of $\boldsymbol{\mu}$ vanish and the only contributions of (6.82) come from terms with $n''' = n$, $n'' = n'$, or

$$\begin{aligned}
&\left\langle 2\pi \int \left\{ \int \tilde{\psi}^\dagger(\mathbf{x})\mathbf{p}_\perp(\mathbf{x}',\mathbf{x})\tilde{\psi}(\mathbf{x})d^3\mathbf{x}\right\}^2 d^3\mathbf{x}' \right\rangle \\
&= \frac{2k_M^3}{9\pi}\sum_{\substack{nn' \\ n\neq n'}} |\,\mu_{nn'}\,|^2 \langle \tilde{c}_n^\dagger \tilde{c}_{n'}\tilde{c}_{n'}^\dagger \tilde{c}_n\rangle \\
&= \frac{2k_M^3}{9\pi}\sum_{\substack{nn' \\ n\neq n'}} |\,\mu_{nn'}\,|^2 \langle \tilde{c}_n^\dagger \tilde{c}_n(1-\tilde{c}_{n'}^\dagger \tilde{c}_{n'})\rangle
\end{aligned} \tag{6.84}$$

For a two-level atom, and in view of (4.103),

$$\tilde{c}_n^\dagger \tilde{c}_n\left(1-\tilde{c}_{n'}^\dagger \tilde{c}_{n'}\right) = \left(\frac{1}{2}\pm S_z\right)^2 \tag{6.85}$$

and

$$\left\langle 2\pi \int \left\{ \int \tilde{\psi}^\dagger(\mathbf{x})\mathbf{p}_\perp(\mathbf{x}',\mathbf{x})\tilde{\psi}(\mathbf{x})d^3\mathbf{x}\right\}^2 d^3\mathbf{x}' \right\rangle = \frac{2\,|\,\mu_{21}\,|^2}{9\pi c^3}\omega_M^3 \tag{6.86}$$

Moreover, for a two-level atom and from (4.103) $\sum_n c_n^\dagger c_n = 1$; consequently use of (2.9) gives

$$\begin{aligned}
&\left\langle \frac{e}{2mc^2}\int \psi^\dagger(\mathbf{x})\mathbf{A}_\perp^2(0)\psi(\mathbf{x})d^3\mathbf{x}\right\rangle \\
&= \frac{e^2}{2mc^2}\langle\{0_{\mathbf{k}j}\},\downarrow|\,\mathbf{A}_\perp^2(0)\,|\,\{0_{\mathbf{k}j}\},\downarrow\rangle \\
&= \frac{\pi e^2\hbar}{mV}\sum_{\mathbf{k}j}\frac{1}{\omega_k}\langle\{0_{\mathbf{k}j}\},\downarrow|\,a_{\mathbf{k}j}a_{\mathbf{k}j}^\dagger\,|\,\{0_{\mathbf{k}j}\},\downarrow\rangle \\
&= \frac{8\pi^2}{(2\pi)^3}\frac{e^2\hbar}{mc^3}\int_0^{\omega_M}\omega_k d\omega_k \\
&= \frac{e^2\hbar}{2\pi mc^3}\omega_M^2 = \frac{|\,\mu_{21}\,|^2}{3\pi c^3}\omega_M^2\omega_0
\end{aligned} \tag{6.87}$$

where in the last step we have used the Thomas-Reiche-Kuhn sum rule (see e.g. Merzbacher 1961) adapted to a two-level atom in the form

$$\frac{2m}{\hbar}\omega_0 \mid \mu_{21} \mid^2 = 3e^2 \tag{6.88}$$

Clearly if one adds (6.86) to (6.72) and (6.87) to (6.71) one obtains the same ground-state shift

$$\Delta = \frac{2 \mid \mu_{21} \mid^2}{3\pi c^3}\left(\frac{1}{2}\omega_M^2\omega_0 - \omega_M\omega_0^2 + \omega_0^3 \ln\frac{\omega_M}{\omega_0}\right) \tag{6.89}$$

This shift is due to self-energy effects. Moreover, it is independent of the coupling scheme adopted, which solves the apparent discrepancy proposed by the form (6.70). We note that use of (6.88) in the first term within brackets in (6.89) yields a contribution to the ground-state shift Δ which can be put in the form

$$\frac{\mid \mu_{21} \mid^2}{3\pi c^3}\omega_M^2\omega_0 = \frac{e^2}{\hbar c}\frac{\hbar^2}{m}\frac{1}{\pi}\int_0^{\omega_M} kdk \tag{6.90}$$

Thus this contribution coincides with (6.48) obtained for a free electron, which was discarded because it does not have any consequence on the electron dynamics. It is obvious that this remains true in the atomic case, and it shows that in the minimal coupling scheme Δ_L as obtained from (6.70) essentially coincides with the self-energy shift of the total (atom+field) ground state.

6.5 Virtual clouds and excited states

It should be remarked that in all cases discussed until now, the only bare states we have considered as candidates for dressing are stable states, that is, states which cannot decay into other bare states by the emission of real quanta. This is evidently the case for bare ground states of the total source+field system, such as the static model of the nucleon or the ground-state two-level atom; but also for the Fröhlich polaron described by Hamiltonian (6.23), where the electron in state $\mid \mathbf{p}, \{0_\mathbf{k}\}\rangle$ has been assumed not to be able to slow down by emission of real phonons in view of condition (6.27), which ensures nonconservation of total energy in one of such emission processes. In this section we shall consider the difficulties one is confronted with when trying to relax this limitation and to apply the ideas of the previous sections to excited states which can decay by emission of real quanta.

We take up again the problem of the two-level atom interacting with the radiation field described by Hamiltonian $H = H_0 + H_{AF}$ where H_0 and H_{AF} are the same as in (6.1) and (6.2). We shall try to apply the perturbative ideas developed in the previous sections to the eigenstate $\mid \phi\rangle = \mid \{0_{\mathbf{k}j}\}, \uparrow\rangle$ of H_0, corresponding to the eigenvalue $E_0 = \frac{1}{2}\hbar\omega_0$. This state $\mid \phi\rangle$ can decay to the set of states $\mid 1_{\mathbf{k}j}, \downarrow\rangle$ by spontaneous emission, which is the energy-conserving process described in Section 5.4, because of the presence of V_1 in H_{AF}. We define the projection operator $P_0 = \mid \{0_{\mathbf{k}j}\}, \uparrow\rangle\langle\{0_{\mathbf{k}j}\}, \uparrow\mid$ and we remark that $V_2 \mid \{0_{\mathbf{k}j}\}, \uparrow\rangle = 0$. Thus straightforward use of (C.10), and remembering $\langle\phi \mid H_{AF} \mid \phi\rangle = 0$, gives a shift Δ_L for the atomic excited state

$$\begin{aligned}\Delta_L &= \langle\phi \mid H_{AF}\frac{1}{E_0 - H_0}(1 - P_0)H_{AF} \mid \phi\rangle \\ &= \langle\{0_{\mathbf{k}j}\}, \uparrow\mid V_1\frac{1}{E_0 - H_0}V_1 \mid \{0_{\mathbf{k}j}\}, \uparrow\rangle \\ &= \sum_{\mathbf{k}j}\frac{\mid \epsilon_{\mathbf{k}j}\mid^2}{\hbar(\omega_0 - \omega_k)}\end{aligned} \tag{6.91}$$

This sum diverges because of the singularity at $\omega_k = \omega_0$, but its principal part coincides with $-\hbar\Omega_-$ as given by (5.71). Thus a picture seems to emerge that fits nicely with the spontaneous emission result of Section 5.4, where it was found that the spectrum of the radiation emitted during spontaneous decay of a two-level atom has a peak shifted by the frequency $\Omega_- - \Omega_+$ from the bare-atom frequency ω_0. In fact in the previous section we have seen that the ground state is shifted by the amount $-\hbar\Omega_+$ by self-energy effects and (6.91) seems to indicate that the same effects lead to a shift $-\hbar\Omega_-$ of the excited level which would account for the shift in the peak of the spontaneous emission spectrum.

In order to show that all is not well, and that the V_1 terms are introducing conceptual difficulties, we simplify the problem by neglecting the counterrotating terms in the Hamiltonian and redefine the zero of the energy of the system, thereby obtaining from (6.1) and (6.2) the RWA Hamiltonian

$$\begin{aligned}H = H_0 + V_1 &= \hbar\omega_0(S_z + 1/2) + \sum_{\mathbf{k}j}\hbar\omega_k\left(a^{\dagger}_{\mathbf{k}j}a_{\mathbf{k}j} + 1/2\right) \\ &+ \sum_{\mathbf{k}j}\left(\epsilon_{\mathbf{k}j}a_{\mathbf{k}j}S_+ + \epsilon^*_{\mathbf{k}j}a^{\dagger}_{\mathbf{k}j}S_-\right)\end{aligned} \tag{6.92}$$

This Hamiltonian is sometimes called the Lee-Friedrichs Hamiltonian (Friedrichs 1948). Early work on (6.92) was performed by Hamilton (1947). The Hilbert space of (6.92) can be factorized into subspaces each characterized by an eigenvalue of the total number of excitations $\mathcal{N} = \sum_{\mathbf{k}j} a^{\dagger}_{\mathbf{k}j} a_{\mathbf{k}j} + S_z + 1/2$, since $[\mathcal{N}, H] = 0$ and $\mathcal{N}$ is a constant of motion like in the single-mode case of Section 5.3. The only state belonging to the eigenvalue 0 of $\mathcal{N}$ is $| \{0_{\mathbf{k}j}\}, \downarrow\rangle$ which is an eigenstate of H_0 as well as of H, corresponding to the same eigenvalue 0. The set of eigenstates $| \{0_{\mathbf{k}j}\}, \uparrow\rangle$, $| 1_{\mathbf{k}j}, \downarrow\rangle$ of H_0 span the subspace of H with eigenvalue 1 of $\mathcal{N}$, corresponding to eigenvalues $\hbar(\omega_0 + \omega_k)$ of H_0. Result (6.91) is obtained within this subspace when the shift of the energy of $| \{0_{\mathbf{k}j}\}, \uparrow\rangle$ is evaluated in second-order perturbation theory. It can be shown, however, that the result diverges in fourth-order perturbation theory, and this divergence cannot be cured by the use of principal parts (Friedrichs 1948, Petrosky *et al.* 1991). Moreover, the normalization constant, accurate to $O(\epsilon^2)$, of the state obtained from $| \{0_{\mathbf{k}j}\}, \uparrow\rangle$ by second-order perturbation theory as in (C.12) is

$$\begin{aligned} Z_{(2)}^{-1} &= 1 - \frac{1}{2}\langle\phi \mid V_1(1 - P_0)\frac{1}{(E_0 - H_0)^2}(1 - P_0)V_1 \mid \phi\rangle \\ &= 1 - \frac{1}{2}\langle\{0_{\mathbf{k}j}\}, \uparrow \mid V_1 \frac{1}{(E_0 - H_0)^2} V_1 \mid \{0_{\mathbf{k}j}\}, \uparrow\rangle \\ &= 1 - \frac{1}{2}\sum_{\mathbf{k}j} \frac{| \epsilon_{\mathbf{k}j} |^2}{\hbar^2(\omega_0 - \omega_k)^2} \end{aligned} \tag{6.93}$$

which diverges badly. Thus perturbation theory fails, which throws some doubts on result (6.91), however reasonable the latter may look in view of agreement with the spectrum of spontaneous emission. The reason for this failure can be ascribed to the near degeneracy of state $| \{0_{\mathbf{k}j}\}, \uparrow\rangle$ with states $| 1_{\mathbf{k}j}, \downarrow\rangle$ having $\omega_k \sim \omega_0$ (atom-field resonances), which gives rise to vanishing energy denominators at all orders of perturbation theory. This is not a serious obstacle if the set of field modes is discrete. In fact, in this case only accidental and countable degeneracies due to the atom-field resonances are present and they can be eliminated by a direct diagonalization in the correspondent subspace, at least in principle. The situation is much more complicated in the case of a continuum embedding the atom's transition frequency. If this is the case, in any arbitrarily small neighbourhood of the atomic energy there is a nonnumerable set of quasi-resonant field states, and this leads to the non-integrability of the system (Antoniou and Prigogine 1993).

Consequently, nonperturbative methods are required to deal with the eigenvalue problem of (6.92) within the subspace with eigenvalue 1 of $\mathcal{N}$. We try to solve the eigenvalue equation for the Hamiltonian (6.92) (in the one-excitation subspace)

$$(H - E) \mid \psi\rangle = 0 \tag{6.94}$$

We will discretize the field, discussing the continuum limit only at the end. Expanding the eigenstate $\mid \psi\rangle$ in terms of the complete set of eigenstates of H_0 in the subspace considered $\{\mid \{0_{\mathbf{k}j}\} \uparrow\rangle, \mid 1_{\mathbf{k}j} \downarrow\rangle\}$, we can obtain the eigenvalues E by equating to zero the following determinant

$$\Delta(E) = \begin{vmatrix} E_0^{(0)} - E & \epsilon_1 & \epsilon_2 & \epsilon_3 & \cdots \\ \epsilon_1^* & E_1^{(0)} - E & 0 & 0 & \cdots \\ \epsilon_2^* & 0 & E_2^{(0)} - E & 0 & \cdots \\ \epsilon_3^* & 0 & 0 & E_3^{(0)} - E & \cdots \\ \vdots & \vdots & \vdots & \vdots & \vdots \end{vmatrix} \tag{6.95}$$

where index $k (k = 1, 2, \cdots)$ represents the pair $\mathbf{k}j$ referring to a free mode. Thus $\epsilon_k \equiv \epsilon_{\mathbf{k}j}$ and $E_k^{(0)} \equiv E_{\mathbf{k}j}^{(0)} = \hbar\omega_k$ is the unperturbed energy of the $\mathbf{k}j$ field mode for $k = 1, 2, \cdots$. $E_0^{(0)} = \hbar\omega_0$ represents the unperturbed atomic energy. For a finite quantization volume V the modes are discrete, and in the presence of dipole approximation the cutoff frequency ω_M implies that the sequence of $E_k^{(0)}$ terminates at some finite value $E_M^{(0)} = \hbar\omega_M$. In these conditions we are faced with a finite-dimensional eigenvalue problem, and the search for the zeros of $\Delta(E)$ reduces to solving a finite set of algebraic equations. By considering the sequence of determinants of the form (6.95) with $M = 2, 3, \cdots$ it is easy to convince oneself that (6.95) can be cast in the form

$$\Delta(E) = \left\{ \prod_{k=1}^{M} \left(E_k^{(0)} - E \right) \right\} \left\{ E_0^{(0)} - E - \sum_{k=1}^{M} \frac{\mid \epsilon_k \mid^2}{E_k^{(0)} - E} \right\} \tag{6.96}$$

The first factor on the RHS of (6.96) does not contain any information on the atom or on the atom-field interaction. Its zeros coincide with the unperturbed field eigenvalues $E_k^{(0)}$. However, these zeros are not the zeros of $\Delta(E)$ because they are cancelled by the poles of the second factor on the RHS of (6.96). This is evident by writing equation (6.96) in the following form

$$\Delta(E) = \left(\prod_{k=1}^{M}\left(E_k^{(0)} - E\right)\right)\left(E_0^{(0)} - E\right)$$

$$- \sum_{k=1}^{M} |\, \epsilon_k \,|^2 \left(\prod_{\substack{k'=1 \\ k' \neq k}}^{M}\left(E_{k'}^{(0)} - E\right)\right) \tag{6.97}$$

This shows that $\Delta(E)$ does not vanish for $E = E_n^{(0)}$ $(n = 0, 1, \ldots, M)$ because of the second term on the RHS of (6.97) which depends on the atom-field coupling. Therefore the $M + 1$ zeros of $\Delta(E)$ must come from the second factor on the RHS of (6.96). Thus the zeros of the second factor

$$\Delta'(E) = E_0^{(0)} - E - F(E)\ ;\ F(E) = \sum_{k=1}^{M} \frac{|\, \epsilon_k \,|^2}{E_k^{(0)} - E} \tag{6.98}$$

yield the eigenvalues E_n $(n = 0, 1, \ldots, M)$ of H within the one-excitation subspace. Equation $\Delta'(E) = 0$ is generally not soluble analytically, but graphically its solutions can be obtained with the help of Figure 6.3, where the spiky function with poles at $E = E_1^{(0)} = \hbar\omega_1 = 0$, $E = E_2^{(0)} = \hbar\omega_2, \ldots$ along the positive E axis and at the origin represents $F(E)$ defined in (6.98). The eigenvalues E_n of H are the values of E for which the straight diagonal line

$$f(E) = E_0^{(0)} - E \tag{6.99}$$

intercepts $F(E)$. The eigenstates $|\, \psi_n\rangle$ of H corresponding to each E_n are obtained as a linear combination of the eigenstates of H_0 belonging to the one-excitation subspace of H_0 in the form

$$|\, \psi_n\rangle = A_0(n) \,|\, \{0_k\}, \uparrow\rangle + \sum_{k=1}^{M} A_k(n) \,|\, 1_k, \downarrow\rangle \tag{6.100}$$

The coefficients $A_j(n)$ $(j = 0, 1, \ldots M)$ are determined from the eigenvalue equation

$$\Delta_{ij}(E_n) A_j(n) = 0 \tag{6.101}$$

where Δ_{ij} is the i, j matrix element in (6.95). Thus (6.101) for a given n is equivalent to a set of coupled equations

$$\left(E_0^{(0)} - E_n\right) A_0(n) + \sum_{k=1}^{M} \epsilon_k A_k(n) = 0\ ;$$

$$\epsilon_k^* A_0(n) + \left(E_k^{(0)} - E_n\right) A_k(n) = 0 \quad (k = 1, 2, \ldots M) \tag{6.102}$$

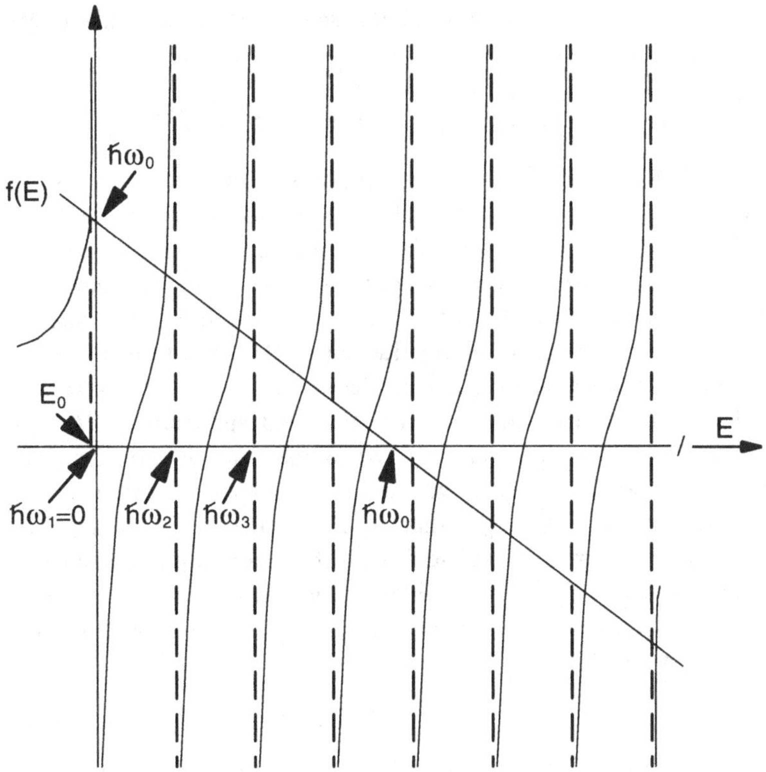

Fig. 6.3 Graphical solution of the eigenvalue problem of the Lee-Friedrichs Hamiltonian. The eigenvalues are the values of E on the horizontal axis corresponding to the points where the straight diagonal line $f(E)$ intercepts the tangent-like curves.

From the second of (6.102) one has

$$A_k(n) = -\frac{\epsilon_k^*}{E_k^{(0)} - E_n} A_0(n) \quad (k = 1, 2, \ldots M) \tag{6.103}$$

while the normalization of $\mid \psi_n \rangle$ yields

$$\mid A_0(n) \mid^2 + \sum_{k=1}^{M} \frac{\mid \epsilon_k \mid^2}{(E_k^{(0)} - E_n)^2} \mid A_0(n) \mid^2 = 1 \; ;$$

$$\mid A_0(n) \mid = \frac{1}{\sqrt{1 + F'(E_n)}} \; ;$$

$$F'(E_n) = \left. \frac{dF}{dE} \right|_{E=E_n} = \sum_{k=1}^{M} \frac{\mid \epsilon_k \mid^2}{(E_k^{(0)} - E_n)^2} \tag{6.104}$$

Consequently, within an arbitrary phase factor (Glaser and Källen 1956/57)

$$|\psi_n\rangle = \frac{1}{\sqrt{1+F'(E_n)}}\left\{|\{0_{\mathbf{k}j}\},\uparrow\rangle + \sum_{k=1}^{M}\frac{\epsilon_k^*}{E_n - E_k^{(0)}}|1_k,\downarrow\rangle\right\} \qquad (6.105)$$

The procedure followed is in principle exact, and the results on the eigenvalue spectrum prompt the following considerations. Since $E_M^{(0)}$ is fixed due to the physical nature of the cutoff frequency ω_M, the system seems to possess at least two important parameters: ϵ_k and the number M of field modes. Increasing M leads to increasing the density of spikes for $E > 0$ in Figure 6.3, but each eigenvalue $E_n (n \geq 1)$ is constrained between $E_n^{(0)}$ and $E_{n+1}^{(0)}$. If we assume ϵ_k independent of M, the $E > 0$ spectrum of H will look more and more like the free-field spectrum of H_F with increasing M (at constant $\epsilon_k \neq 0$), while the only $E < 0$ eigenvalue becomes more and more negative. At the same time, the atomic point eigenvalue of H_0 at $E = E_0^{(0)}$, that is, the excited bare atomic state, becomes more and more "confused" with the quasi-continuum states of the field. This can be seen by considering for example the probability of finding in $|\psi_n\rangle$ the atom excited and no photons; this probability is obtained from (6.105) as

$$|\langle\{0_{\mathbf{k}j}\},\uparrow|\psi_n\rangle|^2 = [1 + F'(E_n)]^{-1} \qquad (6.106)$$

The explicit form of $F'(n)$ in (6.104) shows clearly that $F'(E_n)$ diverges upon increasing M at constant $\epsilon_k \neq 0$, at least if E_n remains finite. Therefore all finite energy eigenstates of H have a vanishingly small $|0_{\mathbf{k}j},\uparrow\rangle$ character in the continuum limit (independently of the value of $\epsilon_k \neq 0$). This seems to show that the discrete state of H_0 dissolves in the continuum spectrum of H. This conclusion is also supported by the fact that in the continuum limit the $E > 0$ spectrum of H looks exactly like that of the free field Hamiltonian H_F and, if the only negative eigenvalue goes to $-\infty$, the same holds for the whole finite spectrum. This equivalence of the spectrum of H_0 with the spectrum of H_F is independent of the magnitude of ϵ_k, provided $\epsilon_k \neq 0$, and it seems to imply a nonanalytical dependence on ϵ_k of the spectrum of H. In fact any arbitrary small value of ϵ_k changes qualitatively the spectrum of H from a continuum plus a discrete (when $\epsilon_k = 0$) to a continuum only (for any $\epsilon_k \neq 0$). Also, all nonzero values of ϵ_k lead to the same spectrum in the continuum limit. In this limit the spectrum of H does not seem to contain any information about the existence in H_0 of a point atomic level at $E = E_0$. On the other hand, one can take $\epsilon_k \to 0$ in each member of the sequence of systems $\Delta(E) = 0$ with increasing $M < \infty$. Then each member of this sequence yields the complete unperturbed spectrum of H_0, including the point eigenvalue at $E = E_0^{(0)}$.

The situation is quite different if it assumed that $| \epsilon_k |^2 \sim 1/M$, which obviously means a "decrease" of the interaction with each single field mode while increasing the number of modes. We only mention that this is the case for the usual atom-field interactions when $| \epsilon_k |^2 \sim 1/V$, V being the quantization volume proportional to the number of modes of the field. We remark that this assumption is necessary to obtain in the continuum limit a finite spontaneous emission rate γ, since γ is proportional to $| \epsilon_k |^2 M$. A discussion of this case can be found in Fano (1961) and in Cohen-Tannoudji *et al.* (1992). In this case the Hamiltonian H can be diagonalized without use of perturbation theory and, analogously to the previous case, it is found that the energy spectrum of H is equal to the purely continuous spectrum of the field H_F. The eigenstates of H, however, when expressed in terms of eigenstates of H_0, retain the memory of the discrete level; this memory is particularly evident in the new continuum in an interval of width $\hbar\gamma$ around $E_0^{(0)}$, as expected from physical considerations.

Until now we have discussed examples of the conceptual difficulties met in the diagonalization of Hamiltonians involving the interaction between discrete levels and a continuum. Another difficulty is the description of decay processes in a consistent way. In fact the states (6.105), being eigenstates of the total Hamiltonian H, are stationary states and do not decay. But decay of excited states is experimentally observed, and this fact should be included in the theory to obtain a nonambiguous definition of a dressed excited state. Recent attemps to introduce spontaneous decay and damping at a fundamental level are based on an extension of the eigenvalue problem (6.94) to the complex E-plane (Petrosky *et al.* 1991, and references therein). This assumption does not contrast with the hermiticity of H if the norm of the eigenstates with complex eigenvalues vanishes. Thus, if we write the complex energy eigenvalue as $\tilde{E} - i\hbar\gamma$ with $\tilde{E}$ and γ real, we have by definition

$$H \mid \psi\rangle = \left(\tilde{E} - i\hbar\gamma\right) \mid \psi\rangle \tag{6.107}$$

Therefore the time evolution of the state $\mid \psi(0)\rangle = \mid \psi\rangle$ is given by

$$\mid \psi(t)\rangle = e^{-iHt} \mid \psi\rangle = e^{-i\tilde{E}t/\hbar - \gamma t} \mid \psi\rangle \tag{6.108}$$

The imaginary part $-i\hbar\gamma$ of the complex eigenvalue yields an exponential damping factor that can be interpreted as the spontaneous decay of the unstable dressed state $|\psi\rangle$. The real part $\tilde{E}$ of the eigenvalue is obviously interpreted as the energy of the unstable state $|\psi\rangle$ inclusive of radiative corrections.

The underlying theory of diagonalization of H in the complex plane has not yet been completed, and its discussion is beyond the purpose of this book. We just mention that its consistent mathematical implementation involves sophisticated contour integrations (Sudarshan *et al.* 1978) or alternatively new rules of analytic continuation in the complex plane. This seems to imply time symmetry breaking and irreversibility in the dynamics of the system (Petrosky *et al.* 1991, 1993).

While these developments are certainly interesting *per se*, the lesson we have to draw from the considerations developed in the present section is that a study of the dress of an excited atomic state starting from the corresponding dressed eigenstate poses rather formidable problems in view of the conceptual difficulties involved in the definition of such a dressed excited eigenstate. In the next section we shall explore a different approach which bypasses an explicit solution of the eigenvalue problem of H.

6.6 Van Hove theory of dressed states

In this section we shall present some of the main results of Van Hove's analysis of the theory of dressed states (Van Hove 1955, 1956). As we shall see, these ideas permit us to side-step some of the difficulties posed by the solution of the eigenvalue problem discussed in the previous section and to obtain a framework useful to describe the main features of the dynamics of the excited states in the presence of interactions leading to dressing effects. It is a fact, however, that a detailed investigation of the virtual cloud for the excited states of a field source, of the sort introduced in this chapter for stable states which we intend to develop more precisely in the rest of this book, has not yet been done on the basis of Van Hove's ideas. Consequently, a fully developed account of these ideas, which would in any case take a disproportionate amount of space, would be out of place here, and our efforts will be aimed at providing a general outline of the main results of Van Hove theory, which may encourage the reader interested in further explorations to consult the relevant references.

The object of Van Hove theory is a classification of the time development of an arbitrary state $|\psi(t)\rangle$ of a general system described by a time-independent Hamiltonian H which can be partitioned as $H = H_0 + V$ into an unperurbed Hamiltonian H_0, whose eigenstates $|\alpha\rangle$

and eigenvalues E_α are known, and into an interaction V. In a typical radiation problem H_0 is the bare source Hamiltonian and V contains in general rotating as well as counterrotating terms. As is well known

$$| \psi(t)\rangle = e^{-(i/\hbar)Ht} | \psi(0)\rangle \tag{6.109}$$

where $| \psi(0)\rangle$ is a prescribed initial state of the total system. The state at time t, as given from (6.109), can be expressed in terms of an operator function of a complex variable, which is called the resolvent of H and which is defined as (see e.g. Hugenholtz 1959)

$$G(z) = \frac{1}{z - H} \tag{6.110}$$

where z is a complex c-variable with the dimensions of an energy. In fact the matrix elements on the eigenstates of H_0 are

$$\langle \alpha | G(z) | \alpha' \rangle = \sum_\lambda \frac{\langle \alpha | \lambda \rangle \langle \lambda | \alpha' \rangle}{z - E_\lambda} \tag{6.111}$$

where $| \lambda \rangle$ are the eigenstates of H and E_λ are the corresponding eigenvalues. Note that the sum in (6.111) becomes an integral for the continuous part of the spectrum of H. Moreover, the form of (6.111) shows that $\langle \alpha | G(z) | \alpha' \rangle$, for any α and α', is analytic everywhere in the complex z-plane except on the real axis, because E_λ must be real in view of the self-adjointness of H. Consider now the matrix element

$$\begin{aligned} \langle \alpha | e^{-(i/\hbar)Ht} | \alpha' \rangle &= \sum_\lambda \langle \alpha | \lambda \rangle \langle \lambda | \alpha' \rangle e^{-(i/\hbar)E_\lambda t} \\ &= -\frac{1}{2\pi i} \sum_\lambda \langle \alpha | \lambda \rangle \langle \lambda | \alpha' \rangle \oint \frac{e^{-(i/\hbar)zt}}{z - E_\lambda} dz \\ &= -\frac{1}{2\pi i} \oint \langle \alpha | G(z) | \alpha' \rangle e^{-(i/\hbar)zt} dz \end{aligned} \tag{6.112}$$

where the closed integration path encircles all of the real z-axis as in Figure 6.4 and where (6.111) has been used. Then from (6.109)

$$\begin{aligned} | \psi(t)\rangle &= \sum_{\alpha\alpha'} | \alpha \rangle \langle \alpha | e^{-(i/\hbar)Ht} | \alpha' \rangle \langle \alpha' | \psi(0)\rangle \\ &= -\frac{1}{2\pi i} \oint \sum_{\alpha\alpha'} | \alpha \rangle \langle \alpha | G(z) | \alpha' \rangle \langle \alpha' | e^{-izt} dz | \psi(0)\rangle \\ &= -\frac{1}{2\pi i} \oint G(z) e^{-(i/\hbar)zt} dz | \psi(0)\rangle \end{aligned} \tag{6.113}$$

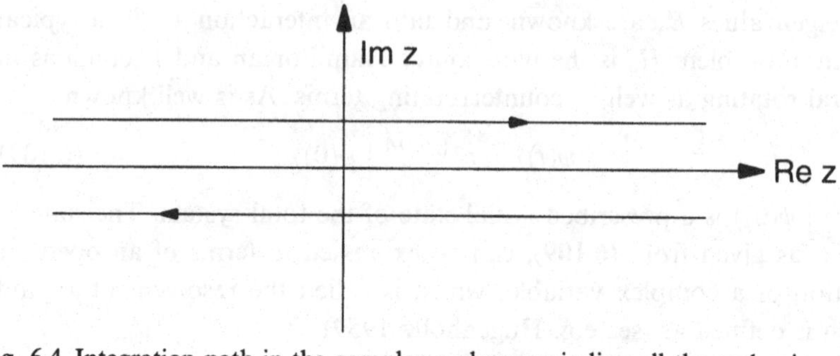

Fig. 6.4 Integration path in the complex z-plane encircling all the real axis.

which proves our statement. For $t > 0$ the lower part of the integration path can be deformed to an infinite half-circle in the lower half-plane, which yields a vanishing contribution to the integral, and the upper part of the integration path can be brought indefinitely near to the real axis without changing the value of the integral. Thus for $t > 0$ (6.113) can be written as

$$| \psi(t)\rangle = -\frac{1}{2\pi i}\int_{-\infty}^{\infty} \frac{e^{-(i/\hbar)zt}}{z + i\eta - H} dz \mid \psi(0)\rangle \quad (\eta \to 0) \tag{6.114}$$

Expressions (6.113) and (6.114) show that the analytic properties of $G(z)$ determine the time evolution of $| \psi(t)\rangle$. For example, a quantity important from the physical point of view is $\langle \alpha \mid e^{-(i/\hbar)Ht} \mid \alpha\rangle$, which gives the probability amplitude of finding the system in a given eigenstate $| \alpha\rangle$ of H_0 after time t. In view of (6.112) one has

$$\langle \alpha \mid e^{-(i/\hbar)Ht} \mid \alpha\rangle = -\frac{1}{2\pi i}\oint D_\alpha(z) e^{-(i/\hbar)zt} dz \; ;$$
$$D_\alpha(z) \equiv \langle \alpha \mid D(z) \mid \alpha\rangle \tag{6.115}$$

where $D(z)$ is an operator defined as the diagonal part of $G(z)$, such that

$$D(z) \mid \alpha\rangle = D_\alpha(z) \mid \alpha\rangle \tag{6.116}$$

for any eigenstate $| \alpha\rangle$ of H_0. Extracting the diagonal part of the resolvent is not a trivial matter in general, but Van Hove succeeded in proving that

$$D(z) = \frac{1}{z - H_0 - \Sigma(z)} \tag{6.117}$$

where $\Sigma(z)$ is an operator diagonal in the α-representation. An expression for the diagonal matrix elements of $\Sigma(z)$ in the α-representation can be obtained in the form (Cohen-Tannoudji *et al.* 1992)

$$P_\alpha \Sigma(z) P_\alpha = P_\alpha V P_\alpha + P_\alpha V \frac{Q_\alpha}{z - H_0 - Q_\alpha V Q_\alpha} V P_\alpha \,;$$
$$Q_\alpha = 1 - P_\alpha \tag{6.118}$$

where P_α is the operator projection onto state $| \alpha\rangle$. Clearly, the properties of $\Sigma(z)$ are important for the analytic properties of $D(z)$. As we have seen, $G(z)$, and consequently $D(z)$, can have singularities only on the real z-axis. These singularities are poles if they correspond to isolated eigenvalues of H, generally indicating bound states. If H has also a continuum of eigenvalues, the corresponding singularities cannot be poles, which are isolated by definition, but they are rather represented by a branch cut along the real z-axis. $D(z)$ takes different values on either side of the cut, and since it can be shown that $\Sigma(z^*) = \Sigma^*(z)$ (Van Hove 1955), the form of this discontinuity can be guessed. In fact, introducing $E \equiv \mathrm{Re}(z)$, one must have for E along the cut

$$\lim_{\eta \to 0} \Sigma(E \pm i\eta) = \Delta(E) \mp \frac{i}{2}\Gamma(E) \tag{6.119}$$

where both Δ and Γ are real diagonal operators. Hence from (6.117)

$$\lim_{\eta \to 0} D_\alpha(E \pm i\eta) = \frac{1}{E - E_\alpha - \Delta_\alpha(E) \pm \frac{i}{2}\Gamma_\alpha(E)} \tag{6.120}$$

where subscript α indicates the diagonal matrix element of the operator on eigenstate $| \alpha\rangle$ of H_0, and where $\Gamma_\alpha(E) \geq 0$ (Davidovich 1975). It should be noted that (6.120) is the quantity appearing in integral (6.115), which is related to the probability of change of state $| \alpha\rangle$. This supports our statement about the importance of the analytic properties of $G(z)$ for the dynamical behaviour of the system, and it encourages us to consider the form (6.120) in some more detail.

Expression (6.120), which is obtained on quite general grounds, shows that for small Δ_α and Γ_α, $D_\alpha(z)$ has a peak of width $\sim \Gamma_\alpha(E_\alpha)$ in the neighbourhood of $E'_\alpha = E_\alpha + \Delta_\alpha(E_\alpha)$. This is the so-called "pole approximation" for $D_\alpha(z)$. If $\Gamma_\alpha(E_\alpha) \neq 0$, there is a finite discontinuity of $D_\alpha(z)$ across the cut. This means that analytic continuation of $D_\alpha(z)$ yields a multivalued function, and that the discontinuity is due to a pole with $\mathrm{Im}(z) \neq 0$ on a second Riemann sheet of $D_\alpha(z)$. Thus, integral (6.115) can be evaluated by deforming the integration contour on this unphysical Riemann, sheet, and the complex pole yields a decay of state $| \alpha\rangle$ with decay time of the order of $\hbar/\Gamma_\alpha(E_\alpha)$. Thus, $\Gamma_\alpha(E_\alpha) \neq 0$ indicates dissipative behaviour of the system in state $|\,\alpha\rangle$. If $\Gamma_\alpha(E_\alpha)=0$ and if $\Gamma_\alpha(E)$ vanishes all the way through the real axis, then according to Van Hove the state $|\,\alpha\rangle$, although not an eigenstate of the complete Hamiltonian H, must be regarded as an asymptotic stationary state with energy E_α. States of this sort are characterized by the property that a wave packet made of these states and of the form

$$|\,\psi(t_0)\rangle = \int c(\alpha)e^{-(i/\hbar)E_\alpha t_0}\,|\,\alpha\rangle d\alpha \tag{6.121}$$

where $c(\alpha)$ is a sufficiently smooth function of α, satisfies asymptotically

$$e^{-(i/\hbar)Ht}\,|\,\psi(t_0)\rangle \sim |\,\psi(t_0+t)\rangle \tag{6.122}$$

in the limit $t_0\to\infty$. The physical meaning of the property of asymptotic stationarity can be understood in terms of V yielding an interaction which can be regarded as localized in space and time, as in the phenomenon of scattering from a fixed centre. In this case, V causes transitions from one eigenstate of H_0 to another. After an infinite time has elapsed, however, and after all scattering processes have taken place, the eigenstates $|\,\alpha\rangle$ of H_0 behave as stable states. Finally, for the case $\Gamma_\alpha(E_\alpha)=0$, but in the presence of a range of E on the real axis for which $\Gamma_\alpha(E)\neq 0$, Van Hove has shown that a wave packet of the form (6.121) is not even asymptotically stationary. This physically corresponds to an interaction brought about by V, which cannot be localized in space and time, but which acts permanently on the unperturbed eigenstates of $|\,\alpha\rangle$. This leads to a qualitative change in the structure of the eigenstates of H with respect to the eigenstates of H_0 and it gives rise to the dressed states.

We shall now illustrate by some practical examples relative to the atom-photon interaction the statements above. Consider first a two-level atom interacting with the free e.m. field in the minimal coupling scheme, in the electric dipole approximation and in the RWA. The Hamiltonian is, from (6.1) and (6.2)

$$\begin{aligned} H &= H_0 + V_1\,; \\ H_0 &= \hbar\omega_0 S_z + \sum_{\mathbf{k}j}\hbar\omega_k a^\dagger_{\mathbf{k}j}a_{\mathbf{k}j}\,; \\ V_1 &= \sum_{\mathbf{k}j}\left(\epsilon_{\mathbf{k}j}a_{\mathbf{k}j}S_+ + \epsilon^*_{\mathbf{k}j}a^\dagger_{\mathbf{k}j}S_-\right) \end{aligned} \tag{6.123}$$

We wish to investigate the properties of state $|\,\alpha\rangle = |\,\{0_{\mathbf{k}j}\},\uparrow\rangle$ using the results of Van Hove theory. This state is an eigenstate of H_0 corresponding to the unperturbed eigenvalue $E_\alpha=\hbar\omega_0/2$. From (6.118) we have

$$P_\alpha \Sigma(z) P_\alpha = P_\alpha V_1 \frac{Q_\alpha}{z - \hbar\omega_0 S_z - \sum_{\mathbf{k}j} \hbar\omega_k a^\dagger_{\mathbf{k}j} a_{\mathbf{k}j}} V_1 P_\alpha + O(\epsilon^3_{\mathbf{k}j}) \quad (6.124)$$

A straightforward algebraic development yields, neglecting $O(\epsilon^3_{\mathbf{k}j})$,

$$\langle \alpha \mid P_\alpha \Sigma(z) P_\alpha \mid \alpha \rangle = \Sigma_{\mathbf{k}j} \frac{\mid \epsilon_{\mathbf{k}j} \mid^2}{z + \frac{1}{2}\hbar\omega_0 - \hbar\omega_k} \quad (6.125)$$

while use of (5.61) leads to

$$\langle \alpha \mid P_\alpha \Sigma(z) P_\alpha \mid \alpha \rangle = \frac{\gamma \hbar^2}{2\pi\omega_0} \int_0^{\omega_M} \frac{\omega_k}{z + \frac{1}{2}\hbar\omega_0 - \hbar\omega_k} d\omega_k$$

$$= -\frac{\gamma}{2\pi\omega_0} \left\{ \hbar\omega_M + \left(z + \frac{1}{2}\hbar\omega_0 \right) \ln \frac{z + \frac{1}{2}\hbar\omega_0 - \hbar\omega_M}{z + \frac{1}{2}\hbar\omega_0} \right\} \quad (6.126)$$

It is not difficult to see that (6.126) has a branch cut along the real z-axis from $E = -\frac{1}{2}\hbar\omega_0$ to $E = \hbar\omega_M - \frac{1}{2}\hbar\omega_0$, where we have set $E = \mathrm{Re}(z)$. Introducing the two variables $z_0 = \mid z_0 \mid e^{i\theta_0}$ and $z_{0M} = \mid z_{0M} \mid e^{i\theta_{0M}}$ instead of z as shown in Figure 6.5, with $-\pi < (\theta_0, \theta_{0M}) < \pi$, one has

$$z_0 = z + \frac{1}{2}\hbar\omega_0 \; ; \; z_{0M} = z + \frac{1}{2}\hbar\omega_0 - \hbar\omega_M \; ;$$

$$\ln \frac{z_{0M}}{z_0} = \ln \mid z_{0M} \mid - \ln \mid z_0 \mid + i(\theta_{0M} - \theta_0) \quad (6.127)$$

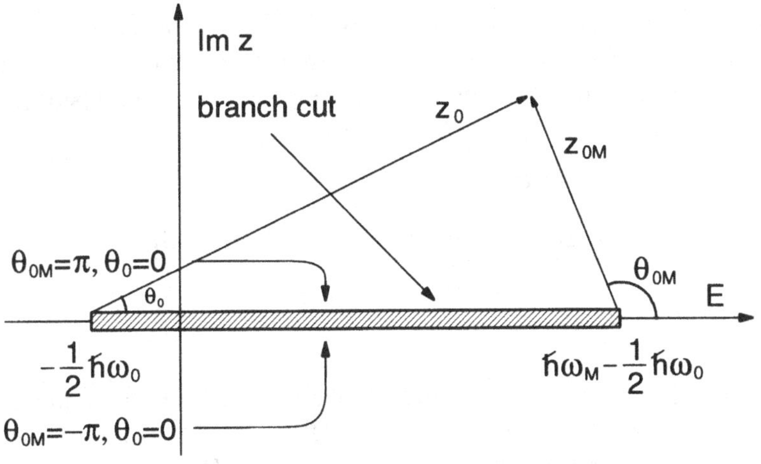

Fig. 6.5 The branch-cut in the complex z-plane for the unstable state where the atom is excited and no photon is present.

Thus, approaching the real axis from above and from below at any point E such that $-\frac{1}{2}\hbar\omega_0 < E < \hbar\omega_M - \frac{1}{2}\hbar\omega_0$ one has from (6.126) and (6.127)

$$\langle\alpha \mid P_\alpha\Sigma(E \pm i\eta)P_\alpha \mid \alpha\rangle = -\frac{\gamma}{2\pi\omega_0}\left\{\hbar\omega_M + \left(E + \frac{1}{2}\hbar\omega_0\right).\right.$$
$$\left.\times\left(\ln\frac{\hbar\omega_M - \frac{1}{2}\hbar\omega_0 - E}{E + \frac{1}{2}\hbar\omega_0} \pm i\pi\right)\right\} \quad (\eta \to 0) \tag{6.128}$$

This shows that (6.126) has indeed a branch cut as shown in Figure 6.5. Moreover, use of (6.119) shows immediately that, for E on this cut,

$$\Delta_\alpha(E) = -\frac{\gamma}{2\pi\omega_0}\left\{\hbar\omega_M + \left(E + \frac{1}{2}\hbar\omega_0\right)\ln\frac{\hbar\omega_M - \frac{1}{2}\hbar\omega_0 - E}{E + \frac{1}{2}\hbar\omega_0}\right\};$$
$$\Gamma_\alpha(E) = \frac{\gamma}{\omega_0}\left(E + \frac{1}{2}\hbar\omega_0\right) \tag{6.129}$$

Since $E_\alpha = \hbar\omega_0/2$ is within the cut, one immediately finds that for the state in question $D_\alpha(z)$ has a peak near $E'_\alpha = E_\alpha + \Delta_\alpha(E_\alpha)$ of width $\Gamma_\alpha(E_\alpha)$ where

$$\Delta_\alpha(E_\alpha) = -\frac{\gamma}{2\pi\omega_0}\left\{\hbar\omega_M + \hbar\omega_0\ln\frac{\omega_M - \omega_0}{\omega_0}\right\};\ \Gamma_\alpha(E_\alpha) = \hbar\gamma \tag{6.130}$$

This characterizes $\mid \{0_{\mathbf{k}j}\}, \uparrow\rangle$ as an unstable state with a decay time γ^{-1} (Davidovich 1975), hardly a surprising result in view of the results of Wigner-Weisskopf theory outlined in Section 5.4.

As a second application of Van Hove theory, we consider the state $\mid \alpha\rangle = \mid 1_{\mathbf{k}j}, \downarrow\rangle$, with the same Hamiltonian (6.123) as in the previous example. This is an eigenstate of H_0 corresponding to the unperturbed eigenvalue $E_\alpha = \hbar\omega_k - \frac{1}{2}\hbar\omega_0$. One has, neglecting $O(\epsilon^3_{\mathbf{k}j})$,

$$\begin{aligned}&\langle\alpha \mid P_\alpha\Sigma(z)P_\alpha \mid \alpha\rangle\\ &= \langle 1_{\mathbf{k}j}, \downarrow\mid V_1\frac{1}{z - \hbar\omega_0 S_z - \sum_{\mathbf{k}j}\hbar\omega_k a^\dagger_{\mathbf{k}j}a_{\mathbf{k}j}}V_1 \mid 1_{\mathbf{k}j}, \downarrow\rangle\\ &= \frac{\mid \epsilon_{\mathbf{k}j}\mid^2}{z - \frac{1}{2}\hbar\omega_0}\end{aligned} \tag{6.131}$$

Since $\mid \epsilon_{\mathbf{k}j}\mid^2$ is inversely proportional to the volume V, in the continuum limit $V \to \infty$ one has trivially $\langle\alpha \mid P_\alpha\Sigma(z)P_\alpha \mid \alpha\rangle = 0$. Thus in this limit $\Gamma_\alpha(E) = 0$ for any E, and according to Van Hove $\mid 1_{\mathbf{k}j}, \downarrow\rangle$ is an asymptotically stable state.

Finally, consider the complete two-level Hamiltonian including the counterrotating terms

$$H = H_0 + V_1 + V_2\ ;\ V_2 = \sum_{\mathbf{k}j}\left(\epsilon_{\mathbf{k}j} a^\dagger_{\mathbf{k}j} S_+ + \epsilon^*_{\mathbf{k}j} a_{\mathbf{k}j} S_-\right) \tag{6.132}$$

and the state $|\,\alpha\rangle = |\,\{0_{\mathbf{k}j}\}, \downarrow\rangle$, which is an eigenstate of H_0 corresponding to the unperturbed eigenvalue $E_\alpha = -\hbar\omega_0/2$. In the presence of V_2 this is not an eigenstate of the total Hamiltonian. Neglecting again $O(\epsilon^3_{\mathbf{k}j})$, and using (5.61) for minimal coupling, one has

$$\begin{aligned}
&\langle\alpha\,|\,P_\alpha\Sigma(z)P_\alpha\,|\,\alpha\rangle \\
&= \langle\{0_{\mathbf{k}j}\}, \downarrow|\, V_2 \frac{1}{z - \hbar\omega_0 S_z - \sum_{\mathbf{k}j}\hbar\omega_k a^\dagger_{\mathbf{k}j} a_{\mathbf{k}j}} V_2\,|\,\{0_{\mathbf{k}j}\}, \downarrow\rangle \\
&= \sum_{\mathbf{k}j} \frac{|\,\epsilon_{\mathbf{k}j}\,|^2}{z - \frac{1}{2}\hbar\omega_0 - \hbar\omega_k} = \frac{\gamma\hbar^2}{2\pi\omega_0}\int_0^{\omega_M} \frac{\omega_k}{z - \frac{1}{2}\hbar\omega_0 - \hbar\omega_k}\, d\omega_k \\
&= -\frac{\gamma}{2\pi\omega_0}\left\{\hbar\omega_M + \left(z - \frac{1}{2}\hbar\omega_0\right)\ln\frac{z - \frac{1}{2}\hbar\omega_0 - \hbar\omega_M}{z - \frac{1}{2}\hbar\omega_0}\right\}
\end{aligned} \tag{6.133}$$

Clearly, (6.133) has a branch cut along $\mathrm{Re}(z) = E$ from $E = \frac{1}{2}\hbar\omega_0$ to $E = \hbar\omega_M + \frac{1}{2}\hbar\omega_0$ as displayed in Figure 6.6. In fact, using the two variables

$$z_0 = z - \frac{1}{2}\hbar\omega_0\ ;\ z_{0M} = z - \frac{1}{2}\hbar\omega_0 - \hbar\omega_M \tag{6.134}$$

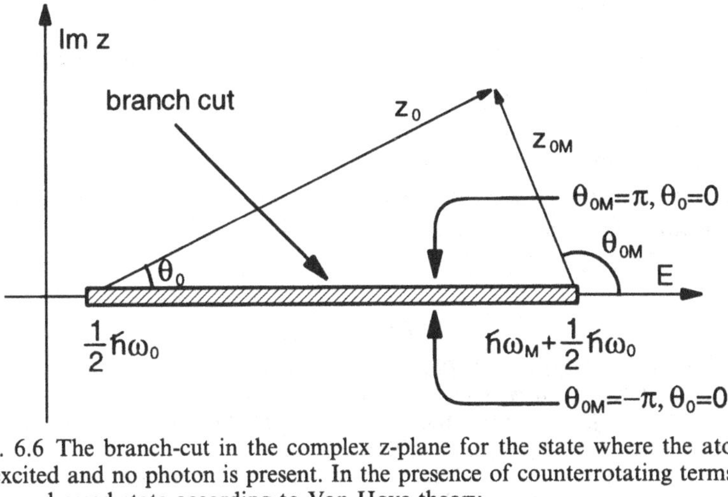

Fig. 6.6 The branch-cut in the complex z-plane for the state where the atom is unexcited and no photon is present. In the presence of counterrotating terms this gives a dressed state according to Van Hove theory.

with $z_0 =| z_0 | e^{i\theta_0}$, $z_{0M} =| z_{0M} | e^{i\theta_{0M}}$ and $-\pi < (\theta_0, \theta_{0M}) < \pi$, one has

$$\ln\frac{z_{0M}}{z_0} = \ln | z_{0M} | - \ln | z_0 | + i(\theta_{0M} - \theta_0) \tag{6.135}$$

Use of (6.135) in (6.133) gives, for E within the branch cut

$$\langle\alpha | P_\alpha \Sigma(E \pm i\eta) P_\alpha | \alpha\rangle = -\frac{\gamma}{2\pi\omega_0}\left\{\hbar\omega_M + \left(E - \frac{1}{2}\hbar\omega_0\right)\cdot \right.$$
$$\left.\times\left(\ln\frac{\hbar\omega_M + \frac{1}{2}\hbar\omega_0 - E}{E - \frac{1}{2}\hbar\omega_0} \pm i\pi\right)\right\} \quad (\eta \to 0) \tag{6.136}$$

In the present case, however, E_α does not lie within the branch cut, but rather in the region to the left of $E = \frac{1}{2}\hbar\omega_0$, where

$$\langle\alpha | P_\alpha \Sigma(E \pm i\eta) P_\alpha | \alpha\rangle$$
$$= -\frac{\gamma}{2\pi\omega_0}\left\{\hbar\omega_M + \left(E - \frac{1}{2}\hbar\omega_0\right)\cdot\right.$$
$$\left.\times \ln\frac{| E - \hbar\omega_M - \frac{1}{2}\hbar\omega_0 |}{E - \frac{1}{2}\hbar\omega_0}\right\} \quad (\eta \to 0) \tag{6.137}$$

has no imaginary part. Thus $| \{0_{\mathbf{k}j}\}, \downarrow\rangle$ is not a decaying state since $\Gamma_\alpha(E_\alpha) = 0$, although $\Gamma_\alpha(E) \neq 0$ within the branch cut on the real axis, where (6.136) shows that

$$\Gamma_\alpha(E) = \frac{\gamma}{\omega_0}\left(E - \frac{1}{2}\hbar\omega_0\right) \tag{6.138}$$

These are the conditions for a dressed state in Van Hove theory, in agreement with the arguments presented in this chapter in favour of the presence of a virtual dressing cloud.

We conclude this section by remarking that although Van Hove theory provides a valid diagnostic means to ascertain the dressed nature of a state, it does not seem to have been used hitherto to calculate the properties of the virtual cloud.

References

I. Antoniou, I. Prigogine (1993). *Physica A* **192,** 443
M. Bolsterli (1991). *J. Math. Phys.* **32,** 254
C. Cohen-Tannoudji, J. Dupont-Roc, G. Grynberg (1989). *Photons and Atoms* (John Wiley and Sons Inc., New York)
C. Cohen-Tannoudji, J. Dupont-Roc, G. Grynberg (1992). *Atom-Photon Interactions* (John Wiley and Sons Inc., New York)

G. Compagno, G.M. Palma, R. Passante, F. Persico (1990). In *New Frontiers in Quantum Electrodynamics and Quantum Optics,* A.O. Barut (ed.) (Plenum Press, New York) p. 129
D.P. Craig, T. Thirunamachandran (1984). *Molecular Quantum Electrodynamics* (Academic Press Inc., London)
L. Davidovich (1975). Ph D Thesis (Rochester)
P.D. Drummond (1986). In *Quantum Optics IV*, J.D. Harvey and D.F. Walls (eds.) (Springer-Verlag, Berlin) p. 90
U. Fano (1961). *Phys. Rev.* **124,** 1866
K.O. Friedrichs (1948). *Commun. Pure Appl. Math.* **1,** 361
H. Fröhlich (1963). In *Polarons and Excitons,* C.G. Kuper and G.D. Whitfield (eds.) (Oliver and Boyd, Edinburgh) p. 1
V. Glaser, G. Källen (1956/57). *Nucl. Phys.* **2,** 706
D. Griffiths (1987). *Introduction to Elementary Particles* (John Wiley and Sons, Singapore)
J. Hamilton (1947). *Proc. Phys. Soc. A* **59,** 917
E.M. Henley, W. Thirring (1962). *Elementary Quantum Field Theory* (McGraw Hill Book Co., New York)
N.M. Hugenholtz (1959). In *The Many-Body Problem* (Dunod, Paris) p. 1
C. Itzykson, J.B. Zuber (1985). *Quantum Field Theory* (McGraw-Hill Book Co., Singapore)
E. Kartheuser (1972). In *Polarons in Ionic Crystals and Polar Semiconductors*, J.T. Devreese (ed.) (North-Holland Publishing Company, Amsterdam), p. 715
I.R. Kenyon (1987). *Elementary Particle Physics* (Routledge and Kegan Paul, London)
K. Kim, E. Wolf (1986). *Opt. Commun.* **59,** 1
E. Merzbacher (1961). *Quantum Mechanics* (John Wiley and Sons, New York)
N. Meyer-Vernet (1989). *Am. J. Phys.* **57,** 1084
T. Petrosky, I Prigogine (1993). *Phys. Lett. A* **182,** 5
T. Petrosky, I. Prigogine, S. Tasaki (1991). *Physica A* **173,** 175
D. Pines (1963). In *Polarons and Excitons,* C.G. Kuper and G.D. Whitfield (eds.) (Oliver and Boyd, Edinburgh) p. 33
E.C.G. Sudarshan, C.B. Chiu, V. Gorini (1978). *Phys. Rev. D* **18,** 2914
L. Van Hove (1955). *Physica* **21,** 901
L. Van Hove (1956). *Physica* **22,** 343

Further reading

The reader interested in the theory of renormalization will find it profitable to consult the book

J. Collins, *Renormalization* (Cambridge University Press, Cambridge 1989)

A clear introduction to the theory of the resolvent operator is given by

P. Roman, *Advanced Quantum Theory* (Addison-Wesley Publishing Co, Reading, Ma. 1965)

A. Messiah, *Quantum Mechanics,* vol. II (North-Holland Publishing Company, Amsterdam 1962)

Supplementary reading on the integration in the complex plane can be found in

L.V. Ahlfors, *Complex Analysis* (McGraw-Hill and Kogakusha Company Ltd, Kyoto 1953)

For a general survey of vacuum fluctuations of the electromagnetic field the reader is referred to the papers contained in
F. Persico, E.A. Power (eds.), *Vacuum in Nonrelativistic Matter-Radiation Systems, Physica Scripta* **T21** (1988)

A theory of the interaction of a discrete level with a continuum was developed by
U. Fano, *Phys. Rev.* **124,** 1866 (1961)
in connection with lineshape and auto-ionizing states. This has recently been used to quantize the electromagnetic field in dielectrics by
B. Huttner, S.M. Barnett, *Phys. Rev. A* **46,** 4306 (1992)

7

Energy density around dressed atoms

Introduction. The main theme of this chapter is the explicit calculation of the shape of the virtual cloud surrounding different kinds of ground-state sources. In Section 7.1 two of the three examples considered in Section 6.1 are taken up again, and it is argued that a convenient description of the shape of the virtual cloud is given by the energy density of the field around the source. This energy density is evaluated in detail for the static source of mesons and for the Fröhlich polaron. Section 7.2 is dedicated to an analogous calculation of the electric energy density around a two-level atom within a perturbation scheme. The virtual cloud around a two-level atom is again the subject of Section 7.3, where we evaluate the magnetic energy density as well as the coarse-grained energy density. From the results of the first three sections we conclude that the qualitative picture of the virtual cloud proposed in Section 6.2 is well founded. Moreover, the discussion of the two-level atom leads us to an important conclusion: the space surrounding an atom can be divided into a near zone and a far zone, where the energy density of the virtual cloud behaves differently as a function of the distance from the atom. In Section 7.4 we evaluate the energy density of the virtual cloud surrounding a slow free electron, separating classical and quantum contributions. Section 7.5 is dedicated to a more realistic atomic model, that is, a dressed nonrelativistic hydrogen atom. We evaluate in detail the various contributions to the energy density of the virtual cloud by second-order perturbation theory. The resulting expressions for these contributions are simplified in Section 7.6 by approximations appropriate to the near-zone and to the far-zone of the hydrogen atom, and the results are compared with those obtained in Section 7.4. In Section 7.6 we calculate the virtual cloud surrounding an atom whose coupling with the electromagnetic field is described by the Craig-Power model introduced at the end of Section 4.5, and we show that this model is able to represent adequately a realistic atom only in the far zone. The relation of the shape of the virtual cloud to the forces between ground-state sources, that is, to observable quantities, is discussed in Section 7.8. We take up again the three examples introduced in Section 6.1, and we show that exchange of virtual mesons between two static sources yields an effective potential of the Yukawa kind, whereas an exchange of virtual phonons between two massive Fröhlich polarons yields a screening of the Coulomb repulsion. Moreover, we show that the exchange of virtual photons between two neutral atoms in the Craig-Power model leads to the correct far-region expression of the van der Waals forces. Finally, we argue that van der Waals forces between an atom and an appropriate test body can be used to measure directly the energy density around the atom both in the near and in the far zone.

7.1 Energy density and virtual quanta

As we have seen in the previous chapter, the problem of dressed excited states of a quantum source interacting with a field presents serious conceptual difficulties. Ultimately, these difficulties are related to the apparent impossibility of separating clearly virtual quanta from real ones, which are emitted by spontaneous decay processes because the source is initially in an excited state. Since these difficulties have not yet been solved, we shall limit our treatment here to source-field systems which are in their ground state, or in states whose lifetime is so long in comparison with the time which is necessary to form or to rearrange the cloud of virtual quanta that they may be considered stable.

First we must address the question of how one can describe in quantitative terms the shape of the cloud of virtual quanta dressing a field source in a stable configuration. The quantities discussed in the previous chapter (e.g. number of quanta in the virtual cloud, source mass or energy renormalization, shifts of the energy levels of the source-field system) are in fact of a global nature, and do not contain information on the shape of the virtual cloud as a function of the distance from the source. This information will be available given an operator functional of the field distribution around the source, whose quantum average on the ground state of the coupled source-field system can be taken as a reliable measure of the form of the virtual cloud.

In the case of the static nucleon coupled to a scalar meson field, discussed in Section 6.1, an obvious choice is the amplitude of the field

$\phi(\mathbf{x})$ defined in (6.9)

$$\phi(\mathbf{x}) = \sum_{\mathbf{k}} \sqrt{\frac{\hbar}{2V\omega_k}} \left(a_{\mathbf{k}} e^{i\mathbf{k}\cdot\mathbf{x}} + a_{\mathbf{k}}^{\dagger} e^{-i\mathbf{k}\cdot\mathbf{x}} \right) \tag{7.1}$$

The ground state of the coupled source-field system is $| \tilde{0}\rangle$ given by (6.12). Use of (6.14) yields

$$\langle \tilde{0} \mid a_{\mathbf{k}} \mid \tilde{0}\rangle = \langle 0 \mid D_{\mathbf{k}}^{-1} a_{\mathbf{k}} D_{\mathbf{k}} \mid 0\rangle = g\left[2\hbar V \omega_k^3\right]^{-1/2} \rho_{\mathbf{k}} \equiv f_{\mathbf{k}} \tag{7.2}$$

and so

$$\begin{aligned} \langle \tilde{0} \mid \phi(\mathbf{x}) \mid \tilde{0}\rangle &= \sum_{\mathbf{k}} \sqrt{\frac{\hbar}{2V\omega_k}} \left(f_{\mathbf{k}} e^{i\mathbf{k}\cdot\mathbf{x}} + f_{\mathbf{k}}^{*} e^{-i\mathbf{k}\cdot\mathbf{x}} \right) \\ &= \frac{1}{2V} g \sum_{\mathbf{k}} \frac{1}{\omega_k^2} \left(\rho_{\mathbf{k}} e^{i\mathbf{k}\cdot\mathbf{x}} + \rho_{\mathbf{k}}^{*} e^{-i\mathbf{k}\cdot\mathbf{x}} \right) \end{aligned} \tag{7.3}$$

To proceed further we have to specify the form $\rho(\mathbf{x})$ of the source, which determines $\rho_{\mathbf{k}}$ by (6.8). The divergences arising from use of a point source do not play a significant role in the present context. For a point source $\rho(\mathbf{x}) = \delta(\mathbf{x})$ and consequently $\rho_{\mathbf{k}} = \rho_{\mathbf{k}}^{*} = 1$. Then with $\kappa \equiv mc/\hbar$, one has from (7.3) after some straightforward elementary angular integrations

$$\begin{aligned} \langle \tilde{0} \mid \phi(\mathbf{x}) \mid \tilde{0}\rangle &= \frac{1}{V} g \sum_{\mathbf{k}} \frac{1}{\omega_k^2} \cos \mathbf{k}\cdot\mathbf{x} = \frac{1}{(2\pi)^3 c^2} g \int \frac{\cos \mathbf{k}\cdot\mathbf{x}}{k^2+\kappa^2} d^3\mathbf{k} \\ &= \frac{1}{2\pi^2 c^2} g \frac{1}{x} \int_0^\infty \frac{k \sin kx}{k^2+\kappa^2} dk = \frac{1}{4\pi c^2} g \frac{e^{-\kappa x}}{x} \end{aligned} \tag{7.4}$$

We see that the field distribution decays in a Yukawa-like way far from the source. The divergence at $x = 0$ is due to the point-like nature of the source (Henley and Thirring 1962).

In many other situations, including the atom-photon case of interest in this book, the choice is not so simple since the quantum average of the field amplitude operator on the ground state of the system vanishes identically. However, nonvanishing averages exist of operators which are quadratic in the field coordinates. Among these quadratic operators a special role is played by the energy-momentum tensor, which we have discussed in connection with various fields in the early chapters of this book, due to simple properties under Lorentz transformations (see e.g. Birrell and Davies 1992). In particular, its time-time component T_{44} is the energy density of the field, which is directly related to observable effects and is thus a plausible candidate for the description of the virtual cloud. We evaluate this quantum average for the static source-meson field problem in order to make an instructive comparison with result (7.4) for the quantum average of the field amplitude before moving on to other cases. The energy-density operator for a real scalar field is immediately obtained from (3.28) as

$$\mathcal{H}(\mathbf{x}) = \frac{1}{2}\left(\dot{\phi}^2 + c^2(\nabla\phi)^2 + \kappa^2 c^2 \phi^2\right) \tag{7.5}$$

where $\phi(\mathbf{x})$ is given by (7.1) and

$$\dot{\phi}(\mathbf{x}) = -i\sum_{\mathbf{k}}\sqrt{\frac{\hbar\omega_k}{2V}}\left(a_{\mathbf{k}}e^{i\mathbf{k}\cdot\mathbf{x}} - a_{\mathbf{k}}^{\dagger}e^{-i\mathbf{k}\cdot\mathbf{x}}\right);$$

$$\nabla\phi(\mathbf{x}) = i\sum_{\mathbf{k}}\sqrt{\frac{\hbar}{2V\omega_k}}\mathbf{k}\left(a_{\mathbf{k}}e^{i\mathbf{k}\cdot\mathbf{x}} - a_{\mathbf{k}}^{\dagger}e^{-i\mathbf{k}\cdot\mathbf{x}}\right) \tag{7.6}$$

We actually need to evaluate

$$\langle\tilde{0} \mid \mathcal{H}(\mathbf{x}) \mid \tilde{0}\rangle = \langle 0 \mid D^{-1}\mathcal{H}(\mathbf{x})D \mid 0\rangle \tag{7.7}$$

where D is defined as $D = \prod_{\mathbf{k}} D_{\mathbf{k}}$. For example, on the basis of (6.14), we have for $\dot{\phi}^2$ in (7.5)

$$\begin{aligned} D^{-1}\dot{\phi}^2(\mathbf{x})D = &-\frac{\hbar}{2V}\sum_{\mathbf{k}\mathbf{k}'}\sqrt{\omega_k\omega_{k'}} \\ &\times\left[(a_{\mathbf{k}} + f_{\mathbf{k}})e^{i\mathbf{k}\cdot\mathbf{x}} - \left(a_{\mathbf{k}}^{\dagger} + f_{\mathbf{k}}^{*}\right)e^{-i\mathbf{k}\cdot\mathbf{x}}\right] \\ &\times\left[(a_{\mathbf{k}'} + f_{\mathbf{k}'})e^{i\mathbf{k}'\cdot\mathbf{x}} - \left(a_{\mathbf{k}'}^{\dagger} + f_{\mathbf{k}'}^{*}\right)e^{-i\mathbf{k}'\cdot\mathbf{x}}\right] \end{aligned} \tag{7.8}$$

from which one easily obtains

$$\begin{aligned} \langle\tilde{0} \mid \dot{\phi}^2(\mathbf{x}) \mid \tilde{0}\rangle &= \frac{\hbar}{2V}\left\{\sum_{\mathbf{k}}\omega_k \right. \\ &\left. - \left[\sum_{\mathbf{k}\mathbf{k}'}\sqrt{\omega_k\omega_{k'}}\left(f_{\mathbf{k}}f_{\mathbf{k}'}e^{i(\mathbf{k}+\mathbf{k}')\cdot\mathbf{x}} - f_{\mathbf{k}}f_{\mathbf{k}'}^{*}e^{i(\mathbf{k}-\mathbf{k}')\cdot\mathbf{x}}\right) + \text{c.c.}\right]\right\} \\ &= \frac{\hbar}{2V}\left\{\sum_{\mathbf{k}}\omega_k - \left[\left(\sum_{\mathbf{k}}\sqrt{\omega_k}f_{\mathbf{k}}e^{i\mathbf{k}\cdot\mathbf{x}}\right)^2 \right.\right. \\ &\left.\left. - \left|\sum_{\mathbf{k}}\sqrt{\omega_k}f_{\mathbf{k}}e^{i\mathbf{k}\cdot\mathbf{x}}\right|^2 + \text{c.c.}\right]\right\} \\ &= \frac{1}{V}\sum_{\mathbf{k}}\frac{\hbar\omega_k}{2} + g^2\left\{\text{Im}\left[\frac{1}{V}\sum_{\mathbf{k}}\frac{1}{\omega_k}\rho_{\mathbf{k}}e^{i\mathbf{k}\cdot\mathbf{x}}\right]\right\}^2 \end{aligned} \tag{7.9}$$

where Im(z) is the imaginary part of z. Proceeding along the same lines one has also

$$\langle\tilde{0} \mid c^2[\nabla\phi(\mathbf{x})]^2 \mid \tilde{0}\rangle$$
$$= \frac{1}{V}\sum_{\mathbf{k}} \frac{\hbar c^2 k^2}{2\omega_k} + g^2c^2\left\{\mathrm{Re}\left[\nabla\left(\frac{1}{V}\sum_{\mathbf{k}}\frac{1}{\omega_k{}^2}\rho_{\mathbf{k}}e^{i\mathbf{k}\cdot\mathbf{x}}\right)\right]\right\}^2 \tag{7.10}$$

$$\langle\tilde{0} \mid c^2\kappa^2\phi^2(\mathbf{x}) \mid \tilde{0}\rangle$$
$$= \frac{1}{V}\sum_{\mathbf{k}} \frac{\hbar c^2 \kappa^2}{2\omega_k} + g^2c^2\kappa^2\left\{\mathrm{Re}\left[\frac{1}{V}\sum_{\mathbf{k}}\frac{1}{\omega_k^2}\rho_{\mathbf{k}}e^{i\mathbf{k}\cdot\mathbf{x}}\right]\right\}^2 \tag{7.11}$$

In the point source limit $\rho_{\mathbf{k}} = 1$ one has

$$\mathrm{Im}\left[\frac{1}{V}\sum_{\mathbf{k}}\frac{1}{\omega_k}e^{i\mathbf{k}\cdot\mathbf{x}}\right] = \frac{1}{V}\sum_{\mathbf{k}}\frac{1}{\omega_k}\sin\mathbf{k}\cdot\mathbf{x} = 0\,;$$
$$\frac{1}{V}\sum_{\mathbf{k}}\frac{1}{\omega_k^2}e^{i\mathbf{k}\cdot\mathbf{x}} = \frac{1}{(2\pi)^3}\int\frac{1}{\omega_k^2}e^{i\mathbf{k}\cdot\mathbf{x}}d^3\mathbf{k}$$
$$= \frac{1}{2\pi^2c^2}\frac{1}{x}\int_0^\infty\frac{k}{k^2+\kappa^2}\sin kx\,dk$$
$$= \frac{1}{4\pi c^2}\frac{e^{-\kappa x}}{x}\,;$$
$$\nabla\left(\frac{1}{V}\sum_{\mathbf{k}}\frac{1}{\omega_k^2}e^{i\mathbf{k}\cdot\mathbf{x}}\right) = -\frac{1}{4\pi c^2}\frac{e^{-\kappa x}}{x}\left(\kappa+\frac{1}{x}\right)\hat{x} \tag{7.12}$$

Substitution of (7.12) into (7.9, 7.10, 7.11) leads to

$$\langle\tilde{0} \mid \dot{\phi}^2(\mathbf{x}) \mid \tilde{0}\rangle = \frac{1}{V}\sum_{\mathbf{k}}\frac{\hbar\omega_k}{2}\,;$$
$$\langle\tilde{0} \mid c^2[\nabla\phi(\mathbf{x})]^2 \mid \tilde{0}\rangle = \frac{1}{V}\sum_{\mathbf{k}}\frac{\hbar c^2k^2}{2\omega_k} + \frac{1}{16\pi^2c^2}g^2\frac{e^{-2\kappa x}}{x^2}\left(\kappa+\frac{1}{x}\right)^2;$$
$$\langle\tilde{0} \mid c^2\kappa^2\phi^2(\mathbf{x}) \mid \tilde{0}\rangle = \frac{1}{V}\sum_{\mathbf{k}}\frac{\hbar c^2\kappa^2}{2\omega_k} + \frac{1}{16\pi^2c^2}g^2\frac{\kappa^2}{x^2}e^{-2\kappa x} \tag{7.13}$$

Finally, use of (7.13) in (7.7) through (7.5) gives

$$\langle\tilde{0} \mid \mathcal{H}(\mathbf{x}) \mid \tilde{0}\rangle = \frac{1}{V}\sum_{\mathbf{k}}\frac{\hbar\omega_k}{2} + \frac{1}{32\pi^2c^2}g^2\frac{e^{-2\kappa x}}{x^2}\left(2\kappa^2+\frac{2\kappa}{x}+\frac{1}{x^2}\right) \tag{7.14}$$

Thus the energy density of the scalar field surrounding the point static source consists of two terms, the first of which is the zero-point energy of the unperturbed meson vacuum $\sum_{\mathbf{k}} \hbar\omega_k/2V$. Actually this term, in the absence of a frequency cutoff, diverges as expected. Since this term is uniform on the whole space, however, it does not contain any information on the source-field interaction and it can be discarded. Hence the part describing the virtual cloud is the second term on the RHS of (7.14), which is due to the virtual mesons continuously emitted and reabsorbed by the source. We remark that the x-dependences of the virtual meson cloud as described by (7.4) and by (7.14) are not terribly different, in the sense that the main features of the cloud are in both cases dominated by the exponential decay for large x with decay constant given by the Compton wavelength of the meson field. In fact the leading term in (7.14) is simply the square of (7.4). These results are in agreement with the qualitative argument presented in Section 6.2 in connection with the linear dimensions of the dressed nucleon.

In the case of the Fröhlich polaron, also introduced in Section 6.1, the energy-momentum tensor of the electric field has not been evaluated, since it has been assumed that the polarization created in the lattice by the presence of the electron at point $\mathbf{r}_e$, or equivalently the electric potential created by this polarization, is an adequate measure of the virtual phonon cloud surrounding the slow electron. The latter can be obtained for the small coupling case starting from the perturbative expansion (6.28) for the state of the dressed electron. In the electron coordinate representation, this perturbed state is

$$\begin{aligned}&\langle \mathbf{x}_e \mid \mathbf{p}_i = \hbar\langle\mathbf{k}\rangle, \{0_{\mathbf{k}}\}\rangle^{(1)}\\&= \frac{1}{\sqrt{V}} e^{i\langle\mathbf{k}\rangle\cdot\mathbf{x}_e}\left\{1 + \sum_{\mathbf{k}} \frac{V_k^*}{\Delta_{\mathbf{k}}} e^{-i\mathbf{k}\cdot\mathbf{x}_e} a_{\mathbf{k}}^\dagger\right\} \mid \{0_{\mathbf{k}}\}\rangle \qquad (7.15)\end{aligned}$$

The operator for the electric potential created at point $\mathbf{x}$ by the presence of the phonons can be shown to be (Fröhlich 1963)

$$\phi(\mathbf{x}) = i\left(\frac{2\pi\hbar\omega}{\bar{\epsilon}V}\right)^{1/2} \sum_{\mathbf{k}} \frac{1}{\mathbf{k}}\left(a_{\mathbf{k}}^\dagger e^{-i\mathbf{k}\cdot\mathbf{x}} - a_{\mathbf{k}} e^{i\mathbf{k}\cdot\mathbf{x}}\right) \qquad (7.16)$$

Thus the polarization potential created at $\mathbf{x}$ by an electron situated at $\mathbf{x}_e$ (with probability $1/V$) is

$$\begin{aligned}&{}^{(1)}\langle \mathbf{p}_i = \hbar\langle\mathbf{K}\rangle, \{0_{\mathbf{k}}\} \mid \mathbf{x}_e\rangle\phi(\mathbf{x})\langle\mathbf{x}_e \mid \mathbf{p}_i = \hbar\langle\mathbf{K}\rangle, \{0_{\mathbf{k}}\}\rangle^{(1)}\\&\equiv \frac{1}{V}\phi(\mathbf{x}, \mathbf{x}_e) \qquad (7.17)\end{aligned}$$

where $\phi(\mathbf{x}, \mathbf{x}_e)$ is the potential at $\mathbf{x}$ created by an electron which is certainly at $\mathbf{x}_e$ (Fröhlich 1963). Hence, up to terms of order q, where q is the electron charge

$$\begin{aligned}
\phi(\mathbf{x}, \mathbf{x}_e) &= \langle\{0_\mathbf{k}\} \mid \left(1 + \sum_\mathbf{k} \frac{V_k}{\Delta_\mathbf{k}} e^{i\mathbf{k}\cdot\mathbf{x}_e} a_\mathbf{k}\right) \phi(\mathbf{x}) \\
&\times \left(1 + \sum_\mathbf{k} \frac{V_k^*}{\Delta_\mathbf{k}} e^{-i\mathbf{k}\cdot\mathbf{x}_e} a_\mathbf{k}^\dagger\right) \mid \{0_\mathbf{k}\}\rangle \\
&= i\left[\frac{2\pi\hbar\omega}{\bar{\epsilon}V}\right]^{1/2} \sum_\mathbf{k} \frac{1}{k}\left(\frac{V_k}{\Delta_\mathbf{k}} e^{i\mathbf{k}\cdot(\mathbf{x}_e - \mathbf{x})} - \text{c.c.}\right) \\
&= -2q\frac{2\pi\hbar\omega}{\bar{\epsilon}V} \sum_\mathbf{k} \frac{\cos \mathbf{k}\cdot(\mathbf{x}_e - \mathbf{x})}{k^2 \Delta_\mathbf{k}} \\
&= \frac{4\pi q}{\bar{\epsilon}V} \sum_\mathbf{k} \frac{\cos \mathbf{k}\cdot(\mathbf{x}_e - \mathbf{x})}{k^2\left(1 + \frac{\hbar}{2m\omega}k^2\right)}
\end{aligned} \tag{7.18}$$

The sum in the last term can be transformed into an integral which is performed easily as follows

$$\begin{aligned}
\phi(\mathbf{x}, \mathbf{x}_e) &= \frac{4\pi q}{\bar{\epsilon}} \frac{1}{(2\pi)^3} \int \frac{\cos \mathbf{k}\cdot(\mathbf{x}_e - \mathbf{x})}{k^2\left(1 + \frac{\hbar}{2m\omega}k^2\right)} d^3\mathbf{k} \\
&= \frac{2}{\pi} \frac{q}{\bar{\epsilon} \mid \mathbf{x}_e - \mathbf{x} \mid} \int_0^\infty \frac{\sin k \mid \mathbf{x}_e - \mathbf{x} \mid}{k\left(1 + \frac{\hbar}{2m\omega}k^2\right)} dk \\
&= \frac{q}{\bar{\epsilon} \mid \mathbf{x}_e - \mathbf{x} \mid}\left(1 - \exp\left(-\left[\frac{2m\omega}{\hbar}\right]^{1/2} \mid \mathbf{x}_e - \mathbf{x} \mid\right)\right)
\end{aligned} \tag{7.19}$$

(7.19) shows that at distances from the electron larger than $\sqrt{\hbar/2m\omega}$ the polarization potential is of the normal screened Coulomb type, whereas at smaller distances the presence of the virtual phonon cloud eliminates the Coulomb singularity (Lee et al. 1953). This is possible because in a polar lattice the ionic displacement due to the optical phonons is associated with a local electric field. Thus $\sqrt{\hbar/2m\omega}$ can be taken as a measure of the radius of the virtual cloud, in agreement with the qualitative argument of Section 6.2. More refined arguments, valid also for finite temperatures and in the presence of external magnetic fields, yield results in agreement with (7.19) in the appropriate limits (Peeters and Devreese 1985).

The lattice polarization charge density around the electron is entirely due to the virtual phonons, and as such it is perhaps a more direct measure of the virtual cloud. This can be obtained directly from Poisson's equation and (7.18) as

$$\rho(\mathbf{x},\mathbf{x}_e) = -\frac{1}{4\pi}\nabla^2\phi(\mathbf{x},\mathbf{x}_e) = \frac{q}{\epsilon V}\sum_{\mathbf{k}}\frac{\cos\mathbf{k}\cdot(\mathbf{x}_e-\mathbf{x})}{1+\frac{\hbar}{2m\omega}k^2}$$
$$= \frac{q}{\epsilon}\frac{1}{(2\pi)^3}\int\frac{\cos\mathbf{k}\cdot(\mathbf{x}_e-\mathbf{x})}{1+\frac{\hbar}{2m\omega}k^2}d^3\mathbf{k}$$
$$= \frac{1}{2\pi^2}\frac{q}{\bar{\epsilon}\mid\mathbf{x}-\mathbf{x}_e\mid}\int_0^\infty\frac{k\sin k\mid\mathbf{x}_e-\mathbf{x}\mid}{1+\frac{\hbar}{2m\omega}k^2}dk$$
$$= \frac{m\omega}{2\pi\hbar}\frac{q}{\bar{\epsilon}\mid\mathbf{x}_e-\mathbf{x}\mid}e^{-[2m\omega/\hbar]^{1/2}|\mathbf{x}_e-\mathbf{x}|} \tag{7.20}$$

In agreement with the results discussed above for the potential, the shape of the virtual cloud as given by the polarization charge density is dominated at large distances from the electron by the exponential in (7.20), which effectively eliminates the virtual phonons at distances larger than $\sqrt{\hbar/2m\omega}$.

Two conclusions can be drawn from the results presented in this section. First, the qualitative arguments used to discuss the virtual cloud in Section 6.2 are supported by the quantitative arguments of this section. This also means that the physics of the virtual clouds can be correctly, albeit qualitatively, described as in Section 6.2. Secondly, the main features of the shape of the virtual cloud may not depend strongly on the operator used to describe it. This is also very reassuring, and it is strengthened by the discussion in the subsequent sections of this chapter.

7.2 Electric energy density around a two-level atom

In this section we shall discuss in some detail the virtual photon cloud surrounding a two-level atom in its dressed ground state. The coupling is taken to be electric dipole in the multipolar coupling scheme with no RWA. So the system is chosen to be that governed by the Hamiltonian (5.51) with $\lambda = -1$

$$H = \hbar\omega_o S_z + \sum_{\mathbf{k}j}\hbar\omega_k\left(a^\dagger_{\mathbf{k}j}a_{\mathbf{k}j}+1/2\right)$$
$$+\sum_{\mathbf{k}j}\left(\epsilon_{\mathbf{k}j}a_{\mathbf{k}j}S_+ + \epsilon^*_{\mathbf{k}j}a^\dagger_{\mathbf{k}j}S_-\right)$$
$$-\sum_{\mathbf{k}j}\left(\epsilon_{\mathbf{k}j}a^\dagger_{\mathbf{k}j}S_+ + \epsilon^*_{\mathbf{k}j}a^\dagger_{\mathbf{k}j}S_-\right)\ ;$$
$$\epsilon_{\mathbf{k}j} = -i\sqrt{\frac{2\pi\hbar\omega_k}{V}}\mu_{21}\cdot\mathbf{e}_{\mathbf{k}j} \tag{7.21}$$

where the polarization vectors $\mathbf{e}_{\mathbf{k}j}$ are taken as real. The dressed ground state of the system is, according to (6.3),

$$\begin{aligned} |\{0_{\mathbf{k}j}\},\downarrow\rangle' = \Big\{ 1 &- \frac{1}{E_0 - H_0}(1-P_0)V_2 \\ &- \frac{1}{E_0-H_0}(1-P_0)V_1\frac{1}{E_0-H_0}(1-P_0)V_2 \\ &- \frac{1}{2}\langle\{0_{\mathbf{k}j}\},\downarrow| V_2 \frac{1}{(E_0-H_0)^2}(1-P_0)V_2 \,|\, \{0_{\mathbf{k}j}\},\downarrow\rangle \Big\} \\ &\times |\{0_{\mathbf{k}j}\},\downarrow\rangle \end{aligned} \tag{7.22}$$

This dressed state has been obtained by perturbation theory discussed in Appendix C, and it is accurate to order $\epsilon^2_{\mathbf{k}j}$ and normalized to the same order. H_0, V_1 and V_2 are defined as in Section 6.1. Throughout the present section we shall take μ_{21} as real, which can always be done by an appropriate definition of the phases of the two atomic states $|\uparrow\rangle$ and $|\downarrow\rangle$.

It is immediate that the quantum average on (7.22) of any operator linear in $a_{\mathbf{k}j}$ and $a^\dagger_{\mathbf{k}j}$ vanishes. Thus, we shall adopt the energy density of the electromagnetic field as a measure of the shape of the virtual cloud. For the multipolar scheme we have, from (4.77)

$$\begin{aligned} \mathcal{H}(\mathbf{x}) &= \mathcal{H}_{el}(\mathbf{x}) + \mathcal{H}_{mag}(\mathbf{x}) \,; \\ \mathcal{H}_{el}(\mathbf{x}) &= \frac{1}{8\pi}\mathbf{D}^2_\perp(\mathbf{x}) \,; \\ \mathcal{H}_{mag}(\mathbf{x}) &= \frac{1}{8\pi}\mathbf{H}^2(\mathbf{x}) \end{aligned} \tag{7.23}$$

We proceed to evaluate $'\langle\{0_{\mathbf{k}j}\},\downarrow|\, \mathbf{D}^2_\perp(\mathbf{x}) \,|\, \{0_{\mathbf{k}j}\},\downarrow\rangle'$. On the basis of expansion (4.81) we immediately obtain

$$\begin{aligned} \mathbf{D}^2_\perp(\mathbf{x}) = -\frac{2\pi\hbar}{V}\sum_{\mathbf{k}\mathbf{k}'jj'} \sqrt{\omega_k\omega_{k'}}\,\mathbf{e}_{\mathbf{k}j}\cdot\mathbf{e}_{\mathbf{k}'j'} \Big\{ & a_{\mathbf{k}j}a_{\mathbf{k}'j'}e^{i(\mathbf{k}+\mathbf{k}')\cdot\mathbf{x}} \\ &+ a^\dagger_{\mathbf{k}j}a^\dagger_{\mathbf{k}'j'}e^{-i(\mathbf{k}+\mathbf{k}')\cdot\mathbf{x}} - a_{\mathbf{k}j}a^\dagger_{\mathbf{k}'j'}e^{i(\mathbf{k}-\mathbf{k}')\cdot\mathbf{x}} \\ &- a^\dagger_{\mathbf{k}j}a_{\mathbf{k}'j'}e^{-i(\mathbf{k}-\mathbf{k}')\cdot\mathbf{x})} \Big\} \end{aligned} \tag{7.24}$$

Moreover, from (7.22) with $E_0 = -\hbar\omega_0/2$ a simple procedure shows that

$$
\begin{aligned}
&{}'\langle\{0_{\mathbf{k}j}\},\downarrow|\, a_{\mathbf{k}j}a_{\mathbf{k}'j'} \mid \{0_{\mathbf{k}j}\},\downarrow\rangle' \\
&\quad = -\langle\{0_{\mathbf{k}j}\},\downarrow|\, a_{\mathbf{k}j}a_{\mathbf{k}'j'} \frac{1}{E_0 - H_0}(1-P_0)V_1 \\
&\quad \times \frac{1}{E_0 - H_0}(1-P_0)V_2 \mid \{0_{\mathbf{k}j}\},\downarrow\rangle \\
&\quad = -\frac{1}{\hbar^2}(1+\delta_{\mathbf{k}\mathbf{k}'}\delta_{jj'})\left\{\frac{\epsilon^*_{\mathbf{k}j}\epsilon_{\mathbf{k}'j'}}{(\omega_o+\omega_{k'})(\omega_k+\omega_{k'})}\right. \\
&\quad \left. + \frac{\epsilon_{\mathbf{k}j}\epsilon^*_{\mathbf{k}'j'}}{(\omega_0+\omega_k)(\omega_k+\omega_{k'})}\right\} \\
&\quad = \left({}'\langle\{0_{\mathbf{k}j}\},\downarrow|\, a^\dagger_{\mathbf{k}j}a^\dagger_{\mathbf{k}'j'} \mid \{0_{\mathbf{k}j}\},\downarrow\rangle'\right)^* ; \\
&{}'\langle\{0_{\mathbf{k}j}\},\downarrow|\, a^\dagger_{\mathbf{k}j}a_{\mathbf{k}'j'} \mid \{0_{\mathbf{k}j}\},\downarrow\rangle' \\
&\quad = {}'\langle\{0_{\mathbf{k}j}\},\downarrow|\, V_2(1-P_0)\frac{1}{E_0 - H_0}a^\dagger_{\mathbf{k}j}a_{\mathbf{k}'j'} \\
&\quad \times \frac{1}{E_0 - H_0}(1-P_0)V_2 \mid \{0_{\mathbf{k}j}\},\downarrow\rangle \\
&\quad = \frac{1}{\hbar^2}\frac{\epsilon^*_{\mathbf{k}j}\epsilon_{\mathbf{k}'j'}}{(\omega_0+\omega_k)(\omega_0+\omega_{k'})} ; \\
&{}'\langle\{0_{\mathbf{k}j}\},\downarrow|\, a_{\mathbf{k}j}a^\dagger_{\mathbf{k}'j'} \mid \{0_{\mathbf{k}j}\},\downarrow\rangle' \\
&\quad = {}'\langle\{0_{\mathbf{k}j}\},\downarrow|\, \left(\delta_{\mathbf{k}\mathbf{k}'}\delta_{jj'} + a^\dagger_{\mathbf{k}'j'}a_{\mathbf{k}j}\right) \mid \{0_{\mathbf{k}j}\},\downarrow\rangle' \\
&\quad = \delta_{\mathbf{k}\mathbf{k}'}\delta_{jj'} + \frac{1}{\hbar^2}\frac{\epsilon_{\mathbf{k}j}\epsilon^*_{\mathbf{k}'j'}}{(\omega_0+\omega_{k'})(\omega_0+\omega_k)}
\end{aligned}
\tag{7.25}
$$

Thus the quantum average of (7.24), on the basis of (7.25), is

$$
\begin{aligned}
&{}'\langle\{0_{\mathbf{k}j}\},\downarrow|\, \mathbf{D}^2_\perp(\mathbf{x}) \mid \{0_{\mathbf{k}j}\},\downarrow\rangle' = \frac{2\pi}{V}\sum_{\mathbf{k}j}\hbar\omega_k + \frac{2\pi}{\hbar V}\sum_{\mathbf{k}\mathbf{k}'jj'}\sqrt{\omega_k\omega_{k'}} \\
&\times \left\{\mathbf{e}_{\mathbf{k}j}\cdot\mathbf{e}_{\mathbf{k}'j'}(1+\delta_{\mathbf{k}\mathbf{k}'}\delta_{jj'})\left[\frac{\epsilon^*_{\mathbf{k}j}\epsilon_{\mathbf{k}'j'}}{(\omega_0+\omega_{k'})(\omega_k+\omega_{k'})}\right.\right. \\
&\left. + \frac{\epsilon_{\mathbf{k}j}\epsilon^*_{\mathbf{k}'j'}}{(\omega_0+\omega_k)(\omega_k+\omega_{k'})}\right] e^{i(\mathbf{k}+\mathbf{k}')\cdot\mathbf{x}} \\
&\left. + \mathbf{e}_{\mathbf{k}j}\cdot\mathbf{e}_{\mathbf{k}'j'}\frac{\epsilon_{\mathbf{k}j}\epsilon^*_{\mathbf{k}'j'}}{(\omega_0+\omega_k)(\omega_0+\omega_{k'})}e^{i(\mathbf{k}-\mathbf{k}')\cdot\mathbf{x}} + \text{c.c.}\right\}
\end{aligned}
\tag{7.26}
$$

We now remark that on the basis of (2.6), and using the multipolar $\epsilon_{\mathbf{k}j}$ as in (7.21) with μ_{21} real,

$$
\begin{aligned}
\sum_{jj'} \mathbf{e}_{\mathbf{k}j} \cdot \mathbf{e}_{\mathbf{k}'j'} \epsilon^*_{\mathbf{k}j} \epsilon_{\mathbf{k}'j'} &= \sum_{jj'} \mathbf{e}_{\mathbf{k}j} \cdot \mathbf{e}_{\mathbf{k}'j'} \epsilon_{\mathbf{k}j} \epsilon^*_{\mathbf{k}'j'} \\
&= \frac{2\pi\hbar}{V} \sqrt{\omega_k \omega_{k'}} (\mu_{21})_m (\mu_{21})_n \sum_{jj'} (\mathbf{e}_{\mathbf{k}j})_\ell (\mathbf{e}_{\mathbf{k}'j'})_\ell (\mathbf{e}_{\mathbf{k}j})_m (\mathbf{e}_{\mathbf{k}'j'})_n \\
&= \frac{2\pi\hbar}{V} \sqrt{\omega_k \omega_{k'}} (\mu_{21})_m (\mu_{21})_n \left(\delta_{\ell m} - \hat{k}_\ell \hat{k}_m\right) \left(\delta_{\ell n} - \hat{k}'_\ell \hat{k}'_n\right) \qquad (7.27)
\end{aligned}
$$

where the symbol $\hat{V}$ indicates the unit vector in the direction of $\mathbf{V}$. Thus the double polarization sums in (7.26) can be easily performed, and the terms contributing to the double sums in (7.26) take the form

$$
\begin{aligned}
A(\mathbf{x}) &\equiv \frac{2\pi}{\hbar V} \sum_{\mathbf{k}\mathbf{k}'jj'} \sqrt{\omega_k \omega_{k'}} \mathbf{e}_{\mathbf{k}j} \cdot \mathbf{e}_{\mathbf{k}'j'} \frac{\epsilon^*_{\mathbf{k}j} \epsilon_{\mathbf{k}'j'}}{(\omega_0 + \omega_{k'})(\omega_k + \omega_{k'})} e^{i(\mathbf{k}+\mathbf{k}')\cdot\mathbf{x}} \\
&= \frac{4\pi^2}{V^2} \sum_{\mathbf{k}\mathbf{k}'} \omega_k \omega_{k'} (\mu_{21})_m (\mu_{21})_n \left(\delta_{\ell m} - \hat{k}_\ell \hat{k}_m\right) \left(\delta_{\ell n} - \hat{k}'_\ell \hat{k}'_n\right) \\
&\quad \times \frac{e^{i(\mathbf{k}+\mathbf{k}')\cdot\mathbf{x}}}{(\omega_0 + \omega_{k'})(\omega_k + \omega_{k'})} ; \\
B(\mathbf{x}) &\equiv \frac{2\pi}{\hbar V} \sum_{\mathbf{k}\mathbf{k}'jj'} \sqrt{\omega_k \omega_{k'}} \mathbf{e}_{\mathbf{k}j} \cdot \mathbf{e}_{\mathbf{k}'j'} \frac{\epsilon_{\mathbf{k}j} \epsilon^*_{\mathbf{k}'j'}}{(\omega_0 + \omega_k)(\omega_k + \omega_{k'})} e^{i(\mathbf{k}+\mathbf{k}')\cdot\mathbf{x}} \\
&= \frac{4\pi^2}{V^2} \sum_{\mathbf{k}\mathbf{k}'} \omega_k \omega_{k'} (\mu_{21})_m (\mu_{21})_n \left(\delta_{\ell m} - \hat{k}_\ell \hat{k}_m\right) \left(\delta_{\ell n} - \hat{k}'_\ell \hat{k}'_n\right) \\
&\quad \times \frac{e^{i(\mathbf{k}+\mathbf{k}')\cdot\mathbf{x}}}{(\omega_0 + \omega_k)(\omega_k + \omega_{k'})} ; \\
C(\mathbf{x}) &\equiv \frac{2\pi}{\hbar V} \sum_{\mathbf{k}\mathbf{k}'jj'} \sqrt{\omega_k \omega_{k'}} \mathbf{e}_{\mathbf{k}j} \cdot \mathbf{e}_{\mathbf{k}'j'} \frac{\epsilon_{\mathbf{k}j} \epsilon^*_{\mathbf{k}'j'}}{(\omega_0 + \omega_k)(\omega_0 + \omega_{k'})} e^{i(\mathbf{k}-\mathbf{k}')\cdot\mathbf{x}} \\
&= \frac{4\pi^2}{V^2} \sum_{\mathbf{k}\mathbf{k}'} \omega_k \omega_{k'} (\mu_{21})_m (\mu_{21})_n \left(\delta_{\ell m} - \hat{k}_\ell \hat{k}_m\right) \left(\delta_{\ell n} - \hat{k}'_\ell \hat{k}'_n\right) \\
&\quad \times \frac{e^{i(\mathbf{k}-\mathbf{k}')\cdot\mathbf{x}}}{(\omega_0 + \omega_k)(\omega_0 + \omega_{k'})} ; \qquad (7.28)
\end{aligned}
$$

The term in $\delta_{\mathbf{k}\mathbf{k}'}\delta_{jj'}$ in (7.26) contributes at order $1/V$, and vanishes for $V \to \infty$. Actually A and B are equal, but it is convenient to keep them in the present form in order to simplify their approximate evaluation (see the near-zone approximation below).

The treatment given so far is rigorous up to terms of order $\epsilon_{\mathbf{k}j}^2$. We shall now approximate (7.28) according to the following argument, which is based on a qualitative description of the virtual cloud of the kind given in Section 6.2. We divide the space around the two-level atom, sitting at the origin $\mathbf{x} = 0$, into two regions: the "near" zone for $x < c/\omega_0$ and the "far" zone for $x > c/\omega_0$. The lifetime of virtual photons with $\omega_k > \omega_0$ is shorter than ω_0^{-1}, and during this lifetime they cannot reach distances greater that c/ω_0 from the atom. Thus they can contribute only to the energy density in the near zone. Also virtual photons with $\omega_k < \omega_0$ contribute to the energy density of the near zone, of course, but this contribution is negligible due to their relatively small energy. Thus in the near zone it is legitimate to neglect ω_0 with respect to ω_k in (7.28) and to approximate $\omega_0 + \omega_k \sim \omega_k$ wherever this quantity appears in $A(\mathbf{x})$, $B(\mathbf{x})$ and $C(\mathbf{x})$. On the contrary, low energy virtual photons with $\omega_k < \omega_0$ have a longer lifetime, and consequently they can reach the far zone, where they give the prevailing contribution to the energy density due to the scarcity of high energy photons at $x > c/\omega_0$. Thus in the far zone we shall approximate $\omega_0 + \omega_k \sim \omega_0$ wherever this quantity appears in $A(\mathbf{x})$, $B(\mathbf{x})$ and $C(\mathbf{x})$. This argument is very qualitative, of course, and it might seem to contradict the qualitative conclusion of Section 6.2 that c/ω_0 is the extreme boundary for low energy photons. That conclusion, however, did not take into account the contribution of two-photon emission with no change of the internal state of the atom, whose importance at large distances has been discussed by Passante and Power (1987), and which is implicitly taken into account in the argument above. Thus we shall use the following approximations

a) near zone ($x < c/\omega_0$)

$$A(\mathbf{x}) = \frac{4\pi^2}{V^2}(\mu_{21})_m(\mu_{21})_n \sum_{\mathbf{k}\mathbf{k}'} \left(\delta_{\ell m} - \hat{k}_\ell \hat{k}_m\right)\left(\delta_{\ell n} - \hat{k}'_\ell \hat{k}'_n\right) \times \frac{\omega_k}{\omega_k + \omega_{k'}} e^{i(\mathbf{k}+\mathbf{k}')\cdot\mathbf{x}} ;$$

$$B(\mathbf{x}) = \frac{4\pi^2}{V^2}(\mu_{21})_m(\mu)_n \sum_{\mathbf{k}\mathbf{k}'} \left(\delta_{\ell m} - \hat{k}_\ell \hat{k}_m\right)\left(\delta_{\ell n} - \hat{k}'_\ell \hat{k}'_n\right) \times \frac{\omega_{k'}}{\omega_k + \omega_{k'}} e^{i(\mathbf{k}+\mathbf{k}')\cdot\mathbf{x}} ;$$

$$C(\mathbf{x}) = \frac{4\pi^2}{V^2}(\mu_{21})_m(\mu_{21})_n \sum_{\mathbf{k}\mathbf{k}'} \left(\delta_{\ell m} - \hat{k}_\ell \hat{k}_n\right) \times \left(\delta_{\ell n} - \hat{k}'_\ell \hat{k}'_n\right) e^{i(\mathbf{k}-\mathbf{k}')\cdot\mathbf{x}} \qquad (7.29)$$

b) far zone ($x > c/\omega_0$)

$$A(\mathbf{x}) = B(\mathbf{x}) = \frac{4\pi^2}{V^2}\frac{(\mu_{21})_m(\mu_{21})_n}{\omega_0}\sum_{\mathbf{k}\mathbf{k}'}\left(\delta_{\ell m} - \hat{k}_\ell\hat{k}_m\right)\left(\delta_{\ell n} - \hat{k}'_\ell\hat{k}'_n\right)$$

$$\times \frac{\omega_k\omega_{k'}}{\omega_k + \omega_{k'}} e^{i(\mathbf{k}+\mathbf{k}')\cdot\mathbf{x}} \;;$$

$$C(\mathbf{x}) = \frac{4\pi^2}{V^2}\frac{(\mu_{21})_m(\mu_{21})_n}{\omega_0^2}\sum_{\mathbf{k}\mathbf{k}'}\left(\delta_{\ell m} - \hat{k}_\ell\hat{k}_m\right)\left(\delta_{\ell n} - \hat{k}'_\ell\hat{k}'_n\right)$$

$$\times\, \omega_k\omega_{k'} e^{i(\mathbf{k}-\mathbf{k}')\cdot\mathbf{x}} \tag{7.30}$$

We concentrate first on the near zone, where we remark that

$$A(\mathbf{x}) + B(\mathbf{x}) = \frac{4\pi^2}{V^2}(\mu_{21})_m(\mu_{21})_n\sum_{\mathbf{k}\mathbf{k}'}\left(\delta_{\ell m} - \hat{k}_\ell\hat{k}_m\right)\left(\delta_{\ell n} - \hat{k}'_\ell\hat{k}'_n\right)$$

$$\times\, e^{i(\mathbf{k}+\mathbf{k}')\cdot\mathbf{x}}$$

$$= 4\pi^2(\mu_{21})_m(\mu_{21})_n\left\{\frac{1}{V}\sum_{\mathbf{k}}\left(\delta_{\ell m} - \hat{k}_\ell\hat{k}_m\right)e^{i\mathbf{k}\cdot\mathbf{x}}\right\}$$

$$\times\left\{\frac{1}{V}\sum_{\mathbf{k}'}\left(\delta_{\ell m} - \hat{k}'_\ell\hat{k}'_n\right)e^{i\mathbf{k}'\cdot\mathbf{x}}\right\};$$

$$C(\mathbf{x}) = (4\pi)^2(\mu_{21})_m(\mu_{21})_n\left\{\frac{1}{V}\sum_{\mathbf{k}}\left(\delta_{\ell m} - \hat{k}_\ell\hat{k}_m\right)e^{i\mathbf{k}\cdot\mathbf{x}}\right\}$$

$$\times\left\{\frac{1}{V}\sum_{\mathbf{k}'}\left(\delta_{\ell n} - \hat{k}'_\ell\hat{k}'_n\right)e^{-i\mathbf{k}'\cdot\mathbf{x}}\right\}\quad (x < c/\omega_0) \tag{7.31}$$

The factorization in (7.31) leads to evaluation of single, rather than double, sums over $\mathbf{k}$. These single summations can be evaluated using the prescription, valid for any reasonable $f(k)$ (Craig and Thirunamachandran 1984),

$$\frac{1}{V}\sum_{\mathbf{k}} f(k)\left(\delta_{\ell m} - \hat{k}_\ell\hat{k}_m\right)e^{i\mathbf{k}\cdot\mathbf{x}} = -\frac{1}{2\pi^2}D^x_{\ell m}\int_0^\infty \frac{1}{k}f(k)\sin kx\,dk \tag{7.32}$$

where $D^x_{\ell m}$ is a differential operator defined as

$$D^x_{\ell m} = \frac{1}{x}\left[(\delta_{\ell m} - \hat{x}_\ell\hat{x}_m)\frac{\partial^2}{\partial x^2} + (\delta_{\ell m} - 3\hat{x}_\ell\hat{x}_m)\left(\frac{1}{x^2} - \frac{1}{x}\frac{\partial}{\partial x}\right)\right] \tag{7.33}$$

On the basis of (7.32)

$$\frac{1}{V}\sum_{\mathbf{k}}\left(\delta_{\ell m}-\hat{k}_\ell\hat{k}_m\right)e^{\pm i\mathbf{k}\cdot\mathbf{x}} = -\frac{1}{2\pi^2}D^x_{\ell m}\int_0^\infty \frac{\sin kx}{k}dk$$
$$= -\frac{1}{2\pi^2}D^x_{\ell m}\frac{\pi}{2}$$
$$= -\frac{1}{4\pi}(\delta_{\ell m}-3\hat{x}_\ell\hat{x}_m)\frac{1}{x^3} \qquad (7.34)$$

Substituting (7.34) into (7.31) and implementing straightforward algebraic procedures yields

$$A(\mathbf{x})+B(\mathbf{x})+C(\mathbf{x}) = \mu_{21}^2\frac{1}{2}\left(1+3\cos^2\theta\right)\frac{1}{x^6} \quad (x < c/\omega_0) \qquad (7.35)$$

where θ is the angle between μ_{21} and $\mathbf{x}$. Thus the electric energy density in the near region of a two-level atom is

$$\langle\mathcal{H}_{el}\rangle = \frac{1}{8\pi}{}'\langle\{0_{\mathbf{k}j}\},\downarrow|\,\mathbf{D}_\perp^2(\mathbf{x})\,|\,\{0_{\mathbf{k}j}\},\downarrow\rangle'$$
$$= \frac{1}{2}\frac{1}{V}\sum_{\mathbf{k}}\frac{\hbar\omega_k}{2}+\frac{1}{8\pi}[A(\mathbf{x})+B(\mathbf{x})+C(\mathbf{x})+\text{c.c.}]$$
$$= \frac{1}{2}\frac{1}{V}\sum_{\mathbf{k}}\frac{\hbar\omega_k}{2}+\frac{1}{8\pi}\mu_{21}^2\left(1+3\cos^2\theta\right)\frac{1}{x^6}$$
$$(x < c/\omega_0) \qquad (7.36)$$

This result lends itself to a simple physical interpretation. In fact the first contribution to (7.36) is simply the infinite contribution to the electric energy density coming from zero-point fluctuations of the unperturbed vacuum e.m. field. This contribution is irrelevant for the present purposes, since it does not depend on $\mathbf{x}$ as indeed expected. As for the second term in (7.36), consider the static electric field generated by a static electric dipole μ_{21} fixed at the origin (see e.g. Becker 1982)

$$\mathbf{E}(\mathbf{x}) = -\frac{1}{x^3}\mu_{21}+\frac{3}{x^3}(\mu_{21})_m(\hat{x})_m\hat{x} \qquad (7.37)$$

The energy density corresponding to this field is

$$\frac{1}{8\pi}\mathbf{E}^2(\mathbf{x}) = \frac{1}{8\pi}\mu_{21}^2\left(1+3\cos^2\theta\right)\frac{1}{x^6} \qquad (7.38)$$

coinciding with the second term in (7.36). Thus in the near region the energy density of the displacement field coincides with the energy density of the electric field of a static dipole.

We now turn to the far zone. Where necessary, double sums over $\mathbf{k}$ and $\mathbf{k}'$ in (7.30) can be disentangled, putting $\omega_k = ck$ and using

$$\frac{1}{k+k'} = \int_0^\infty e^{-(k+k')y}dy \tag{7.39}$$

Thus from (7.30) we get

$$\begin{aligned}
A(\mathbf{x}) = B(\mathbf{x}) &= 4\pi^2 c \frac{(\mu_{21})_m(\mu_{21})_n}{\omega_0} \\
&\times \int_0^\infty \left\{ \frac{1}{V}\sum_{\mathbf{k}} k e^{-ky}\left(\delta_{\ell m} - \hat{k}_\ell \hat{k}_m\right) e^{i\mathbf{k}\cdot\mathbf{x}} \right\} \\
&\times \left\{ \frac{1}{V}\sum_{\mathbf{k}'} k' e^{-k'y}\left(\delta_{\ell n} - \hat{k}'_\ell \hat{k}'_n\right) e^{i\mathbf{k}'\cdot\mathbf{x}} \right\} dy\ ; \\
C(\mathbf{x}) &= 4\pi^2 c^2 \frac{(\mu_{21})_m(\mu_{21})_n}{\omega_0^2} \left\{ \frac{1}{V}\sum_{\mathbf{k}} k\left(\delta_{\ell m} - \hat{k}_\ell \hat{k}_m\right) e^{i\mathbf{k}\cdot\mathbf{x}} \right\} \\
&\times \left\{ \frac{1}{V}\sum_{\mathbf{k}'} k'\left(\delta_{\ell m} - \hat{k}'_\ell \hat{k}'_m\right) e^{i\mathbf{k}'\cdot\mathbf{x}} \right\} \quad (x > c/\omega_0)
\end{aligned} \tag{7.40}$$

This leads to evaluation of

$$\begin{aligned}
&\frac{1}{V}\sum_{\mathbf{k}} k e^{-ky}\left(\delta_{\ell m} - \hat{k}_\ell \hat{k}_m\right) e^{i\mathbf{k}\cdot\mathbf{x}} \\
&= -\frac{1}{2\pi^2} D^x_{\ell m} \int_0^\infty e^{-ky} \sin kx dk = -\frac{1}{2\pi^2} D^x_{\ell m} \frac{x}{x^2+y^2} \\
&= -\frac{2}{\pi^2}\frac{1}{(x^2+y^2)^3}\left[(\delta_{\ell m} - 2\hat{x}_\ell \hat{x}_m)x^2 - \delta_{\ell m} y^2\right]
\end{aligned} \tag{7.41}$$

which for $y = 0$ yields

$$\frac{1}{V}\sum_{\mathbf{k}} k\left(\delta_{\ell m} - \hat{k}_\ell \hat{k}_m\right) e^{i\mathbf{k}\cdot\mathbf{x}} = -\frac{2}{\pi^2}\frac{1}{x^4}(\delta_{\ell m} - 2\hat{x}_\ell \hat{x}_m) \tag{7.42}$$

Substitution of (7.41) and (7.42) into (7.40) gives, after some algebra,

$$\begin{aligned}
A(\mathbf{x}) = B(\mathbf{x}) &= \frac{16}{\pi^2} c \frac{(\mu_{21})_m(\mu_{21})_n}{\omega_0} \int_0^\infty \frac{1}{(x^2+y^2)^6} \\
&\times \left[\delta_{mn}x^4 - (2\delta_{mn} - 4\hat{x}_m\hat{x}_n)x^2y^2 + \delta_{mn}y^4\right] dy \\
&= \frac{1}{8\pi} c \frac{(\mu_{21})_m(\mu_{21})_n}{\omega_0} (13\delta_{mn} + 7\hat{x}_m\hat{x}_n)\frac{1}{x^7}\ ; \\
C(\mathbf{x}) &= \frac{16}{\pi^2} c^2 \frac{(\mu_{21})_m(\mu_{21})_n}{\omega_0{}^2} \delta_{mn}\frac{1}{x^8} \quad (x > c/\omega_0)
\end{aligned} \tag{7.43}$$

Clearly, in the limit of large x, C can be neglected with the following result

$$\begin{aligned}\langle \mathcal{H}_{el} \rangle &= \frac{1}{2}\frac{1}{V}\sum_{\mathbf{k}j}\frac{\hbar\omega_k}{2} + \frac{1}{8\pi}[A(\mathbf{x}) + B(\mathbf{x}) + \text{c.c.}] \\ &= \frac{1}{2}\frac{1}{V}\sum_{\mathbf{k}j}\frac{\hbar\omega_k}{2} + \frac{1}{32\pi^2}\hbar c\frac{2(\mu_{21})_m(\mu_{21})_n}{\hbar\omega_0}(13\delta_{mn} + 7\hat{x}_m\hat{x}_n)\frac{1}{x^7} \\ &= \frac{1}{2}\frac{1}{V}\sum_{\mathbf{k}j}\frac{\hbar\omega_k}{2} + \frac{1}{16\pi^2}\frac{c}{\omega_0}\mu_{21}^2(13 + 7\cos^2\theta)\frac{1}{x^7} \\ &\quad (x > c/\omega_0) \end{aligned} \tag{7.44}$$

Introducing the static anisotropic polarizability tensor of the two-level atom

$$\alpha_{mn} = \frac{2(\mu_{21})_m(\mu_{21})_n}{\hbar\omega_0} \tag{7.45}$$

the electric energy density in the far zone takes the form

$$\langle \mathcal{H}_{el} \rangle = \frac{1}{2}\frac{1}{V}\sum_{\mathbf{k}}\frac{\hbar\omega_k}{2} + \frac{1}{32\pi^2}\hbar c\alpha_{mn}(13\ \delta_{mn} + 7\hat{x}_m\hat{x}_n)\frac{1}{x^7} \quad (x > c/\omega_0) \tag{7.46}$$

We may summarize our results as follows. In the near zone around a ground-state dressed two-level atom the cloud of virtual photons yields an electric energy density identical to that of a static dipole. Thus in the near zone the contribution of the transverse part of the electric field is negligible, and the dominant contribution to the electric energy density comes from the longitudinal unretarded electric field. Consequently the x-dependence of the energy density is of the form of $(x^3)^{-2} = x^{-6}$. In the far zone, on the other hand, we find no x^{-6} term, the leading term being at x^{-7}. This means that the contribution of the longitudinal field is cancelled exactly by part of the energy density of the transverse field, leaving only the x^{-7} contribution, and this shows that in the far region the longitudinal and transverse electric energy densities due to the virtual photons are equally important. Finally, the cancellation in the far zone leads to a further decrease of the thickness of the virtual cloud for $x > c/\omega_0$, as measured by $\langle \mathcal{H}_{el} \rangle$, in agreement with the qualitative considerations of Section 6.2. In Section 7.8 we will discuss how this energy density can be measured.

7.3 Other energy densities around a two-level atom

The space-dependence of the electric energy density of the virtual cloud is not the only possible description of the virtual field around a two-level atom. In this section we shall consider other descriptions, with the aim of showing that they give qualitatively similar results.

First we consider the magnetic energy density defined in (7.23). On the basis of (2.9) one has

$$
\begin{aligned}
\mathbf{H}^2(\mathbf{x}) = -\frac{2\pi\hbar}{V}\sum_{\mathbf{k}\mathbf{k}'jj'} \sqrt{\omega_k\omega_{k'}}\mathbf{b}_{\mathbf{k}j}\cdot\mathbf{b}_{\mathbf{k}'j'}\Big\{ & a_{\mathbf{k}j}a_{\mathbf{k}'j'}e^{i(\mathbf{k}+\mathbf{k}')\cdot\mathbf{x}} \\
& + a^\dagger_{\mathbf{k}j}a^\dagger_{\mathbf{k}'j'}e^{-i(\mathbf{k}+\mathbf{k}')\cdot\mathbf{x}} - a_{\mathbf{k}j}a^\dagger_{\mathbf{k}'j'}e^{i(\mathbf{k}-\mathbf{k}')\cdot\mathbf{x}} \\
& - a^\dagger_{\mathbf{k}j}a_{\mathbf{k}'j'}e^{-i(\mathbf{k}-\mathbf{k}')\cdot\mathbf{x}}\Big\}
\end{aligned}
\tag{7.47}
$$

where $\mathbf{b}_{\mathbf{k}j} = \hat{\mathbf{k}} \times \mathbf{e}_{\mathbf{k}j}$ is taken as real. Use of (7.25) thus yields

$$
\begin{aligned}
'\langle\{0_{\mathbf{k}j}\},\downarrow|\,\mathbf{H}^2(\mathbf{x})\,|\,\{0_{\mathbf{k}j}\},\downarrow\rangle' &= \frac{2\pi}{V}\sum_{\mathbf{k}j}\hbar\omega_k + \frac{2\pi}{\hbar V}\sum_{\mathbf{k}\mathbf{k}'jj'}\sqrt{\omega_k\omega_{k'}} \\
\times\Bigg\{\mathbf{b}_{\mathbf{k}j}\cdot\mathbf{b}_{\mathbf{k}'j'}\left(1+\delta_{\mathbf{k}\mathbf{k}'}\delta_{jj'}\right) & \left[\frac{\epsilon^*_{\mathbf{k}j}\epsilon_{\mathbf{k}'j'}}{(\omega_0+\omega_{k'})(\omega_k+\omega_{k'})}\right. \\
& \left. + \frac{\epsilon_{\mathbf{k}j}\epsilon^*_{\mathbf{k}'j'}}{(\omega_0+\omega_k)(\omega_k+\omega_{k'})}\right] e^{i(\mathbf{k}+\mathbf{k}')\cdot\mathbf{x}} \\
& + \mathbf{b}_{\mathbf{k}j}\cdot\mathbf{b}_{\mathbf{k}'j'}\frac{\epsilon_{\mathbf{k}j}\epsilon^*_{\mathbf{k}'j'}}{(\omega_0+\omega_k)(\omega_0+\omega_{k'})}e^{i(\mathbf{k}-\mathbf{k}')\cdot\mathbf{x}} + \text{c.c.}\Bigg\}
\end{aligned}
\tag{7.48}
$$

As in the previous section, the contribution of the $\delta_{\mathbf{k}\mathbf{k}'}\delta_{jj'}$ terms in (7.48) can be neglected for $V \to \infty$. For the remaining terms, the polarization sums can be performed with the help of (2.6). Taking μ_{21} real, one has in fact

$$
\begin{aligned}
\sum_{jj'}\mathbf{b}_{\mathbf{k}j}\cdot\mathbf{b}_{\mathbf{k}'j'}\epsilon^*_{\mathbf{k}j}\epsilon_{\mathbf{k}'j'} &= \sum_{jj'}\mathbf{b}_{\mathbf{k}j}\cdot\mathbf{b}_{\mathbf{k}'j'}\epsilon_{\mathbf{k}j}\epsilon^*_{\mathbf{k}'j'} \\
&= \frac{2\pi\hbar}{V}\sqrt{\omega_k\omega_{k'}}(\mu_{21})_m(\mu_{21})_n \\
&\quad \times\sum_{jj'}(\mathbf{b}_{\mathbf{k}j})_\ell(\mathbf{b}_{\mathbf{k}'j'})_\ell(\mathbf{e}_{\mathbf{k}j})_m(\mathbf{e}_{\mathbf{k}'j'})_n \\
&= \frac{2\pi\hbar}{V}\sqrt{\omega_k\omega_{k'}}(\mu_{21})_m(\mu_{21})_n e_{\ell mp}\hat{k}_p e_{\ell nq}\hat{k}'_q
\end{aligned}
\tag{7.49}
$$

where $e_{\ell mp}$ is the complete antisymmetric pseudotensor defined in (1.39). From the properties of this pseudotensor it is possible to prove

$$e_{\ell mp}e_{\ell nq} = \delta_{mn}\delta_{pq} - \delta_{mq}\delta_{np} \tag{7.50}$$

Hence

$$\begin{aligned}
\sum_{jj'} \mathbf{b}_{\mathbf{k}j} \cdot \mathbf{b}_{\mathbf{k}'j'} \epsilon^*_{\mathbf{k}j} \epsilon_{\mathbf{k}'j'} &= \sum_{jj'} \mathbf{b}_{\mathbf{k}j} \cdot \mathbf{b}_{\mathbf{k}'j'} \epsilon_{\mathbf{k}j} \epsilon^*_{\mathbf{k}'j'} \\
&= \frac{2\pi\hbar}{V} \sqrt{\omega_k \omega_{k'}} (\mu_{21})_m (\mu_{21})_n \left(\delta_{mn}\delta_{pq} - \delta_{mq}\delta_{np}\right) \hat{k}_p \hat{k}'_q \\
&= -\frac{2\pi\hbar}{V} \sqrt{\omega_k \omega_{k'}} \left[\left(\mu_{21} \cdot \hat{\mathbf{k}}\right)\left(\mu_{21} \cdot \hat{\mathbf{k}}'\right) - \mu_{21}^2 \hat{\mathbf{k}} \cdot \hat{\mathbf{k}}'\right\}
\end{aligned} \tag{7.51}$$

and the terms contributing to (7.48) can be written as

$$\begin{aligned}
A(\mathbf{x}) &= \frac{2\pi}{\hbar V} \sum_{\mathbf{k}\mathbf{k}'j'j} \sqrt{\omega_k \omega_{k'}} \mathbf{b}_{\mathbf{k}j} \cdot \mathbf{b}_{\mathbf{k}'j'} \frac{\epsilon^*_{\mathbf{k}j} \epsilon_{\mathbf{k}'j'}}{(\omega_0 + \omega_{k'})(\omega_k + \omega_{k'})} e^{i(\mathbf{k}+\mathbf{k}')\cdot\mathbf{x}} \\
&= -\frac{4\pi^2}{V^2} \sum_{\mathbf{k}\mathbf{k}'} \omega_k \omega_{k'} \left[\left(\mu_{21} \cdot \hat{\mathbf{k}}\right)\left(\mu_{21} \cdot \hat{\mathbf{k}}'\right) - \mu_{21}^2 \hat{\mathbf{k}} \cdot \hat{\mathbf{k}}'\right] \\
&\quad \times \frac{e^{i(\mathbf{k}+\mathbf{k}')\cdot\mathbf{x}}}{(\omega_0 + \omega_{k'})(\omega_k + \omega_{k'})}; \\
B(\mathbf{x}) &= \frac{2\pi}{\hbar V} \sum_{\mathbf{k}\mathbf{k}'j'j} \sqrt{\omega_k \omega_{k'}} \mathbf{b}_{\mathbf{k}j} \cdot \mathbf{b}_{\mathbf{k}'j'} \frac{\epsilon_{\mathbf{k}j} \epsilon^*_{\mathbf{k}'j'}}{(\omega_0 + \omega_{k'})(\omega_k + \omega_{k'})} e^{i(\mathbf{k}+\mathbf{k}')\cdot\mathbf{x}} \\
&= -\frac{4\pi^2}{V^2} \sum_{\mathbf{k}\mathbf{k}'} \omega_k \omega_{k'} \left[\left(\mu_{21} \cdot \hat{\mathbf{k}}\right)\left(\mu_{21} \cdot \hat{\mathbf{k}}'\right) - \mu_{21}^2 \hat{\mathbf{k}} \cdot \hat{\mathbf{k}}'\right] \\
&\quad \times \frac{e^{i(\mathbf{k}+\mathbf{k}')\cdot\mathbf{x}}}{(\omega_0 + \omega_k)(\omega_k + \omega_{k'})}; \\
C(\mathbf{x}) &= \frac{2\pi}{\hbar V} \sum_{\mathbf{k}\mathbf{k}'jj'} \sqrt{\omega_k \omega_{k'}} \mathbf{b}_{\mathbf{k}j} \cdot \mathbf{b}_{\mathbf{k}'j'} \frac{\epsilon_{\mathbf{k}j} \epsilon^*_{\mathbf{k}'j'}}{(\omega_0 + \omega_k)(\omega_0 + \omega_{k'})} e^{i(\mathbf{k}-\mathbf{k}')\cdot\mathbf{x}} \\
&= -\frac{4\pi^2}{V^2} \sum_{\mathbf{k}\mathbf{k}'} \omega_k \omega_{k'} \left[\left(\mu_{21} \cdot \hat{\mathbf{k}}\right)\left(\mu_{21} \cdot \hat{\mathbf{k}}'\right) - \mu_{21}^2 \hat{\mathbf{k}} \cdot \hat{\mathbf{k}}'\right] \\
&\quad \times \frac{e^{i(\mathbf{k}-\mathbf{k}')\cdot\mathbf{x}}}{(\omega_0 + \omega_k)(\omega_0 + \omega_{k'})}
\end{aligned} \tag{7.52}$$

Following the development of the previous section, we now obtain approximate expressions valid in the near zone ($x < c/\omega_0$) where we assume $\omega_k/\omega_0 \gg 1$, and in the far zone ($x > c/\omega_0$) where we assume $\omega_0/\omega_k \gg 1$.

a) Near zone ($x < c/\omega_0$). Here a preliminary remark is appropriate. If one approximates $\omega_0 + \omega_k \sim \omega_k$ one has exact cancellation of the x^{-6} terms in $A(\mathbf{x}) + B(\mathbf{x}) + C(\mathbf{x})$. In fact with this approximation (7.52) yields

$$
\begin{aligned}
A(\mathbf{x}) + B(\mathbf{x}) &= -\frac{4\pi^2}{V^2}\sum_{\mathbf{k}\mathbf{k}'}\left[\left(\mu_{21}\cdot\hat{\mathbf{k}}\right)\left(\mu_{21}\cdot\hat{\mathbf{k}}'\right) - \mu_{21}^2\hat{\mathbf{k}}\cdot\hat{\mathbf{k}}'\right] \\
&\quad \times e^{i(\mathbf{k}+\mathbf{k}')\cdot\mathbf{x}} \\
&= -4\pi^2\left\{\left(\frac{1}{V}\sum_{\mathbf{k}}\mu_{21}\cdot\hat{\mathbf{k}}e^{i\mathbf{k}\cdot\mathbf{x}}\right)\right. \\
&\quad \times\left(\frac{1}{V}\sum_{\mathbf{k}'}\mu_{21}\cdot\hat{\mathbf{k}}'e^{i\mathbf{k}'\cdot\mathbf{x}}\right) \\
&\quad \left. - \mu_{21}^2\left(\frac{1}{V}\sum_{\mathbf{k}}\hat{\mathbf{k}}e^{i\mathbf{k}\cdot\mathbf{x}}\right)\cdot\left(\frac{1}{V}\sum_{\mathbf{k}'}\hat{\mathbf{k}}'e^{i\mathbf{k}'\cdot\mathbf{x}}\right)\right\}; \\
C(\mathbf{x}) &= -\frac{4\pi^2}{V^2}\sum_{\mathbf{k}\mathbf{k}'}\left[\left(\mu_{21}\cdot\hat{\mathbf{k}}\right)\left(\mu_{21}\cdot\hat{\mathbf{k}}'\right) - \mu_{21}^2\hat{\mathbf{k}}\cdot\hat{\mathbf{k}}'\right] \\
&\quad \times e^{i(\mathbf{k}-\mathbf{k}')\cdot\mathbf{x}} \\
&= -4\pi^2\left\{\left(\frac{1}{V}\sum_{\mathbf{k}}\mu_{21}\cdot\hat{\mathbf{k}}e^{i\mathbf{k}\cdot\mathbf{x}}\right)\right. \\
&\quad \times\left(\frac{1}{V}\sum_{\mathbf{k}'}\mu_{21}\cdot\hat{\mathbf{k}}'e^{-i\mathbf{k}'\cdot\mathbf{x}}\right) \\
&\quad \left. - \mu_{21}^2\left(\frac{1}{V}\sum_{\mathbf{k}}\hat{\mathbf{k}}e^{i\mathbf{k}\cdot\mathbf{x}}\right)\cdot\left(\frac{1}{V}\sum_{\mathbf{k}'}\hat{\mathbf{k}}'e^{-i\mathbf{k}'\cdot\mathbf{x}}\right)\right\}
\end{aligned}
\tag{7.53}
$$

Now, transforming sums into integrals as usual and performing elementary angular integration, it is easy to prove that

$$
\frac{1}{V}\sum_{\mathbf{k}}\hat{\mathbf{k}}e^{\pm i\mathbf{k}\cdot\mathbf{x}} = \mp i\nabla^x\frac{1}{V}\sum_{\mathbf{k}}e^{i\mathbf{k}\cdot\mathbf{x}} = \mp i\frac{1}{\pi^2}\frac{\hat{x}}{x^3} \tag{7.54}
$$

When (7.54) is substituted into (7.53) one has immediately

$$
A(\mathbf{x}) + B(\mathbf{x}) = -C(\mathbf{x}) = -\frac{4}{\pi^2}\mu_{21}^2\left(1-\cos^2\theta\right)\frac{1}{x^6} \tag{7.55}
$$

which yields the cancellation mentioned above. Thus in order to find a nonzero result in the near zone one has to proceed to the next order in ω_0/ω_k, using

$$\frac{1}{\omega_0+\omega_k} \sim \frac{1}{\omega_k} - \frac{\omega_0}{\omega_k^2} ;$$
$$\frac{1}{(\omega_0+\omega_k)(\omega_0+\omega_{k'})} \sim \frac{1}{\omega_k\omega_{k'}} - \omega_0\frac{\omega_k+\omega_{k'}}{\omega_k^2\omega_{k'}^2} \tag{7.56}$$

Substituting (7.56) into (7.52) and neglecting terms of order zero in ω_0, which we have shown to cancel, one has

$$A(\mathbf{x})+B(\mathbf{x}) = \frac{4\pi^2}{V^2}\frac{\omega_0}{c}\sum_{\mathbf{kk}'}\frac{k^2+k'^2}{kk'(k+k')}\left[\left(\mu_{21}\cdot\hat{\mathbf{k}}\right)\left(\mu_{21}\cdot\hat{\mathbf{k}}'\right)\right]$$
$$\times\, e^{i(\mathbf{k}+\mathbf{k}')\cdot\mathbf{x}}$$
$$= \lim_{\mathbf{x}'\to\mathbf{x}} -4\pi^2\frac{\omega_0}{c}G^{\mathbf{x},\mathbf{x}'}\frac{1}{V^2}\sum_{\mathbf{kk}'}\frac{k^2+k'^2}{k^2k'^2(k+k')}e^{i(\mathbf{k}\cdot\mathbf{x}+\mathbf{k}'\cdot\mathbf{x}')} \tag{7.57}$$

where $G^{\mathbf{x},\mathbf{x}'}$ is a differential operator defined as

$$G^{\mathbf{x},\mathbf{x}'} = (\mu_{21}\cdot\nabla^x)(\mu_{21}\cdot\nabla^{x'}) - \mu_{21}^2\nabla^x\cdot\nabla^{x'} \tag{7.58}$$

This leads to the evaluation of

$$\frac{1}{V^2}\sum_{\mathbf{kk}'}\frac{k^2+k'^2}{k^2k'^2(k+k')}e^{i(\mathbf{k}\cdot\mathbf{x}+\mathbf{k}'\cdot\mathbf{x}')}$$
$$= \int_0^\infty\left\{\left(\frac{1}{V}\sum_{\mathbf{k}}\frac{1}{k^2}e^{-ky}e^{i\mathbf{k}\cdot\mathbf{x}}\right)\left(\frac{1}{V}\sum_{\mathbf{k}'}e^{-k'y}e^{i\mathbf{k}'\cdot\mathbf{x}'}\right)\right.$$
$$\left.+\left(\frac{1}{V}\sum_{\mathbf{k}}e^{-ky}e^{i\mathbf{k}\cdot\mathbf{x}}\right)\left(\frac{1}{V}\sum_{\mathbf{k}'}\frac{1}{k'^2}e^{-k'y}e^{i\mathbf{k}'\cdot\mathbf{x}'}\right)\right\}dy \tag{7.59}$$

where (7.39) has been used to achieve factorization of the double sums on $\mathbf{k}$ and $\mathbf{k}'$. The following results are easily obtained transforming the sums into integrals and after elementary integration

$$\frac{1}{V}\sum_{\mathbf{k}}e^{-ky}e^{i\mathbf{k}\cdot\mathbf{x}} = \frac{1}{\pi^2}\frac{y}{(x^2+y^2)^2} ;$$
$$\frac{1}{V}\sum_{\mathbf{k}}\frac{1}{k^2}e^{-ky}e^{i\mathbf{k}\cdot\mathbf{x}} = \frac{1}{2\pi^2}\frac{1}{x}\arctan\frac{x}{y} \tag{7.60}$$

Substitution of (7.60) into (7.59) gives

$$
\begin{aligned}
&\frac{1}{V^2}\sum_{\mathbf{k}\mathbf{k}'}\frac{k^2+k'^2}{k^2k'^2(k+k')}e^{i(\mathbf{k}\cdot\mathbf{x}+\mathbf{k}'\cdot\mathbf{x}')}\\
&=\frac{1}{2\pi^4}\int_0^\infty\left\{\frac{y/x}{(x'^2+y^2)^2}\arctan\frac{x}{y}+\frac{y/x'}{(x^2+y^2)^2}\arctan\frac{x'}{y}\right\}dy\\
&=\frac{1}{(2\pi)^3}\frac{x^2+x'^2}{x^2x'^2(x+x')}
\end{aligned}
\tag{7.61}
$$

after partial integration and some simple algebra. Inserting (7.61) in (7.57) and using (7.58) yields

$$
A(\mathbf{x})+B(\mathbf{x})=\frac{1}{\pi}\frac{\omega_0}{c}\mu_{21}^2\frac{3}{4}\sin^2\theta\frac{1}{x^5}\quad (x<c/\omega_0)
\tag{7.62}
$$

where θ is the angle between μ_{21} and $\mathbf{x}$ as usual. Moreover, substitution of (7.56) into (7.52) gives also, following the same procedure as above

$$
C(\mathbf{x})=\lim_{\mathbf{x}'\to\mathbf{x}}4\pi^2\frac{\omega_0}{c}G^{\mathbf{x},\mathbf{x}'}\frac{1}{V^2}\sum_{\mathbf{k}\mathbf{k}'}\left(\frac{1}{k}+\frac{1}{k'}\right)\frac{1}{kk'}e^{i(\mathbf{k}\cdot\mathbf{x}-\mathbf{k}'\cdot\mathbf{x}')}
\tag{7.63}
$$

Thus one needs to evaluate

$$
\begin{aligned}
&\frac{1}{V^2}\sum_{\mathbf{k}\mathbf{k}'}\left(\frac{1}{k}+\frac{1}{k'}\right)\frac{1}{kk'}e^{i(\mathbf{k}\cdot\mathbf{x}-\mathbf{k}'\cdot\mathbf{x}')}\\
&=\left(\frac{1}{V}\sum_{\mathbf{k}}\frac{1}{k^2}e^{i\mathbf{k}\cdot\mathbf{x}}\right)\left(\frac{1}{V}\sum_{\mathbf{k}'}\frac{1}{k'}e^{-i\mathbf{k}'\cdot\mathbf{x}'}\right)\\
&+\left(\frac{1}{V}\sum_{\mathbf{k}}\frac{1}{k}e^{i\mathbf{k}\cdot\mathbf{x}}\right)\left(\frac{1}{V}\sum_{\mathbf{k}'}\frac{1}{k'^2}e^{-i\mathbf{k}'\cdot\mathbf{x}'}\right)
\end{aligned}
\tag{7.64}
$$

The sums appearing in (7.64) can be easily evaluated as

$$
\begin{aligned}
&\frac{1}{V}\sum_{\mathbf{k}}\frac{1}{k}e^{\pm i\mathbf{k}\cdot\mathbf{x}}=\frac{1}{2\pi^2}\frac{1}{x^2};\\
&\frac{1}{V}\sum_{\mathbf{k}}\frac{1}{k^2}e^{\pm i\mathbf{k}\cdot\mathbf{x}}=\frac{1}{4\pi}\frac{1}{x}
\end{aligned}
\tag{7.65}
$$

These results can be substituted in (7.64), which can be expressed as

$$
\frac{1}{V^2}\sum_{\mathbf{k}\mathbf{k}'}\left(\frac{1}{k}+\frac{1}{k'}\right)\frac{1}{kk'}e^{i(\mathbf{k}\cdot\mathbf{x}-\mathbf{k}'\cdot\mathbf{x}')}=\frac{1}{(2\pi)^3}\left(\frac{1}{x}+\frac{1}{x'}\right)\frac{1}{xx'}
\tag{7.66}
$$

Substitution in turn of (7.66) into (7.63) leads to

$$C(\mathbf{x}) = -\frac{1}{\pi}\frac{\omega_0}{c}\mu_{21}^2 2\sin^2\theta\frac{1}{x^5} \tag{7.67}$$

which, together with (7.62), yields

$$A(\mathbf{x}) + B(\mathbf{x}) + C(\mathbf{x}) = -\frac{1}{\pi}\frac{\omega_0}{c}\mu_{21}^2\frac{5}{4}\sin^2\theta\frac{1}{x^5} \quad (x < c/\omega_0) \tag{7.68}$$

When this result is used in (7.48) one finds

$$\begin{aligned}\langle\mathcal{H}_{mag}\rangle &= \frac{1}{8\pi}\ {}'\langle\{0_{\mathbf{k}j}\},\downarrow|\,\mathbf{H}^2(\mathbf{x})\mid\{0_{\mathbf{k}j}\},\downarrow\rangle' \\ &= \frac{1}{2}\frac{1}{V}\sum_{\mathbf{k}j}\frac{\hbar\omega_k}{2} - \frac{1}{16\pi^2}\frac{\omega_0}{c}\mu_{21}^2 5\sin^2\theta\frac{1}{x^5} \\ &\quad (x < c/\omega_0)\end{aligned} \tag{7.69}$$

The first term on the RHS of (7.69) is the zero-point magnetic contribution to the energy density. This is infinite and equal to the electric contribution obtained in (7.36), and we are going to neglect it for the same reasons as previously. The second part is smaller by a factor $x\omega_0/c$ than its electric counterpart at any point $\mathbf{x}$ of the near zone.

b) Far zone ($x > c/\omega_0$). Putting $\omega_0 + \omega_k \sim \omega_0$ in (7.52) and disentangling as in (7.39) leads to

$$\begin{aligned}A(\mathbf{x}) = B(\mathbf{x}) &= -\frac{4\pi^2}{V^2}\frac{c}{\omega_0}\sum_{\mathbf{k}\mathbf{k}'}\frac{kk'}{k+k'} \\ &\times\left[\left(\mu_{21}\cdot\hat{\mathbf{k}}\right)\left(\mu_{21}\cdot\hat{\mathbf{k}}'\right) - \mu_{21}^2\hat{\mathbf{k}}\cdot\hat{\mathbf{k}}'\right]e^{i(\mathbf{k}+\mathbf{k}')\cdot\mathbf{x}} \\ &= 4\pi^2\frac{c}{\omega_0}\int_0^\infty\left\{\left(\mu_{21}\cdot\nabla^x\frac{1}{V}\sum_{\mathbf{k}}e^{-ky}e^{i\mathbf{k}\cdot\mathbf{x}}\right)^2\right. \\ &\quad\left. - \mu_{21}^2\left(\nabla^x\frac{1}{V}\sum_{\mathbf{k}}e^{-ky}e^{i\mathbf{k}\cdot\mathbf{x}}\right)^2\right\}dy\end{aligned} \tag{7.70}$$

From (7.60) one has

$$\nabla^x\frac{1}{V}\sum_{\mathbf{k}}e^{-ky}e^{i\mathbf{k}\cdot\mathbf{x}} = -\frac{4}{\pi^2}\frac{xy}{(x^2+y^2)^3}\hat{x} \tag{7.71}$$

and after elementary integration

$$A(\mathbf{x}) = B(\mathbf{x}) = -\frac{1}{\pi}\frac{c}{\omega_0}\frac{7}{8}\sin^2\theta\frac{1}{x^7} \tag{7.72}$$

Moreover

$$\begin{aligned} C(\mathbf{x}) &= -\frac{4\pi^2}{V^2}\frac{c^2}{\omega_0^2}\sum_{\mathbf{kk}'} kk'\left[\left(\mu_{21}\cdot\hat{\mathbf{k}}\right)\left(\mu_{21}\cdot\hat{\mathbf{k}}'\right) - \mu_{21}^2\hat{\mathbf{k}}\cdot\hat{\mathbf{k}}'\right]e^{i(\mathbf{k}-\mathbf{k}')\cdot\mathbf{x}} \\ &= 4\pi^2\frac{c^2}{\omega_0^2}\left\{\left|\mu_{21}\cdot\nabla^x\frac{1}{V}\sum_{\mathbf{k}} e^{i\mathbf{k}\cdot\mathbf{x}}\right|^2 - \mu_{21}^2\left|\nabla^x\frac{1}{V}\sum_{\mathbf{k}} e^{i\mathbf{k}\cdot\mathbf{x}}\right|^2\right\} \end{aligned} \tag{7.73}$$

This vanishes, as is evident from the first of (7.60) evaluated at $y = 0$, and

$$\langle\mathcal{H}_{mag}\rangle = \frac{1}{2}\frac{1}{V}\sum_{\mathbf{k}j}\frac{\hbar\omega_k}{2} - \frac{7}{16\pi^2}\frac{c}{\omega_0}\mu_{21}^2\sin^2\theta\frac{1}{x^7} \quad (x > c/\omega_0) \tag{7.74}$$

Expressions (7.69) and (7.74) show that the virtual photons tend to decrease the magnetic energy density around the two-level atom, while they tend to increase the electric energy density, as is evident from (7.36) and (7.44). The total energy density can be obtained as the sum of these two energies.

Other descriptions of the virtual cloud are possible. As an example, consider the following partition of the displacement operator (4.81)

$$\begin{aligned} \mathbf{D}_\perp(\mathbf{x}) &= \mathbf{D}_\perp^{(-)}(\mathbf{x}) + \mathbf{D}_\perp^{(+)}(\mathbf{x})\ ; \\ \mathbf{D}_\perp^{(-)}(\mathbf{x}) &= -i\sum_{\mathbf{k}j}\sqrt{\frac{2\pi\hbar\omega_k}{V}}\mathbf{e}_{\mathbf{k}j}a_{\mathbf{k}j}^\dagger e^{-i\mathbf{k}\cdot\mathbf{x}}\ ; \\ \mathbf{D}_\perp^{(+)}(\mathbf{x}) &= i\sum_{\mathbf{k}j}\sqrt{\frac{2\pi\hbar\omega_k}{V}}\mathbf{e}_{\mathbf{k}j}a_{\mathbf{k}j}e^{i\mathbf{k}\cdot\mathbf{x}} \end{aligned} \tag{7.75}$$

and the quadratic form

$$\begin{aligned} W_{el}(\mathbf{x}) &= \frac{1}{4\pi}\mathbf{D}_\perp^{(-)}(\mathbf{x})\cdot\mathbf{D}_\perp^{(+)}(\mathbf{x}) \\ &= \frac{\hbar}{2V}\sum_{\mathbf{kk}'jj'}\sqrt{\omega_k\omega_{k'}}\mathbf{e}_{\mathbf{k}j}\cdot\mathbf{e}_{\mathbf{k}'j'}a_{\mathbf{k}j}^\dagger a_{\mathbf{k}'j'}e^{-i(\mathbf{k}-\mathbf{k}')\cdot\mathbf{x}} \end{aligned} \tag{7.76}$$

This operator has an interesting interpretation (Mandel 1966). In fact, integrating over a volume $\ell^3 \ll V$ centred at a point $\mathbf{x}_0$ yields

$$\int_{\ell^3} e^{-i(\mathbf{k}-\mathbf{k}')\cdot\mathbf{x}} d^3\mathbf{x} = \ell^3 e^{-i(\mathbf{k}-\mathbf{k}')\cdot\mathbf{x}_0} f(\mathbf{k}-\mathbf{k}')$$

$$f(\mathbf{k}-\mathbf{k}') = \prod_j \frac{\sin(\mathbf{k}-\mathbf{k}')_j \ell/2}{(\mathbf{k}-\mathbf{k}')_j \ell/2} \tag{7.77}$$

Clearly, $f(\mathbf{k}-\mathbf{k}') \sim 0$ except for $\mathbf{k}$ and $\mathbf{k}'$ satisfying

$$(\mathbf{k}-\mathbf{k}')_j \ell/2 \ll 1 \quad (j = 1, 2, 3) \tag{7.78}$$

when $f(\mathbf{k}-\mathbf{k}') \sim 1$. Choosing e.g. $\ell \sim 10^{-1}$ cm, (7.78) implies near equality of $\mathbf{k}$ and $\mathbf{k}'$ for photons in the optical range. Although the same is not true for photons in the microwave range or in ranges of lower frequency, these latter photons play a role in the virtual cloud only at distances from the atom so large that the cloud is practically negligible. Thus from (7.76) one is entitled to approximate

$$\frac{1}{\ell^3}\int_{\ell^3} W_{el}(\mathbf{x}) d^3\mathbf{x} \sim \frac{1}{2}\frac{1}{V}\sum_{\mathbf{k}j} \hbar\omega_k a^\dagger_{\mathbf{k}j} a_{\mathbf{k}j} \tag{7.79}$$

This shows that W_{el} can be interpreted approximately as the electric energy density carried by the photons, when it is averaged (or "coarse-grained") within a volume $\ell^3 \ll V$. A more sophisticated definition of this averaged energy density can be found in Passante et al. (1985).

From (7.76) use of (7.25) and (7.27) leads to

$$\begin{aligned} &'\langle\{0_{\mathbf{k}j}\}, \downarrow | \, W_{el}(\mathbf{x}) \, | \{0_{\mathbf{k}j}\}, \downarrow\rangle' \\ &= \pi(\mu_{21})_m(\mu_{21})_n \left[\frac{1}{V}\sum_{\mathbf{k}} \frac{k}{k_0+k}\left(\delta_{\ell m} - \hat{k}_\ell \hat{k}_m\right) e^{-i\mathbf{k}\cdot\mathbf{x}}\right] \\ &\times \left[\frac{1}{V}\sum_{\mathbf{k}'} \frac{k'}{k_0+k'}\left(\delta_{\ell n} - \hat{k}'_\ell \hat{k}'_n\right) e^{i\mathbf{k}\cdot\mathbf{x}}\right] \end{aligned} \tag{7.80}$$

whereas from (7.32) we find

$$\begin{aligned} &\frac{1}{V}\sum_{\mathbf{k}} \frac{k}{k_0+k}\left(\delta_{\ell m} - \hat{k}_\ell \hat{k}_m\right) e^{-i\mathbf{k}\cdot\mathbf{x}} = -\frac{1}{2\pi^2} D^x_{\ell m} \int_0^\infty \frac{\sin kx}{k_0+k} dk \\ &= -\frac{1}{2\pi^2} D^x_{\ell m} f(k_0 x) \, ; \\ &f(k_0 x) = Ci(k_0 x)\sin k_0 x - si(k_0 x)\cos k_0 x \end{aligned} \tag{7.81}$$

where

$$Ci(z) = C + \ln z + \int_0^z \frac{\cos t - 1}{t} dt \,;$$
$$si(z) = \int_0^z \frac{\sin t}{t} dt - \frac{\pi}{2} \tag{7.82}$$

are the usual definitions of cosine and sine integrals (see e.g. Abramowitz and Stegun 1965), C being the Euler-Mascheroni constant. Use of the well-known limiting expressions for $f(z)$

$$f(z) \simeq \pi/2 \ \ (z \ll 1) \ ; \ f(z) \simeq 1/z \ \ (z \gg 1) \tag{7.83}$$

and of some straightforward algebra gives

$$'\langle\{0_{\mathbf{k}j}\}, \downarrow | \ W_{el}(\mathbf{x}) \ | \{0_{\mathbf{k}j}\}, \downarrow\rangle' = \begin{cases} \frac{2}{\pi^3} \mu_{21}^2 \frac{c^2}{\omega_0^2} \frac{1}{x^8} & (k_0 x \gg 1) \\ \frac{1}{16\pi} \mu_{21}^2 \left(1 + 3\cos^2\theta\right) \frac{1}{x^6} & (k_0 x \ll 1) \end{cases} \tag{7.84}$$

in the far and in the near zone respectively.

We shall conclude the section by remarking that all the descriptions of the virtual photon cloud around a two-level atom, in the electric dipole approximation and up to order e^2, indicate that the density of the cloud decreases according to a power law with increasing distance from the atom. Moreover, at distances $\sim c/\omega_0$, the power law changes and the decrease of the density becomes faster. This shows that the virtual electromagnetic field around the atom carries information on the internal structure of the source, since the change in the power law depends on ω_0 which is a fundamental parameter of the bare two-level atom. This information is also carried in part by the anisotropy of the virtual cloud. This anisotropy, given by the dependence of (7.84) on θ, which is the angle between the atomic transition dipole μ_{21} and the observation point at $\mathbf{x}$, displays very clearly the fact that a two-level atom is a fundamentally anisotropic object.

7.4 Energy density around a slow free electron

The next model we shall consider is that of a spinless free electron of momentum $\mathbf{p}$ interacting with the vacuum fluctuations of the e.m. field. This model has been recently investigated by Compagno and Salamone

(1991). This section is based essentially on the work of these authors. As discussed in the course of this book, the process responsible for the cloud around this electron is the emission and reabsorption of virtual photons in the course of recoil events. In a relativistic context one must consider also relativistic photons inducing fluctuation of the total (electric+magnetic) bare energy $\Delta E \geq mc^2$. These high-energy photons and the associated positron cloud have been discussed long ago mainly from a global point of view (Weisskopf 1939). They contribute a cloud of virtual particles extending up to a distance from the electron which is of the order of the electron Compton wavelength $\lambda = \hbar/mc$. Thus a nonrelativistic approach like the one we shall describe here is valid only at distances from the electron larger than λ_c, where the positron cloud can be neglected and the electric charge is the renormalized one $-e$. At these relatively large distances only low-frequency photons contribute to the cloud around the electron, so that approximations of the kind (6.43) can be assumed to be valid for a slow electron.

We will evaluate the structure of the electromagnetic field around such a slow electron *in vacuo* much in the same way as we have evaluated the cloud of virtual phonons surrounding a weakly coupled Fröhlich polaron in Section 7.1. We shall use perturbation theory starting from the minimal-coupling Hamiltonian (6.39). If the free electron has momentum $\mathbf{p} \ll mc$, perturbation expansion accurate up to terms of order e^2 leads to a perturbed state given by (6.41). Thus

$$| \mathbf{p}, \{0_{\mathbf{k}j}\}\rangle' = | \phi\rangle + | \phi_1\rangle + | \phi_2\rangle + | \phi_3\rangle + | \phi_4\rangle \tag{7.85}$$

where the various terms are easily calculated from (6.39) and approximation (6.43) as

$$\begin{aligned}
| \phi\rangle &= | \mathbf{p}, \{0_{\mathbf{k}j}\}\rangle \,; \\
| \phi_1\rangle &= \frac{1}{E_0 - H_0}(1 - P_0) V_1 \,| \phi\rangle \\
&= -\frac{e}{m}\sum_{\mathbf{k}j}\sqrt{\frac{2\pi\hbar}{V\omega_k}}\mathbf{e}^*_{\mathbf{k}j}\cdot\mathbf{p}\frac{1}{\hbar\omega_k}\,| \mathbf{p} - \hbar\mathbf{k}, 1_{\mathbf{k}j}\rangle \,; \\
| \phi_2\rangle &= \frac{1}{E_0 - H_0}(1 - P_0) V_2 \,| \phi\rangle \\
&= -\frac{\pi e^2}{mV}\Bigg\{\sum_{\substack{\mathbf{k}\mathbf{k}'jj' \\ \mathbf{k}j\neq\mathbf{k}'j'}}\frac{1}{\sqrt{\omega_k\omega_{k'}}}\mathbf{e}^*_{\mathbf{k}j}\cdot\mathbf{e}^*_{\mathbf{k}'j'}\frac{1}{\omega_k + \omega_{k'}}\cdot \\
&\quad\times | \mathbf{p} - \hbar(\mathbf{k} + \mathbf{k}'), 1_{\mathbf{k}j}, 1_{\mathbf{k}'j'}\rangle + \frac{1}{\sqrt{2}}\sum_{\mathbf{k}j}\frac{\mathbf{e}^{*\,2}_{\mathbf{k}j}}{\omega_k^2}\,| \mathbf{p} - 2\hbar\mathbf{k}, 2_{\mathbf{k}j}\rangle\Bigg\} \,;
\end{aligned}$$

$$
\begin{aligned}
| \phi_3\rangle &= \frac{1}{E_0 - H_0}(1 - P_0)V_1 \frac{1}{E_0 - H_0}(1 - P_0)V_1 \mid \phi\rangle \\
&= \frac{2\pi e^2}{m^2\hbar V}\left\{ \sum_{\substack{\mathbf{k}\mathbf{k}'j'j \\ \mathbf{k}j \neq \mathbf{k}'j'}} \frac{1}{\sqrt{\omega_k \omega_{k'}}} \frac{1}{\omega_k(\omega_k + \omega_{k'})} \left(\mathbf{e}^*_{\mathbf{k}j} \cdot \mathbf{p}\right) . \right. \\
&\times \left[e^*_{\mathbf{k}'j'} \cdot (\mathbf{p} - \hbar\mathbf{k}) \right] \mid \mathbf{p} - \hbar(\mathbf{k} + \mathbf{k}'), 1_{\mathbf{k}j}, 1_{\mathbf{k}'j'}\rangle \\
&\left. + \frac{1}{\sqrt{2}} \sum_{\mathbf{k}j} \frac{1}{\omega_k^3} \left(\mathbf{e}^*_{\mathbf{k}j} \cdot \mathbf{p}\right)^2 \mid \mathbf{p} - 2\hbar\mathbf{k}, 2_{\mathbf{k}j}\rangle \right\} \\
| \phi_4\rangle &= -\frac{1}{2}\langle \phi \mid V_1 \frac{1}{(E_0 - H_0)^2}(1 - P_0)V_1 \mid \phi\rangle \mid \phi\rangle \\
&= -\frac{\pi e^2}{m^2\hbar V}\left(\sum_{\mathbf{k}j} \frac{1}{\omega_k^3} \mid \mathbf{e}_{\mathbf{k}j} \cdot \mathbf{p} \mid^2 \right) \mid \mathbf{p}, \{0_{\mathbf{k}j}\}\rangle
\end{aligned} \tag{7.86}
$$

Here the symbols used are the same as in Section 6.3. It is evident from (7.86) that both one- and two-photon states contribute to the virtual cloud around the slow electron at order e^2.

Proceeding as in Section 7.1 for the polaron case, we write the state (7.85) in the electron coordinate representation. Thus we require

$$
\begin{aligned}
\langle \mathbf{x}_e \mid \mathbf{p}, \{0_{\mathbf{k}j}\}\rangle' &= \langle \mathbf{x}_e \mid \phi\rangle + \langle \mathbf{x}_e \mid \phi_1\rangle \\
&+ \langle \mathbf{x}_e \mid \phi_2\rangle + \langle \mathbf{x}_e \mid \phi_3\rangle + \langle \mathbf{x}_e \mid \phi_4\rangle
\end{aligned} \tag{7.87}
$$

where $\mathbf{x}_e$ is the electron coordinate. We have from (7.86)

$$
\begin{aligned}
\langle \mathbf{x}_e \mid \phi\rangle &= \frac{1}{\sqrt{V}} e^{(i/\hbar)\mathbf{p}\cdot\mathbf{x}_e} \mid \{0_{\mathbf{k}j}\}\rangle \; ; \\
\langle \mathbf{x}_e \mid \phi_1\rangle &= -\frac{e}{mV}\sqrt{\frac{2\pi}{\hbar}} \sum_{\mathbf{k}j} \frac{1}{\omega_k^{3/2}} \mathbf{e}^*_{\mathbf{k}j} \cdot \mathbf{p} e^{i((1/\hbar)\mathbf{p} - \mathbf{k})\cdot\mathbf{x}_e} \mid 1_{\mathbf{k}j}\rangle \; ; \\
\langle \mathbf{x}_e \mid \phi_2\rangle &= \frac{\pi e^2}{mV^{3/2}} \sum_{\mathbf{k}\mathbf{k}'jj'} \frac{\mathbf{e}^*_{\mathbf{k}j} \cdot \mathbf{e}^*_{\mathbf{k}'j'}}{\sqrt{\omega_k \omega_{k'}}(\omega_k + \omega_{k'})} \\
&\times \left[1 + \left(\sqrt{2} - 1\right)\delta_{\mathbf{k}\mathbf{k}'}\delta_{jj'} \right] e^{i((1/\hbar)\mathbf{p} - \mathbf{k} - \mathbf{k}')\cdot\mathbf{x}_e} \mid 1_{\mathbf{k}j}, 1_{\mathbf{k}'j'}\rangle \; ; \\
\langle \mathbf{x}_e \mid \phi_3\rangle &= \frac{2\pi e^2}{m^2\hbar V^{3/2}} \sum_{\mathbf{k}\mathbf{k}'jj'} \frac{1}{\sqrt{\omega_k \omega_{k'}}\omega_k(\omega_k + \omega_{k'})} \left(\mathbf{e}^*_{\mathbf{k}j} \cdot \mathbf{p}\right) \\
&\times \left[\mathbf{e}^*_{\mathbf{k}'j'} \cdot (\mathbf{p} - \hbar\mathbf{k}) \right] \left[1 + \left(\sqrt{2} - 1\right)\delta_{\mathbf{k}\mathbf{k}'}\delta_{jj'} \right] \\
&\times e^{i((1/\hbar)\mathbf{p} - \mathbf{k} - \mathbf{k}')\cdot\mathbf{x}_e} \mid 1_{\mathbf{k}j}, 1_{\mathbf{k}'j'}\rangle \; ; \\
\langle \mathbf{x}_e \mid \phi_4\rangle &= -\frac{\pi e^2}{m^2\hbar V^{3/2}} \left(\sum_{\mathbf{k}j} \frac{1}{\omega_k^3} \left|\mathbf{e}_{\mathbf{k}j} \cdot \mathbf{p}\right|^2 \right) e^{(i/\hbar)\mathbf{p}\cdot\mathbf{x}_e} \mid \{0_{\mathbf{k}j}\}\rangle
\end{aligned} \tag{7.88}
$$

For any local field operator $F(\mathbf{x})$, the expression

$$'\langle \mathbf{p}, \{0_{\mathbf{k}j}\} \mid \mathbf{x}_e\rangle F(\mathbf{x})\langle \mathbf{x}_e \mid \mathbf{p}, \{0_{\mathbf{k}j}\}\rangle' = \frac{1}{V}\langle F(\mathbf{x}, \mathbf{x}_e)\rangle \tag{7.89}$$

gives the quantum average of F at point $\mathbf{x}$ if the electron is at point $\mathbf{x}_e$ with probability density $1/V$. Thus $F(\mathbf{x}, \mathbf{x}_e)$ is the quantum average of F at point $\mathbf{x}$ if the electron is at $\mathbf{x}_e$ with certainty. The simplest local field quantities are the electric and magnetic fields around the electron. It is convenient to partition the electric field into longitudinal and transverse components as in (4.24)

$$\mathbf{E} = \mathbf{E}_{\perp} + \mathbf{E}_{\parallel} \ ; \ \mathbf{E}_{\perp} = -\frac{1}{c}\dot{\mathbf{A}} \ ; \ \mathbf{E}_{\parallel} = -\nabla V \tag{7.90}$$

where from (2.9)

$$\mathbf{E}_{\perp}(\mathbf{x}) = i\sum_{\mathbf{k}j}\sqrt{\frac{2\pi\hbar\omega_k}{V}}\left(\mathbf{e}_{\mathbf{k}j}a_{\mathbf{k}j}e^{i\mathbf{k}\cdot\mathbf{x}} - \mathbf{e}^*_{\mathbf{k}j}a^\dagger_{\mathbf{k}j}e^{-i\mathbf{k}\cdot\mathbf{x}}\right) \tag{7.91}$$

and from (4.13) for an electron at point $\mathbf{x}_e$

$$V(\mathbf{x}) = \int \frac{\rho(\mathbf{x}')}{\mid \mathbf{x} - \mathbf{x}' \mid} d^3\mathbf{x}' = -\frac{e}{\mid \mathbf{x} - \mathbf{x}_e \mid} \ ; \ \mathbf{E}_{\parallel} = -e\frac{\mathbf{x} - \mathbf{x}_e}{\mid \mathbf{x} - \mathbf{x}_e \mid^3} \tag{7.92}$$

The magnetic field is only transverse and can be expressed as in (2.9)

$$\mathbf{H}(\mathbf{x}) = i\sum_{\mathbf{k}j}\sqrt{\frac{2\pi\hbar\omega_k}{V}}\left(\mathbf{b}_{\mathbf{k}j}a_{\mathbf{k}j}e^{i\mathbf{k}\cdot\mathbf{x}} - \mathbf{b}^*_{\mathbf{k}j}a^\dagger_{\mathbf{k}j}e^{-i\mathbf{k}\cdot\mathbf{x}}\right) \tag{7.93}$$

Up to terms of order e^2 and from (7.92)

$$\begin{aligned} &'\langle \mathbf{p}, \{0_{\mathbf{k}j}\} \mid \mathbf{x}_e\rangle \mathbf{E}_{\parallel}\langle \mathbf{x}_e \mid \mathbf{p}, \{0_{\mathbf{k}j}\}\rangle' \\ &= \langle \phi \mid \mathbf{x}_e\rangle \mathbf{E}_{\parallel}\langle \mathbf{x}_e \mid \phi\rangle = -\frac{e}{V}\frac{\mathbf{x} - \mathbf{x}_e}{\mid \mathbf{x} - \mathbf{x}_e \mid^3} \end{aligned} \tag{7.94}$$

Moreover, within the same approximation,

$$\begin{aligned} &'\langle \mathbf{p}, \{0_{\mathbf{k}j}\} \mid \mathbf{x}_e\rangle a_{\mathbf{k}j}\langle \mathbf{x}_e \mid \mathbf{p}, \{0_{\mathbf{k}j}\}\rangle' \\ &= \left('\langle \mathbf{p}, \{0_{\mathbf{k}j}\} \mid \mathbf{x}_e\rangle a^\dagger_{\mathbf{k}j}\langle \mathbf{x}_e \mid \mathbf{p}, \{0_{\mathbf{k}j}\}\rangle'\right)^* \\ &= \langle \phi \mid \mathbf{x}_e\rangle a_{\mathbf{k}j}\langle \mathbf{x}_e \mid \phi_1\rangle \\ &= -\frac{e}{mV^{3/2}}\sqrt{\frac{2\pi}{\hbar}}\frac{1}{\omega_k^{3/2}}\mathbf{e}^*_{\mathbf{k}j}\cdot\mathbf{p}e^{-i\mathbf{k}\cdot\mathbf{x}_e} \end{aligned} \tag{7.95}$$

From (7.85) and (2.6) one easily obtains

$$'\langle \mathbf{p}, \{0_{\mathbf{k}j}\} \mid \mathbf{x}_e\rangle E_{\perp m}(\mathbf{x})\langle \mathbf{x}_e \mid \mathbf{p}, \{0_{\mathbf{k}j}\}\rangle'$$

$$= -i\frac{2\pi e}{mV^2}\sum_{\mathbf{k}j}\frac{1}{\omega_k}\left(\mathbf{e}_{\mathbf{k}j}\right)_m\left(\mathbf{e}^*_{\mathbf{k}j}\right)_n p_n e^{i\mathbf{k}\cdot(\mathbf{x}-\mathbf{x}_e)} + \text{c.c.}$$

$$= \frac{4\pi e}{mV^2}p_n\sum_{\mathbf{k}}\frac{1}{\omega_k}\left(\delta_{mn} - \hat{k}_m\hat{k}_n\right)\sin \mathbf{k}\cdot(\mathbf{x}-\mathbf{x}_e) = 0 \qquad (7.96)$$

since the summand is an even function of $\mathbf{k}$. On the other hand, summing over the polarization index according to (2.6),

$$'\langle \mathbf{p}, \{0_{\mathbf{k}j}\} \mid \mathbf{x}_e\rangle H_m(\mathbf{x})\langle \mathbf{x}_e \mid \mathbf{p}, \{0_{\mathbf{k}j}\}\rangle'$$

$$= -i\frac{2\pi e}{mV^2}\sum_{\mathbf{k}j}\frac{1}{\omega_k}\left(\mathbf{b}_{\mathbf{k}j}\right)_m\left(\mathbf{e}^*_{\mathbf{k}j}\right)_n p_n e^{i\mathbf{k}\cdot(\mathbf{x}-\mathbf{x}_2)} + \text{c.c.}$$

$$= -i\frac{2\pi e}{mV^2}\sum_{\mathbf{k}}\frac{1}{\omega_k}e_{nm\ell}\hat{k}_\ell p_n e^{i\mathbf{k}\cdot(\mathbf{x}-\mathbf{x}_e)} + \text{c.c.}$$

$$= -\frac{2\pi e}{mV^2c}e_{mn\ell}p_n\nabla_\ell\sum_{\mathbf{k}}\frac{1}{k^2}e^{i\mathbf{k}\cdot(\mathbf{x}-\mathbf{x}_e)} + \text{c.c.}$$

$$= -\frac{e}{mVc}e_{mn\ell}p_n(\mathbf{x}-\mathbf{x}_e)_\ell\frac{1}{\mid \mathbf{x}-\mathbf{x}_e\mid^3} \qquad (7.97)$$

where summation over $\mathbf{k}$ has been performed according to (7.65) and where the symmetry properties of the tensor $e_{nm\ell}$ have been exploited. Collecting results (7.94), (7.96) and (7.97) one has the electromagnetic field at point $\mathbf{x}$ if the electron is certainly at point $\mathbf{x}_e$ in the form (Craig and Thirunamachandran 1984)

$$\langle \mathbf{E}(\mathbf{x}-\mathbf{x}_e)\rangle = -e\frac{\mathbf{x}-\mathbf{x}_e}{\mid \mathbf{x}-\mathbf{x}_e\mid^3}\ ;$$

$$\langle \mathbf{H}(\mathbf{x}-\mathbf{x}_e)\rangle = -\frac{e}{mc}\mathbf{p}\times\frac{\mathbf{x}-\mathbf{x}_e}{\mid \mathbf{x}-\mathbf{x}_e\mid^3} \qquad (7.98)$$

This result is exactly equal to that for a classical particle moving at sufficiently small velocity (see e.g. Becker 1982). Thus QED at order e^2 does not add anything new to classical electrodynamics as far as the field amplitudes are concerned.

Within the same approximation, however, the energy density around a slow free electron presents novel interesting features when calculated by QED. The operator corresponding to the T_{44} component of the momentum-energy tensor is

$$\begin{aligned}\mathcal{H}(\mathbf{x}) &= \mathcal{H}_{el}(\mathbf{x}) + \mathcal{H}_{mag}(\mathbf{x})\ ;\\ \mathcal{H}_{el}(\mathbf{x}) &= \frac{1}{8\pi}\left[\mathbf{E}_{\parallel}^2(\mathbf{x}) + \mathbf{E}_{\perp}^2(\mathbf{x}) + 2\mathbf{E}_{\parallel}(\mathbf{x})\cdot\mathbf{E}_{\perp}(\mathbf{x})\right]\ ;\\ \mathcal{H}_{mag}(\mathbf{x}) &= \frac{1}{8\pi}\mathbf{H}^2(\mathbf{x})\end{aligned} \tag{7.99}$$

Concentrating first on the transverse electric part, use of the expansion (7.91) shows that $\mathbf{E}_{\perp}^2(\mathbf{x})$ is the same operator as (7.24). Thus to evaluate $'\langle \mathbf{p}, \{0_{\mathbf{k}j}\} \mid \mathbf{x}_e\rangle E_{\perp}^2(\mathbf{x})\langle \mathbf{x}_e \mid \mathbf{p}, \{0_{\mathbf{k}j}\}\rangle'$, the following quantum averages must be first obtained from (7.88), to order e^2,

$$\begin{aligned}&'\langle \mathbf{p},\{0_{\mathbf{k}j}\} \mid \mathbf{x}_e\rangle a_{\mathbf{k}j}a_{\mathbf{k}'j'}\langle \mathbf{x}_e \mid \mathbf{p}, \{0_{\mathbf{k}j}\}\rangle'\\ &= \left('\langle \mathbf{p}, \{0_{\mathbf{k}j}\} \mid \mathbf{x}_e\rangle a_{\mathbf{k}j}^{\dagger}a_{\mathbf{k}'j'}^{\dagger}\langle \mathbf{x}_e \mid \mathbf{p}, \{0_{\mathbf{k}j}\}\rangle'\right)^*\\ &= \langle \phi \mid \mathbf{x}_e\rangle a_{\mathbf{k}j}a_{\mathbf{k}'j'}\langle \mathbf{x}_e \mid \phi_2\rangle + \langle \phi \mid \mathbf{x}_e\rangle a_{\mathbf{k}j}a_{\mathbf{k}'j'}\langle \mathbf{x}_e \mid \phi_3\rangle\\ &= \frac{2\pi e^2}{mV^2}\frac{1}{\sqrt{\omega_k\omega_{k'}}(\omega_k+\omega_{k'})}\left\{ -\mathbf{e}_{\mathbf{k}j}^*\cdot\mathbf{e}_{\mathbf{k}'j'}^*\right.\\ &+\frac{1}{m\hbar}\frac{1}{\omega_{k'}}\left(\mathbf{e}_{\mathbf{k}j}^*\cdot\mathbf{p}\right)\left[\mathbf{e}_{\mathbf{k}j}^*\cdot(\mathbf{p}-\hbar\mathbf{k}')\right]\\ &\left.+\frac{1}{m\hbar}\frac{1}{\omega_{k'}}\left(\mathbf{e}_{\mathbf{k}'j'}^*\cdot\mathbf{p}\right)\left[\mathbf{e}_{\mathbf{k}j}^*\cdot(\mathbf{p}-\hbar\mathbf{k}')\right]\right\}\\ &\times\left[1+\left(\sqrt{2}-1\right)\delta_{\mathbf{k}\mathbf{k}'}\delta_{jj'}\right]e^{-i(\mathbf{k}+\mathbf{k}')\cdot\mathbf{x}_e}\ ;\\ &'\langle \mathbf{p},\{0_{\mathbf{k}j}\} \mid \mathbf{x}_e\rangle a_{\mathbf{k}j}^{\dagger}a_{\mathbf{k}'j'}\langle \mathbf{x}_e \mid \mathbf{p}, \{0_{\mathbf{k}j}\}\rangle'\\ &= \left('\langle \mathbf{p}, \{0_{\mathbf{k}j}\} \mid \mathbf{x}_e\rangle a_{\mathbf{k}j}a_{\mathbf{k}'j'}^{\dagger}\langle \mathbf{x}_e \mid \mathbf{p}, \{0_{\mathbf{k}j}\}\rangle'\right)^* - \frac{1}{V}\delta_{\mathbf{k}\mathbf{k}'}\delta_{jj'}\\ &= \langle \phi_1 \mid \mathbf{x}_e\rangle a_{\mathbf{k}j}^{\dagger}a_{\mathbf{k}'j'}\langle \mathbf{x}_e \mid \phi_1\rangle\\ &= \frac{2\pi e^2}{m^2\hbar V^2}\frac{1}{(\omega_k\omega_{k'})^{3/2}}\left(\mathbf{e}_{\mathbf{k}j}\cdot\mathbf{p}\right)\left(\mathbf{e}_{\mathbf{k}'j'}^*\cdot\mathbf{p}\right)e^{i(\mathbf{k}-\mathbf{k}')\cdot\mathbf{x}_e}\end{aligned} \tag{7.100}$$

The result (7.100) and extensive use of polarization sums (2.6) leads to

$$
\begin{aligned}
&{}'\langle \mathbf{p}, \{0_{\mathbf{k}j}\} \mid \mathbf{x}_e\rangle E_\perp^2(\mathbf{x})\langle \mathbf{x}_e \mid \mathbf{p}, \{0_{\mathbf{k}j}\}\rangle' \\
&= \frac{4\pi}{V^2}\sum_{\mathbf{k}j}\frac{1}{2}\hbar\omega_k - \frac{2\pi\hbar}{V}\sum_{\mathbf{k}\mathbf{k}'jj'}\sqrt{\omega_k\omega_{k'}} \\
&\times \Big\{\left(\mathbf{e}_{\mathbf{k}j}\cdot\mathbf{e}_{\mathbf{k}'j'}\right)'\langle \mathbf{p}, \{0_{\mathbf{k}j}\} \mid \mathbf{x}_e\rangle a_{\mathbf{k}j}a_{\mathbf{k}'j'}\langle \mathbf{x}_e \mid \mathbf{p}, \{0_{\mathbf{k}j}\}\rangle' e^{i(\mathbf{k}+\mathbf{k}')\cdot\mathbf{x}} \\
&- \left(\mathbf{e}^*_{\mathbf{k}j}\cdot\mathbf{e}_{\mathbf{k}j}\right)'\langle \mathbf{p}, \{0_{\mathbf{k}j}\} \mid \mathbf{x}_e\rangle a^\dagger_{\mathbf{k}j}a_{\mathbf{k}'j'}\langle \mathbf{x}_e \mid \mathbf{p}, \{0_{\mathbf{k}j}\}\rangle' \\
&\times e^{-i(\mathbf{k}-\mathbf{k}')\cdot\mathbf{x}} + \text{c.c.}\Big\} \\
&= \frac{4\pi}{V^2}\sum_{\mathbf{k}j}\frac{1}{2}\hbar\omega_k + (A+B+C+D+\text{c.c.})
\end{aligned}
\tag{7.101}
$$

where

$$
\begin{aligned}
A &= \frac{4\pi^2e^2\hbar}{mV^3c}\sum_{\mathbf{k}\mathbf{k}'}\frac{1}{k+k'}\left(\delta_{mn}-\hat{k}_m\hat{k}_n\right)\left(\delta_{mn}-\hat{k}'_m\hat{k}'_n\right) \\
&\quad\times e^{i(\mathbf{k}+\mathbf{k}')\cdot(\mathbf{x}-\mathbf{x}_e)}\ ; \\
B &= -\frac{4\pi^2e^2}{m^2V^3c^2}p_np_\ell\sum_{\mathbf{k}\mathbf{k}'}\frac{1}{kk'}\left(\delta_{mn}-\hat{k}_m\hat{k}_n\right)\left(\delta_{m\ell}-\hat{k}'_m\hat{k}'_\ell\right) \\
&\quad\times e^{-i(\mathbf{k}+\mathbf{k}')\cdot(\mathbf{x}-\mathbf{x}_e)}\ ; \\
C &= \frac{4\pi^2e^2\hbar}{m^2V^3c^2}p_n\sum_{\mathbf{k}\mathbf{k}'}\frac{1}{k+k'}\Big[\hat{k}_\ell\left(\delta_{mn}-\hat{k}_m\hat{k}_n\right)\left(\delta_{m\ell}-\hat{k}'_m\hat{k}'_\ell\right) \\
&\quad+\hat{k}'_\ell\left(\delta_{mn}-\hat{k}'_m\hat{k}'_n\right)\left(\delta_{m\ell}-\hat{k}_m\hat{k}_\ell\right)\Big]e^{-i(\mathbf{k}+\mathbf{k}')\cdot(\mathbf{x}-\mathbf{x}_e)}\ ; \\
D &= \frac{4\pi^2e^2}{m^2V^3c^2}p_np_\ell\sum_{\mathbf{k}\mathbf{k}'}\frac{1}{kk'}\left(\delta_{mn}-\hat{k}_m\hat{k}_n\right)\left(\delta_{m\ell}-\hat{k}'_m\hat{k}'_\ell\right) \\
&\quad\times e^{-i(\mathbf{k}-\mathbf{k}')\cdot(\mathbf{x}-\mathbf{x}_e)}
\end{aligned}
\tag{7.102}
$$

Various cancellations in (7.102)) can be exploited in order to simplify (7.101). In fact, putting $\mathbf{k}' = -\mathbf{k}'$ in D immediately yields $B + D = 0$. Moreover, $C + C^*$ vanishes because

$$
\begin{aligned}
C + C^* &= \frac{8\pi^2e^2\hbar}{m^2V^3c^2}p_n\sum_{\mathbf{k}\mathbf{k}'}\frac{1}{k+k'}\Big[\hat{k}_\ell\left(\delta_{mn}-\hat{k}_m\hat{k}_n\right)\left(\delta_{m\ell}-\hat{k}'_m\hat{k}'_\ell\right) \\
&\quad+\hat{k}'_\ell\left(\delta_{mn}-\hat{k}'_m\hat{k}'_n\right)\left(\delta_{m\ell}-\hat{k}_m\hat{k}_\ell\right)\Big] \\
&\quad\times\cos(\mathbf{k}+\mathbf{k}')\cdot(\mathbf{x}-\mathbf{x}_e)
\end{aligned}
\tag{7.103}
$$

and the summand in this expression is an odd function of $\mathbf{k}+\mathbf{k}'$. Finally, the double sum appearing in A can be disentangled with the help of (7.39) as

$$\frac{1}{V^2}\sum_{\mathbf{k}\mathbf{k}'}\frac{1}{k+k'}\left(\delta_{mn}-\hat{k}_m\hat{k}_n\right)\left(\delta_{mn}-\hat{k}'_m\hat{k}'_n\right)e^{i(\mathbf{k}+\mathbf{k}')\cdot\mathbf{r}}$$
$$=\int_0^\infty\left(\frac{1}{V}\sum_{\mathbf{k}}e^{-ky}\left(\delta_{mn}-\hat{k}_m\hat{k}_n\right)e^{i\mathbf{k}\cdot\mathbf{r}}\right)^2 dy \tag{7.104}$$

where $\mathbf{r}=\mathbf{x}-\mathbf{x}_e$. Evaluation of A proceeds through the following steps.

i) Use of (7.32) gives

$$\frac{1}{V}\sum_{\mathbf{k}}e^{-ky}\left(\delta_{mn}-\hat{k}_m\hat{k}_n\right)e^{i\mathbf{k}\cdot\mathbf{r}}$$
$$=-\frac{1}{2\pi^2}D^r_{mn}\int_0^\infty\frac{1}{k}e^{-ky}\sin kr dk=-\frac{1}{2\pi^2}D^r_{mn}\arctan\frac{r}{y}$$
$$=-\frac{1}{2\pi^2}\frac{1}{r}\left[(\delta_{mn}-\hat{r}_m\hat{r}_n)\frac{-2ry}{(r^2+y^2)^2}\right.$$
$$\left.+(\delta_{mn}-3\hat{r}_m\hat{r}_n)\left(\frac{1}{r^2}\arctan\frac{r}{y}-\frac{1}{r}\frac{y}{r^2+y^2}\right)\right] \tag{7.105}$$

ii) The square of (7.105) is obtained using the following products

$$(\delta_{mn}-\hat{r}_m\hat{r}_n)^2=2\quad;\quad(\delta_{mn}-3\hat{r}_m\hat{r}_n)^2=6\,;$$
$$(\delta_{mn}-\hat{r}_m\hat{r}_n)(\delta_{mn}-3\hat{r}_m\hat{r}_n)=2 \tag{7.106}$$

and it is

$$\left(\frac{1}{V}\sum_{\mathbf{k}}e^{-ky}\left(\delta_{mn}-\hat{k}_m\hat{k}_n\right)e^{i\mathbf{k}\cdot\mathbf{r}}\right)^2$$
$$=\frac{1}{4\pi^4}\frac{1}{r^2}\left\{\frac{8r^2y^2}{(r^2+y^2)^4}-\frac{8ry}{(r^2+y^2)^2}\left(\frac{1}{r^2}\arctan\frac{r}{y}-\frac{1}{r}\frac{y}{r^2+y^2}\right)\right.$$
$$+6\left(\frac{1}{r^4}\arctan^2\frac{r}{y}-\frac{2}{r^3}\frac{y}{r^2+y^2}\arctan\frac{r}{y}\right.$$
$$\left.\left.+\frac{1}{r^2}\frac{y^2}{(r^2+y^2)^2}\right)\right\} \tag{7.107}$$

iii) Extensive use of partial integration and the evaluation of elementary integrals as in Sections 7.2 and 7.3 leads to

$$\int_0^\infty \left(\frac{1}{V} \sum_{\mathbf{k}} e^{-ky} \left(\delta_{mn} - \hat{k}_m \hat{k}_n \right) e^{i\mathbf{k}\cdot\mathbf{r}} \right)^2 dy = \frac{5}{16\pi^3} \frac{1}{r^5} \tag{7.108}$$

From (7.104) and (7.101) we have

$$A = \frac{1}{V} \frac{5}{4\pi} \frac{e^2\hbar}{mc} \frac{1}{r^5};$$

$$'\langle \mathbf{p}, \{0_{\mathbf{k}j}\} \mid \mathbf{x}_e \rangle \mathbf{E}_\perp^2(\mathbf{x}) \langle \mathbf{x}_e \mid \mathbf{p}, \{0_{\mathbf{k}j}\} \rangle'$$
$$= \frac{4\pi}{V^2} \sum_{\mathbf{k}j} \frac{1}{2} \hbar\omega_k + \frac{1}{V} \frac{5}{2\pi} \frac{e^2\hbar}{mc} \frac{1}{|\mathbf{x} - \mathbf{x}_e|^5} \tag{7.109}$$

Hence the transverse electric energy density at point $\mathbf{x}$, if the electron is at point $\mathbf{x}_e$ with certainty, is

$$\langle \frac{1}{8\pi} \mathbf{E}_\perp^2 (\mathbf{x} - \mathbf{x}_e) \rangle = \frac{1}{2} \frac{1}{V} \sum_{\mathbf{k}j} \frac{1}{2} \hbar\omega_k + \frac{5}{16\pi^2} \frac{e^2\hbar}{mc} \frac{1}{|\mathbf{x} - \mathbf{x}_e|^5} \tag{7.110}$$

where the first term on the RHS is the (infinite) contribution of the zero-point photons of the unperturbed vacuum to the electric transverse energy density. The second term must be attributed to the cloud of virtual photons dressing the electron. It is accurate to order e^2 and is valid at distances larger than λ_c, as discussed previously. This term is purely quantum in nature, and exists in spite of the fact that the average amplitude of the virtual field vanishes, as shown in (7.96).

As for the energy density due to the longitudinal field, from (7.92) we have at order e^2

$$'\langle \mathbf{p}, \{0_{\mathbf{k}j}\} \mid \mathbf{x}_e \rangle \mathbf{E}_\parallel^2(\mathbf{x}) \langle \mathbf{x}_e \mid \mathbf{p}, \{0_{\mathbf{k}j}\} \rangle' = \langle \phi \mid \mathbf{x}_e \rangle \mathbf{E}_\parallel^2(\mathbf{x}) \langle \mathbf{x}_e \mid \phi \rangle$$
$$= \frac{e^2}{V} \frac{1}{|\mathbf{x} - \mathbf{x}_e|^4} \tag{7.111}$$

and this contribution, if the electron is certainly at $\mathbf{x}_e$, gives

$$\langle \frac{1}{8\pi} \mathbf{E}_\parallel^2 (\mathbf{x} - \mathbf{x}_e) \rangle = \frac{1}{8\pi} e^2 \frac{1}{|\mathbf{x} - \mathbf{x}_e|^4} \tag{7.112}$$

The interference term between the longitudinal and transverse field in (7.99) vanishes for the same mathematical reasons that led to the vanishing of (7.96), as is easy to see. Thus the total electric energy density around the electron obtained from (7.110) and (7.112) is

$$\langle \mathcal{H}_{el}(\mathbf{x}-\mathbf{x}_e)\rangle = \frac{1}{2}\frac{1}{V}\sum_{\mathbf{k}j}\frac{1}{2}\hbar\omega_k + \frac{5}{16\pi^2}\frac{e^2\hbar}{mc}\frac{1}{|\mathbf{x}-\mathbf{x}_e|^5}$$
$$+\frac{1}{8\pi}e^2\frac{1}{|\mathbf{x}-\mathbf{x}_e|^4} \quad (|\mathbf{x}-\mathbf{x}_e| > \lambda_c) \tag{7.113}$$

Turning to the magnetic energy density, from (7.93), (7.100) and by repeated use of polarization sums (2.6) we have

$$'\langle \mathbf{p},\{0_{\mathbf{k}j}\} \mid \mathbf{x}_e\rangle \mathbf{H}^2(\mathbf{x})\langle \mathbf{x}_e \mid \mathbf{p},\{0_{\mathbf{k}j}\}\rangle'$$
$$= \frac{4\pi}{V^2}\sum_{\mathbf{k}j}\frac{1}{2}\hbar\omega_k - \frac{2\pi\hbar}{V}\sum_{\mathbf{k}\mathbf{k}'jj'}\sqrt{\omega_k\omega_{k'}}$$
$$\times\Big\{(\mathbf{b}_{\mathbf{k}j}\cdot\mathbf{b}_{\mathbf{k}'j'})'\langle \mathbf{p},\{0_{\mathbf{k}j}\} \mid \mathbf{x}_e\rangle' a_{\mathbf{k}j}a_{\mathbf{k}'j'}\langle \mathbf{x}_e \mid \mathbf{p},\{0_{\mathbf{k}j}\}\rangle' e^{i(\mathbf{k}+\mathbf{k}')\cdot\mathbf{x}}$$
$$-(\mathbf{b}^*_{\mathbf{k}j}\cdot\mathbf{b}_{\mathbf{k}'j'})'\langle \mathbf{p},\{0_{\mathbf{k}j}\} \mid \mathbf{x}_e\rangle a^\dagger_{\mathbf{k}j}a_{\mathbf{k}'j'}\langle \mathbf{x}_e \mid \mathbf{p},\{0_{\mathbf{k}j}\}\rangle'$$
$$\times e^{-i(\mathbf{k}-\mathbf{k}')\cdot\mathbf{x}} + c.c.\Big\}$$
$$= \frac{4\pi}{V^2}\sum_{\mathbf{k}}\frac{1}{2}\hbar\omega_k + (A+B+C+D) \tag{7.114}$$

where

$$A = \frac{8\pi^2e^2\hbar}{mV^3c}\sum_{\mathbf{k}\mathbf{k}'}\frac{\hat{\mathbf{k}}\cdot\hat{\mathbf{k}}'}{k+k'}e^{i(\mathbf{k}+\mathbf{k}')\cdot(\mathbf{x}-\mathbf{x}_e)};$$

$$B = \frac{-4\pi^2e^2\hbar}{m^2V^3c^2}p^2\sum_{\mathbf{k}\mathbf{k}'}\frac{1}{kk'}\left[\hat{\mathbf{k}}\cdot\hat{\mathbf{k}}' - \left(\hat{p}\cdot\hat{\mathbf{k}}\right)\left(\hat{p}\cdot\hat{\mathbf{k}}'\right)\right]e^{-i(\mathbf{k}+\mathbf{k}')\cdot(\mathbf{x}-\mathbf{x}_e)};$$

$$C = \frac{4\pi^2e^2\hbar}{m^2V^3c^2}\sum_{\mathbf{k}\mathbf{k}'}\frac{1}{k+k'}\mathbf{p}\cdot\left(\hat{\mathbf{k}}+\hat{\mathbf{k}}'\right)\left(\hat{\mathbf{k}}\cdot\hat{\mathbf{k}}'-1\right)e^{-i(\mathbf{k}+\mathbf{k}')\cdot(\mathbf{x}-\mathbf{x}_e)};$$

$$D = \frac{4\pi^2e^2}{m^2V^3c^2}p^2\sum_{\mathbf{k}\mathbf{k}'}\frac{1}{kk'}\left[\hat{\mathbf{k}}\cdot\hat{\mathbf{k}}' - \left(\hat{p}\cdot\hat{\mathbf{k}}\right)\left(\hat{p}\cdot\hat{\mathbf{k}}'\right)\right]$$
$$\times e^{-i(\mathbf{k}-\mathbf{k}')\cdot(\mathbf{x}-\mathbf{x}_e)} \tag{7.115}$$

Substituting $\mathbf{k}'$ with $-\mathbf{k}'$ in D shows that $B = D$. Moreover, $C + C^*$ vanishes because the summand is an odd function of $\mathbf{k}+\mathbf{k}'$. The quantities contributing to (7.114) are calculated as follows.

$$A = \lim_{\mathbf{r}'\to\mathbf{r}} -\frac{8\pi^2 e^2\hbar}{mVc}\nabla^r\cdot\nabla^{r'}\frac{1}{V^2}\sum_{\mathbf{k}\mathbf{k}'}\frac{1}{kk'(k+k')}e^{i(\mathbf{k}\cdot\mathbf{r}+\mathbf{k}'\cdot\mathbf{r}')}$$
$$= \lim_{\mathbf{r}'\to\mathbf{r}} -\frac{8\pi^2 e^2\hbar}{mVc}\nabla^r\cdot\nabla^{r'}\int_0^\infty\left(\frac{1}{V}\sum_{\mathbf{k}}\frac{1}{k}e^{-ky}e^{i\mathbf{k}\cdot\mathbf{r}}\right)$$
$$\times\left(\frac{1}{V}\sum_{\mathbf{k}'}\frac{1}{k'}e^{-k'y}e^{i\mathbf{k}'\cdot\mathbf{r}'}\right)dy \tag{7.116}$$

where $\mathbf{r} = \mathbf{x} - \mathbf{x}_e$. Elementary integration gives

$$\frac{1}{V}\sum_{\mathbf{k}}\frac{1}{k}e^{-ky}e^{i\mathbf{k}\cdot\mathbf{r}} = \frac{1}{2\pi^2}\frac{1}{r^2+y^2} \tag{7.117}$$

Substitution of (7.117) into (7.116) and further elementary integration yields

$$A = \lim_{\mathbf{r}'\to\mathbf{r}} -\frac{e^2\hbar}{\pi mVc}\nabla^r\cdot\nabla^{r'}\frac{1}{rr'(r+r')} = -\frac{1}{V}\frac{5}{4\pi}\frac{e^2\hbar}{mc}\frac{1}{|\mathbf{x}-\mathbf{x}_e|^5} \tag{7.118}$$

Moreover

$$B = \lim_{\mathbf{r}'\to\mathbf{r}}\frac{4\pi^2 e^2}{m^2Vc^2}p^2\left[\nabla^r\cdot\nabla^{r'} - (\hat{p}\cdot\nabla^r)\left(\hat{p}\cdot\nabla^{r'}\right)\right]$$
$$\times\frac{1}{V^2}\sum_{\mathbf{k}\mathbf{k}'}\frac{1}{k^2k'^2}e^{-i(\mathbf{k}\cdot\mathbf{r}+\mathbf{k}'\cdot\mathbf{r}')}$$
$$= \lim_{\mathbf{r}'\to\mathbf{r}}\frac{e^2}{4m^2Vc^2}p^2\left[\nabla^r\cdot\nabla^{r'} - (\hat{p}\cdot\nabla^r)\left(\hat{p}\cdot\nabla^{r'}\right)\right]\frac{1}{rr'}$$
$$= \frac{1}{V}\frac{e^2p^2}{4m^2c^2}\frac{\sin^2\theta}{|\mathbf{x}-\mathbf{x}_e|^4} \tag{7.119}$$

where use has been made of result (7.65) and where θ is the angle between the electron momentum $\mathbf{p}$ and $\mathbf{x}-\mathbf{x}_e$. Substituting (7.118) and (7.119) into (7.114) and use of (7.89) leads to

$$\langle\mathcal{H}_{mag}(\mathbf{x}-\mathbf{x}_e)\rangle = \frac{1}{2}\frac{1}{V}\sum_{\mathbf{k}j}\frac{1}{2}\hbar\omega_k - \frac{5}{16\pi^2}\frac{e^2\hbar}{mc}\frac{1}{|\mathbf{x}-\mathbf{x}_e|^5}$$
$$+\frac{1}{8\pi}\frac{e^2p^2}{m^2c^2}\frac{\sin^2\theta}{|\mathbf{x}-\mathbf{x}_e|^4}\quad(|\mathbf{x}-\mathbf{x}_e|>\lambda_c) \tag{7.120}$$

Like its electric counterpart (7.113), this expression, which is accurate to order e^2 and valid at distances from the electron larger than λ_c, contains more terms than might have been anticipated from the quantum average of the magnetic field amplitude (7.98). Indeed, while the third term on the RHS of (7.120) originates simply from the square of the field amplitude, the first term is the contribution of the zero-point photons of the unperturbed vacuum to the magnetic energy density and the second term is due to the cloud of virtual photons dressing the electron. The latter term is equal and opposite to its electric counterpart in (7.113), so that the total energy density around the slow electron takes the form

$$\begin{aligned}\langle \mathcal{H}(\mathbf{x}-\mathbf{x}_e)\rangle &= \langle \mathcal{H}_{el}(\mathbf{x}-\mathbf{x}_e)\rangle + \langle \mathcal{H}_{mag}(\mathbf{x}-\mathbf{x}_e)\rangle \\ &= \frac{1}{V}\sum_{\mathbf{k}j}\frac{1}{2}\hbar\omega_k + \frac{1}{8\pi}e^2\frac{1}{|\mathbf{x}-\mathbf{x}_e|^4} \\ &+ \frac{1}{8\pi}e^2\frac{e^2p^2}{m^2c^2}\frac{\sin^2\theta}{|\mathbf{x}-\mathbf{x}_e|^4} \end{aligned} \qquad (7.121)$$

In spite of the cancellation in the total energy density of the electric and magnetic virtual clouds, each of the two clouds separately is capable of giving rise to observable effects (Compagno and Salamone 1991).

7.5 Energy density around a nonrelativistic hydrogen atom

We consider a spinless nonrelativistic electron of mass m and charge $-e$ bound to a proton fixed at the origin of the coordinate system. The energy spectrum consists, as is well known, of bound as well as of ionized levels. The former are distributed according to (4.37) with $Z=1$, while the latter form a continuum starting at $E=0$. Since there is no electron-electron interaction, the Hamiltonian of the system is of the form (4.40) with $D=0$. The electron is coupled to the radiation field, which we expand in spherical rather than in plane waves, in order to take advantage of the spherical symmetry of the system. Thus the unperturbed Hamiltonian is of the form

$$H_0 = \sum_{NLM} E_N c^{\dagger}_{NLM} c_{NLM} + \sum_{\lambda\ell m}\int \hbar\omega_k a^{\dagger}(k,\lambda,\ell,m)a(k,\lambda,\ell,m)dk \qquad (7.122)$$

where the sum over N includes an integration over the continuous part of the atomic spectrum. The eigenstates of H_0 are of the form $|u_{NLM}, \{n(k,\lambda,\ell,m)\}\rangle$ where $\{n(k,\lambda,\ell,m)\}$ represents a distribution of photon population among the field modes and

$$u_{NLM}(\mathbf{x}_e) \equiv \langle \mathbf{x}_e \mid u_{NLM} \rangle = R_{NL}(x_e) Y_L^M(\theta_e, \varphi_e) \tag{7.123}$$

In (7.123), x_e, θ_e, φ_e are the electron spherical coordinates and Y_L^M are the spherical harmonics described in Appendix A, whereas R_{NL} is the radial part of the hydrogenic wavefunction as discussed in detail in many textbooks (see e.g. Landau and Lifshitz 1958, Merzbacher 1961). The ground state of H_0 is $\mid u_{100}, \{0\}\rangle$, describing the hydrogen atom in its lowest possible bare state and the field devoid of photons. We recall that

$$u_{100}(x_e) = \frac{1}{\sqrt{4\pi}} R_{10}(x_e) = \frac{1}{\sqrt{4\pi}} \frac{2}{a^{3/2}} e^{-x_e/a} \tag{7.124}$$

where a is the Bohr radius. Excited one- and two-photon eigenstates of H_0, together with respective eigenvalues, are indicated as

$$\begin{aligned} &\mid u_{NLM}, 1_{k,\ell,m}\rangle \quad (E = E_N + \hbar\omega_k) \; ; \\ &\mid u_{NLM}, 1_{k,\lambda,\ell,m}, 1_{k',\lambda',\ell',m'}\rangle \quad (E = E_N + \hbar\omega_k + \hbar\omega_{k'}) \end{aligned} \tag{7.125}$$

For the field part of these states, we shall adopt Schweber's normalization (1964), in such a way that

$$\begin{aligned} &a(k,\lambda,\ell,m) \mid 1_{k',\lambda',\ell',m'}, 1_{k'',\lambda'',\ell'',m''}\rangle \\ &= \frac{1}{\sqrt{2}} a(k,\lambda,\ell,m) a^\dagger(k',\lambda',\ell',m') a^\dagger(k'',\lambda'',\ell'',m'') \mid \{0\}\rangle \; ; \\ &a^\dagger(k,\lambda,\ell,m) \mid 1_{k',\lambda',\ell',m'}\rangle = \sqrt{2} \mid 1_{k,\lambda,\ell,m}, 1_{k',\lambda',\ell',m'}\rangle \end{aligned} \tag{7.126}$$

We shall discuss the problem in the minimal coupling scheme and in the electric dipole approximation. Thus the interaction Hamiltonian is given by (4.49) with $R = 0$. The term linear in $\mathbf{A}$ has been obtained in (4.107) as

$$\begin{aligned} H_{AF}^{(1)} = -2\sqrt{\frac{2}{3}} \frac{e\hbar^{1/2}}{mc^{1/2}} \sum_m \sum_{\substack{NLM \\ N'L'M'}} \int u_{NLM}^*(\mathbf{x}_e) \mathbf{Y}_{1m-}(\theta_e, \varphi_e) \cdot \mathbf{p}_e \\ \times u_{N'L'M'}(\mathbf{x}_e) d^3\mathbf{x}_e \int c_{NLM}^\dagger c_{N'L'M'} a(k, \mathcal{E}, 1, m) k^{1/2} dk \\ + \text{h.c.} \end{aligned} \tag{7.127}$$

The expression for the term quadratic in **A** is readily obtained introducing (2.22) into (4.49) and taking the limit $kx_e \to 0$. The result is

$$H_{AF}^{(2)} = \frac{4e^2\hbar}{3mc}\sum_{mm'}\sum_{\substack{NLM\\N'L'M'}}\Big\{\int u^*_{NLM}(\mathbf{x}_e)\mathbf{Y}_{1m-}(\theta_e,\varphi_e).$$
$$\cdot\mathbf{Y}_{1m'-}(\theta_e,\varphi_e)u_{N'L'M'}(\mathbf{x}_e)d^3\mathbf{x}_e$$
$$\times\int\int(kk')^{1/2}c^\dagger_{NLM}c_{N'L'M'}a(k,\mathcal{E},1,m)a(k',\mathcal{E},1,m')dkdk'$$
$$+\int u^*_{NLM}(\mathbf{x}_e)\mathbf{Y}_{1m-}(\theta_e,\varphi_e)\cdot\mathbf{Y}^*_{1m'-}(\theta_e,\varphi_e)u_{N'L'M'}(\mathbf{x}_e)d^3\mathbf{x}_e$$
$$\times\int\int(kk')^{1/2}c^\dagger_{NLM}c_{N'L'M'}a(k,\mathcal{E},1,m)a^\dagger(k',\mathcal{E},1,m')dkdk'$$
$$+\int u^*_{NLM}(\mathbf{x}_e)\mathbf{Y}^*_{1m-}(\theta_e,\varphi_e)\cdot\mathbf{Y}_{1m'-}(\theta_e,\varphi_e)u_{N'L'M'}(\mathbf{x}_e)d^3\mathbf{x}_e$$
$$\times\int\int(kk')^{1/2}c^\dagger_{NLM}c_{N'L'M'}a^\dagger(k,\mathcal{E},1,m)a(k',\mathcal{E},1,m')dkdk'$$
$$+\int u^*_{NLM}(\mathbf{x}_e)\mathbf{Y}^*_{1m-}(\theta_e,\varphi_e)\cdot\mathbf{Y}^*_{1m'-}(\theta_e,\varphi_e)u_{N'L'M'}(\mathbf{x}_e)d^3\mathbf{x}_e$$
$$\times\int\int(kk')^{1/2}c^\dagger_{NLM}c_{N'L'M'}$$
$$\times\, a^\dagger(k,\mathcal{E},1,m)a^\dagger(k',\mathcal{E},1,m')dkdk'\Big\} \qquad (7.128)$$

The ground state of the coupled atom-radiation system, accurate to order e^2, is obtained from (C.13)

$$|\,\phi\rangle' = \Big\{1+\frac{1}{E_0-H_0}(1-P_0)\Big(H_{AF}^{(1)}+H_{AF}^{(2)}\Big).$$
$$+\frac{1}{E_0-H_0}(1-P_0)H_{AF}^{(1)}\frac{1}{E_0-H_0}(1-P_0)H_{AF}^{(1)}$$
$$-\frac{1}{2}\langle\phi\,|\,H_{AF}^{(1)}\frac{1}{(E_0-H_0)^2}(1-P_0)H_{AF}^{(1)}\,|\,\phi\rangle\Big\}\,|\,\phi\rangle \qquad (7.129)$$

where $|\,\phi\rangle \equiv |\,u_{100},\{0\}\rangle$ and $E_0 \equiv E_1$. Some straightforward algebra leads to the following expressions for the various contributions to $|\,\phi\rangle'$ appearing in (7.129)

$$\frac{1}{E_0-H_0}(1-P_0)H_{AF}^{(1)}\,|\,\phi\rangle$$
$$= 2\sqrt{\frac{2}{3}}\frac{e\hbar^{1/2}}{mc^{1/2}}\sum_{NLM}\sum_{m}\int u^*_{NLM}(\mathbf{x}_e)\mathbf{Y}^*_{1m-}(\theta_e,\varphi_e)\cdot\mathbf{p}u_{100}(\mathbf{x}_e)d^3\mathbf{x}_e$$

$$\times \int \frac{k^{1/2}}{\hbar(\omega_N + \omega_k)} \mid u_{NLM}, 1_{k,\mathcal{E},1,m}\rangle dk \; ;$$

$$\frac{1}{E_0 - H_0}(1 - P_0)H_{AF}^{(2)} \mid \phi\rangle = -\frac{4\sqrt{2}}{3}\frac{e^2\hbar}{mc}\sum_{mm'}\sum_{NLM}\int u_{NLM}^*(\mathbf{x}_e)$$

$$\times \mathbf{Y}_{1m-}^*(\theta_e, \varphi_e) \cdot \mathbf{Y}_{1m'-}^*(\theta_e, \varphi_e) u_{100}(\mathbf{x}_e) d^3\mathbf{x}_e$$

$$\times \int\int \frac{(kk')^{1/2}}{\hbar(\omega_N + \omega_k + \omega_{k'})} \mid u_{NLM}, 1_{k,\mathcal{E},1,m}, 1_{k',\mathcal{E},1,m'}\rangle dk dk' \; ;$$

$$\frac{1}{E_0 - H_0}(1 - P_0)H_{AF}^{(1)}\frac{1}{E_0 - H_0}(1 - P_0)H_{AF}^{(1)} \mid \phi\rangle$$

$$= \frac{8}{3}\frac{e^2\hbar}{m^2 c}\Bigg\{\sum_m \sum_{\substack{NLM \\ N'L'M'}}{}'$$

$$\times \int u_{NLM}^*(\mathbf{x}_e)\mathbf{Y}_{1m-}^*(\theta_e, \varphi_e) \cdot \mathbf{p} u_{100}(\mathbf{x}_e) d^3\mathbf{x}_e$$

$$\times \int u_{N'L'M'}^* \mathbf{Y}_{1m-}(\theta_e, \varphi_e) \cdot \mathbf{p} u_{NLM}(\mathbf{x}_e) d^3\mathbf{x}_e$$

$$\times \int \frac{k dk}{\hbar^2 \omega_{N'}(\omega_N + \omega_k)} \mid u_{N'L'M'}, \{0\}\rangle$$

$$+ \sqrt{2}\sum_{mm'}\sum_{\substack{NLM \\ N'L'M'}} \int u_{NLM}^*(\mathbf{x}_e)\mathbf{Y}_{1m-}^*(\theta_e, \varphi_e) \cdot \mathbf{p} u_{100}(\mathbf{x}_e) d^3\mathbf{x}_e$$

$$\times \int u_{N'L'M'}^*(\mathbf{x}_e)\mathbf{Y}_{1m'-}^*(\theta_e, \varphi_e) \cdot \mathbf{p} u_{NLM}(\mathbf{x}_e) d^3\mathbf{x}_e$$

$$\times \int\int \frac{(kk')^{1/2}}{\hbar^2(\omega_N + \omega_k)(\omega_{N'} + \omega_k + \omega_{k'})}$$

$$\times \mid u_{N'L'M'}, 1_{k,\mathcal{E},1,m}, 1_{k',\mathcal{E},1,m'}\rangle dk dk' \Bigg\} \; ;$$

$$\langle\phi \mid H_{AF}^{(1)} \frac{1}{(E_0 - H_0)^2}(1 - P_0)H_{AF}^{(1)} \mid \phi\rangle$$

$$= \frac{8}{3}\frac{e^2\hbar}{m^2 c}\sum_m \sum_{NLM}\left| \int u_{NLM}^*(\mathbf{x}_e)\mathbf{Y}_{1m-}^*(\theta_e, \varphi_e) \cdot \mathbf{p} u_{100}(\mathbf{x}_e) d^3\mathbf{x}_e \right|^2$$

$$\times \int \frac{k dk}{\hbar^2(\omega_N + \omega_k)^2}$$

$$= \frac{2e^2\hbar}{3\pi m^2 a^2 c^3}\sum_N |\langle R_{N1} \mid R_{10}\rangle|^2 \int \frac{k dk}{(k_N + k)^2} \quad \left(k_N = \frac{\omega_N}{c}\right) \quad (7.130)$$

In these expressions Σ' indicates a sum of NLM and $N'L'M'$ with $NLM \neq N'L'M'$ and $\omega_N = (E_N - E_1)/\hbar$. Moreover, in the final step of (7.130) use has been made of the result

$$\int u^*_{NLM}(\mathbf{x}_e)\mathbf{Y}^*_{1m-}(\theta_e, \varphi_e) \cdot \mathbf{p} u_{100}(\mathbf{x}_e) d^3\mathbf{x}_e$$
$$= -i\frac{(-1)^M}{\sqrt{12\pi}}\frac{\hbar}{a}\langle R_{N1} \mid R_{10}\rangle \delta_{L,1}\delta_{M,-m} \tag{7.131}$$

which follows from the properties of vector spherical harmonics derived in Appendix A. The notation $\langle R_{N1} \mid R_{10}\rangle$ means

$$\langle R_{N1} \mid R_{10}\rangle \equiv \int R_{N1}(x_e)R_{10}(x_e)x_e^2 dx_e \tag{7.132}$$

which we shall generalize in the future to

$$\langle R_{N1} \mid x_e^\ell \mid R_{10}\rangle \equiv \int R_{N1}(x_e)R_{10}(x_e)x_e^{\ell+2} dx_e \tag{7.133}$$

Our aim is to obtain an analytic expression for the energy density of the e.m. field around the atom in its dressed ground state $\mid \phi\rangle'$ as given by (7.129). As in the two-level atom case, this energy density can be partitioned into an electric and a magnetic energy density. Unlike the two-level atom case discussed in Section 7.2, however, but similar to the free-electron case discussed in the previous section, we shall use the minimal coupling scheme. Consequently it is convenient here to sub-partition the electric energy density into contributions from the transverse and longitudinal components of the electric field. Thus the total energy density around the ground-state hydrogen atom will be partitioned as in (7.99)

$$\mathcal{H}(\mathbf{x}) = \mathcal{H}_{el}(\mathbf{x}) + \mathcal{H}_{mag}(\mathbf{x}) \ ;$$
$$\mathcal{H}_{el}(\mathbf{x}) = \frac{1}{8\pi}\left[E_\parallel^2(\mathbf{x}) + E_\perp^2(\mathbf{x}) + 2\mathbf{E}_\parallel(\mathbf{x}) \cdot \mathbf{E}_\perp(\mathbf{x})\right] \ ;$$
$$\mathcal{H}_{mag}(\mathbf{x}) = \frac{1}{8\pi}H^2(\mathbf{x}) \tag{7.134}$$

We first consider the transverse electric part. In spherical coordinates the expression for $\mathbf{E}_\perp(\mathbf{x})$ can be obtained from (2.22). Such an expression, however, contains many terms which can be seen immediately to give a vanishing contribution to $'\langle\phi \mid \mathbf{E}_\perp^2(\mathbf{x}) \mid \phi\rangle'$. In fact, because of the electric dipole approximation, $\mid \phi\rangle'$ corresponds to the vacuum of $\mathcal{M}$-photons, as is evident from the explicit expressions (7.130). Thus the nonvanishing contributions to the transverse electric energy density are

$$
\begin{aligned}
&'\langle\phi \mid E_\perp^2(\mathbf{x}) \mid \phi\rangle' \\
&= 4\hbar c \sum_{\ell\ell' mm'} \int\int \Big\{ '\langle\phi \mid a(k,\mathcal{M},\ell,m)a(k',\mathcal{M},\ell',m') \mid \phi\rangle' \\
&\times j_\ell(kx) j_{\ell'}(k'x) \mathbf{Y}_{\ell m0}(\theta,\varphi)\cdot \mathbf{Y}^*_{\ell' m'0}(\theta,\varphi) \\
&- '\langle\phi \mid a(k,\mathcal{E},\ell,m)a(k',\mathcal{E},\ell',m') \mid \phi\rangle' \frac{1}{\sqrt{(2\ell+1)(2\ell'+1)}} \\
&\times \left[\sqrt{\ell} j_{\ell+1}(kx)\mathbf{Y}_{\ell m+}(\theta,\varphi) - \sqrt{\ell+1} j_{\ell-1}(kx)\mathbf{Y}_{\ell m-}(\theta,\varphi)\right] \\
&\cdot \left[\sqrt{\ell'} j_{\ell'+1}(k'x)\mathbf{Y}_{\ell' m'+}(\theta,\varphi) - \sqrt{\ell'+1} j_{\ell'-1}(k'x)\mathbf{Y}_{\ell' m'-}(\theta,\varphi)\right] \\
&+ '\langle\phi \mid a(k,\mathcal{E},\ell,m)a^\dagger(k',\mathcal{E},\ell',m') \mid \phi\rangle' \frac{1}{\sqrt{(2\ell+1)(2\ell'+1)}} \\
&\times \left[\sqrt{\ell} j_{\ell+1}(kx)\mathbf{Y}_{\ell m+}(\theta,\varphi) - \sqrt{\ell+1} j_{\ell-1}(kx)\mathbf{Y}_{\ell m-}(\theta,\varphi)\right] \\
&\cdot \left[\sqrt{\ell'} j_{\ell'+1}(k'x)\mathbf{Y}^*_{\ell' m+}(\theta,\varphi) - \sqrt{\ell'+1} j_{\ell'-1}(k'x)\mathbf{Y}^*_{\ell' m-}(\theta,\varphi)\right] \\
&+ '\langle\phi \mid a^\dagger(k,\mathcal{E},\ell,m)a(k',\mathcal{E},\ell',m') \mid \phi\rangle' \frac{1}{\sqrt{(2\ell+1)(2\ell'+1)}} \\
&\times \left[\sqrt{\ell} j_{\ell+1}(kx)\mathbf{Y}^*_{\ell m+}(\theta,\varphi) - \sqrt{\ell+1} j_{\ell-1}(kx)\mathbf{Y}^*_{\ell m-}(\theta,\varphi)\right] \\
&\cdot \left[\sqrt{\ell'} j_{\ell'+1}(k'x)\mathbf{Y}_{\ell' m'+}(\theta,\varphi) - \sqrt{\ell'+1} j_{\ell'-1}(k'x)\mathbf{Y}_{\ell' m'-}(\theta,\varphi)\right] \\
&- '\langle\phi \mid a^\dagger(k,\mathcal{E},\ell,m)a^\dagger(k',\mathcal{E},\ell',m') \mid \phi\rangle' \frac{1}{\sqrt{(2\ell+1)(2\ell'+1)}} \\
&\times \left[\sqrt{\ell} j_{\ell+1}(kx)\mathbf{Y}^*_{\ell m+}(\theta,\varphi) - \sqrt{\ell+1} j_{\ell-1}(kx)\mathbf{Y}^*_{\ell m-}(\theta,\varphi)\right] \\
&\cdot \left[\sqrt{\ell'} j_{\ell'+1}(k'x)\mathbf{Y}^*_{\ell' m'+}(\theta,\varphi) - \sqrt{\ell'+1} j_{\ell'-1}(k'x)\mathbf{Y}^*_{\ell' m'-}(\theta,\varphi)\right]\Big\} \\
&\times (kk')^{3/2} dk dk' \\
&= 4\hbar c\{A(\mathbf{x}) + B(\mathbf{x}) + C(\mathbf{x}) + D(\mathbf{x}) + E(\mathbf{x})\}
\end{aligned}
\tag{7.135}
$$

Using the commutation relations for a and $a^\dagger$ obtained in Section 2.1, a certain amount of algebra leads to

$$'\langle\phi\,|a(k,\mathcal{M},\ell,m)a^\dagger(k',\mathcal{M},\ell',m')\mid\phi\rangle' = \delta(k-k')\delta_{\ell\ell'}\delta_{mm'}\,;$$

$$'\langle\phi\,|a(k,\mathcal{E},\ell,m)a(k',\mathcal{E},\ell',m')\mid\phi\rangle'$$

$$= \langle\phi\mid a(k,\mathcal{E},\ell,m)a(k',\mathcal{E},\ell',m')\frac{1}{E_0-H_0}(1-P_0)H_{AF}^{(2)}\mid\phi\rangle$$

$$+ \langle\phi\mid a(k,\mathcal{E},\ell,m)a(k',\mathcal{E},\ell',m')\frac{1}{E_0-H_0}(1-P_0)$$

$$\times H_{AF}^{(1)}\frac{1}{E_0-H_0}H_{AF}^{(1)}\mid\phi\rangle$$

$$= (-1)^{m+1}\frac{2e^2\hbar}{3\pi mc}\frac{(kk')^{1/2}}{\hbar(\omega_N+\omega_{k'})}\delta_{\ell,1}\delta_{\ell',1}\delta_{m',-m}$$

$$+ (-1)^m\frac{2e^2\hbar^3}{9\pi m^2a^2c}\frac{(kk')^{1/2}}{\hbar(\omega_N+\omega_{k'})}$$

$$\times \sum_N(\langle R_{N1}\mid R_{10}\rangle)^2\left(\frac{1}{\hbar(\omega_N+\omega_k)}+\frac{1}{\hbar(\omega_N+\omega_{k'})}\right)$$

$$\times\ \delta_{\ell,1}\delta_{\ell',1}\delta_{m',-m}\,;$$

$$'\langle\phi\,|a^\dagger(k,\mathcal{E},\ell,m)a(k',\mathcal{E},\ell',m')\mid\phi\rangle'$$

$$= \langle\phi\mid H_{AF}^{(1)}(1-P_0)\frac{1}{E_0-H_0}a^\dagger(k,\mathcal{E},\ell,m)a(k',\mathcal{E},\ell',m')$$

$$\times \frac{1}{E_0-H_0}(1-P_0)H_{AF}^{(1)}\mid\phi\rangle$$

$$= \frac{2e^2\hbar^3}{9\pi m^2a^2c}\sum_N(\langle R_{N1}\mid R_{10}\rangle)^2\frac{(kk')^{1/2}}{\hbar^2(\omega_N+\omega_k)(\omega_N+\omega_{k'})}$$

$$\times\ \delta_{\ell,1}\delta_{\ell',1}\delta_{m',m}\,;$$

$$'\langle\phi\,|a(k,\mathcal{E},\ell,m)a^\dagger(k',\mathcal{E},\ell',m')\mid\phi\rangle'$$

$$= \delta(k-k')\delta_{\ell\ell'}\delta_{mm'} + \,'\langle\phi\mid a^\dagger(k',\mathcal{E},\ell',m')a(k,\mathcal{E},\ell,m)\mid\phi\rangle'\,;$$

$$'\langle\phi\,|a^\dagger(k,\mathcal{E},\ell,m)a^\dagger(k',\mathcal{E},\ell',m')\mid\phi\rangle'$$

$$= \,'\langle\phi\mid a(k,\mathcal{E},\ell,m)a(k',\mathcal{E},\ell',m')\mid\phi\rangle' \tag{7.136}$$

Use of (7.136) and extensive use of the algebra of vector spherical harmonics presented in Appendix A permits evaluation of the double sums over ℓ and m in (7.135) and leads to the following relations

$$A(\mathbf{x}) = \frac{1}{4\pi}\sum_\ell (2\ell+1)\int j_\ell^2(kx)k^3 dk \ ;$$

$$B(\mathbf{x}) = \frac{e^2\hbar}{6\pi^2 mc}\int\int \frac{(kk')^2}{\hbar(\omega_k+\omega_{k'})}$$
$$\times\left\{1-\frac{\hbar^2}{3ma^2}\sum_N(\langle R_{N1}\mid R_{10}\rangle)^2\right.$$
$$\left.\times\left[\frac{1}{\hbar(\omega_N+\omega_k)}+\frac{1}{\hbar(\omega_N+\omega_{k'})}\right]\right\}$$
$$\times\,[j_2(kx)j_2(k'x)+2j_0(kx)j_0(k'x)]dkdk' \ ;$$

$$D(\mathbf{x}) = \frac{e^2\hbar^3}{18\pi^2 m^2 a^2 c}\int\int\left\{\sum_N(\langle R_{N1}\mid R_{10}\rangle)^2\right.$$
$$\left.\times\left[\frac{(kk')^2}{\hbar^2(\omega_N+\omega_k)(\omega_N+\omega_{k'})}\right]\right\}$$
$$\times\,[j_2(kx)j_2(k'x)+2j_0(kx)j_0(k'x)]dkdk' \ ;$$

$$C(\mathbf{x}) = D(\mathbf{x})$$
$$+\frac{1}{4\pi}\sum_\ell\int[\ell j_{\ell+1}^2(kx)+(\ell+1)j_{\ell-1}^2(kx)]k^3dk \ ;$$
$$E(\mathbf{x}) = B(\mathbf{x}) \ ;$$

$$C(\mathbf{x})+A(\mathbf{x}) = D(\mathbf{x})+\frac{1}{4\pi}\sum_\ell\int[\ell j_{\ell+1}^2(kx)+(2\ell+1)j_\ell^2(kx)$$
$$+(\ell+1)j_{\ell-1}^2(kx)]k^3dk$$
$$= D(\mathbf{x})+\frac{1}{2\pi}\int k^3dk$$
$$\equiv D(\mathbf{x})+2\pi\frac{1}{(2\pi)^3}\int\frac{1}{2}kd^3\mathbf{k} \qquad (7.137)$$

Substitution of (7.137) into (7.135) yields

$$\frac{1}{8\pi}{}'\langle\phi \mid E_\perp^2(\mathbf{x}) \mid \phi\rangle' = \frac{1}{(2\pi)^3}\int \frac{1}{2}\hbar\omega_k d^3\mathbf{k} + \frac{e^2\hbar}{6\pi^3 mc}\int\int \frac{(kk')^2}{k+k'}$$
$$\times\left\{1 - \frac{\hbar^2}{3ma^2}\sum_N(\langle R_{N1} \mid R_{10}\rangle)^2\left[\frac{1}{\hbar(\omega_N+\omega_k)} + \frac{1}{\hbar(\omega_N+\omega_{k'})}\right]\right\}$$
$$\times [j_2(kx)j_2(k'x) + 2j_0(kx)j_0(k'x)]dkdk'$$
$$+\frac{e^2\hbar^2}{18\pi^3 m^2a^2}\int\int (kk')^2$$
$$\times\left\{\sum_N(\langle R_{N1} \mid R_{10}\rangle)^2\frac{1}{(\omega_N+\omega_k)(\omega_N+\omega_{k'})}\right\}$$
$$\times [j_2(kx)j_2(k'x) + 2j_0(kx)j_0(k'x)]dkdk' \tag{7.138}$$

An analogous procedure, starting from the expressions for $\mathbf{H}(\mathbf{x})$ given in (2.22), leads to the following expression for the magnetic contribution to the energy density

$$\frac{1}{8\pi}{}'\langle\phi \mid H^2(\mathbf{x}) \mid \phi\rangle' = \frac{1}{(2\pi)^3}\int \frac{1}{2}\hbar\omega_k d^3\mathbf{k}$$
$$-\frac{e^2\hbar}{2\pi^3 mc}\int\int \frac{(kk')^2}{k+k'}\left\{1 - \frac{\hbar^2}{3ma^2}\sum_N(\langle R_{N1} \mid R_{10}\rangle)^2.\right.$$
$$\left.\times\left[\frac{1}{\hbar(\omega_N+\omega_k)} + \frac{1}{\hbar(\omega_N+\omega_{k'})}\right]\right\}j_1(kx)j_1(k'x)dkdk' + \frac{e^2\hbar^2}{6\pi^3 m^2a^2}$$
$$\times\int\int (kk')^2\{\sum_N(\langle R_{N1} \mid R_{10}\rangle)^2\frac{1}{(\omega_N+\omega_k)(\omega_N+\omega_{k'})}$$
$$\times j_1(kx)j_1(k'x)\}dkdk' \tag{7.139}$$

It should be noted that the first term on the RHS of both expressions (7.138) and (7.139) represents the infinite contribution of the unperturbed vacuum e.m. field, which does not depend on $\mathbf{x}$. The other terms are due to the presence of the atom at the origin of the reference frame.

We now turn to the contribution to (7.134) stemming from the presence of the longitudinal electric field $E_\parallel(\mathbf{x})$. It is convenient to use the well-known spherical harmonic expansion (see e.g. Barton 1989), valid in the region outside the atom,

$$\frac{1}{\mid \mathbf{x}-\mathbf{x}_e \mid} = 4\pi\sum_{\ell=0}^{\infty}\frac{1}{2\ell+1}\frac{x_e^\ell}{x^{\ell+1}}\sum_{m=-\ell}^{\ell}Y_\ell^{m*}(\theta_e,\varphi_e)Y_\ell^m(\theta,\varphi) \tag{7.140}$$

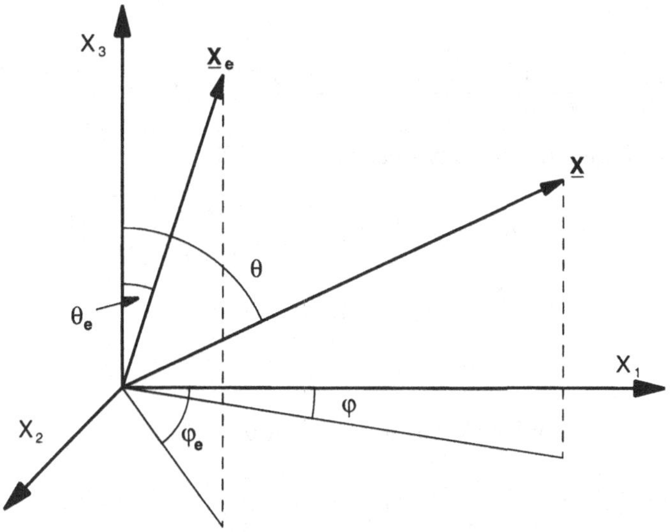

Fig. 7.1 Polar coordinates for the position of the electron $\mathbf{x}_e$ and the observation point $\mathbf{x}$. The nucleus of the hydrogen atom is taken at the origin of the reference frame.

where the meaning of the symbols is illustrated in Figure 7.1. Thus the potential created at point $\mathbf{x}$ by the electron at point $\mathbf{x}_e$ and by the proton fixed at the origin can be expressed as

$$\begin{aligned} V(\mathbf{x}) &= \frac{e}{x} - \frac{e}{\mid \mathbf{x} - \mathbf{x}_e \mid} \\ &= -4\pi e \sum_{\ell=1}^{\infty} \frac{1}{2\ell+1} \frac{x_e^{\ell}}{x^{\ell+1}} \sum_{m=-\ell}^{\ell} Y_\ell^{m*}(\theta_e, \varphi_e) Y_\ell^m(\theta, \varphi) \end{aligned} \tag{7.141}$$

The longitudinal electric field operator for the hydrogen atom can then be obtained as from (7.92)

$$\begin{aligned} \mathbf{E}_{\|}(\mathbf{x}) = -\nabla V(\mathbf{x}) = 4\pi e \sum_{\ell=1}^{\infty} \sum_{m=-\ell}^{\ell} \frac{1}{2\ell+1} x_e^{\ell} Y_\ell^{m*}(\theta_e, \varphi_e) \\ \times \nabla \left[\frac{1}{x^{\ell+1}} Y_\ell^m(\theta, \varphi) \right] \end{aligned} \tag{7.142}$$

A certain amount of algebra involving the properties of vector spherical harmonics summarized in Appendix A leads to the following form for

$\mathbf{E}_{\parallel}(\mathbf{x})$

$$\mathbf{E}_{\parallel}(\mathbf{x}) = 4\pi e \sum_{\ell=1}^{\infty} (i)^{\ell+1} \frac{x_e^{\ell}}{x^{\ell+2}} \sqrt{\frac{\ell+1}{2\ell+1}} \sum_m Y_{\ell}^{m*}(\theta_e, \varphi_e) \mathbf{Y}_{\ell m+}(\theta, \varphi) \quad (7.143)$$

from which one immediately obtains

$$\begin{aligned} \mathbf{E}_{\parallel}^2(\mathbf{x}) = 16\pi^2 e^2 \sum_{\ell\ell'} i^{\ell+\ell'+2} \frac{x_e^{\ell+\ell'}}{x^{\ell+\ell'+4}} \sqrt{\frac{(\ell+1)(\ell'+1)}{(2\ell+1)(2\ell'+1)}} \\ \times \sum_{mm'} Y_{\ell}^{m*}(\theta_e, \varphi_e) Y_{\ell'}^{m'*}(\theta_e, \varphi_e) \\ \times \mathbf{Y}_{\ell m+}(\theta, \varphi) \cdot \mathbf{Y}_{\ell' m'}(\theta, \varphi) \end{aligned} \quad (7.144)$$

Since $\mathbf{E}_{\parallel}^2(\mathbf{x})$ is of order e^2, it is not difficult to evaluate

$$\begin{aligned} \frac{1}{8\pi}\,{}'\langle \phi \mid \mathbf{E}_{\parallel}^2(\mathbf{x}) \mid \phi \rangle' &= \frac{1}{8\pi} \langle \phi \mid \mathbf{E}_{\parallel}^2(\mathbf{x}) \mid \phi \rangle \\ &= \frac{1}{32\pi^2} e^2 \sum_{\ell=1}^{\infty} \frac{1}{x^{2\ell+4}} (\ell+1) \langle R_{10} \mid x_e^{2\ell} \mid R_{10} \rangle \\ &\sim \frac{e^2}{16\pi^2} \langle R_{10} \mid x_e^2 \mid R_{10} \rangle \frac{1}{x^6} \end{aligned} \quad (7.145)$$

where multipoles of order higher than $\ell = 1$ have been neglected. From (7.143) and (2.22) one has

$$\begin{aligned} \mathbf{E}_{\parallel}(\mathbf{x}) \cdot \mathbf{E}_{\perp}(\mathbf{x}) &= i8\pi e \sqrt{\hbar c} \sum_{\ell\ell'} \sum_{mm'} i^{\ell+1} \sqrt{\frac{\ell+1}{2\ell+1}} \frac{x_e^{\ell}}{x^{\ell+2}} Y_{\ell}^{m*}(\theta_e, \varphi_e) \\ &\times \int \Big\{ a(k, \mathcal{M}, \ell', m') j_{\ell'}(kx) \mathbf{Y}_{\ell' m' 0}(\theta, \varphi) \cdot \mathbf{Y}_{\ell m+}(\theta, \varphi) \\ &+ a(k, \mathcal{E}, \ell', m') \frac{1}{\sqrt{2\ell'+1}} \Big[\sqrt{\ell'} j_{\ell'+1}(kx) \mathbf{Y}_{\ell' m'+}(\theta, \varphi) \cdot \mathbf{Y}_{\ell m+}(\theta, \varphi) \\ &- \sqrt{\ell'+1} j_{\ell'-1}(kx) \mathbf{Y}_{\ell' m'-}(\theta, \varphi) \cdot \mathbf{Y}_{\ell m+}(\theta, \varphi) \Big] \Big\} k^{3/2} dk \\ &- i8\pi e \sqrt{\hbar c} \sum_{\ell\ell'} \sum_{mm'} i^{\ell+1} \sqrt{\frac{\ell+1}{2\ell+1}} \frac{x_e^{\ell}}{x^{\ell+2}} Y_{\ell}^{m*}(\theta_e, \varphi_e) \\ &\times \int \Big\{ a^{\dagger}(k, \mathcal{M}, \ell', m') j_{\ell'}(kx) \mathbf{Y}_{\ell' m' 0}^{*}(\theta, \varphi) \cdot \mathbf{Y}_{\ell m+}(\theta, \varphi) \\ &+ a^{\dagger}(k, \mathcal{E}, \ell', m') \frac{1}{\sqrt{2\ell'+1}} \Big[\sqrt{\ell'} j_{\ell'+1}(kx) \mathbf{Y}_{\ell' m'+}^{*}(\theta, \varphi) \cdot \mathbf{Y}_{\ell m+}(\theta, \varphi) \\ &- \sqrt{\ell'+1} j_{\ell'-1}(kx) \mathbf{Y}_{\ell' m'-}^{*}(\theta, \varphi) \cdot \mathbf{Y}_{\ell' m'+}(\theta, \varphi) \Big] \Big\} k^{3/2} dk \end{aligned} \quad (7.146)$$

Thus (7.146) is of order e, and it is easy to convince ourselves that at order e^2 one has

$$
\begin{aligned}
&{}'\langle\phi \mid \mathbf{E}_{\|}(\mathbf{x}) \cdot \mathbf{E}_{\perp}(\mathbf{x}) \mid \phi\rangle' \\
&= \langle\phi \mid \mathbf{E}_{\|}(\mathbf{x}) \cdot \mathbf{E}_{\perp}(\mathbf{x}) \frac{1}{E_0 - H_0}(1 - P_0) H_{AF}^{(1)} \mid \phi\rangle \\
&+ \langle\phi \mid H_{AF}^{(1)}(1 - P_0) \frac{1}{E_0 - H_0} \mathbf{E}_{\|}(\mathbf{x}) \cdot \mathbf{E}_{\perp}(\mathbf{x}) \mid \phi\rangle
\end{aligned}
\tag{7.147}
$$

Use of (7.130) and of (7.146) in (7.147) gives, after extensive algebra,

$$
\begin{aligned}
&\frac{1}{4\pi}{}'\langle\phi \mid \mathbf{E}_{\|}(\mathbf{x}) \cdot \mathbf{E}_{\perp}(\mathbf{x}) \mid \phi\rangle' \\
&= -\frac{e^2 \hbar^2}{3\pi^2 m a} \sum_N \langle R_{N1} \mid x_e \mid R_{10}\rangle \langle R_{N1} \mid R_{10}\rangle \\
&\times \int \frac{k^2 j_2(kx)}{\hbar(\omega_N + \omega_k)} dk \frac{1}{x^3}
\end{aligned}
\tag{7.148}
$$

Expressions (7.138), (7.139), (7.145) and (7.148) are the desired results which, apart from the dipole approximation in (7.145), are exact at order e^2 and which can be substituted in principle in (7.134) to obtain the total energy density around the ground-state hydrogen atom. These expressions, however, are not very transparent and easily manageable, apart from the unperturbed zero-point contributions appearing in the first two of them. Thus in the next section we will make appropriate approximations which lead to a simpler physical interpretation of these results.

7.6 Approximate energy density around a hydrogen atom

In the two-level atom case discussed earlier in this chapter we have found it useful to distinguish two regions of space around the atom. In the first region the virtual cloud is mainly made up of photons whose frequency is higher than the atomic natural frequency, whereas in the second the opposite is true. The two regions have been called near and far zone respectively, because high-frequency photons are not likely to reach far from the atom in view of their relatively short lifetime, whereas low-frequency photons can reach relatively large distances since they live longer. These features have also been discussed qualitatively in Section 6.2. This suggests that we should use the same distinction for the present case between a near and a far zone characterized by $\omega_N \ll \omega_k$ and

$\omega_N \gg \omega_k$ respectively, where ω_N is a typical atomic frequency in an optical range corresponding to a virtual transition leading the atom from the ground to an excited state. Naturally in both zones the purely longitudinal contribution to the energy density is the same and it is given by (7.145), but the other contributions involving the transverse electric or magnetic field are different.

In the near zone, in fact, one approximates

$$\frac{1}{\omega_N + \omega_k} \sim \frac{1}{\omega_k} \tag{7.149}$$

and (7.138), (7.139) and (7.148) take the form

$$\begin{aligned}
&\frac{1}{8\pi}\,'\langle\phi \mid \mathbf{E}_\perp^2(\mathbf{x}) \mid \phi\rangle' = \frac{1}{(2\pi)^3}\int \frac{1}{2}\hbar\omega_k d^3\mathbf{k} \\
&+ \frac{e^2\hbar}{6\pi^3 mc}\int\int \frac{(kk')^2}{k+k'}[j_2(kx)j_2(k'x) + 2j_0(kx)j_0(k'x)] \\
&\times dkdk' \, ;
\end{aligned} \tag{7.150}$$

$$\begin{aligned}
&\frac{1}{8\pi}\,'\langle\phi \mid \mathbf{H}^2(\mathbf{x}) \mid \phi\rangle' = \frac{1}{(2\pi)^3}\int \frac{1}{2}\hbar\omega_k d^3\mathbf{k} \\
&- \frac{e^2\hbar}{2\pi^3 mc}\int\int \frac{(kk')^2}{k+k'} j_1(kx)j_1(k'x)dkdk' \\
&+ \frac{e^2\hbar^2}{3\pi^3 m^2 a^2 c^2}\sum_N (\langle R_{N1} \mid R_{10}\rangle)^2 \int\int kk' j_1(kx)j_1(k'x)dkdk' \\
&\sim \frac{1}{(2\pi)^3}\int \frac{1}{2}\hbar\omega_k d^3\mathbf{k} \\
&- \frac{e^2\hbar}{2\pi^3 mc}\int\int \frac{(kk')^2}{k+k'} j_1(kx)j_1(k'x)dkdk' ;
\end{aligned} \tag{7.151}$$

$$\begin{aligned}
&\frac{1}{4\pi}\,'\langle\phi \mid \mathbf{E}_\parallel(\mathbf{x}) \cdot \mathbf{E}_\perp(\mathbf{x}) \mid \phi\rangle' \\
&= -\frac{e^2\hbar}{3\pi^2 mac}\sum_N \langle R_{N1} \mid x_e \mid R_{10}\rangle\langle R_{N1} \mid R_{10}\rangle \\
&\times \int kj_2(kx)dk \frac{1}{x^3}
\end{aligned} \tag{7.152}$$

It should be noted that the magnetic energy density has been approximated by neglecting one of the double integrations, which, on purely dimensional grounds, yields a contribution behaving as x^{-4}. In contrast, the double integration which has been retained yields a contribution behaving as x^{-5}, which is prevalent at short distances. The integrals appearing in (7.150, 7.151) can be factorized as in (7.39) and then evaluated using partial integration and a certain amount of algebraic labour. The result is

$$\int\int \frac{(kk')^2}{k+k'}[j_2(kx)j_2(k'x) + 2j_0(kx)j_0(k'x)]dkdk' = \frac{15\pi}{8}\frac{1}{x^5} ;$$
$$\int\int \frac{(kk')^2}{k+k'}j_1(kx)j_1(k'x)dkdk' = \frac{5\pi}{8}\frac{1}{x^5} \tag{7.153}$$

The integral appearing in (7.152) can be regularized at the upper limit of integration and then evaluated as a Laplace transform (Erdelyi *et al.* 1954) as follows

$$\int kj_2(kx)dk = \sqrt{\frac{\pi}{2x}}\int k^{1/2}J_{5/2}(kx)dk$$
$$= \lim_{\alpha\to 0}\sqrt{\frac{\pi}{2x}}\int_0^\infty e^{-\alpha k}k^{1/2}J_{5/2}(kx)dk = \frac{2}{3}\frac{1}{x^2} \tag{7.154}$$

Finally, subtitution of (7.153) and (7.154) in (7.150, 7.151, 7.152), together with use of (D.17) yields

$$\frac{1}{8\pi}\,'\langle\phi \mid \mathbf{E}_\perp^2(\mathbf{x}) \mid \phi\rangle' = \frac{1}{(2\pi)^3}\int \frac{1}{2}\hbar\omega_k d^3\mathbf{k} + \frac{1}{8\pi}\frac{5e^2\hbar}{2\pi mc}\frac{1}{x^5} ;$$
$$\frac{1}{8\pi}\,'\langle\phi \mid \mathbf{H}^2(\mathbf{x}) \mid \phi\rangle' = \frac{1}{(2\pi)^3}\int \frac{1}{2}\hbar\omega_k d^3\mathbf{k} - \frac{1}{8\pi}\frac{5e^2\hbar}{2\pi mc}\frac{1}{x^5} ;$$
$$\frac{1}{4\pi}\,'\langle\phi \mid \mathbf{E}_\parallel(\mathbf{x})\cdot\mathbf{E}_\perp(\mathbf{x}) \mid \phi\rangle' = -\frac{e^2\hbar}{3\pi^2 mc}\frac{1}{x^5} \tag{7.155}$$

It is interesting to remark that the purely transverse part of the electric energy density around a ground-state hydrogen atom in the near zone, as given by the first of (7.155), coincides with that of a free electron obtained in (7.110), and that the same is true for the magnetic energy density terms behaving as x^{-5}. These similarities originate from the fact that in the near region the dominant contribution to the virtual cloud around the atom stems from high-frequency photons, which are not influenced by the atomic structure. In this sense one is led to conclude that in the near region of the ground-state hydrogen atom the virtual transverse energy density is essentially the same as for a free electron. It should be noted, however, that for the hydrogen atom the transverse and the mixed contributions to the electric energy densities in the near zone, the first and third expressions of (7.155), are dominated by the longitudinal contribution (7.145)

$$\frac{1}{8\pi}{}'\langle\phi \mid \mathbf{E}_{\|}^2(\mathbf{x}) \mid \phi\rangle' = \frac{e^2}{16\pi^2}\langle R_{10} \mid x_e^2 \mid R_{10}\rangle \frac{1}{x^6} \tag{7.156}$$

which behaves as x^{-6}. Obviously the latter contribution is different from the longitudinal contribution (7.112) for a free electron, because of the presence of the proton. In fact (7.156) is of the same dipolar form as that of a two-level atom in the near zone, expression (7.36), although the latter is strongly influenced by the intrinsic anisotropic structure of the two-level atom which is rooted in the form of the atom-photon coupling constant $\epsilon_{\mathbf{k}j}$. It must be emphasized that the prevalence of the longitudinal over the transverse electric energy density characterizes the near zone around a neutral atom, both two-level and multilevel, as that region of space where retardation effects are negligible. In the two-level atom case there is an x^{-5} contribution to the electric energy density in the near zone which corresponds to the first and third terms in (7.155) for the hydrogen atom. Also (7.69) for the magnetic energy density in the near zone of a two-level atom has its counterpart in the second of (7.155) for the hydrogen atom.

We now turn to the far zone of the hydrogen atom, characterized by $\omega_N \gg \omega_k$. Consequently, the energy denominators appearing in the relevant expressions of Section 7.5 will be approximated according to

$$\begin{aligned} \frac{1}{\omega_N + \omega_k} &\sim \frac{1}{\omega_N}\left(1 - \frac{\omega_k}{\omega_N} + \frac{\omega_k^2}{\omega_N^2}\right); \\ \frac{1}{(\omega_N + \omega_k)(\omega_N + \omega_{k'})} &\sim \frac{1}{\omega_N^2}\left(1 - \frac{\omega_k}{\omega_N} - \frac{\omega_{k'}}{\omega_N}\right) \end{aligned} \tag{7.157}$$

where inclusion of terms up to order ω_N^{-3} is necessary in view of various cancellations taking place among different contributions to the energy density. Using (7.157) the purely transverse electric energy density (7.138) can be approximated as

$$\frac{1}{8\pi}{}'\langle\phi \mid \mathbf{E}_{\perp}^2(\mathbf{x}) \mid \phi\rangle' = \frac{1}{(2\pi)^3}\int \frac{1}{2}\hbar\omega_k d^3\mathbf{k} + O(x^{-5}) + O(x^{-6}) + O(x^{-7}) \tag{7.158}$$

where

$$O(x^{-5}) = \frac{e^2\hbar}{6\pi^3 mc}\left\{1 - \frac{2\hbar}{3ma^2}\sum_N \frac{1}{\omega_N}(\langle R_{N1} \mid R_{10}\rangle)^2\right\}$$
$$\times \int\int \frac{(kk')^2}{k+k'}[j_2(kx)j_2(k'x) + 2j_0(kx)j_0(k'x)]dkdk' \;;$$
$$O(x^{-6}) = \frac{e^2\hbar}{9\pi^3 m^2 a^2}\sum_N \frac{1}{\omega_N^2}(\langle R_{N1} \mid R_{10}\rangle)^2$$
$$\times \int\int (kk')^2[j_2(kx)j_2(k'x) + 2j_0(kx)j_0(k'x)]dkdk' \;;$$
$$O(x^{-7}) = -\frac{e^2\hbar^2 c}{9\pi^3 m^2 a^2}\sum_N \frac{1}{\omega_N^3}(\langle R_{N1} \mid R_{10}\rangle)^2$$
$$\times \int\int \left[(kk')^2(k+k') - \frac{(kk')^3}{k+k'}\right]$$
$$\times [j_2(kx)j_2(k'x) + 2j_0(kx)j_0(k'x)]dkdk' \qquad (7.159)$$

It should be noted that $O(x^{-i})$ in (7.159) yields an x-dependence of the form x^{-i}, as is immediately obtained by dimensional considerations on the double integrals on k and k' appearing in each term. Analogous treatment for the magnetic contribution to the energy density in the far zone yields, from (7.139),

$$\frac{1}{8\pi}{}'\langle\phi \mid \mathbf{H}^2(\mathbf{x}) \mid \phi\rangle' = \frac{1}{(2\pi)^3}\int \frac{1}{2}\hbar\omega_k d^3\mathbf{k} + O(x^{-5}) + O(x^{-7}) \qquad (7.160)$$

where $O(x^{-6})$ vanishes exactly and

$$O(x^{-5}) = -\frac{e^2\hbar}{2\pi^3 mc}\left\{1 - \frac{2\hbar}{3ma^2}\sum_N(\langle R_{N1} \mid R_{10}\rangle)^2\right\}$$
$$\times \int\int \frac{(kk')^2}{k+k'}j_1(kx)j_1(k'x)dkdk' \;;$$
$$O(x^{-7}) = -\frac{e^2\hbar^2 c}{3\pi^3 m^2 a^2}\sum_N \frac{1}{\omega_N^3}(\langle R_{N1} \mid R_{10}\rangle)^2$$
$$\times \int\int \frac{(kk')^3}{k+k'}j_1(kx)j_1(k'x)dkdk' \qquad (7.161)$$

Finally, the cross term (7.148) gives

$$\frac{1}{4\pi}{}'\langle\phi \mid \mathbf{E}_\parallel(\mathbf{x})\cdot\mathbf{E}_\perp(\mathbf{x}) \mid \phi\rangle' = O(x^{-6}) + O(x^{-7}) \qquad (7.162)$$

where

$$O(x^{-6}) = -\frac{e^2\hbar}{3\pi^2 mac}\sum_N \frac{1}{\omega_N}\langle R_{N1} \mid x_e \mid R_{10}\rangle\langle R_{N1} \mid R_{10}\rangle$$
$$\times \int k^2 j_2(kx)dk \frac{1}{x^3} \ ;$$
$$O(x^{-7}) = \frac{e^2\hbar c}{3\pi^2 ma}\sum_N \frac{1}{\omega_N^2}\langle R_{N1} \mid x_e \mid R_{10}\rangle\langle R_{N1} \mid R_{10}\rangle$$
$$\times \int k^3 j_2(kx)dk \frac{1}{x^3} \tag{7.163}$$

Use of (D.10) leads to vanishing of $O(x^{-5})$ in (7.159) and in (7.161). Moreover the method described by Passante and Power (1987) leads to the following results for the relevant double integrations over k and k'

$$\int\int (kk')^2[j_2(kx)j_2(k'x) + 2j_0(kx)j_0(k'x)]dkdk'$$
$$= \left(\int k^2 j_2(kx)dk\right)^2 + 2\left(\int k^2 j_0(kx)dk\right)^2 = \frac{9\pi^2}{4}\frac{1}{x^6} \ ;$$
$$\int\int (kk')^2(k+k')[j_2(kx)j_2(k'x) + 2j_0(kx)j_0(k'x)]dkdk'$$
$$= 2\int k^3 j_2(kx)dk \int k^2 j_2(kx)dk$$
$$+ 4\int k^3 j_0(kx)dk \int k^2 j_0(kx)dk = 24\pi\frac{1}{x^7} \ ;$$
$$\int\int \frac{(kk')^3}{k+k'}[j_2(kx)j_2(k'x) + 2j_0(kx)j_0(k'x)]dkdk'$$
$$= \frac{1}{2}\int\int (kk')^2(k+k')[j_2(kx)j_2(k'x) + 2j_0(kx)j_0(k'x)]dkdk'$$
$$-\frac{1}{2}\int\int \frac{(kk')^2(k^2+k'^2)}{k+k'}[j_2(kx)j_2(k'x) + 2j_0(kx)j_0(k'x)]dkdk'$$
$$= \frac{69\pi}{8}\frac{1}{x^7} \ ;$$
$$\int\int \frac{(kk')^3}{k+k'} j_1(kx)j_1(k'x)dkdk'$$
$$= \frac{1}{2}\int\int (kk')^2(k+k')j_1(kx)j_1(k'x)dkdk'$$
$$-\frac{1}{2}\int\int \frac{(kk')^2(k^2+k'^2)}{k+k'} j_1(kx)j_1(k'x)dkdk' = \frac{7\pi}{8}\frac{1}{x^7} \ ;$$
$$\int k^2 j_2(kx)dk = \frac{3\pi}{2}\frac{1}{x^3} \ ; \ \int k^3 j_2(kx)dk = 8\frac{1}{x^4} \tag{7.164}$$

Substitution of (7.164) in (7.158), (7.160) and (7.162) and use of the sum rules evaluated in Appendix D leads to the following results, valid in the far zone

$$\frac{1}{8\pi}\,'\langle\phi \mid \mathbf{E}_{\perp}^{2}(\mathbf{x}) \mid \phi\rangle' = \frac{1}{(2\pi)^3}\int \frac{1}{2}\hbar\omega_k d^3\mathbf{k}$$
$$+\frac{e^2}{16\pi^2}\langle R_{10} \mid x_e^2 \mid R_{10}\rangle \frac{1}{x^6} - \frac{41}{16\pi^2}\hbar c\alpha_H \frac{1}{x^7} \tag{7.165}$$

$$\frac{1}{8\pi}\,'\langle\phi \mid \mathbf{H}^{2}(\mathbf{x}) \mid \phi\rangle'$$
$$= \frac{1}{(2\pi)^3}\int \frac{1}{2}\hbar\omega_k d^3\mathbf{k} - \frac{7}{16\pi^2}\hbar c\alpha_H \frac{1}{x^7} \tag{7.166}$$

$$\frac{1}{4\pi}\,'\langle\phi \mid \mathbf{E}_{\parallel}(\mathbf{x}) \cdot \mathbf{E}_{\perp}(\mathbf{x}) \mid \phi\rangle'$$
$$= -\frac{e^2}{8\pi^2}\langle R_{10} \mid x_e^2 \mid R_{10}\rangle \frac{1}{x^6} + \frac{4}{\pi^2}\hbar c\alpha_H \frac{1}{x^7} \tag{7.167}$$

where α_H is the ground-state electric polarizability of the hydrogen atom defined as in (D.13).

The influence of the internal structure of the source, which did not play any role in the transverse part of the energy density in the near zone, is now evident in all contributions (7.165, 7.166, 7.167) relative to the far zone. Thus all these terms are very different from those of the free electron. In particular, there is no cancellation between the transverse electric and magnetic energy densities as in the near-zone case. An important cancellation, however, takes place in the total electric energy density, which is of the form

$$\frac{1}{8\pi}\,'\langle\phi \mid \left\{\mathbf{E}_{\perp}^{2}(\mathbf{x}) + \mathbf{E}_{\parallel}^{2}(\mathbf{x}) + 2\mathbf{E}_{\parallel}(\mathbf{x}) \cdot \mathbf{E}_{\perp}(\mathbf{x})\right\} \mid \phi\rangle'$$
$$= \frac{1}{(2\pi)^3}\int \frac{1}{2}\hbar\omega_k d^3\mathbf{k} + \frac{23}{16\pi^2}\hbar c\alpha_H \frac{1}{x^7} \tag{7.168}$$

and which shows that nonretarded effects, yielding terms behaving as x^{-6} in the energy density, are absent in the far zone of the hydrogen atom. Also, it should be noted that the total energy density in the far zone takes the particularly simple form

$$\frac{1}{8\pi}\,'\langle\phi \mid \{\mathbf{E}^2(\mathbf{x}) + \mathbf{H}^2(\mathbf{x})\} \mid \phi\rangle' = \frac{1}{(2\pi)^3}\int \hbar\omega_k d^3\mathbf{k} + \frac{1}{\pi^2}\hbar c\alpha_H \frac{1}{x^7} \tag{7.169}$$

Finally, it is interesting to compare the results obtained in the far zone of the hydrogen atom with those in the corresponding zone of the two-level atom. The far zone total electric energy density of the two-level atom

(7.46) is anisotropic, but if one smears the anisotropy out by assuming a polarizability $\alpha_{mn} = \alpha\delta_{mn}$, (7.46) is seen to reduce to (7.168). The same is true, using the two-level atom electric polarizability defined in (7.45), for the far-zone magnetic energy density (7.74) which reduces to (7.166) when the anisotropy is smeared out.

7.7 Energy density in the Craig-Power model

It is instructive to evaluate the energy density of the virtual cloud surrounding a ground-state neutral atom or molecule whose electrodynamical behaviour can be described in terms of its static polarizability. It is convenient to start from the effective field Hamiltonian (4.120) obtained for any "rigid" electronic configuration of such an object. Coherently with the general assumptions of this chapter, we take $\{n_i\}$ in (4.120) to be the ground-state electronic configuration of the field source. Then (4.120) reduces to

$$H = H_F^{mul} - \frac{1}{2}\alpha_{\ell m} D_{\perp\ell}(0) D_{\perp m}(0) \tag{7.170}$$

In this expression, which is usually called the Craig-Power Hamiltonian (Craig and Power 1969), the zero of the energy has been chosen to eliminate the atomic (or molecular) ground state energy E_A. Moreover, $\alpha_{\ell m}$ in (7.170) is the ground state polarizability of the source, which we assume as point-like and placed at the origin of the coordinates. We emphasize that H here is a purely field operator which does not act on the internal degrees of freedom of the source.

In second-quantized form and using a plane-wave set of field normal modes, H_F^{mul} is given by (4.109) and $D_\perp$ by (4.81). Thus, dropping superflous indices, (7.170) takes the form

$$\begin{aligned} H = \sum_{\mathbf{k}j} \hbar\omega_k \left(a_{\mathbf{k}j}^\dagger a_{\mathbf{k}j} + 1/2\right) + \frac{\pi\hbar}{V}\alpha_{\ell m} \sum_{\mathbf{k}\mathbf{k}'jj'} \sqrt{\omega_k \omega_{k'}} \\ \times \Big\{ (\mathbf{e}_{\mathbf{k}j})_\ell (\mathbf{e}_{\mathbf{k}'j'})_m a_{\mathbf{k}j} a_{\mathbf{k}'j'} - (\mathbf{e}_{\mathbf{k}j})_\ell \left(\mathbf{e}_{\mathbf{k}'j'}^*\right)_m a_{\mathbf{k}j} a_{\mathbf{k}'j'}^\dagger \\ - \left(\mathbf{e}_{\mathbf{k}j}^*\right)_\ell (\mathbf{e}_{\mathbf{k}'j'})_m a_{\mathbf{k}j}^\dagger a_{\mathbf{k}'j'} \\ + \left(\mathbf{e}_{\mathbf{k}j}^*\right)_\ell \left(\mathbf{e}_{\mathbf{k}'j'}^*\right)_m a_{\mathbf{k}j}^\dagger a_{\mathbf{k}'j'}^\dagger \Big\} \end{aligned} \tag{7.171}$$

We will treat this Hamiltonian by perturbation theory, taking advantage of the fact that the double sum in (7.171) is of the order e^2, as is evident from (4.118), hence presumably small with respect to the first sum which we take as the unperturbed Hamiltonian H_0. Moreover, since we limit overall accuracy to order e^2, we can restrict our treatment to straightforward first-order perturbation theory. The ground state of

$$H_0 = \sum_{\mathbf{k}j} \hbar\omega_k \left(a^\dagger_{\mathbf{k}j} a_{\mathbf{k}j} + 1/2 \right) \tag{7.172}$$

is the unperturbed vacuum of the e.m. field $| \{0_{\mathbf{k}j}\}\rangle$ in the absence of the source. From (C.12) the first-order perturbed ground state is

$$| \{0_{\mathbf{k}j}\}\rangle' = \left\{ 1 + \frac{1}{E_0 - H_0}(1 - P_0)H_{int} \right\} | \{0_{\mathbf{k}j}\}\rangle \tag{7.173}$$

where $E_0 = \sum_{\mathbf{k}j} \hbar\omega_k/2$, P_0 is the projector on $| \{0_{\mathbf{k}j}\}\rangle$ and H_{int} is the double sum operator appearing in (7.171). A simple algebra shows that

$$\begin{aligned} | \{0_{\mathbf{k}j}\}\rangle' &= | \{0_{\mathbf{k}j}\}\rangle - \frac{\pi}{V}\alpha_{\ell m} \\ &\times \left\{ \sum_{\substack{\mathbf{k}\mathbf{k}'jj' \\ \mathbf{k}j \neq \mathbf{k}'j'}} \left(\mathbf{e}^*_{\mathbf{k}j}\right)_\ell \left(\mathbf{e}^*_{\mathbf{k}'j'}\right)_m \frac{\sqrt{\omega_k \omega_{k'}}}{\omega_k + \omega_{k'}} | 1_{\mathbf{k}j} 1_{\mathbf{k}'j'}\rangle . \right. \\ &\left. + \frac{1}{\sqrt{2}} \sum_{\mathbf{k}j} \left(\mathbf{e}^*_{\mathbf{k}j}\right)_\ell \left(\mathbf{e}^*_{\mathbf{k}j}\right)_m | 2_{\mathbf{k}j}\rangle \right\} \end{aligned} \tag{7.174}$$

Thus the presence of the source yields admixture of the old vacuum with pairs of virtual photons. These pairs give rise to the virtual cloud surrounding the electrical neutral source fixed at the origin of the reference frame.

Using (7.174) one obtains, up to terms quadratic in e, the following quantum averages

$$\begin{aligned} {}'\langle\{0_{\mathbf{k}j}\} \mid a_{\mathbf{k}j} a_{\mathbf{k}'j'} \mid \{0_{\mathbf{k}j}\}\rangle' &= \left({}'\langle\{0_{\mathbf{k}j}\} \mid a^\dagger_{\mathbf{k}j} a^\dagger_{\mathbf{k}'j'} \mid \{0_{\mathbf{k}j}\}\rangle' \right)^* \\ &= -\frac{\pi}{V}\alpha_{\ell m} \left[\left(\mathbf{e}^*_{\mathbf{k}j}\right)_\ell \left(\mathbf{e}^*_{\mathbf{k}'j'}\right)_m + \left(\mathbf{e}^*_{\mathbf{k}'j'}\right)_\ell \left(\mathbf{e}^*_{\mathbf{k}j}\right)_m \right] \frac{\sqrt{\omega_k \omega_{k'}}}{\omega_k + \omega_{k'}} \; ; \\ {}'\langle\{0_{\mathbf{k}j}\} \mid a^\dagger_{\mathbf{k}j} a_{\mathbf{k}'j'} \mid \{0_{\mathbf{k}j}\}\rangle' &= 0 \; ; \\ {}'\langle\{0_{\mathbf{k}j}\} \mid a_{\mathbf{k}j} a^\dagger_{\mathbf{k}'j'} \mid \{0_{\mathbf{k}j}\}\rangle' &= \delta_{\mathbf{k}\mathbf{k}'}\delta_{jj'} \end{aligned} \tag{7.175}$$

the last of which represents the contribution of the unperturbed vacuum. As we have seen previously in this chapter, this contribution, albeit infinite, cannot give rise to a space-dependent distribution of virtual energy around the atom. Thus we will simply discard it as irrelevant to our purposes.

We are now in the position to evaluate the energy density of the virtual electromagnetic field as a function of the position $\mathbf{x}$ in space. With the help of (2.6) and (7.39) one finds

$$
\begin{aligned}
&{}'\langle\{0_{\mathbf{k}j}\} \mid \mathbf{D}_\perp^2(\mathbf{x}) \mid \{0_{\mathbf{k}j}\}\rangle' = \frac{2\pi^2\hbar}{V^2}\alpha_{\ell m}\sum_{\mathbf{k}\mathbf{k}'jj'} \mathbf{e}_{\mathbf{k}j}\cdot\mathbf{e}_{\mathbf{k}'j'} \\
&\times\left[\left(\mathbf{e}^*_{\mathbf{k}j}\right)_\ell\left(\mathbf{e}^*_{\mathbf{k}'j'}\right)_m+\left(\mathbf{e}^*_{\mathbf{k}'j'}\right)_\ell\left(\mathbf{e}^*_{\mathbf{k}j}\right)_m\right]\frac{\omega_k\omega_{k'}}{\omega_k+\omega_{k'}}e^{i(\mathbf{k}+\mathbf{k}')\cdot\mathbf{x}}+\text{c.c.} \\
&=\frac{2\pi^2}{V^2}\hbar c\alpha_{\ell m}\int_0^\infty\left\{\sum_{\mathbf{k}}\left(\delta_{n\ell}-\hat{k}_n\hat{k}_\ell\right)ke^{-ky}e^{i\mathbf{k}\cdot\mathbf{x}}\right. \\
&\times\sum_{\mathbf{k}'}\left(\delta_{nm}-\hat{k}'_n\hat{k}'_m\right)k'e^{-k'y}e^{i\mathbf{k}'\cdot\mathbf{x}} \\
&+\sum_{\mathbf{k}}\left(\delta_{nm}-\hat{k}_n\hat{k}_m\right)ke^{-ky}e^{i\mathbf{k}\cdot\mathbf{x}} \\
&\times\left.\sum_{\mathbf{k}'}\left(\delta_{n\ell}-\hat{k}'_n\hat{k}'_\ell\right)k'e^{-k'y}e^{i\mathbf{k}'\cdot\mathbf{x}}\right\}dy+\text{c.c.}
\end{aligned}
\tag{7.176}
$$

Use of (7.41) and some algebra yields

$$
\begin{aligned}
{}'\langle\{0_{\mathbf{k}j}\} \mid \mathbf{D}_\perp^2(\mathbf{x}) \mid \{0_{\mathbf{k}j}\}\rangle' &= \frac{32}{\pi^2}\hbar c\alpha_{\ell m}\int_0^\infty\frac{1}{(x^2+y^2)^6} \\
&\times\left\{x^4\delta_{\ell m}-2(\delta_{\ell m}-2\hat{x}_\ell\hat{x}_m)x^2y^2+y^4\delta_{\ell m}\right\}dy
\end{aligned}
\tag{7.177}
$$

We have already performed in (7.43) the integral appearing in (7.177). Substitution yields

$$
{}'\langle\{0_{\mathbf{k}j}\} \mid \mathbf{D}_\perp^2(\mathbf{x}) \mid \{0_{\mathbf{k}j}\}\rangle' = \frac{1}{4\pi}\hbar c\alpha_{\ell m}(13\delta_{\ell m}+7\hat{x}_\ell\hat{x}_m)\frac{1}{x^7}
\tag{7.178}
$$

Expressions (7.175) can also be used to evaluate the quantum average of $\mathbf{H}^2(\mathbf{x})$, as given in second quantized form by (7.47). This leads to

$$
\begin{aligned}
&{}'\langle\{0_{\mathbf{k}j}\} \mid \mathbf{H}^2(\mathbf{x}) \mid \{0_{\mathbf{k}j}\}\rangle' = \frac{2\pi^2\hbar}{V^2}\chi_{\ell m}\sum_{\mathbf{k}\mathbf{k}'jj'}\mathbf{b}_{\mathbf{k}j}\cdot\mathbf{b}_{\mathbf{k}'j'} \\
&\times\left[\left(\mathbf{e}^*_{\mathbf{k}j}\right)_\ell\left(\mathbf{e}^*_{\mathbf{k}'j'}\right)_m+\left(\mathbf{e}^*_{\mathbf{k}'j'}\right)_\ell\left(\mathbf{e}^*_{\mathbf{k}j}\right)_m\right]\frac{\omega_k\omega_{k'}}{\omega_k+\omega_{k'}}e^{i(\mathbf{k}+\mathbf{k}')\cdot\mathbf{x}} \\
&+\text{c.c.}
\end{aligned}
\tag{7.179}
$$

From the sum rules (2.6) one has

$$\alpha_{\ell m}\sum_{jj'}\mathbf{b}_{\mathbf{k}j}\cdot\mathbf{b}_{\mathbf{k}'j'}\left[\left(\mathbf{e}^*_{\mathbf{k}j}\right)_\ell\left(\mathbf{e}^*_{\mathbf{k}'j'}\right)_m+\left(\mathbf{e}^*_{\mathbf{k}'j'}\right)_\ell\left(\mathbf{e}^*_{\mathbf{k}j}\right)_m\right]$$

$$\times\, e^{i(\mathbf{k}+\mathbf{k}')\cdot\mathbf{x}}$$

$$=\lim_{\mathbf{x}'\to\mathbf{x}}\left\{2\alpha_{\ell\ell}\hat{\mathbf{k}}\cdot\hat{\mathbf{k}}'e^{i\mathbf{k}\cdot\mathbf{x}+i\mathbf{k}'\cdot\mathbf{x}'}-\alpha_{\ell m}\left(\hat{k}_\ell\hat{k}'_m+\hat{k}_m\hat{k}'_\ell\right)e^{i\mathbf{k}\cdot\mathbf{x}+i\mathbf{k}'\cdot\mathbf{x}'}\right\}$$

$$=-\lim_{\mathbf{x}'\to\mathbf{x}}\left\{2\alpha_{\ell\ell}\nabla^x\cdot\nabla^{x'}-\alpha_{\ell m}\left(\nabla^x_\ell\nabla^{x'}_m+\nabla^x_m\nabla^{x'}_\ell\right)\right\}$$

$$\times\,\frac{1}{kk'}e^{i(\mathbf{k}\cdot\mathbf{x}+\mathbf{k}'\cdot\mathbf{x}')} \tag{7.180}$$

Substitution of this result into (7.179) and use of (7.41) gives

$${}'\langle\{0_{\mathbf{k}j}\}\mid\mathbf{H}^2(\mathbf{x})\mid\{0_{\mathbf{k}j}\}\rangle'$$

$$=-\frac{2\pi^2}{V^2}\hbar c\lim_{\mathbf{x}'\to\mathbf{x}}\left\{2\alpha_{\ell\ell}\nabla^x\cdot\nabla^{x'}-\alpha_{\ell m}\left(\nabla^x_\ell\nabla^{x'}_m+\nabla^x_m\nabla^{x'}_\ell\right)\right\}$$

$$\times\int_0^\infty\left(\sum_{\mathbf{k}}e^{-ky}e^{i\mathbf{k}\cdot\mathbf{x}}\right)\left(\sum_{\mathbf{k}'}e^{-k'y}e^{i\mathbf{k}'\cdot\mathbf{x}'}\right)dy+\text{c.c.} \tag{7.181}$$

Use of (7.60) to evaluate the sum over **k** appearing in (7.181) yields

$$\frac{1}{V^2}\left(\sum_{\mathbf{k}}e^{-ky}e^{i\mathbf{k}\cdot\mathbf{x}}\right)\left(\sum_{\mathbf{k}'}e^{-k'y}e^{i\mathbf{k}'\cdot\mathbf{x}'}\right)$$

$$=\frac{1}{\pi^4x^4x'^4}\frac{y^2}{(1+y^2/x^2)^2(1+y^2/x'^2)^2} \tag{7.182}$$

and with the help of integration tables (Gradshteyn and Ryzhik 1973)

$$\frac{1}{V^2}\int_0^\infty\left(\sum_{\mathbf{k}}e^{-ky}e^{i\mathbf{k}\cdot\mathbf{x}}\right)\left(\sum_{\mathbf{k}'}e^{-k'y}e^{i\mathbf{k}'\cdot\mathbf{x}'}\right)dy$$

$$=\frac{1}{2\pi^4xx'^4}B\left(\frac{3}{2},\frac{5}{2}\right)\,{}_2F_1(2,3/2;4;1-x^2/x'^2)$$

$$=\frac{1}{4\pi^3}\frac{1}{xx'(x+x')^3} \tag{7.183}$$

where B is the beta-function and ${}_2F_1$ is the hypergeometric function. Result (7.183) can be substituted in (7.182) which, after some straightforward algebra leads to (Persico and Power 1986)

$${}'\langle\{0_{\mathbf{k}j}\}\mid\mathbf{H}^2(\mathbf{x})\mid\{0_{\mathbf{k}j}\}\rangle'=-\frac{7}{4\pi}\hbar c\alpha_{\ell m}(\delta_{\ell m}-\hat{x}_\ell\hat{x}_m)\frac{1}{x^7} \tag{7.184}$$

The electric and magnetic energy densities around the source can be immediately obtained from (7.178) and (7.184) as

$$\langle \mathcal{H}_{el}(\mathbf{x}) \rangle = \frac{1}{32\pi^2} \hbar c \alpha_{\ell m} (13\delta_{\ell m} + 7\hat{x}_\ell \hat{x}_m) \frac{1}{x^7} ;$$

$$\langle \mathcal{H}_{mag}(\mathbf{x}) \rangle = -\frac{7}{32\pi^2} \hbar c \alpha_{\ell m} (\delta_{\ell m} - \hat{x}_\ell \hat{x}_m) \frac{1}{x^7} \tag{7.185}$$

These expressions are easily seen to be of the same form as their respective counterparts obtained for the virtual energy density of a two-level atom in the far zone, namely expressions (7.46) and (7.74), although here the (infinite) zero-point contributions are missing simply because we have discarded them as explained after (7.175). The reason why we fail to find a near-zone expression in the Craig-Power model is the approximation performed to obtain expression (4.117) on which (7.170) is based. In fact, this approximation consists of neglecting the contributions to the Hamiltonian which come from the high-energy virtual photons, whose energy is larger than the average spacing between atomic energy levels. These are the photons which are the main contributors to the virtual cloud in the near zone. While these photons were properly treated in our two-level atom calculations, they have been discarded in the Craig-Power model. Thus one cannot expect to find a near-zone expression in the latter model, which can consequently be legitimately used only to investigate the far zone energy density of a neutral atomic or molecular source. In other words, one may conclude that the Craig-Power Hamiltonian can be expected to treat in an appropriate way only the low-frequency photons, which indeed are the main contributors to the virtual field in the far zone. If one is not interested, however, in the details of the virtual cloud in the near zone of a given source, the technical advantages of the Craig-Power model are evident because the mathematical treatment is rather simple as compared, for example, with the treatment presented for the two-level atom in Sections 7.2 and 7.3.

Finally, we remark that for an isotropic electrically polarizable source with $\alpha_{\ell m} = \alpha\delta_{\ell m}$, the energy density expressions (7.185) take the form

$$\langle \mathcal{H}_{el}(\mathbf{x}) \rangle = \frac{23}{16\pi^2} \hbar c \alpha \frac{1}{x^7} ;$$

$$\langle \mathcal{H}_{mag}(\mathbf{x}) \rangle = -\frac{7}{16\pi^2} \hbar c \alpha \frac{1}{x^7} \tag{7.186}$$

Apart from the missing zero-point contributions of the unperturbed vacuum, these expressions coincide with their counterparts (7.168) and

(7.166) that we have obtained for the electric and magnetic energy density in the far zone of a hydrogen atom. This is also to be expected, since the ground state of a hydrogen atom is endowed with spherical symmetry and it could hardly be described by an anisotropic static polarizability tensor. Consequently, its virtual cloud must also reflect this spherical symmetry, as is confirmed by the similarity of the isotropic forms (7.186) to (7.168) and (7.166). It is also interesting to remark that for the isotropic case under discussion, the Craig-Power Hamiltonian (7.170) takes the particularly simple form

$$H = H_F^{mul} - \frac{1}{2}\alpha \mathbf{D}_\perp^2(0) \tag{7.187}$$

This is suggestive of the fact that an electrically isotropic source has an influence on the far zone field which can be calculated up to terms quadratic in e, by assuming that the coupling between source and field is proportional to the total electric energy density of the field at the location of the source.

7.8 Van der Waals forces and virtual energy density

In this and in the previous chapter we have seen that the interaction of a source with a field leads in general to the creation of a cloud of virtual field quanta which dress the source also in the ground state of the source-field system. These virtual quanta can be visualized as being continuously emitted and reabsorbed by the source. In this scheme one immediately envisages the possibility that, in the presence of more than one source coupled to the field, a virtual quantum emitted by one of the sources is not reabsorbed by the same source but by another, if the second source is not too far away. We may ask what are the physical effects to which such a process can give rise. We shall show here that such an exchange of virtual quanta gives rise to forces between the various sources. To fix the ideas, we shall consider only two sources at distance R large enough that no overlap between the sources exists, each source being coupled to the field at two different points in space.

In order to emphasize that the phenomenon under study is very general and in no way limited to the case of QED, we shall consider first two static meson sources at points 0 and $\mathbf{R}$, as in Figure 7.2. Thus the Hamiltonian of the system can be taken to be the same as (6.10), the information about the presence of two sources, rather than of only one, being contained in the form of the source density $\rho(\mathbf{x})$. Indeed the latter is here assumed to have two similarly shaped, sharp and non-overlapping peaks of the form

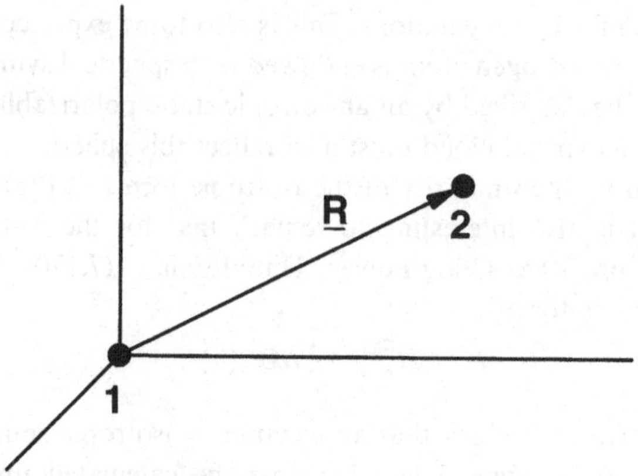

Fig. 7.2 Two sources at distance R.

$\rho_1(\mathbf{x})$ and $\rho_2(\mathbf{x})$ centred respectively at points 0 and $\mathbf{R}$. Hence one can reliably approximate

$$\rho_{\mathbf{k}} = \rho_{1\mathbf{k}}\left(1 + e^{-i\mathbf{k}\cdot\mathbf{R}}\right) \tag{7.188}$$

where $\rho_{1\mathbf{k}}$ is the Fourier transform of $\rho_1(\mathbf{x})$. We have already seen that the ground-state shift due to the source-meson coupling depends on $\rho_{\mathbf{k}}$ as in (6.68). Thus substitution of (7.188) in the latter expression yields the shift

$$\Delta = -g^2 \frac{1}{(2\pi)^3} \int \frac{1}{\omega_k^2} \rho_{1\mathbf{k}}^2 (1 + \cos \mathbf{k}\cdot\mathbf{R}) d^3\mathbf{k} \tag{7.189}$$

If the two peaks are sharp enough one can approximate $\rho_{1\mathbf{k}} \sim 1$ and Δ can be put in the form

$$\Delta = 2\Delta_0 - g^2 \frac{1}{(2\pi)^3} \int \frac{\cos \mathbf{k}\cdot\mathbf{R}}{\omega_k^2} d^3\mathbf{k} = 2\Delta_0 - \frac{1}{4\pi c^2} g^2 \frac{e^{-\kappa R}}{R} \tag{7.190}$$

where the integration has been performed using (7.12). The contribution $2\Delta_0$ to the total ground-state shift (7.190) can be understood as being the shift caused by each of the two sources as if the other were absent. This can be interpreted as stemming from processes in which each source emits one meson and successively reabsorbs it, and it can be pictorically represented by Feynman graphs a) and b) in Figure 7.3. The third contribution is also proportional to g^2, since it must originate from a one-meson process like the first two. Its R-dependence, however, indicates that this meson is not reabsorbed by the same source which emitted it, but by the other at distance R. Consequently, it must be pictorically represented

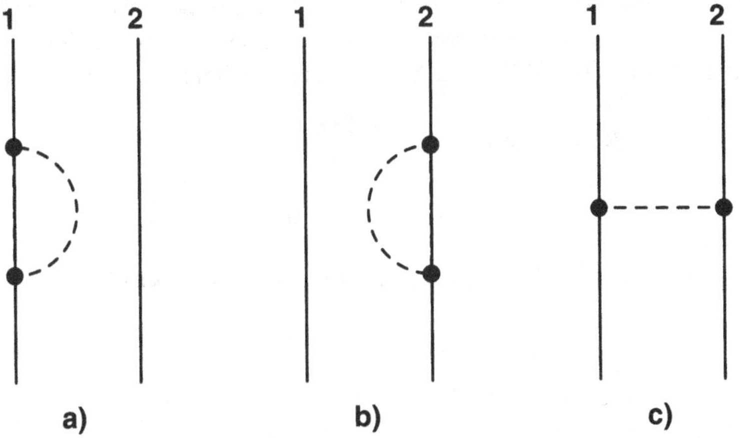

Fig. 7.3 Diagrams a) and b) for two static meson sources yield energy shifts which are independent of the intersource distance R. Diagram c) yields a contribution to the energy of the system which is R-dependent and which gives rise to the Yukawa potential. Broken lines represent virtual mesons which in c) are exchanged between the two sources.

by graph c) in (7.3). Moreover its R-dependence indicates the presence of a force $\partial\Delta/\partial R$ acting between the two sources, which in the case under discussion is attractive. This is called the Yukawa force (see e.g. Henley and Thirring 1962), and it can be thought of as originating from an attractive potential, the Yukawa potential,

$$V(\mathbf{R}) = -\frac{1}{4\pi c}g^2\frac{e^{-\kappa R}}{R} \tag{7.191}$$

It is the kind of potential giving rise to the strong force acting between nucleons and is due to virtual meson exchange. We remark that the exponential behaviour of $V(\mathbf{R})$ makes it a short-range potential, the range being measured by κ^{-1} which is the Compton wavelength of the meson.

We now consider two polarons in a polar semiconductor at $T = 0\text{K}$. The situation here is more complicated than in the previous example because the electron, contrary to the usual hadronic sources of strong force, cannot really be considered infinitely massive because of recoil in the processes of absorption and emission of phonons, and this is indeed reflected in the momentum-conserving structure of the polaron Hamiltonian (6.23). One can, however, derive a phenomenological electron-phonon Hamiltonian for infinitely massive electrons in the following way (Kittel 1963). In a polar semiconductor the electric potential generated at point $\mathbf{x}$ in the crystal by the optical phonons is given by (7.16). This potential interacts with two infinitely massive charges $-e$ placed at points 0 and $\mathbf{R}$ in such a way that the interaction energy is

$$V(\mathbf{R}) = -e\phi(0,\mathbf{R}) = -ie\left(\frac{2\pi\hbar\omega}{\bar{\epsilon}V}\right)^{1/2}$$

$$\times\sum_{\mathbf{k}}\frac{1}{k}\left\{a_{\mathbf{k}}^{\dagger}\left(1+e^{-i\mathbf{k}\cdot\mathbf{R}}\right)-a_{\mathbf{k}}\left(1+e^{i\mathbf{k}\cdot\mathbf{R}}\right)\right\}$$

$$=\sum_{\mathbf{k}}\left\{V_k^{*}a_{\mathbf{k}}^{\dagger}\left(1+e^{-i\mathbf{k}\cdot\mathbf{R}}\right)+V_k a_{\mathbf{k}}\left(1+e^{i\mathbf{k}\cdot\mathbf{R}}\right)\right\};$$

$$V_k = ie\left(\frac{2\pi\hbar\omega}{\bar{\epsilon}V}\right)^{1/2}\frac{1}{k} \qquad (7.192)$$

where V_k is the same as the electron-phonon coupling constant defined in (6.23). This can be added to the unperturbed optical phonon Hamiltonian to give the phenomenological total Hamiltonian

$$H = \hbar\omega\sum_{\mathbf{k}}a_{\mathbf{k}}^{\dagger}a_{\mathbf{k}}+\sum_{\mathbf{k}}\left\{V_k^{*}a_{\mathbf{k}}^{\dagger}\left(1+e^{-i\mathbf{k}\cdot\mathbf{R}}\right)+V_k a_{\mathbf{k}}\left(1+e^{i\mathbf{k}\cdot\mathbf{R}}\right)\right\} \qquad (7.193)$$

useful for our purposes (Compagno *et al.* 1990). The perturbed ground state energy of (7.193) is obtained by second-order perturbation theory as

$$\Delta = 2\Delta_0 - \frac{2}{\hbar\omega}\sum_{\mathbf{k}}\mid V_k\mid^2 \cos\mathbf{k}\cdot\mathbf{R}\ ;\ \Delta_0 = -\frac{1}{\hbar\omega}\sum_{\mathbf{k}}\mid V_k\mid^2 \qquad (7.194)$$

In the expression (7.194) for Δ, Δ_0 represents the energy shift which would be generated by each of the two charges in the absence of the other. This can be attributed to virtual processes corresponding to Feynman graphs a) and b) in Figure 7.4. Consequently

$$V(\mathbf{R}) = -\frac{2}{\hbar\omega}\sum_{\mathbf{k}}\mid V_k\mid^2\cos\mathbf{k}\cdot\mathbf{R} = -\frac{1}{\bar{\epsilon}}\frac{e^2}{R} \qquad (7.195)$$

must be the contribution of process c) in Figure 7.4 in which a virtual phonon generated by one electron is absorbed by the other, giving rise to the screening potential represented by (7.195). Contrarily to the Yukawa potential (7.191), this screening potential is not short-range. In fact, it is of the Coulomb form, and thus it leads to an effective reduction of the electronic charge. Account of the dynamical features of the problem, which have been described in the oversimplified treatment given here, leads to many remarkable effects, including BCS superconductivity, whose discussion falls outside of the scope of this book.

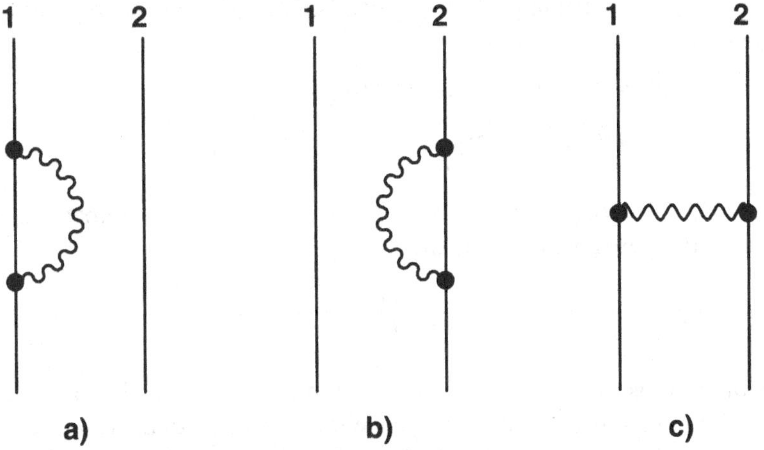

Fig. 7.4 Diagrams a) and b) for two infinitely massive electrons in a polar semiconductor yield energy shifts which are independent of the interelectron distance **R**. Diagram c) yields a contribution to the energy of the system which is R-dependent and which gives rise to a screening potential. Wiggly lines represent virtual phonons which in case c) are exchanged between the electrons.

Returning to the main theme of atom-photon interaction, we consider two neutral atoms coupled to the vacuum electromagnetic field. We describe these two atoms by their ground-state static electric polarizabilities α_S and α_T. The reason for the use of indices S and T instead of 1 and 2 respectively will become clear later on. We adopt the Craig-Power Hamiltonian with atom S at the origin of our reference frame and atom T at point $\mathbf{R}$ in space. Assuming electrically isotropic sources, this Hamiltonian is given by (7.187) and so for the present problem we have

$$\begin{aligned}
H &= H_F^{mul} + H(\mathbf{R}) \; ; \\
H_F^{mul} &= \sum_{\mathbf{k}j} \hbar\omega_k \left(a^\dagger_{\mathbf{k}j} a_{\mathbf{k}j} + 1/2\right) \; ; \\
H(\mathbf{R}) &= -\frac{1}{2}\alpha_S D_\perp^2(0) - \frac{1}{2}\alpha_T D_\perp^2(\mathbf{R}) \\
&= \frac{\pi\hbar}{V}\alpha_S \sum_{\mathbf{k}\mathbf{k}'jj'} \sqrt{\omega_k\omega_{k'}} \Big\{ \mathbf{e}_{\mathbf{k}j}\cdot\mathbf{e}_{\mathbf{k}'j'} a_{\mathbf{k}j} a_{\mathbf{k}'j'} \left[1 + \beta e^{i(\mathbf{k}+\mathbf{k}')\cdot\mathbf{R}}\right] \\
&- \mathbf{e}_{\mathbf{k}j}\cdot\mathbf{e}^*_{\mathbf{k}'j'} a_{\mathbf{k}j} a^\dagger_{\mathbf{k}'j'} \left[1 + \beta e^{i(\mathbf{k}-\mathbf{k}')\cdot\mathbf{R}}\right] \\
&- \mathbf{e}^*_{\mathbf{k}j}\cdot\mathbf{e}_{\mathbf{k}'j'} a^\dagger_{\mathbf{k}j} a_{\mathbf{k}'j'} \left[1 + \beta e^{-i(\mathbf{k}-\mathbf{k}')\cdot\mathbf{R}}\right] \\
&+ \mathbf{e}^*_{\mathbf{k}j}\cdot\mathbf{e}^*_{\mathbf{k}'j'} a^\dagger_{\mathbf{k}j} a^\dagger_{\mathbf{k}'j'} \left[1 + \beta e^{-i(\mathbf{k}+\mathbf{k}')\cdot\mathbf{R}}\right] \Big\} \qquad (7.196)
\end{aligned}$$

where $\beta = \alpha_T/\alpha_S$. From (C.10) the perturbed ground-state energy can be obtained at order e^4 as

$$E^{(2)} = E_0 + \langle\{0_{\mathbf{k}j}\} \mid H(\mathbf{R}) \mid \{0_{\mathbf{k}j}\}\rangle + \langle\{0_{\mathbf{k}j}\} \mid H(\mathbf{R}) \frac{1}{E_0 - H_0}(1 - P_0) H(\mathbf{R}) \mid \{0_{\mathbf{k}j}\}\rangle \quad (7.197)$$

where $E_0 = \sum_{\mathbf{k}j} \hbar\omega_k/2$ is the zero-point energy of the unperturbed vacuum. It is immediate to obtain

$$\langle\{0_{\mathbf{k}j}\} \mid H(\mathbf{R}) \mid \{0_{\mathbf{k}j}\}\rangle = -\frac{2\pi}{V}(\alpha_S + \alpha_T) \sum_{\mathbf{k}} \hbar\omega_k \quad (7.198)$$

This quantity is of order e^2 and infinite, but it cannot contribute to any interatomic force because it is independent of the interatomic distance **R**. As for the third term on the RHS of (7.197), we find after a simple calculation

$$\langle\{0_{\mathbf{k}j}\} \mid H(\mathbf{R}) \frac{1}{E_0 - H_0}(1 - P_0) H(\mathbf{R}) \mid \{0_{\mathbf{k}j}\}\rangle = -\frac{2\pi^2}{V^2} \hbar c \alpha_S^2 \sum_{\mathbf{k}\mathbf{k}'jj'} \left(\mathbf{e}_{\mathbf{k}j} \cdot \mathbf{e}_{\mathbf{k}'j'}\right)\left(\mathbf{e}^*_{\mathbf{k}j} \cdot \mathbf{e}^*_{\mathbf{k}'j'}\right) \times \left[1 + \beta^2 + 2\beta \cos(\mathbf{k} + \mathbf{k}') \cdot \mathbf{R}\right] \frac{kk'}{k + k'} \quad (7.199)$$

The only quantity of interest of (7.199) in the present context is the **R**-dependent part

$$V(\mathbf{R}) = -\frac{4\pi^2}{V^2} \hbar c \alpha_S \alpha_T \sum_{\mathbf{k}\mathbf{k}'jj'} \left(\mathbf{e}_{\mathbf{k}j} \cdot \mathbf{e}_{\mathbf{k}'j'}\right)\left(\mathbf{e}^*_{\mathbf{k}j} \cdot \mathbf{e}^*_{\mathbf{k}'j'}\right) \frac{kk'}{k + k'} \times \cos(\mathbf{k} + \mathbf{k}') \cdot \mathbf{R} \quad (7.200)$$

Given the two-photon nature of $H(\mathbf{R})$ clearly displayed by (7.196), a little thought will show that $V(\mathbf{R})$ originates from processes in which virtual photon pairs are exchanged between the atoms and which can be pictorically represented as in Figure 7.5. With the help of (2.6) one can evaluate the polarization sums

$$\sum_{jj'} \left(\mathbf{e}_{\mathbf{k}j} \cdot \mathbf{e}_{\mathbf{k}'j'}\right)\left(\mathbf{e}^*_{\mathbf{k}j} \cdot \mathbf{e}^*_{\mathbf{k}'j'}\right) = \left(\delta_{mn} - \hat{k}_m \hat{k}_n\right)\left(\delta_{mn} - \hat{k}'_m \hat{k}'_n\right) \quad (7.201)$$

Moreover, use of (7.32) and of (7.39) yields

$$V(\mathbf{R}) = -\frac{1}{\pi^2} \hbar c \alpha_S \alpha_T \int_0^\infty \left\{ D^R_{mn} \int_0^\infty e^{-ky} \sin kR dk \right\}^2 dy \quad (7.202)$$

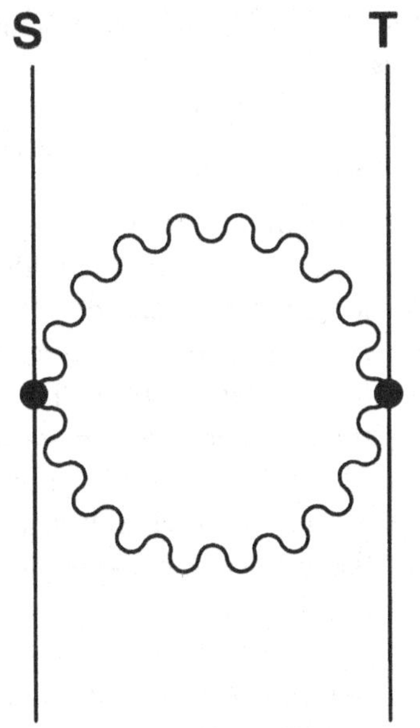

Fig. 7.5 Diagram representing the energy shift of a system of two neutral atoms S and T coupled to the electromagnetic field in a Craig-Power fashion. Virtual photons, corresponding to wiggly lines, are exchanged in pairs. The energy shift is R-dependent and gives rise to the long distance van der Waals potential.

From (7.41) we know that

$$D^R_{mn} \int_0^\infty e^{-ky} \sin kR dk = \frac{4}{(R^2+y^2)^3} \left[\left(\delta_{mn} - 2\hat{R}_m \hat{R}_n \right) R^2 - \delta_{mn} y^2 \right] \tag{7.203}$$

and since

$$\left(\delta_{mn} - 2\hat{R}_m \hat{R}_n \right)^2 = 3 \; ; \; \left(\delta_{mn} - 2\hat{R}_m \hat{R}_n \right) \delta_{mn} = 1 \; ; \; \delta^2_{mn} = 3 \tag{7.204}$$

one finds

$$\left\{ D^R_{mn} \int_0^\infty e^{-ky} \sin kR dk \right\}^2 = \frac{16}{(R^2+y^2)^6} \left(3R^4 - 2R^2 y^2 + 3y^4 \right)$$
$$= \frac{48}{(R^2+y^2)^4} - \frac{128 R^2 y^2}{(R^2+y^2)^6} \tag{7.205}$$

Substituting this quantity in (7.202) and performing elementary integrations, one obtains the result

$$V(\mathbf{R}) = -\frac{23}{4\pi}\hbar c \alpha_S \alpha_T \frac{1}{R^7} = -4\pi\alpha_T \langle \mathcal{H}_{el}^S(\mathbf{R})\rangle \tag{7.206}$$

where $\langle \mathcal{H}_{el}^S(\mathbf{R})\rangle$ is the ground-state quantum average of the virtual electric energy density generated by atom S at point $\mathbf{R}$ in the absence of atom T. $V(\mathbf{R})$ in (7.206) is the long-distance van der Waals potential acting between two electrically polarizable sources interacting with the vacuum field as described by Hamiltonian (7.196). This coincides with the Casimir-Polder potential at large intersource separation (Power 1964). We remark that on the basis of the discussion of the previous section on the limits of validity of the Craig-Power Hamiltonian, one should expect (7.206) to be valid only if the distance R between the sources is larger than the typical natural wavelength of each of the two sources. This is so because, as we have seen, the Craig-Power Hamiltonian can be expected to treat correctly only the long-wavelength virtual photons, which prevail in the virtual photon exchanges leading to the van der Waals potential (7.206). Thus one should expect to obtain a more general $V(\mathbf{R})$, valid also when the sources are rather close to each other, and which reduces to (7.206) in the limit of large R, if one starts from the multipolar Hamiltonian for two neutral sources derived from (4.109). This is indeed the case as shown by the very lengthy calculations described in detail in the book by Craig and Thirunamachandran (1984). The result of this approach is the general van der Waals potential between two ground-state isotropic electrically polarizable sources

$$\begin{aligned} V(\mathbf{R}) &= -\frac{4}{9\pi\hbar c R^2}\sum_{t,s} |\mu_{t0}^T|^2 |\mu_{s0}^S|^2 \int_0^\infty \frac{k_t k_s u^4 e^{-2uR}}{(k_t^2+u^2)(k_s^2+u^2)} \\ &\quad \times \left\{1+\frac{2}{uR}+\frac{5}{u^2R^2}+\frac{6}{u^3R^3}+\frac{3}{u^4R^4}\right\} du\,, \\ k_{t,s} &= \frac{\omega_{t,s}}{c} \end{aligned} \tag{7.207}$$

In this expression μ_{t0}^T (μ_{s0}^S) is the electric dipole matrix element between the ground state of source T (S) and its t (s) excited state. Moreover, the sum over t (s) runs over all excited states of source T (S), and ω_t (ω_s) is the frequency of the $t \leftrightarrow 0$ ($s \leftrightarrow 0$) transition of source T (S). Introducing average values $\bar{k}_t$ and $\bar{k}_s$ for each of the sources and taking $R \gg \bar{k}_t^{-1}, \bar{k}_s^{-1}$ in such a way that each source is in the far zone of the other, the exponential within the integral in (7.207) will cut all contributions for

$u > k_t, k_s$. In this condition one can assume $u \ll k_t, k_s$ in the integrand of (7.207) and perform the integration analytically. This yields (7.206) as expected, using the normal definition of the isotropic static electric polarizabilities (Craig and Thirunamachandran 1984)

$$\alpha_T = \frac{2}{3}\sum_t \frac{1}{\hbar\omega_t} \mid \mu_{t0}^T \mid^2 ; \; \alpha_S = \frac{2}{3}\sum_s \frac{1}{\hbar\omega_s} \mid \mu_{s0}^S \mid^2 \tag{7.208}$$

In the appropriate limit $R \ll \bar{k}_t^{-1}, \bar{k}_s^{-1}$, when each source is in the near zone of the other, the largest contribution to the integral in (7.207) will come from the range of u such that $uR < 1$. Thus one can approximate the integrand by putting the exponential equal to 1 and by keeping only the term $3/u^4R^4$ which dominates the others. In this limit

$$\begin{aligned} V(\mathbf{R}) &= -\frac{4}{3\pi\hbar c R^6}\sum_{t,s} \mid \mu_{t0}^T \mid^2 \mid \mu_{s0}^S \mid^2 \int_0^\infty \frac{k_t k_s}{(k_t^2+u^2)(k_s^2+u^2)} du \\ &= -\frac{2}{3}\hbar c \sum_{t,s} \frac{1}{\hbar(\omega_t+\omega_s)} \mid \mu_{t0}^T \mid^2 \mid \mu_{s0}^S \mid^2 \frac{1}{R^6} \end{aligned} \tag{7.209}$$

which shows that the near-zone van der Waals potential is proportional to R^{-6}, differently from the far-zone potential which is proportional to R^{-7}.

A more interesting situation arises when one of the sources, say S, is in the far zone of T but T is in the near zone of S (Compagno *et al.* 1987). Thus one has $\bar{k}_s^{-1} > R > \bar{k}_t^{-1}$ and one can guess that in (7.207) the presence of the exponential will substantially decrease only the contributions coming from the range $u > k_t$ and not those coming from the range $u > k_s$. Consequently, one is entitled to approximate

$$\begin{aligned} V(\mathbf{R}) = &-\frac{2}{3\pi R^2}\alpha_T \sum_s \mid \mu_{s0}^S \mid^2 k_s \int_0^\infty \frac{u^4 e^{-2uR}}{k_s^2+u^2} \\ &\times \left\{1+\frac{2}{uR}+\frac{5}{u^2R^2}+\frac{6}{u^3R^3}+\frac{3}{u^4R^4}\right\} du \end{aligned} \tag{7.210}$$

Moreover, the largest contribution to this integral will presumably come from the region $uR < 1$, where the exponential is ~ 1. Thus $V(\mathbf{R})$ can be further approximated as

$$\begin{aligned} V(\mathbf{R}) &= -\frac{2}{\pi R^6}\alpha_T \sum_s \mid \mu_{s0}^S \mid^2 k_s \int_0^\infty \frac{1}{k_s^2+u^2} du \\ &= -\alpha_T \sum_s \mid \mu_{s0}^S \mid^2 \frac{1}{R^6} \end{aligned} \tag{7.211}$$

Using completeness of the set of states $|\psi_s^S\rangle$ of source S

$$\sum_s |\mu_{s0}^S|^2 = \sum_s \langle\phi_0^S|\mu^S|\phi_s^S\rangle \cdot \langle\phi_s^S|\mu^S|\phi_0^S\rangle$$
$$= \langle\phi_0^S|(\mu^S)^2|\phi_0^S\rangle \qquad (7.212)$$

one can write

$$\frac{1}{R^6}\sum_s |\mu_{s0}^S|^2 = \frac{1}{R^6}\langle\phi_0^S|(\mu^S)^2|\phi_0^S\rangle \qquad (7.213)$$

Comparison with (7.156) obtained for the hydrogen atom shows that (7.213) is simply proportional to the electric energy density in the near zone of source S. Thus (7.211) can be written in the form

$$V(\mathbf{R}) = -4\pi\alpha_T\langle\mathcal{H}_{el}^S(\mathbf{R})\rangle \qquad (7.214)$$

also when T is in the near zone of S. Expression (7.206) and (7.214) show that, provided one chooses a test source T with $\hat{k}_t$ large enough, the van der Waals potential felt by this test source is proportional to the energy density created by the source S both in the far and in the near zone of S. This explains the reason for labelling the two sources as S and T. Thus the van der Waals force provides a means of measuring directly the electric energy density of a source both in the near and in the far region. It should be mentioned that recently the rate of spontaneous emission of a photon by an excited atom has been suggested to be determined by the local electric energy density (Khosravi and Loudon 1991, Barnett *et al.* 1992). It should also be mentioned that more general expressions of the van der Waals force, including the effect of magnetic ground-state polarizability, have been obtained by Feinberg and Sucher (1970). Thus in the appropriate limits the same procedure of measuring the van der Waals force provides a way of measuring the ground-state magnetic energy-density of a magnetically polarizable source.

The question of the actual experimental detection of van der Waals forces deserves some further comments. Direct measurement of these forces between two isolated neutral atoms is very difficult to perform, but the transition from a near-zone R^{-6} regime for the potential to a R^{-7} far-zone regime was found necessary by Overbeek *et al.*, in order to explain the stability properties of some colloidal solutions. These papers are cited by Casimir and Polder (1948) and by Milonni and Shih (1992). Also scarce are experiments on the van der Waals interaction between a neutral atom and a macroscopic body, which we can here rather optimistically regard as an additive aggregate of many neutral atoms, although we would expect that many-body forces might play an important role. A summary of recent experimental results can be found in Sandoghdar *et al.* (1992) who measured the near-zone force exerted by cavity walls on Rydberg Na atoms. The far-zone force exerted by cavity walls on ground-state Na atoms has been measured by Sukenik *et al.* (1993). Another recent summary of experimental results on van der Waals forces between macroscopic bodies can be found in a paper by Elizalde and Romeo (1991). A general review of problems connected with van der Waals forces has been given by Barash and Ginzburg (1984).

References

M. Abramowitz, I.A. Stegun (eds.) (1965). *Handbook of Mathematical Functions* (Dover Publications Inc., New York)

Yu.S. Barash, V.L. Ginzburg (1984). *Usp. Fiz. Nauk.* **143,** 345 (*Sov. Phys. Usp.* **27,** 467 (1984))

S.M. Barnett, B. Huttner, R. Loudon (1992). *Phys. Rev. Lett.* **68,** 3698

G. Barton (1989). *Elements of Green's Functions and Propagation* (Oxford University Press, Oxford)

R. Becker (1982). *Electromagnetic Fields and Interactions* vol. 1 (Dover Publications Inc., New York)

N.D. Birrell, P.C.W. Davies (1992). *Quantum Fields in Curved Space* (Cambridge University Press, Cambridge)

H.B.G. Casimir, D. Polder (1948). *Phys. Rev.* **73,** 360

G. Compagno, G.M. Palma, R. Passante, F. Persico (1990). In *New Frontiers in Quantum Electrodynamics and Quantum Optics,* A.O. Barut (ed.) (Plenum Press, New York), p. 129

G. Compagno, R. Passante, F. Persico (1987). *Phys. Lett. A* **121,** 19

G. Compagno, G. Salamone (1991). *Phys. Rev. A* **44,** 5390

D.P. Craig, E.A. Power (1969). *Int. J. Quantum Chem.* **3,** 903

D.P. Craig, T. Thirunamachandran (1984). *Molecular Quantum Electrodynamics* (Academic Press Inc., London)

E. Elizalde, A. Romeo (1991). *Am. J. Phys.* **59,** 711

A. Erdelyi, W. Magnus, F. Oberhettinger, F.G. Tricomi (1954). *Tables of Integral Transforms,* vol. I (McGraw-Hill Book Company Inc., New York)

G. Feinberg, J. Sucher (1970). *Phys. Rev. A* **2,** 2395

H. Fröhlich (1963). In *Polarons and Excitons,* C.G. Kuper and G.D. Whitfield (eds.) (Oliver and Boyd, Edinburgh), p. 1

I.S. Gradshteyn, I.M. Ryzhik (1973). *Tables of Integrals, Series and Products* (Academic Press Inc., New York)

E.M. Henley, W. Thirring (1962). *Elementary Quantum Field Theory* (McGraw-Hill Book Co., New York)

H. Khosravi, R. Loudon (1991). *Proc. R. Soc. London A* **433,** 337

C. Kittel (1963). *Quantum Theory of Solids* (John Wiley and Sons Inc., New York)

L.D. Landau, E.M. Lifshitz (1958). *Quantum Mechanics* (Pergamon Press, London)
T.D. Lee, F. Low, D. Pines (1953). *Phys. Rev.* **90,** 297
L. Mandel (1966). *Phys. Rev.* **144,** 1071
E. Merzbacher (1961). *Quantum Mechanics* (John Wiley and Sons, New York)
P.W. Milonni, M.L. Shih (1992). *Phys. Rev. A* **45,** 4241
R. Passante, G. Compagno, F. Persico (1985). *Phys. Rev. A* **31,** 2827
R. Passante, E.A. Power (1987). *Phys. Rev. A* **35,** 188
F.M. Peeters, J.T. Devreese (1985). *Phys. Rev. B* **31,** 1985
F. Persico, E.A. Power (1986). *Phys. Lett. A* **114,** 309
E.A. Power (1964). *Introductory Quantum Electrodynamics* (Longmans, Green and Co. Ltd., London)
E.A. Power, T. Thirunamachandran (1983). *Phys. Rev. A* **28,** 2663
V. Sandoghdar, C.I. Sukenik, E.A. Hinds, S. Haroche (1992). *Phys. Rev. Lett.* **68,** 3432
S.S. Schweber (1964). *An Introduction to Relativistic Quantum Field Theory* (John Weatherhill, Tokyo)
C.I. Sukenik, M.G. Boshier, D. Cho, V. Sandoghdar, E.A. Hinds (1993). *Phys. Rev. Lett.* **70,** 560
V.F. Weisskopf (1939). *Phys. Rev.* **56,** 72

Further reading

A thorough discussion of the photon density operator is given by
R.J. Cook, *Phys. Rev. A* **25,** 2164 (1982) and *Phys. Rev. A* **26,** 2754 (1982).

The following paper discusses van der Waals forces in connection with interference of virtual photons
G. Compagno, F. Persico, R. Passante, *Phys. Lett. A* **112,** 215 (1985).

Elementary discussions of Casimir potentials are given by
L. Spruch, *Physics Today,* November 1986, p. 37.
P.W. Milonni, M.L. Shih, in *Contemporary Physics* (1993).

A clear account of BCS superconductivity in connection with electron-phonon interaction can be found in
H. Haken, *Quantum Field Theory of Solids* (North-Holland Publishing Company, Amsterdam 1988).

The connection between the Casimir effect and van der Waals forces has been examined in
G. Plunien, B. Müller, W. Greiner, *Phys. Rep.* **134,** 87 (1986).

The general theory of dispersion forces is discussed in
G. Feinberg, J. Sucher, C.K. Au, *Phys. Rep.* **180,** 83 (1989).

An example of the use of the concept of virtual cloud in QCD can be found in
W. Feilmair, M. Faber, H. Markum, *Phys. Rev. D* **39,** 1409 (1989).

The role of the energy density in determining intermolecular potentials has been illustrated by
E.A. Power, T. Thirunamachandran, *Chem. Phys.* **171,1** (1993).

A theory of Casimir effects in dissipative media has been recently presented by
D. Kupiszewska, *Phys. Rev. A* **46,** 2286 (1992).

8

Further considerations on the nature of dressed states

Introduction. In this final chapter we review two topics of the literature concerning dressed atoms which are of conceptual relevance and capable of shedding some light on the physical meaning and significance of atomic dressing. The first section is devoted to some recent work in connection with the quantum theory of measurement. We include the measuring apparatus in the Hamiltonian along with an appropriate apparatus-atom coupling. We argue that the theory of measurement of finite duration provides us with a tool for detecting the spectral composition of the virtual cloud surrounding an atom. In fact, we show that in a measurement of duration τ on a two-level atom fully dressed by the vacuum fluctuations, as discussed in Chapters 6 and 7, the apparatus perceives the atom as dressed only by photons of frequency larger than τ^{-1}. In the case of a two-level atom dressed by a single-mode field populated by real photons, discussed in Chapter 5, we show that if τ is smaller than the inverse Rabi frequency $\hbar\Delta^{-1}$, the atom is perceived by the apparatus as bare; on the contrary, if τ is larger than $\hbar\Delta^{-1}$, the atom is perceived as dressed. The time scales for the two cases of dressing, by vacuum fluctuations or by a real single-mode field, are very different, but the similarity of the effects indicates a common physical aspect of the two kinds of dressing. The second section of the chapter is dedicated to the vexed question of the interpretation of radiative effects in terms of vacuum fluctuations or of radiation reaction field. We discuss in turn spontaneous emission, the Lamb shift and van der Waals forces. For each of these phenomena we give both the vacuum fluctuation and the radiation reaction intepretation, and we argue in favour of the legitimacy of both interpretations in terms of complementarity. First we consider spontaneous emission, and we introduce partitioning of the field into a vacuum fluctuation and a radiation reaction operator. We show that different ordering of synchronous commuting operators leads to interpreting spontaneous emission as due to to the effect of vacuum fluctuations or of radiation reaction in different proportions. Then we consider Welton's and Power's interpretations of the Lamb shift, in which the existence of vacuum fluctuations plays a central role, and we compare these with Milonni's interpretation in terms of radiation reaction for which the zero-point field seems superfluous. Finally, we describe in a semiquantitative way Spruch and Kelsey's interpretation of van der Waals forces in terms of vacuum fluctuations and, after mentioning briefly Casimir's, Boyer's and Power's interpretations, which are also based on the existence of the zero-point field, we discuss Milonni's interpretation of these forces in terms of the radiation reaction field. The three physical examples considered lead to the conclusion that both interpretations, complementary to a large extent like "two sides of a coin" are useful and even necessary for a full understanding of the physics of radiative effects in QED. We conclude the section by discussing the nature of the cloud of virtual photons from the point of view of radiation reaction and from that of vacuum fluctuations.

8.1 Dressed atoms and the quantum theory of measurement

This section is divided into two parts. In the first we apply the concepts developed in the quantum theory of measurement of finite duration (Peres and Wootters 1985, Peres 1989) to a fully dressed ground-state neutral atom. In the second part the same theory is applied to investigate the nature of an atom dressed by a field of real photons of the kind discussed in Chapter 5.

Concentrating on the first of the two situations, we recall that in Section 6.2 we have attributed the presence of the virtual cloud around a neutral atom to energy nonconservation. More precisely, we have argued that the energy of the ground-state bare atom+field system is not conserved during a quantum fluctuation, due to the coupling of the unexcited atom with the vacuum electromagnetic field, since bare excited states of energy δE are attained which contain one or more photons. These virtual photons contribute to the ground-state cloud, and we have visualized them as having a finite lifetime τ due to the Heisenberg uncertainty principle $\delta E \tau \sim \hbar$. Although we have shown that this apparently naive time-dependent representation of the ground-state dressed atom yields the main qualitative features of the virtual cloud, one may well doubt the soundness of a dynamic picture which refers to bare (hence normally unobservable) quantities. On the other hand, the dressed atom-field ground state is an eigenstate of the total radiation Hamiltonian, and as such endowed with static rather than with dynamic properties. Thus a certain degree of scepticism might seem justified, with respect to the qualitative picture of the virtual cloud proposed in Section 6.2. In what follows, however, we shall discuss a gedanken experiment which shows that it is possible, at least in principle, to detect directly the quantum fluctuations contributing to the virtual cloud surrounding a ground-state two-level atom. This lends support to the qualitative picture of Section 6.2 which is ultimately based on the existence of these fluctuations. Indeed detecting these fluctuations amounts to detecting the virtual cloud in equilibrium conditions, which seems more direct than inferring its existence from that of the van der Waals forces or of the Lamb shift. Moreover we shall discuss the role of the duration τ of these fluctuations and how τ influences the outcome of the gedanken experiment, which is a paradigm of a finite-duration measurement process.

To this aim we have to introduce briefly the main ideas of the theory of measurements of finite duration in the version given by Peres and co-workers. According to this theory, before the measurement process starts the system to be measured is in state $|\varphi\rangle$, the measuring apparatus (or 'pointer') is in state $|\psi\rangle$, the state of the total system being the uncorrelated state $|\Psi\rangle = |\psi\rangle\,|\varphi\rangle$. States $|\psi\rangle$ and $|\varphi\rangle$ are defined in two different Hilbert spaces pertaining to two different Hamiltonians H_M and H_R respectively. The eigenstates of H_M are denoted by $|\psi_k\rangle$ and those of H_R by $|\varphi_k\rangle$; it is assumed that one can estabilish a one to one correspondence between the $|\psi_k\rangle$ and the $|\varphi_k\rangle$.

The measurement process starts at $t = 0$ and it is described in terms of an appropriate time-dependent Hamiltonian $H_{MR}(t)$ containing both object and apparatus variables. Thus in this theory the measurement is a quantum-mechanical process describable in terms of a unitary operator $U(t)$ such that during the period of the measurement the total state $|\Psi(t)\rangle$ is given by

$$|\Psi(t)\rangle = U(t)\,|\Psi(0)\rangle \tag{8.1}$$

The interaction Hamiltonian H_{MR} must be such as to yield eventually correlations between the object being measured and the apparatus, in such a way that

$$|\Psi(t)\rangle = \sum_k c_k\,|\psi_k\rangle\,|\varphi_k\rangle \tag{8.2}$$

where c_k are c-numbers. The measurement is taken to end abruptly at time τ_m, when an observation takes place which induces collapse of (8.2) into one of the possible states $| \psi_k\rangle | \varphi_k\rangle$. This observation is a non-unitary process. The main difference between this theory and the conventional, or "textbook", theory of measurement is the distinction made between measurement and observation. Thus this theory seems to provide a reasonable generalization of the conventional quantum theory of measurement. The basic concepts for such a generalization can be found in the book by von Neumann (1955), whereas the criteria for an appropriate choice of H_{MR} have been discussed more recently (Grabowski 1990).

In preparation for the application of these ideas to the theory of dressed atoms, we assume that the object to be measured is simply an isolated two-level atom, described by the Hamiltonian

$$H_R = \hbar\omega_0 S_z \tag{8.3}$$

We model the pointer as a body of large mass M and momentum p, free to move in one dimension, such that

$$H_M = \frac{1}{2M}p^2 \tag{8.4}$$

and for the atom-pointer Hamiltonian we assume the form

$$H_{MR} = g(t)pS_z \; ;$$
$$g(t) = \frac{1}{\tau_m}g_0[\theta(t) - \theta(t - \tau_m)] \; ;$$
$$\int_0^\infty g(t)dt = g_0 \tag{8.5}$$

where $g(t)$ is a function of time represented by Figure 8.1. We now proceed to show that H_{MR} in (8.5) is of the appropriate form to measure S_z, which is an operator directly related to the bare atomic population distribution. On the basis of the general ideas on the theory of measurement of finite duration, one expects that during the time in the interval $0, \tau_m$ the pointer gets correlated with the atom, and one postulates that the wavefunction collapses at $t = \tau_m$ as a result of an observation (Peres and Wootters 1985). This bare two-level atom problem can be solved exactly. In fact the total Hamiltonian is

$$H = H_M + H_R + H_{MR} = \frac{1}{2M}p^2 + \hbar\omega_0 S_z + g(t)pS_z \tag{8.6}$$

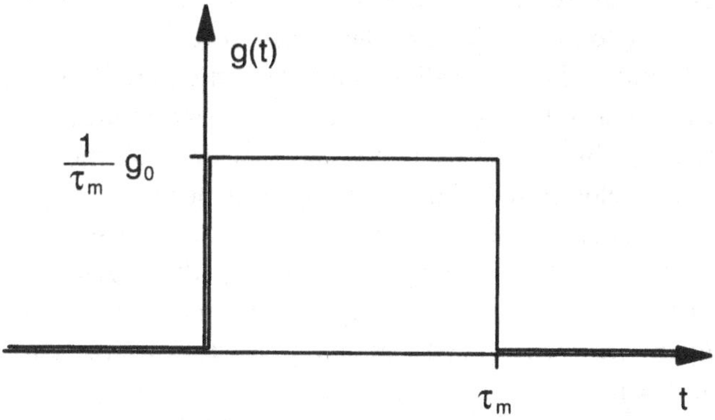

Fig. 8.1 Time-dependence of the atom-pointer coupling.

and the equations of motion of the relevant dynamical variables are

$$\dot{q} = -\frac{i}{\hbar}[q, H] = \frac{1}{M}p + g(t)S_z \; ; \; \dot{p} = 0 \; ; \; \dot{S}_z = 0 \tag{8.7}$$

where q is the position of the pointer. Thus p and S_z are constants of motion, and for $t \geq \tau_m$ integration of (8.7) gives

$$q(t) = g_0 S_z(0) + \frac{1}{M}p(0)t + q(0) \; ;$$
$$q^2(t) = \frac{1}{4}g_0^2 + 2g_0 S_z(0)\left[\frac{1}{M}p(0)t + q(0)\right] + \left[\frac{1}{M}p(0)t + q(0)\right]^2 \tag{8.8}$$

Coherently with the previous discussion, we take $|\Psi(0)\rangle = |\psi\rangle|\varphi\rangle$. Moreover we take advantage of the large mass M of the pointer to choose $|\psi\rangle$ in such a way that

$$\langle\psi | q(0) | \psi\rangle \sim \langle\psi | p(0) | \psi\rangle \sim 0 \tag{8.9}$$

and quite generally we write for the atomic state

$$|\varphi\rangle = \begin{vmatrix} a \\ b \end{vmatrix} \quad \left(|a|^2 = \frac{1}{2} + \langle S_z(0)\rangle \; ; \right.$$
$$\left. |b|^2 = \frac{1}{2} - \langle S_z(0)\rangle \right) \tag{8.10}$$

Thus from (8.8) at $t = \tau_m$

$$\langle\Psi(0) | q(\tau_m) | \Psi(0)\rangle \equiv \langle q(\tau_m)\rangle = g_0\langle S_z(0)\rangle \; ;$$
$$\langle q^2(\tau_m)\rangle - \langle q(\tau_m)\rangle^2 = g_0^2\left(\frac{1}{4} - \langle S_z(0)\rangle^2\right) \tag{8.11}$$

On the basis of this result it is not difficult to convince ourselves that the probability distribution $P(q)$ of the position of the pointer at the observation time τ_m is that qualitatively represented in Figure 8.2. Thus, in view of the first of (8.8) and of (8.9), in a bare atom measurement the position of the pointer is perfectly correlated with $S_z(0)$ and observation of q at $t = \tau_m$ yields either of the two positions of the pointer at $\pm g_0/2$. The relative frequency of either result is related to the variance of $q(\tau_m)$ given by the second of (8.11), whereas the first of (8.11) gives the average of these results. Thus the gedanken experiment modelled by Hamiltonian (8.6) is capable of measuring $|a|^2$ and $|b|^2$, that is, the bare atomic population distribution.

We shall apply the same kind of measurement described above, using the same apparatus, to an atom dressed by the zero-point fluctuations of a single-mode electromagnetic field (Compagno *et al.* 1990). Accordingly we take H_R to be of the same form as in (5.22)

$$H_R = \hbar\omega_0 S_z + \hbar\omega a^\dagger a + \epsilon a S_+ + \epsilon^* a^\dagger S_- - \epsilon a^\dagger S_+ - \epsilon^* a S_- \tag{8.12}$$

and the total Hamiltonian takes the form

$$\begin{aligned} H = \frac{1}{2M}p^2 + \hbar\omega_0 S_z + g(t) S_z p + \hbar\omega a^\dagger a \\ + \epsilon a S_+ + \epsilon^* a^\dagger S_- - \epsilon a^\dagger S_+ - \epsilon^* a S_- \end{aligned} \tag{8.13}$$

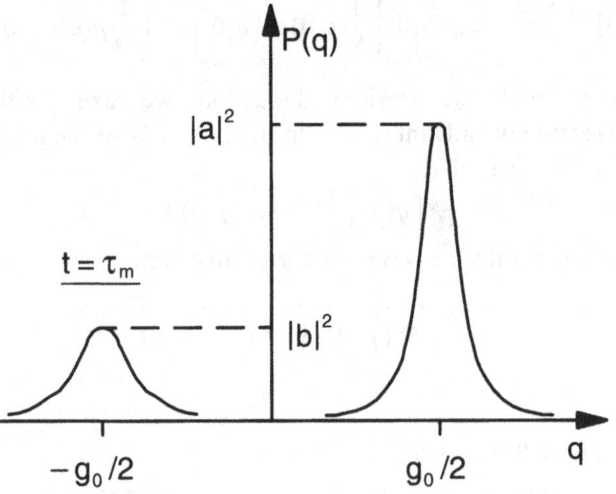

Fig. 8.2 Qualitative representation of the probability distribution of the position q of the pointer in a bare two-level atom measurement of S_z. The observation takes place at $t = \tau_m$. The peak at $q = g_0/2$ corresponds to the atom excited and the peak at $q = -g_0/2$ to the ground-state atom.

The Heisenberg equations of motion for q and p are the same as in (8.7), but now $\dot{S}_z \neq 0$ because S_z does not commute with the atom-field interaction Hamiltonian in H_R. This is clearly a feature that in the ground state of the atom-field system will give rise to virtual transitions between bare atomic levels induced by the zero-point fluctuations of the field, inducing in turn fluctuations of S_z. For this reason direct integration of the first of (8.7) yields

$$\begin{aligned} q(t) &= \int_0^t S_z(t')g(t')dt' + \frac{1}{M}p(0)t + q(0) \, ; \\ q^2(t) &= \int_0^t \int_0^t S_z(t')S_z(t'')g(t')g(t'')dt'dt'' \\ &+ \int_0^t S_z(t')g(t')dt' \left[\frac{1}{M}p(0)t + q(0)\right] \\ &+ \left[\frac{1}{M}p(0)t + q(0)\right] \int_0^t S_z(t')g(t')dt' \\ &+ \left[\frac{1}{M}p(0)t + q(0)\right]^2 \end{aligned} \tag{8.14}$$

which does not reduce to (8.8) even for $t \geq \tau_m$. Up to this point the procedure has been exact, but in order to integrate (8.14) one has to solve the Heisenberg equation for $S_z(t)$, which can be done only approximately. More precisely it can be shown that, neglecting terms $O(\epsilon^3)$,

$$\begin{aligned} S_z(t) = &-\frac{i}{\hbar}\epsilon\Big\{(aS_+)_0 \int_0^t e^{i(\omega_0-\omega)t' + ip(0)G(t')/\hbar}dt' \\ &- (a^\dagger S_+)_0 \int_0^t e^{i(\omega_0+\omega)t' + ip(0)G(t')/\hbar}dt'\Big\} \\ &- \frac{1}{\hbar^2}|\epsilon|^2 \Big\{\int_0^t \int_0^{t'} e^{i(\omega_0-\omega)(t'-t'')+ip(0)[G(t')-G(t'')]/\hbar} \, . \\ &\times \left[S_z + \frac{1}{2} + 2(a^\dagger a - a^2 e^{-2i\omega t''})S_z\right]_0 dt'dt'' \\ &+ \int_0^t \int_0^{t'} e^{i(\omega_0+\omega)(t'-t'')+ip(0)[G(t')-G(t'')]/\hbar} \\ &\times \left[S_z - \frac{1}{2} + 2(a^\dagger a - a^{\dagger 2} e^{2i\omega t''})S_z\right]_0 dt'dt''\Big\} \\ &+ \text{h.c.} + S_z(0) \end{aligned} \tag{8.15}$$

where, for any operator A, $(A)_0$ or $[A]_0$ indicates $A(t=0)$ and where $G(t) = \int_0^t g(t')dt'$. The rather complicated expression for $S_z(t')S_z(t'')$ can be obtained directly from (8.15). For reasons of space it will not be reproduced here, but it can obviously be substituted into (8.14) together with $S_z(t')$ in preparation for the final integrations over t' and t''.

The mathematics is somewhat simplified, however, by preliminarily taking the quantum average of the relevant operators on the initial state $|\Psi(0)\rangle = |\psi\rangle|\varphi\rangle$. For the initial state of the pointer $|\psi\rangle$ we assume the same properties (8.9) as for the case of the bare atom problem previously discussed. For the initial state $|\varphi\rangle$ of the atom-field system we take the dressed ground state of (8.12), which we have already explicitly obtained in Section 5.3, accurate at order ϵ^2, as

$$|\varphi\rangle = \left\{1 - \frac{1}{2}\frac{|\epsilon|^2}{\hbar^2(\omega_0+\omega)^2}\right\}|0,\downarrow\rangle + \frac{\epsilon}{\hbar(\omega_0+\omega)}|1,\uparrow\rangle - \frac{1}{\sqrt{2}\hbar^2}\frac{|\epsilon|^2}{\omega(\omega_0+\omega)}|2,\downarrow\rangle \tag{8.16}$$

where $|n,\uparrow\rangle$ or $|n,\downarrow\rangle$ are states of the atom-field system with n photons and the atom excited or unexcited respectively. A lengthy but straightforward calculation leads to

$$\langle S_z(t)\rangle \equiv \langle S_z(0)\rangle = -\frac{1}{2} + \frac{|\epsilon|^2}{\hbar^2(\omega_0+\omega)^2};$$
$$\langle S_z(t')S_z(t'')\rangle = \frac{1}{4} + \left(\langle S_z(0)\rangle + \frac{1}{2}\right)\left(e^{-i(\omega_0+\omega)(t'-t'')} - 1\right) \tag{8.17}$$

where $\langle A\rangle = \langle\Psi(0)|A|\Psi(0)\rangle$. Expressions (8.17) are accurate up to terms of order ϵ^2, and the same accuracy is attained for $\langle q(t)\rangle$ and $\langle q^2(t)\rangle$ by taking the quantum averages of (8.14) on $|\Psi(0)\rangle$ and by substituting (8.17) in the appropriate places. The final result at $t=\tau_m$ is (Compagno *et al.* 1990, 1991)

$$\langle q(\tau_m)\rangle = g_0\langle S_z(0)\rangle;$$
$$\langle q^2(\tau_m)\rangle - \langle q(\tau_m)\rangle^2 = g_0^2\left(\frac{1}{4} - \langle S_z(0)\rangle^2\right) \times \frac{2[1-\cos(\omega_0+\omega)\tau_m]}{(\omega_0+\omega)^2\tau_m^2} \tag{8.18}$$

Result (8.18) for the dressed ground-state atom should be compared with result (8.11) for the bare atom. In the limit of a "short" measurement, such that $\tau_m \ll (\omega_0+\omega)^{-1}$, the variance of the distribution in the pointer positions $P(q)$, given by (8.18), coincides with the variance of the bare atom in (8.11). In particular, $P(q)$ is two-peaked in the bare as well as in the dressed atom case. Thus the measuring apparatus perceives the atom as essentially bare and the measurement is not influenced much by the coupling with the single-mode vacuum fluctuations. On the contrary, in the limit $\tau_m \gg (\omega_0 + \omega)^{-1}$ the variance (8.18) of the dressed atom case is seen to vanish. This indicates the presence of a single peak in $P(q)$, as shown in Figure 8.3, rather than of two peaks as in the bare atom case. Thus "long" measurements such that $\tau_m \gg (\omega_0 + \omega)^{-1}$ detect a new object, namely the dressed atom. This shows that long measurements are strongly influenced by the coupling with the vacuum fluctuations. Since, as discussed in Section 6.2, $(\omega_0 + \omega)^{-1}$ is the duration τ of a quantum fluctuation leading to the appearance of a virtual photon of frequency ω, this result can also be interpreted as follows. A measurement which lasts for a time τ_m can only detect the effects of vacuum fluctuations which last for a time $\tau \ll \tau_m$. These are rapid fluctuations on the time scale set by the apparatus, whereas the effects of slow fluctuations which last for a time $\tau \gg \tau_m$ cannot be detected. Thus the gedanken experiment which we have described is capable of detecting the vacuum fluctuations of the single-mode field. Yet another way of describing the same effect is in terms of energy. A measurement of the kind we have described lasting for a time τ_m is capable of transferring to the object which is being measured (in our

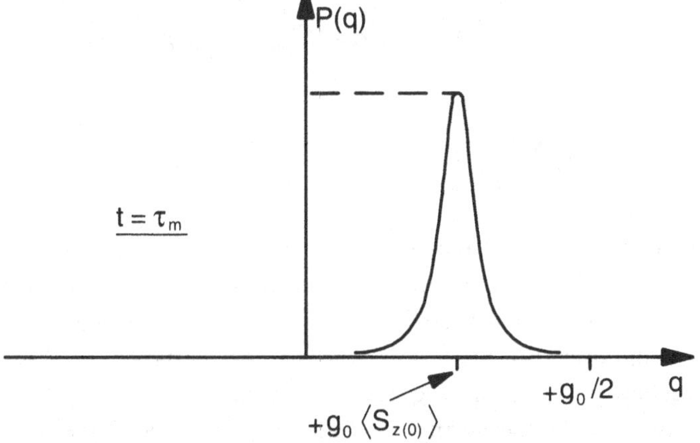

Fig. 8.3 Qualitative representation of the probability distribution of the position q of the pointer in a dressed ground-state two-level atom measurement of S_z. The observation takes place at $t = \tau_m \gg (\omega_0 + \omega)^{-1}$. A single peak is found in contrast with the bare-atom case.

case the atom-field system) an energy $\hbar\tau_m^{-1}$, which is sufficient to create real atom-field excitations of the same energy. In our system the minimum energy required to create such real excitations is $\hbar(\omega_0 + \omega)$ since virtual two-photon excitations cannot be detected by measuring only S_z. Thus if $\tau_m < (\omega_0 + \omega)^{-1}$ the virtual excitations are transformed into real ones and the photon is free to leave the atom which is then measured as bare. If on the contrary $\tau_m > (\omega_0 + \omega)^{-1}$, the energy transferred to the atom-field system during the measurement is not sufficient to untrap the virtual photons of the cloud surrounding the atom, and the latter is then detected as dressed.

The single-mode model discussed above can be generalized to the many-mode case described by the Hamiltonian

$$H = \frac{1}{2M}p^2 + \hbar\omega_0 S_z + g(t)S_z p + \sum_k \hbar\omega_k a_k^\dagger a_k$$
$$+ \sum_k \left(\epsilon_k a_k S_+ + \epsilon_k^* a_k^\dagger S_- - \epsilon_k a_k^\dagger S_+ - \epsilon_k^* a_k S_- \right) \tag{8.19}$$

where subscript k stands for $\mathbf{k}j$. The mathematics gets rather more involved, although no serious conceptual difficulties arise at order ϵ^2. Thus we shall report here only the final result, where quantum averages $\langle\ \rangle$ are taken on the state $|\Psi(0)\rangle = |\psi\rangle|\varphi\rangle$, with $|\psi\rangle$ being the same as in the previous examples and with $|\varphi\rangle$ corresponding to the ground-state atom dressed by the many-mode field. One has

$$\langle q(\tau_m)\rangle = g_0 \left(-\frac{1}{2} + \sum_k \frac{|\epsilon_k|^2}{\hbar^2(\omega_0 + \omega_k)^2} \right);$$

$$\langle q^2(\tau_m)\rangle - \langle q(\tau_m)\rangle^2$$
$$= g_0^2 \sum_k \frac{|\epsilon_k|^2}{\hbar^2(\omega_0 + \omega_k)^2} \frac{2[1 - \cos(\omega_0 + \omega_k)\tau_m]}{(\omega_0 + \omega_k)^2 \tau_m^2} \tag{8.20}$$

From the form of the variance in (8.20) one deduces that a measurement of duration τ_m detects the atom as if it were dressed only by virtual photons of high frequency such that $\omega_0 + \omega_k \gg \tau_m^{-1}$, since these high-frequency photons do not contribute to the variance of $P(q)$. The low-frequency photons of frequency ω_k such that $\omega_0 + \omega_k \ll \tau_m^{-1}$ are not perceived by the measuring apparatus. Thus it is as if the measuring apparatus perceived an atom which is dressed by the high-frequency virtual photons only, or equivalently, as if the virtual cloud around the atom were deficient in low-frequency components. These last considerations seem to lead to the concept of half-dressed, or partially dressed atoms, which was first introduced by E. Feinberg (1980). Since this subject is out of the scope of this book we shall not pursue it further.

We now turn to the second of the two situations briefly described at the beginning of this section, namely that of an atom dressed by a field of real photons. We will consider only a single-mode case of the kind described in Section 5.3. We will also assume that the counterrotating terms in the atom-field interaction Hamiltonian can be neglected, and we consequently adopt the RWA. Thus we take

$$H_R = \hbar\omega_0 S_z + \hbar\omega a^\dagger a + \epsilon a S_+ + \epsilon^* a^\dagger S_- \tag{8.21}$$

This Hamiltonian has been shown in Section 5.3 to be diagonalized by the dressed atomic states of the kind defined in (5.31) or in (5.39). Here the theory of measurement of finite duration described earlier in this section will be applied in order to investigate the physical nature of these dressed states. The total Hamiltonian, including the pointer and the pointer-atom interaction, is

$$H = \frac{1}{2M}p^2 + \hbar\omega_0 S_z + g(t)S_z p + \hbar\omega a^\dagger a + \epsilon a S_+ + \epsilon^* a^\dagger S_- \tag{8.22}$$

while result (8.14) is obviously recovered from direct integration of the Heisenberg equations for q. Thus the problem is once more reduced to solving the dynamics of S_z, which has already been obtained in (5.42) for the RWA Hamiltonian (8.21). During the measurement, however, the dynamics of S_z is determined by (8.22) rather than by (8.21) alone. Thus $\hbar\omega_0$ in (5.42) must be substituted by $\hbar\omega_0 + g_0 p(0)/\tau_m$ and

$$\begin{aligned} S_z(t) &= \left(S_z(0) - \frac{1}{4}\delta_m K_m^{-1}\right)e^{-(i/\hbar)2K_m t} + \frac{1}{4}\delta_m K_m^{-1}\ ; \\ \delta_m &= \hbar(\omega_0 - \omega) + \frac{1}{\tau_m} g_0 p(0)\ ; \\ K_m &= \delta_m S_z + \epsilon a S_+ + \epsilon^* a^\dagger S_- \end{aligned} \tag{8.23}$$

From this expression one easily obtains also

$$\begin{aligned} S_z(t')S_z(t'') &= \frac{1}{4}\left(1 - \frac{1}{4}\delta_m^2 K_m^{-1}\right)e^{(i/\hbar)2K_m(t'-t'')} \\ &+ \frac{1}{8}[S_z(0), K_m]\delta_m K_m^{-2}\left(e^{-(i/\hbar)2K_m t'} - e^{-(i/\hbar)2K_m t''}\right) \\ &+ \frac{1}{16}\delta_m^2 K_m^{-2} \end{aligned} \tag{8.24}$$

As for $| \Psi(0)\rangle = | \psi\rangle | \varphi\rangle$, we choose $| \psi\rangle$ to be the same as in the other cases considered in this section, whereas we choose $| \varphi\rangle$ to be one of the dressed states (5.31) or (5.39). In particular, taking $| \varphi\rangle \equiv | u_n^{(+)}\rangle$ one has

$$\begin{aligned} \langle S_z(t)\rangle &\equiv \langle \Psi(0) | S_z(t) | \Psi(0)\rangle = \frac{1}{2}\frac{\delta}{\Delta}; \\ \langle S_z(t')S_z(t'')\rangle &- \langle S_z(t')\rangle\langle S_z(t'')\rangle \\ &= \frac{1}{4}\left(1 - \frac{\delta^2}{\Delta^2}\right)e^{(i/\hbar)\Delta(t'-t'')} \end{aligned} \tag{8.25}$$

since it is easy to prove by a direct calculation that

$$\langle S_z(0)\rangle = \frac{1}{2}\frac{\delta}{\Delta} \tag{8.26}$$

Substituting (8.26) in (8.14), taking (8.9) into account and performing appropriate manipulations one gets (Lo Cascio and Persico 1991)

$$\begin{aligned} \langle q(\tau_m)\rangle &= g_0\langle S_z(0)\rangle ; \\ \langle q^2(\tau_m)\rangle - \langle q(\tau_m)\rangle^2 &= g_0^2\left(\frac{1}{4} - \langle S_z(0)\rangle^2\right) \\ &\times \frac{2[1 - \cos \Delta\tau_m/\hbar]}{\Delta^2\tau_m^2/\hbar^2} \end{aligned} \tag{8.27}$$

Expression (8.27) should be compared with the bare-atom case (8.11). The dressed-atom variance reduces to the corresponding bare-atom quantity for short measurements such that $\Delta\tau_m/\hbar \ll 1$. In this case one has two peaks in the probability distribution $P(q)$, and the situation is similar to that represented in Figure 8.2. On the other hand, for long measurements such that $\Delta\tau_m/\hbar \gg 1$, the variance in (8.27) vanishes and $P(q)$ displays a single peak. Clearly, the measuring apparatus in a short measurement is not able to perceive the influence of the driving field on the atom, which is consequently detected as bare, whereas a long measurement is not capable of disentangling the atom-field correlations, and consequently the apparatus perceives a single atom-field object, which is what we have called a dressed atom. It should be noted that in the case of dressing by real photons under consideration, the time scale is set by Δ which is the Rabi frequency of the atom-field system, while in the case of dressing by vacuum fluctuations the time scale is set by the sum of the bare atom and the bare photon frequencies. The fact that the orders of magnitude of these two scales are very different indicates that the nature of the dressing process in the two cases is also different.

8.2 The physical interpretation of vacuum radiative effects

We wish to take up again three important observable QED effects discussed in the previous chapters and to discuss in detail their physical interpretation. These radiative effects are spontaneous emission, which we have dealt with in Section 5.4, the Lamb shift which we have discussed in Sections 5.4, 6.4 and in Appendix J, and van der Waals forces between neutral atoms which we have considered in Section 7.8. In presenting the theory of all these effects the concept of dressed atom has been used to smaller or larger extent, and it has been shown to play a role in determining prominent features of the phenomenon. This is another way of saying that we have taken proper account of the existence of the electromagnetic field surrounding the atoms when we have discussed the theory of these effects. In fact, this is the field which gives rise to the virtual cloud of the dressed atom and which we have discussed in some detail in Chapters 6 and 7.

Clearly, this dressing field originates from the atomic source, which is constituted by electric charges. An overall neutral set of electric charges can be described as affecting the electromagnetic field in two complementary ways. One can define a surface surrounding the sources and impose boundary conditions at this surface, which take into account the field generated by the sources. These boundary conditions change the structure of the normal modes of the field and/or of its frequency spectrum. One can, however, also consider the total field in space as a superposition of the field generated by the source and of that which would exist independently of the source. In the case of a localized QED source, such as a neutral atom or molecule, the first line of thought leads one to consider a localized distortion of the structure of the zero-point field modes and possibly a modification of their frequency spectrum. In this picture the dressing field is due to this distortion; moreover, radiative effects must be attributed in part to the action of the distorted zero-point field on the source and in part to the changes induced in the zero-point field by the presence of the source. On the other hand, in the second line of thought the field dressing the source is generated by the source itself on top of the background of an unchanged zero-point field. In this picture the zero-point field is superfluous, the radiative effects are attributed to the interaction of the atom or molecule with the field which the atom or molecule creates (radiation reaction) and this field is called the self-reaction field.

One should not expect that these two ways of thinking lead to different observable results: provided the mathematics is done correctly, the results cannot change if one point of view or the other is taken. One might expect, however, to gain physical insight if one of the two pictures, or an appropriate combination of both, were shown to yield a more convincing physical interpretation of radiative effects. In what follows we will show that this is not the case, and that phenomena like spontaneous emission, the Lamb shift and retarded van der Waals forces can be indifferently and equally convincingly regarded as arising from the zero-point field, from radiation reaction or from a combination of both. This apparently paradoxical situation indicates that the two extreme interpretations, in terms of vacuum fluctuations or in terms of radiation reaction, are indeed "two sides of a coin", in the sense that they seem complementary explanations (Milonni 1988). In order to illustrate this point of view we now proceed to examine each of three effects in turn.

i) Spontaneous emission

The traditional interpretation of spontaneous emission, discussed in Section 5.4, is in terms of vacuum fluctuations (see e.g. Weisskopf 1981). Zero-point fluctuations are visualized as acting on excited atomic electrons and inducing downwards transitions only, since any net amount of energy cannot be extracted from the vacuum. Since zero-point fluctuations are of purely quantum origin, so is the phenomenon of spontaneous emission. Interesting considerations on technical and historical aspects of this phenomenon can be found in Ginzburg (1983). More recently, it has been suggested that radiation reaction can also be considered as a source of spontaneous emission (see e.g. Ackerhalt *et al.* 1973, Milonni *et al.* 1973, Milonni and Smith 1975, Milonni 1980). Here we will show that for a two-level atom spontaneous emission can be attributed equivalently to zero-point fluctuations and to radiation reaction. We anticipate that, from the technical point of view, this equivalence can be traced to different but equivalent orderings of the product of operators which commute at equal times and which appear in the Heisenberg equations of motion of the atom-radiation system. The starting point is Hamiltonian (5.51) in the minimal coupling scheme ($\lambda = 1$)

$$H = \hbar\omega_0 S_z + \sum_{\mathbf{k}j} \hbar\omega_k a^\dagger_{\mathbf{k}j} a_{\mathbf{k}j} + \sum_{\mathbf{k}j} \epsilon_{\mathbf{k}j}(a_{\mathbf{k}j} + a^\dagger_{\mathbf{k}j})(S_+ + S_-) \tag{8.28}$$

where $\epsilon_{\mathbf{k}j}$ is real because the field polarization vectors have been taken as real and the dipole matrix elements as imaginary for simplicity. From the

Heisenberg equations we have

$$\dot{a}_{\mathbf{k}j} = -i\omega_k a_{\mathbf{k}j} - \frac{i}{\hbar}\epsilon_{\mathbf{k}j}(S_+ + S_-) \tag{8.29}$$

We now depart from the procedure followed in Section 5.4. The solution of (8.29) can be formally partitioned as

$$a_{\mathbf{k}j}(t) = a^0_{\mathbf{k}j}(t) + a^{RR}_{\mathbf{k}j}(t) \tag{8.30}$$

where

$$a^0_{\mathbf{k}j}(t) = a_{\mathbf{k}j}(0)e^{-i\omega_k t} \tag{8.31}$$

$$a^{RR}_{\mathbf{k}j}(t) = -\frac{i}{\hbar}\epsilon_{\mathbf{k}j}\int_0^t [S_+(t') + S_-(t')]e^{-i\omega_k(t-t')}dt' \tag{8.32}$$

Clearly, $a^0_{\mathbf{k}j}$ represents the contribution of the zero-point field, since it is present also for $\epsilon = 0$ and it does not contain atomic variables. In contrast, $a^{RR}_{\mathbf{k}j}$ is a radiation reaction term, since it does not contain the vacuum contribution and it is expressed in terms of atomic operators. At the lowest order in ϵ we may use approximation (5.68) in (8.32). Then (8.32) can be integrated as in (5.69), yielding for large t

$$\begin{aligned} a^{RR}_{\mathbf{k}j}(t) = -\frac{1}{\hbar}\epsilon_{\mathbf{k}j}\Bigg\{ & S_+(t)\frac{1}{\omega_0 + \omega_k} \\ & -S_-(t)\left[P\frac{1}{\omega_0 - \omega_k} - i\pi\delta(\omega_0 - \omega_k)\right]\Bigg\} \end{aligned} \tag{8.33}$$

This expression shows explicitly the atomic origin of $a^{RR}_{\mathbf{k}j}(t)$. Now consider the equations of motion of an atomic variable, for example S_+. We can write (5.65) as

$$\dot{S}_+ = i\omega_0 S_+ - \frac{2i}{\hbar}\sum_{\mathbf{k}j}\epsilon_{\mathbf{k}j}\left(S_z a_{\mathbf{k}j} + a^\dagger_{\mathbf{k}j}S_z\right) \tag{8.34}$$

where we adopted the so called "normal ordering" in which the field annihilation operators are placed at the right end and the field creation operators at the left end of each operator product. We may also write (5.65) adopting the "antinormal ordering" in which the field annihilation operators are placed at the left end and the creation operators at the right end of each product, that is

$$\dot{S}_+ = i\omega_0 S_+ - \frac{2i}{\hbar}\sum_{\mathbf{k}j}\epsilon_{\mathbf{k}j}\left(a_{\mathbf{k}j}S_z + S_z a^\dagger_{\mathbf{k}j}\right) \tag{8.35}$$

The two forms (8.34) and (8.35) are equivalent, since $a_{\mathbf{k}j}(t)$ and $S_z(t)$ commute. If we introduce the partition (8.30), however, (8.34) takes the form

$$\dot{S}_+ = i\omega_0 S_+ - \frac{2i}{\hbar}\sum_{\mathbf{k}j} \epsilon_{\mathbf{k}j}\left(S_z a^0_{\mathbf{k}j} + a^{0\dagger}_{\mathbf{k}j} S_z\right)$$
$$-\frac{2i}{\hbar}\sum_{\mathbf{k}j} \epsilon_{\mathbf{k}j}\left(S_z a^{RR}_{\mathbf{k}j} + a^{RR\dagger}_{\mathbf{k}j} S_z\right) \tag{8.36}$$

where the normal order of the operators in each product is significant, since from (8.31) and (8.33) it is clear that $S_z(t)$ does not commute with either $a^0_{\mathbf{k}j}(t)$ or $a^{RR}_{\mathbf{k}j}(t)$ separately. When we take the quantum average of (8.36) on an initial state $| \{0_{\mathbf{k}j}\}, \uparrow\rangle$ (atom excited and no photon) in order to work out the dynamics of S_+ in spontaneous emission, from (8.31) we see that the vacuum part of the field arising from $a^0_{\mathbf{k}j}$ does not contribute to the atomic dynamics, since

$$\langle\{0_{\mathbf{k}j}\}, \uparrow| \left(S_z a^0_{\mathbf{k}j} + a^{0\dagger}_{\mathbf{k}j} S_z\right) | \{0_{\mathbf{k}j}\}, \uparrow\rangle$$
$$\equiv \langle\left(S_z a^0_{\mathbf{k}j} + a^{0\dagger}_{\mathbf{k}j} S_z\right)\rangle = 0 \tag{8.37}$$

and we have

$$\langle\dot{S}_+\rangle = i\omega_0\langle S_+\rangle - \frac{2i}{\hbar}\sum_{\mathbf{k}j} \epsilon_{\mathbf{k}j}\langle S_z a^{RR}_{\mathbf{k}j} + a^{RR\dagger}_{\mathbf{k}j} S_z\rangle \tag{8.38}$$

Thus, adopting normal ordering and starting from state $| \{0_{\mathbf{k}j}\}, \uparrow\rangle$, it appears that the vacuum fluctuations do not contribute to the dynamics of S_+ in spontaneous decay, which is driven only by radiation reaction. This conclusion is reminiscent of the classical theory of radiative damping of a charged harmonic oscillator (see e.g. Jackson 1962, Rohrlich 1965). If we adopt antinormal ordering, however, starting from (8.35) we have

$$\dot{S}_+ = i\omega_0 S_+ - \frac{2i}{\hbar}\sum_{\mathbf{k}j} \epsilon_{\mathbf{k}j}\left(a^0_{\mathbf{k}j} S_z + a^{0\dagger}_{\mathbf{k}j} S_z\right)$$
$$-\frac{2i}{\hbar}\sum_{\mathbf{k}j} \epsilon_{\mathbf{k}j}\left(a^{RR}_{\mathbf{k}j} S_z + S_z a^{RR\dagger}_{\mathbf{k}j}\right) \tag{8.39}$$

Proceeding as in the previous case and taking the quantum average of (8.39) on the initial state $| \{0_{\mathbf{k}j}\}, \uparrow\rangle$, we now see that

$$\langle\{0_{\mathbf{k}j}\}, \uparrow| \left(a^0_{\mathbf{k}j} S_z + S_z a^{0\dagger}_{\mathbf{k}j}\right) | \{0_{\mathbf{k}j}\}, \uparrow\rangle$$
$$\equiv \langle\left(a^0_{\mathbf{k}j} S_z + S_z a^{0\dagger}_{\mathbf{k}j}\right)\rangle \neq 0 \tag{8.40}$$

and consequently the quantum average of (8.39) is

$$\langle \dot{S}_+ \rangle = i\omega_0 \langle S_+ \rangle - \frac{2i}{\hbar} \sum_{\mathbf{k}j} \epsilon_{\mathbf{k}j} \left\langle \left(a^0_{\mathbf{k}j} S_z + S_z a^{0\dagger}_{\mathbf{k}j} \right) \right\rangle$$

$$- \frac{2i}{\hbar} \sum_{\mathbf{k}j} \epsilon_{\mathbf{k}j} \left\langle \left(a^{RR}_{\mathbf{k}j} S_z + S_z a^{RR\dagger}_{\mathbf{k}j} \right) \right\rangle \tag{8.41}$$

which is different from (8.38) and shows that if the antinormal ordering is adopted, the vacuum fluctuations do seem to play a role in the dynamics of S_+ in spontaneous emission. In fact, the conclusion is general, since the argument can be extended to the equations of motion of all atomic variables. Naturally, if the calculations are performed explicitly (Milonni 1976), both (8.38) and (8.41) will give the same result, equal to that found in Section 5.4. We note, however, that the physical interpretation of spontaneous emission in the two points of view is quite different. Furthermore, other recipes for mixed orderings of the operator products are possible and correspondingly other physical interpretations exist which attribute different roles and different relative weight to vacuum fluctuations and radiative reaction in spontaneous emission. In conclusion, depending on which ordering of commuting operators is assumed (and all possible orderings are correct) spontaneous emission can be attributed to the action of vacuum fluctuations or to the action of radiation reaction or to a combination of both. It should be mentioned, however, that Dalibard *et al.* (1982) have suggested that there are reasons to prefer a particular ordering which is called "completely symmetric". This ordering leads one to consider only Hermitian variables of the system and to require that their respective rates of variation be separately Hermitian (Dalibard *et al.* 1982). Moreover, it should be emphasized that no ordering seems to exist that attributes the radiative decay entirely to the action of the zero-point field (see e.g. Milonni 1988).

ii) Lamb shift

Radiative shifts of the atomic levels have been discussed in Section 6.4. These shifts combine to yield the Lamb shift, as shown in Section 5.4 for a two-level atom in the context of spontaneous emission and in Appendix J for a hydrogen atom. With reference to the hydrogen atom, the radiative shift of state n is given by (J.8), which, using (J.9) and (J.10), can be put in the form

$$\Delta_n = \frac{4e^2}{3}\alpha\left(\frac{\hbar}{mc}\right)^2 |\, u_n(0)\,|^2 \ln\frac{\hbar\omega_M}{\langle E_{n'} - E_n\rangle_{\text{Av}}} \tag{8.42}$$

An elegant physical interpretation of (8.42) in terms of vacuum fluctuations was proposed by Welton (1948). We shall begin our discussion of the Lamb shift by presenting a simplified account of this interpretation.

The Heisenberg equation for the position operator $\boldsymbol{\xi}(t)$ of a free electron of mass m in an electric field $\mathbf{E}(t)$ is the same as the classical equation

$$m\ddot{\boldsymbol{\xi}} = -e\mathbf{E}(t) \tag{8.43}$$

Taking the electric field operator in the electric dipole approximation, keeping only zero-order terms in the electric charge, and using (2.9) we can expand

$$\mathbf{E}(t) = \sum_{\mathbf{k}j}\left[\mathbf{e}_{\mathbf{k}j}E(\mathbf{k}j)e^{-i\omega_k t} + \mathbf{e}^*_{\mathbf{k}j}E^\dagger(\mathbf{k}j)e^{i\omega_k t}\right];$$

$$E(\mathbf{k}j) = i\sqrt{\frac{2\pi\hbar\omega_k}{V}}\,a_{\mathbf{k}j}(0) \tag{8.44}$$

It is convenient to expand $\boldsymbol{\xi}(t)$ accordingly as

$$\boldsymbol{\xi} = \sum_{\mathbf{k}j}\left[\mathbf{e}_{\mathbf{k}j}\xi(\mathbf{k}j)e^{-i\omega_k t} + \mathbf{e}^*_{\mathbf{k}j}\xi^\dagger(\mathbf{k}j)e^{i\omega_k t}\right] \tag{8.45}$$

Substituting (8.44) and (8.45) in (8.43) and equating components with equal $\mathbf{k}$ and j leads to

$$m\omega_k^2\xi(\mathbf{k}j) = eE(\mathbf{k}j)\,,\quad \xi(\mathbf{k}j) = \frac{e}{m\omega_k^2}E(\mathbf{k}j) \tag{8.46}$$

Thus a solution of (8.43) in the dipole approximation is

$$\boldsymbol{\xi} = i\frac{e}{m}\sqrt{\frac{2\pi\hbar}{V}}\sum_{\mathbf{k}j}\left[\mathbf{e}_{\mathbf{k}j}a_{\mathbf{k}j}(0)e^{-i\omega_k t} - \mathbf{e}^*_{\mathbf{k}j}a^\dagger_{\mathbf{k}j}(0)e^{i\omega_k t}\right] \tag{8.47}$$

and the quantum average of ξ^2 on the ground state of the field is

$$\langle\{0_{\mathbf{k}j}\}\,|\,\xi^2(t)\,|\,\{0_{\mathbf{k}j}\}\rangle = \frac{2\pi}{V}\frac{e^2\hbar}{m^2}\sum_{\mathbf{k}j}\frac{1}{\omega_k^3}$$

$$= \frac{2}{(2\pi)^2}\frac{e^2\hbar}{m^2c^3}\int\frac{1}{k^3}d^3\mathbf{k} = \frac{2}{\pi}\frac{e^2\hbar}{m^2c^3}\int_0^\infty\frac{1}{\omega}d\omega \tag{8.48}$$

The last integral in (8.48) diverges, but the divergence at the upper limit is fictitious because the nonrelativistic approximation limits the range of integration to $\omega < \omega_M = c\kappa$ where $\kappa = mc/\hbar$ is the electron Compton wavelength. The divergence of the same integral at the lower limit arises from slow low-frequency fluctuations. These fluctuations, however, are suppressed by any kind of binding which introduces a natural cut-off at a frequency $\bar{\omega}$, given by an appropriate average of the transition frequencies of the bound electron. Thus we may conclude that the effect of the zero-point field on a bound electron is to introduce a spread of its position, whose mean square value is

$$\langle \xi^2(t) \rangle = \frac{2}{\pi} \alpha \left(\frac{\hbar}{mc} \right)^2 \ln \frac{\omega_M}{\bar{\omega}} \tag{8.49}$$

where $\bar{\omega}$ depends on the details of the binding potential. Consider now a binding potential $V(\mathbf{x})$. In the absence of the interaction with the vacuum field the potential energy operator of the electron would be $V(\mathbf{x}_e)$. We shall assume that the interaction with an external field changes $\mathbf{x}_e$ into $\mathbf{x}_e + \boldsymbol{\xi}$; consequently the operator $V(\mathbf{x}_e)$ is changed to $V(\mathbf{x}_e + \boldsymbol{\xi})$. Thus, taking quantum averages on the ground state of the field $| \{0_{\mathbf{k}j}\} \rangle$, we are able to introduce an effective potential of the electron, which we express as

$$\langle V(\mathbf{x}_e + \boldsymbol{\xi}) \rangle = [1 + \langle \boldsymbol{\xi} \rangle \cdot \nabla + \frac{1}{2!} \langle (\boldsymbol{\xi} \cdot \nabla)^2 \rangle + \ldots] V(\mathbf{x}_e) \tag{8.50}$$

($\nabla = \partial / \partial \mathbf{x}_e$). From (8.47) we have $\langle \boldsymbol{\xi} \rangle = 0$ and

$$\langle (\boldsymbol{\xi} \cdot \nabla)^2 \rangle = \langle \xi_m \xi_n \rangle \nabla_m \nabla_n = \langle \xi_m^2 \rangle \nabla_m^2 = \frac{1}{3} \langle \xi^2 \rangle \nabla^2 \tag{8.51}$$

since $\langle \xi_m \xi_n \rangle$ is expected to vanish for $m \neq n$ in view of the isotropy of the zero-point fluctuations. Thus from (8.50) we find approximately

$$\langle V(\mathbf{x}_e + \boldsymbol{\xi}) \rangle = V(\mathbf{x}_e) + \frac{1}{6} \langle \xi^2 \rangle \nabla^2 V(\mathbf{x}_e) \tag{8.51}$$

We see that because of the interaction with vacuum fluctuations, which induce a spread of its position, the electron can be described as moving in an effective potential which differs from $V(\mathbf{x}_e)$ by a small term

$$\Delta V(\mathbf{x}_e) = \frac{4}{3} e\alpha \left(\frac{\hbar}{mc} \right)^2 \delta(\mathbf{x}_e) \ln \frac{\omega_M}{\bar{\omega}} \tag{8.53}$$

where we have used $\nabla^2 V = -4\pi\rho$, ρ being the nuclear point-like charge distribution at the origin. The quantum average of $\Delta V(\mathbf{x}_e)$ on an atomic state $u_n(\mathbf{x}_e)$ leads to an energy shift

$$\Delta_n = e\langle u_n \mid \Delta V(\mathbf{x}_e) \mid u_n\rangle$$
$$= \frac{4e^2}{3}\alpha\left(\frac{\hbar}{mc}\right)^2 \mid u_n(0) \mid^2 \ln\frac{\omega_M}{\bar{\omega}} \tag{8.54}$$

This coincides with (8.42), provided we identify $\bar{\omega}$ with $\langle E_{n'} - E_n\rangle_{\mathrm{Av}}/\hbar$, which takes into account the suppression of fluctuations of frequency smaller than $\bar{\omega}$ by the binding of the electron to the nucleus.

We remark that in Welton's interpretation the zero-point field plays an essential role, by causing a spread of the electron position which changes the average potential energy of the electron. This change is the radiative shift of the atomic level, which gives the Lamb shift.

Another interesting physical interpretation of the Lamb shift was proposed by Power (1966) following a suggestion due to Feynman (1961). Power evaluates atomic radiative level shifts as due to the change of the zero-point energy when an atom is introduced in the field. A dilute gas of N hydrogenic atoms per unit volume in state u_ℓ has a refractive index $n(\ell)$. The velocity of light in the gas is $c/n(\ell)$ and the dispersion relation in the medium is

$$\omega_k(\ell) = \frac{c}{n(\ell)}k \tag{8.55}$$

This shows that the frequency of the normal modes of the field depend on the atomic state u_ℓ. Consequently also, the zero-point energy of the field is ℓ-dependent. In fact

$$E_{\mathrm{ZP}}(\ell) = \frac{1}{2}\sum_{\mathbf{k}j} \hbar\omega_k(\ell) = \frac{1}{2}\hbar c \sum_{\mathbf{k}j} \frac{k}{n(\ell)} \tag{8.56}$$

The difference in zero-point energy between a system where the N atoms are in the 2s state and a system where they are in the 2p state is thus

$$E_{\mathrm{ZP}}(2\mathrm{s}) - E_{\mathrm{ZP}}(2\mathrm{p}) = \frac{1}{2}\hbar c \sum_{\mathbf{k}j} k\left[\frac{1}{n(2\mathrm{s})} - \frac{1}{n(2\mathrm{p})}\right] \tag{8.57}$$

The refractive index in the medium constituted by the gas is related to the atomic structure by the relationship (see e.g. Becker 1982)

$$n(\ell) = 1 + \frac{4\pi}{3}N\sum_m \frac{\mid \langle \ell \mid \mu \mid m\rangle \mid^2 (E_m - E_\ell)}{(E_m - E_\ell)^2 - (\hbar ck)^2} \tag{8.58}$$

where m runs over all the atomic states connected by the field to state u_ℓ in the electric dipole approximation, whose energies are E_m. Expression (8.58) shows that $n(\ell)$ is k-dependent and displays dispersion effects. In view of the fact that the density of the field modes increases as k^2, the largest contribution to $n(\ell)$ is likely to come from the high-frequency modes for which $(\hbar ck)^2 \gg (E_m - E_\ell)^2$. Thus we approximate

$$\frac{1}{n(\ell)} \sim 1 + \frac{4\pi}{3} N \sum_m \frac{|\langle \ell | \mu | m \rangle|^2}{(\hbar ck)^2} \left[(E_m - E_\ell) + \frac{(E_m - E_\ell)^3}{(\hbar ck)^2} \right] \tag{8.59}$$

and

$$\begin{aligned} \frac{1}{n(2s)} - \frac{1}{n(2p)} &\sim \frac{4\pi}{3} \frac{N}{(\hbar ck)^2} \left\{ \sum_m |\langle 2s | \mu | m \rangle|^2 (E_m - E_{2s}) \right. \\ &- \sum_m |\langle 2p | \mu | m \rangle|^2 (E_m - E_{2p}) \\ &+ \frac{1}{(\hbar ck)^2} \sum_m |\langle 2s | \mu | m \rangle|^2 (E_m - E_{2s})^3 \\ &\left. - \frac{1}{(\hbar ck)^2} \sum_m |\langle 2p | \mu | m \rangle|^2 (E_m - E_{2p})^3 \right\} \end{aligned} \tag{8.60}$$

We now use the Thomas-Reiche-Kuhn sum rule (see e.g. Merzbacher 1961) in the form

$$\sum_m |\langle \ell | \mu | m \rangle|^2 (E_m - E_\ell) = \frac{3e^2\hbar^2}{2m} \tag{8.61}$$

and

$$\begin{aligned} \langle \ell | \mathbf{p} | m \rangle &= \frac{m}{i\hbar} \langle \ell | [\mathbf{x}_e, H_P] | m \rangle \\ &= \frac{m}{i\hbar} (E_m - E_\ell) \langle \ell | \mathbf{x}_e | m \rangle \end{aligned} \tag{8.62}$$

where H_P is the atomic hydrogen Hamiltonian (4.36). From (8.61) and (8.62) we find

$$\begin{aligned} &(E_m - E_n)^2 |\langle \ell | \mu | m \rangle|^2 = \frac{e^2\hbar^2}{m^2} |\langle \ell | \mathbf{p} | m \rangle|^2 ; \\ &\sum_m |\langle \ell | \mu | m \rangle|^2 (E_m - E_\ell)^3 \\ &= \frac{e^2\hbar^2}{m^2} \sum_m |\langle \ell | \mathbf{p} | m \rangle|^2 (E_m - E_\ell) \end{aligned} \tag{8.63}$$

Substituting (8.62) and (8.63) in (8.60) leads to

$$\frac{1}{n(2s)} - \frac{1}{n(2p)} = \frac{4\pi}{3}\frac{N}{(\hbar ck)^4}\frac{e^2\hbar^2}{m^2} \times \left\{ \sum_m |\langle 2s|\mathbf{p}|m\rangle|^2 (E_m - E_{2s}) - \sum_m |\langle 2p|\mathbf{p}|m\rangle|^2 (E_m - E_{2p}) \right\} \tag{8.64}$$

Using (J.10) the first sum over m in (8.64) yields $2\pi\hbar^2 e^2 \mid u_{2s}(0) \mid^2$ and the second sum vanishes. Thus

$$\frac{1}{n(2s)} - \frac{1}{n(2p)} = \frac{8\pi^2}{3} N \frac{e^2\hbar^4}{m^2} \frac{1}{(\hbar ck)^4} \mid u_{2s}(0) \mid^2 \tag{8.65}$$

Substitution of (8.65) in (8.57) yields the contribution to the zero-point energy shift per atom in the form

$$\begin{aligned} \frac{E_{ZP}(2s) - E_{ZP}(2p)}{N} &= \frac{8\pi^2}{3}\frac{e^2\hbar}{m^2c^3}\sum_{\mathbf{k}}\frac{1}{k^3} \mid u_{2s}(0) \mid^2 \\ &= \frac{4}{3}\frac{e^4}{\hbar c}\left(\frac{\hbar}{mc}\right)^2 \int_{\bar{\omega}}^{\omega_M} \frac{d\omega}{\omega} \mid u_{2s}(0) \mid^2 \\ &= \frac{4}{3} e^2\alpha \left(\frac{\hbar}{mc}\right)^2 \mid u_{2s}(0) \mid^2 \ln\frac{\omega_M}{\bar{\omega}} \end{aligned} \tag{8.66}$$

Comparison of (8.66) with $\Delta_{2s} - \Delta_{2p}$ which can be obtained from (8.42) shows that the Lamb shift can be obtained also as the shift of the zero-point energy of the field modes (of frequency within the band between $\bar{\omega}$ and ω_M) induced by each atom. This is different from Welton's interpretation, where the shift was attributed to the atomic levels rather than to the field. It should be mentioned that Power's approach has been successfully used by Barton (1987) to obtain radiative shifts at finite temperature.

We wish to stress that Welton's and Power's interpretations of Lamb shift, although both based on the vacuum field, are complementary in some aspects. In Welton's case, the origin of the shift is just the change of the average Coulomb nucleus-electron potential energy because of the vacuum fluctuations and no changes in the transverse fields are considered. On the other hand, in Power's case the origin of the shift is just the change of the (transverse) vacuum fluctuations due to the presence of the atom acting as a dielectric body and there is no change of the atomic quantities. In other words, the Lamb shift, which is indeed an energy shift of the complete system (atom + field + interaction; see e.g. Equation (J.4)) can be also qualitatively attributed to an energy shift of just a part of the system ("atom" in the Welton case and "field" in the Power case).

Both in Welton's and in Power's interpretations the existence of vacuum fluctuations is essential for understanding the Lamb shift. Another interpretation of the same effect, however, has been proposed which does not take into account vacuum fluctuations, but which is based entirely on the radiation reaction field. An abundant literature on this subject exists (see e.g. Ackerhalt *et al.* 1973, Ackerhalt and Eberly 1974, Milonni and Smith 1975, Milonni 1976). Here we will follow the approach used by Milonni (1982) and we refer the reader to his papers for more detail (see also Milonni 1988). The main idea is that the Lamb shift is due to the interaction of the atom with the reaction field created by the atom itself. For a hydrogen atom in state u_n Milonni evaluates the quantum average

$$\Delta_n' = -\frac{1}{2}\langle\{0_{\mathbf{k}j}\}, n \mid \boldsymbol{\mu}\cdot\mathbf{E}_\perp \mid \{0_{\mathbf{k}j}\}, n\rangle \tag{8.67}$$

where $\boldsymbol{\mu}$ is the electron dipole operator and $\mathbf{E}_\perp$ is the transverse electric field at the position of the atom, located at the origin of the reference frame. Milonni takes the Hamiltonian of the system in the minimal coupling form and in the electric dipole approximation, which from Section 4.5 is

$$H = \sum_n E_n c_n^\dagger c_n + \sum_{\mathbf{k}j} \hbar\omega_k \left(a_{\mathbf{k}j}^\dagger a_{\mathbf{k}j} + 1/2\right) + i\sum_{nm}\sum_{\mathbf{k}j}\sqrt{\frac{2\pi\hbar}{V\omega_k}}\left(\omega_{mn}\boldsymbol{\mu}_{nm}\cdot\mathbf{e}_{\mathbf{k}j}c_n^\dagger c_m a_{\mathbf{k}j} - \text{h.c.}\right) \tag{8.68}$$

where we take $\boldsymbol{\mu}_{mn}$ and $\mathbf{e}_{\mathbf{k}j}$ to be real vectors for simplicity. The relevant Heisenberg equation is

$$\dot{a}_{\mathbf{k}j} = -i\omega_k a_{\mathbf{k}j} - \sum_{nm}\sqrt{\frac{2\pi\hbar}{V\omega_k}}\frac{1}{\hbar}\omega_{mn}\boldsymbol{\mu}_{nm}\cdot\mathbf{e}_{\mathbf{k}j}c_m^\dagger c_n \tag{8.69}$$

and its formal solution is

$$a_{\mathbf{k}j}(t) = a_{\mathbf{k}j}(0)e^{-i\omega_k t} - \sum_{nm} \sqrt{\frac{2\pi\hbar}{V\omega_k}} \frac{1}{\hbar} \omega_{mn} \boldsymbol{\mu}_{nm} \cdot \mathbf{e}_{\mathbf{k}j} \times e^{-i\omega_k t} \int_0^t e^{i\omega_k t'} c_m^\dagger(t') c_n(t') dt' \tag{8.70}$$

Partitioning (8.70) into vacuum and radiation reaction operators as in (8.31) and (8.32), and using (2.9), we obtain the transverse electric field at the position of the atom as

$$\begin{aligned} \mathbf{E}_\perp(t) &= \mathbf{E}_0(t) + \mathbf{E}_{RR}(t) \,; \\ \mathbf{E}_0(t) &= i\sqrt{\frac{2\pi\hbar\omega_k}{V}} \mathbf{e}_{\mathbf{k}j} a_{\mathbf{k}j}(0) e^{-i\omega_k t} + \text{h.c.} \,; \\ \mathbf{E}_{RR}(t) &= -i\frac{2\pi}{V} \sum_{\mathbf{k}j} \sum_{nm} \mathbf{e}_{\mathbf{k}j} \omega_{mn} \boldsymbol{\mu}_{nm} \cdot \mathbf{e}_{\mathbf{k}j} \\ &\quad \times \int_0^t e^{-i\omega_k (t-t')} c_m^\dagger(t') c_n(t') dt' + \text{h.c.} \end{aligned} \tag{8.71}$$

A different partition of $\mathbf{E}_\perp$ is $\mathbf{E}_\perp(t) = \mathbf{E}_\perp^{(+)}(t) + \mathbf{E}_\perp^{(-)}(t)$ where

$$\begin{aligned} \mathbf{E}_\perp^{(+)}(t) &= i\sum_{\mathbf{k}j} \sqrt{\frac{2\pi\hbar\omega_k}{V}} \mathbf{e}_{\mathbf{k}j} a_{\mathbf{k}j}(t) \,; \\ \mathbf{E}_\perp^{(-)}(t) &= -i\sum_{\mathbf{k}j} \sqrt{\frac{2\pi\hbar\omega_k}{V}} \mathbf{e}_{\mathbf{k}j} a_{\mathbf{k}j}^\dagger(t) \end{aligned} \tag{8.72}$$

Since $\boldsymbol{\mu}(t)$ commutes with $\mathbf{E}_\perp^{(\pm)}(t)$, it is possible to adopt the normal ordering and to write

$$\begin{aligned} \Delta'_n &= -\frac{1}{2}\langle\{0_{\mathbf{k}j}\}, n \mid \left[\boldsymbol{\mu}(t) \cdot \mathbf{E}_\perp^{(+)}(t) + \mathbf{E}_\perp^{(-)}(t) \cdot \boldsymbol{\mu}(t)\right] \mid \{0_{\mathbf{k}j}\}, n\rangle \\ &= -\frac{1}{2}\langle\{0_{\mathbf{k}j}\}, n \mid \left[\boldsymbol{\mu}(t) \cdot \mathbf{E}_{RR}^{(+)}(t) + \mathbf{E}_{RR}^{(-)}(t) \cdot \boldsymbol{\mu}(t)\right] \mid \{0_{\mathbf{k}j}\}, n\rangle \end{aligned} \tag{8.73}$$

where $\mathbf{E}_{RR}^{(+)}$ is the part of $\mathbf{E}_{RR}(t)$ explicitly shown in (8.71) and $\mathbf{E}_{RR}^{(-)}$ is its h.c. We note that the choice of normal ordering leads to disappearance of the vacuum part of $\mathbf{E}_\perp(t)$. Thus Δ'_n contains only the radiation reaction part of the field. Now we have

$$\begin{aligned}\mu &= \int \psi^\dagger(\mathbf{x})\mu(\mathbf{x})\psi(\mathbf{x})d^3\mathbf{x} \\ &= \sum_{pq} \int u_p^\dagger(\mathbf{x})\mu(\mathbf{x})u_q(\mathbf{x})d^3\mathbf{x}c_p^\dagger c_q \\ &= \sum_{pq} \mu_{pq}c_p^\dagger c_q \end{aligned} \tag{8.74}$$

and

$$\begin{aligned}\mu(t)\cdot\mathbf{E}_{RR}^{(+)}(t) = -i\frac{2\pi}{V}\sum_{\mathbf{k}j}\sum_{\ell mpq}(\mu_{pq}\cdot\mathbf{e}_{\mathbf{k}j})c_p^\dagger(t)c_q(t)\omega_{m\ell} \\ \times(\mu_{\ell m}\cdot\mathbf{e}_{\mathbf{k}j})\int_0^t e^{-i\omega_k(t-t')}c_m^\dagger(t')c_\ell(t')dt'\end{aligned} \tag{8.75}$$

This quantity is of order e^2. Thus at this order we may adopt the simplest possible approximation for the various c operators appearing in (8.75), that is

$$c_\ell(t) = e^{-(i/\hbar)E_\ell t}c_\ell(0) \tag{8.76}$$

Substituting (8.76) in (8.75) yields

$$\begin{aligned}\mu(t)\cdot\mathbf{E}_{RR}^{(+)}(t) = -i\frac{2\pi}{V}\sum_{\mathbf{k}j}\sum_{\ell mpq}\omega_{m\ell}(\mu_{pq}\cdot\mathbf{e}_{\mathbf{k}j})(\mu_{\ell m}\cdot\mathbf{e}_{\mathbf{k}j})e^{i(\omega_{pq}-\omega_k)t} \\ \times\int_0^t e^{i(\omega_{m\ell}+\omega_k)t'}dt'\,c_p^\dagger(0)c_q(0)c_m^\dagger(0)c_\ell(0)\end{aligned} \tag{8.77}$$

where $\omega_{pq} = (E_p - E_q)/\hbar$. Thus the relevant matrix elements, after performing the time integration, give

$$\begin{aligned}&\langle\{0_{\mathbf{k}j}\}, n \mid \mu(t)\cdot\mathbf{E}_{RR}^{(+)}(t) \mid \{0_{\mathbf{k}j}\}, n\rangle \\ &= \frac{2\pi}{V}\sum_{\mathbf{k}j}\sum_m(\mu_{nm}\cdot\mathbf{e}_{\mathbf{k}j})^2\frac{\omega_{nm}}{\omega_k-\omega_{nm}}\left[1-e^{-i(\omega_k-\omega_{nm})t}\right]\end{aligned} \tag{8.78}$$

and, for large t,

$$\begin{aligned}\Delta_n' &= \lim_{t\to\infty} -\frac{2\pi}{V}\sum_{\mathbf{k}j}\sum_m(\mu_{nm}\cdot\mathbf{e}_{\mathbf{k}j})^2\omega_{nm}\frac{1-\cos(\omega_k-\omega_{nm})t}{\omega_k-\omega_{nm}} \\ &= -\frac{2\pi}{V}\sum_{\mathbf{k}j}\sum_m(\mu_{nm}\cdot\mathbf{e}_{\mathbf{k}j})^2\omega_{nm}P\frac{1}{\omega_k-\omega_{nm}}\end{aligned} \tag{8.79}$$

Using the fact that $\mathbf{e}_{\mathbf{k}j}$ and $\hat{\mathbf{k}}$ are three orthogonal unit vectors and converting sums into integrals, we have for any function $f(k)$

$$\begin{aligned}
&\frac{1}{V}\sum_{\mathbf{k}j} f(k) \mid \mathbf{e}_{\mathbf{k}j} \cdot \mu_{nm} \mid^2 \\
&= \frac{1}{V}\sum_{\mathbf{k}} f(k)\left[\mid \mu_{nm} \mid^2 - (\mu_{nm} \cdot \hat{\mathbf{k}})^2\right] \\
&= \frac{1}{V}\sum_{\mathbf{k}} f(k) \mid \mu_{nm} \mid^2 \left[1 - (\hat{\mu} \cdot \hat{\mathbf{k}})^2\right] \\
&= \frac{1}{(2\pi)^2} \mid \mu_{nm} \mid^2 \int_0^{k_M} f(k) k^2 dk \int_0^{\pi} \sin^3\theta d\theta \\
&= \frac{1}{3\pi^2} \mid \mu_{nm} \mid^2 \int_0^{k_M} f(k) k^2 dk
\end{aligned} \tag{8.80}$$

Introducing this result in (8.79) we find

$$\Delta'_n = -\frac{2}{3\pi c^3}\sum_m \mid \mu_{nm} \mid^2 \omega_{nm} \int_0^{\omega_M} \frac{\omega^2}{\omega - \omega_{nm}} d\omega \tag{8.81}$$

where the integral is to be calculated as a principal part. At this point Milonni subtracts from each Δ'_n the quantity

$$\Delta_{\text{free}} = -\frac{2}{3\pi c^3}\sum_m \mid \mu_{nm} \mid^2 \omega_{nm} \int_0^{\omega_M} \omega d\omega \tag{8.82}$$

which, in view of (8.61), is independent of n and consequently it is irrelevant if Δ'_n is related to a physically measurable shift. This leads us to consider the quantity

$$\begin{aligned}
\Delta''_n = \Delta'_n - \Delta_{\text{free}} &= -\frac{2}{3\pi c^3}\sum_m \mid \mu_{nm} \mid^2 \omega_{nm}^2 \int_0^{\omega_M} \frac{\omega}{\omega - \omega_{nm}} d\omega \\
&= -\frac{2}{3\pi}\frac{e^2}{m^2 c^3}\sum_m \mid \mathbf{p}_{nm} \mid^2 \int_0^{\omega_M} \frac{\omega}{\omega - \omega_{nm}} d\omega
\end{aligned} \tag{8.83}$$

where we have used (8.62). The last step performed by Milonni (1988) is to remark that, if Δ''_n is a physical shift of the nth atomic level, it should be subjected to renormalization . This means that the quantum average of the counterterm (J.5) should be added to Δ''_n. Since (8.83) coincides with (J.4), this last step yields (J.8) and, after the usual mathematics, we obtain the radiative shift Δ_n of (8.42) and Bethe's expression for the Lamb shift. We remark that (8.83) is due entirely to the radiation reaction field, and this implies that Milonni's interpretation of the Lamb shift is basically different from those by Welton and by Power.

iii) van der Waals forces

The van der Waals forces have been discussed in Section 7.8, where we have shown that they are related to the virtual photon cloud surrounding an atom. The so-called "far zone" part of the force is particularly important from a quantum electrodynamical point of view; it is often called the Casimir force. Here we shall see that, analogously to spontaneous emission and to the Lamb shift, the physical origin of the Casimir force can be traced to the vacuum fluctuations or to the reaction field, or to a combination of them.

The correct space-dependence of the far-zone van der Waals force (apart from numerical factors) as originating from vacuum fluctuations only can be easily obtained (Spruch and Kelsey 1978). Let us assume that the zero-point field is a physical reality, in the sense that, within a dipole approximation framework, it gives rise to a fluctuating field $\mathbf{E}_V(\mathbf{r}_A, t)$ at the position $\mathbf{r}_A$ of an atom assumed to be in an eigenstate $| n\rangle$ of the atomic part of the Hamiltonian. $\mathbf{E}_V$ can be expanded in plane wave amplitudes $\mathbf{E}_V(\mathbf{k}j, \mathbf{r}_A)$ as in (8.44), but here $\mathbf{E}_V(\mathbf{k}j, \mathbf{r}_A)$ is given by

$$\mathbf{E}_V(\mathbf{k}j, \mathbf{r}_A) = i\sqrt{\frac{2\pi\hbar\omega_k}{V}} a_{\mathbf{k}j}(0) e^{i\mathbf{k}\cdot\mathbf{r}_A} \tag{8.84}$$

to account for the fact that the atom is not necessarily at the origin. Each of these field components will give rise to a contribution $\mathbf{p}(\mathbf{k}j, \mathbf{r}_A)$ to the atomic polarization operator, according to

$$\mathbf{p}(\mathbf{k}j, \mathbf{r}_A) = \alpha_A(\mathbf{k}j)\mathbf{E}_V(\mathbf{k}j, \mathbf{r}_A) \tag{8.85}$$

where $\alpha_A(\mathbf{k}j)$ is the atomic dynamic polarizability of atom A, defined as (see e.g. Friedrich 1991)

$$\alpha_A(\mathbf{k}j) = \frac{2}{3}\sum_m \frac{|\langle n | \boldsymbol{\mu} | m\rangle|^2 \omega_{mn}}{\hbar(\omega_{mn}^2 - c^2k^2)} \tag{8.86}$$

For $k \to 0$, $\alpha_A(\mathbf{k}j)$ tends to the static polarizability α_A. The dipole moment $\mathbf{p}(\mathbf{k}j, \mathbf{r}_A)$ in turn generates an electric field at a point $\mathbf{r}_B$ which, neglecting angular dependence and numerical factors, has an amplitude given by (see e.g. (7.37))

$$\mathbf{E}_{A\to B}(\mathbf{k}j,\mathbf{r}_A) \sim \alpha_A(\mathbf{k}j)\mathbf{E}_V(\mathbf{k}j,\mathbf{r}_A)\frac{1}{R^3} \quad (\mathbf{R}=\mathbf{r}_A-\mathbf{r}_B) \tag{8.87}$$

The field at point $\mathbf{r}_B$ stemming from the $\mathbf{k}j$ contribution of the vacuum field in the presence of atom A is then $\mathbf{E}_V(\mathbf{k}j,\mathbf{r}_B)+\mathbf{E}_{A\to B}(\mathbf{k}j,\mathbf{r}_B)$. A second atom at $\mathbf{r}_B$ gets polarized by this field, and the $\mathbf{k}j$ contribution to its interaction with the field is

$$\mathcal{E}_B(\mathbf{k}j,\mathbf{R}) \sim \alpha_B(\mathbf{k}j)(\mathbf{E}_V(\mathbf{k}j,\mathbf{r}_B)+\mathbf{E}_{A\to B}(\mathbf{k}j,\mathbf{r}_B)+\text{h.c.})^2 \tag{8.88}$$

At the lowest possible order in α and neglecting terms which do not depend on $\mathbf{R}$, use of (8.87) gives

$$\mathcal{E}_B(\mathbf{k}j,\mathbf{R}) \sim \alpha_A(\mathbf{k}j)\alpha_B(\mathbf{k}j)\mathbf{E}_V(\mathbf{k}j,\mathbf{r}_A)\cdot\mathbf{E}_V^\dagger(\mathbf{k}j,\mathbf{r}_B)\frac{1}{R^3} \tag{8.89}$$

Summing over all $\mathbf{k}j$, transforming this sum into an integral and taking the quantum average on the vacuum of the field $\mid\{0_{\mathbf{k}j}\}\rangle$, the interaction term (8.89) leads to a total interaction energy of the form

$$\begin{aligned} V(\mathbf{R}) &\sim \int \langle\{0_{\mathbf{k}j}\}\mid \mathcal{E}_B(\omega,\mathbf{R})\mid\{0_{\mathbf{k}j}\}\rangle\omega^2 d\omega \\ &\sim \frac{1}{R^3}\int \alpha_A(\omega)\alpha_B(\omega) \\ &\quad\times \langle\{0_{\mathbf{k}j}\}\mid \mathbf{E}_V(\omega,\mathbf{r}_A)\cdot\mathbf{E}_V^\dagger(\omega,\mathbf{r}_B)\mid\{0_{\mathbf{k}j}\}\rangle\omega^2 d\omega \end{aligned} \tag{8.90}$$

Since V is $\mathbf{R}$-dependent, it plays the role of an interatomic interaction potential. The similarity with the situation discussed in Section 7.8 leads us to investigate if, using the static rather the dynamic polarizability in (8.90), $V(\mathbf{R})$ yields the far-zone expression for the van der Waals potential. In these conditions (8.90) yields

$$V(\mathbf{R}) \sim \frac{1}{R^3}\alpha_A\alpha_B\int\langle\{0_{\mathbf{k}j}\mid \mathbf{E}_V(\omega,\mathbf{r}_A)\cdot\mathbf{E}_V^\dagger(\omega,\mathbf{r}_B)\mid\{0_{\mathbf{k}j}\}\rangle\omega^2 d\omega \tag{8.91}$$

The quantity $\langle\{0_{\mathbf{k}j}\mid \mathbf{E}_V(\omega,\mathbf{r}_A)\cdot\mathbf{E}_V^\dagger(\omega,\mathbf{r}_B)\mid\{0_{\mathbf{k}j}\}\rangle$ in (8.91) is a fundamental quantity representing the correlations of three-dimensional vacuum fluctuations in the absence of charges and currents. Thus its explicit expression can only contain the fundamental constants $\hbar$ and c. Moreover, from the dimensional point of view its integral is an energy density, and because of translational invariance of the vacuum zero-point field it must depend on $\mathbf{R}=\mathbf{r}_B-\mathbf{r}_A$. The only scalar quantity satisfying these requirements in three-dimensional space is $\hbar c/R^4$. Thus, apart from angular dependences and numerical factors, we expect

$$V(R) \sim \alpha_A \alpha_B \hbar c \frac{1}{R^7} \tag{8.92}$$

This results agrees with the interpretation of $V(R)$ in terms of the Casimir-Polder potential, as comparison with (7.206) shows. A complete calculation confirms this interpretation. Thus van der Waals forces, at least in the far zone, can be seen as originating from vacuum fluctuations. More precisely, van der Waals forces seem to originate from the interaction of the atomic dipoles which are induced and correlated by the zero-point field (see also Power and Thirunamachandran 1993).

An alternative description of the Casimir force that invokes vacuum fluctations can be formulated; it has been formulated by Casimir (1949), Boyer (1969) and Power (1972). It has many similarities with Power's description of the Lamb shift that we have discussed previously in this section. In fact the idea is that the presence of two atoms changes the dispersion relation (i.e. the relationship between wavelength and frequency) of the normal modes of the field. As a consequence the zero-point energy E_{ZP} changes because of the presence of the atoms. The difference between the zero-point energy $E_{\text{ZP}}(R)$ when the separation of the two ground-state atoms is R and the zero-point energy $E_{\text{ZP}}(\infty)$ when the atoms are at infinite distance can be evaluated, and we find (for more detail see Boyer 1969)

$$E_{\text{ZP}}(R) - E_{\text{ZP}}(\infty) = -\frac{23}{4\pi} \hbar c \frac{\alpha_A \alpha_B}{R^7} \tag{8.93}$$

This R-dependent energy of the vacuum state yields an interatomic potential that coincides with the Casimir potential. Therefore the van der Waals force can be also considered as the R-dependent part of the change of the energy of the zero-point fluctuations due to the presence of the two atoms.

Analogously to spontaneous emission and to the Lamb shift, van der Waals forces can also be obtained without any reference to vacuum fluctuations, but in terms of source fields only (Milonni 1982, Milonni 1988, Milonni and Shih 1992). Here we only outline the calculations, referring the reader interested in more details to the paper by Milonni (1982). The idea used by Milonni is the following. The atom B at $\mathbf{r}_B$ changes the modes of the field (through its polarizability) and this yields a change in the radiation reaction field of the atom A at $\mathbf{r}_A$ from its free-space form: therefore the radiative level shifts of the energy levels of atom

A change too. The R-dependent part of the ground-state level shift should then give the van der Waals potential.

The field modes in the presence of the ground-state atom B can be obtained from the refractive index of the medium consisting of the vacuum plus the atom, similarly to how field modes are obtained in the presence of a dielectric medium. In our case the field modes when the atom B is present consist of the plane waves of free space plus the dipole field produced by the atom. These new modes are (Milonni 1982)

$$\begin{aligned} \mathbf{F}_{\mathbf{k}j}(\mathbf{x}) &= \mathbf{e}_{\mathbf{k}j} e^{i\mathbf{k}\cdot\mathbf{x}} \\ &+ \alpha_B(\omega_k) k^3 e^{i\mathbf{k}\cdot\mathbf{r}_B} e^{ikr} \left(\mathbf{e}_{\mathbf{k}j} \left(\frac{1}{kr} + \frac{i}{k^2 r^2} - \frac{1}{k^3 r^3} \right) \right. \\ &\left. - (\mathbf{e}_{\mathbf{k}j} \cdot \hat{r}) \hat{r} \left(\frac{1}{kr} + \frac{3i}{k^2 r^2} - \frac{3}{k^3 r^3} \right) \right) \end{aligned} \tag{8.94}$$

where $\mathbf{r} = \mathbf{x} - \mathbf{r}_B = r\hat{r}$ and $\alpha_B(\omega_k)$ is the ground-state polarizability of atom B. The first term on the RHS represents the free field modes and the second term the dipolar field scattered by atom B.

Using the mode functions (8.94) we can calculate the radiation reaction field of atom A in the presence of atom B, similarly to the calculation performed for the Lamb shift. The essential differences are that we now use the mode functions (8.94) rather than the free field modes and also that we use the multipolar coupling scheme, more convenient for this calculation. The Hamiltonian describing the interaction of the atom A with the radiation field is obtained from (4.98) as

$$\begin{aligned} H &= \sum_n E_n c_n^\dagger c_n + \sum_{\mathbf{k}j} \hbar\omega_k \left(a_{\mathbf{k}j}^\dagger a_{\mathbf{k}j} + \frac{1}{2} \right) \\ &- i \sum_{nm} \sum_{\mathbf{k}j} \sqrt{\frac{2\pi\hbar\omega_k}{V}} (\mu_{mn} \cdot \mathbf{F}_{\mathbf{k}j}(\mathbf{x}_A) c_m^\dagger c_n a_{\mathbf{k}j} - \text{h.c.}) \end{aligned} \tag{8.95}$$

where μ_{mn} is real and all atomic quantities refer to atom A.

We now proceed along the same lines as in the calculation of the Lamb shift. We consider the transverse electric field evaluated at the position of atom A

$$\mathbf{E}_\perp(\mathbf{x}_A, t) = i \sum_{\mathbf{k}j} \sqrt{\frac{2\pi\hbar\omega_k}{V}} \left(a_{\mathbf{k}j}(t) \mathbf{F}_{\mathbf{k}j}(\mathbf{x}_A) - a_{\mathbf{k}j}^\dagger(t) \mathbf{F}_{\mathbf{k}j}^*(\mathbf{x}_A) \right) \tag{8.96}$$

We then split the solution of the Heisenberg equation of motion for $a_{\mathbf{k}j}(t)$ into a vacuum part and a radiation reaction part, and from this we obtain a similar splitting for the positive frequency part of the electric field (8.96)

$$\mathbf{E}_{\perp}^{(+)} = \mathbf{E}_0^{(+)}(t) + \mathbf{E}_{RR}^{(+)}(t)$$
$$\mathbf{E}_0^{(+)}(t) = i\sum_{\mathbf{k}j}\sqrt{\frac{2\pi\hbar\omega_k}{V}}\, a_{\mathbf{k}j}(0)e^{-i\omega_k t}\mathbf{F}_{\mathbf{k}j}(\mathbf{x}_A)$$
$$\mathbf{E}_{RR}^{(+)}(t) = i\frac{2\pi}{V}\sum_{\mathbf{k}j}\sum_{m\ell}\omega_k\left(\boldsymbol{\mu}_{\ell m}\cdot\mathbf{F}_{\mathbf{k}j}^*(\mathbf{x}_A)\right)\mathbf{F}_{\mathbf{k}j}(\mathbf{x}_A)$$
$$\times\int_0^t e^{-i\omega_k(t-t')}c_\ell^\dagger(t')c_m(t')dt' \tag{8.97}$$

Milonni proposes to calculate the Lamb shift of the level n of the atom A (using the mode functions (8.94) that take into account the presence of the atom B) as (see Equation (8.67))

$$\Delta_n = -\frac{1}{2}\langle\{0_{\mathbf{k}j}\}, n \mid \boldsymbol{\mu}(t)\cdot\mathbf{E}_{\perp}(\mathbf{x}_A, t) \mid \{0_{\mathbf{k}j}\}, n\rangle \tag{8.98}$$

Adopting the normal ordering, when (8.97) is substituted into (8.98) the vacuum part vanishes and only radiation reaction gives a contribution

$$\Delta_n = -\frac{1}{2}\langle\{0_{\mathbf{k}j}\}, n \mid \left(\boldsymbol{\mu}(t)\cdot\mathbf{E}_{RR}^{(+)}(t) + \mathbf{E}_{RR}^{(-)}(t)\cdot\boldsymbol{\mu}(t)\right) \mid \{0_{\mathbf{k}j}\}, n\rangle \tag{8.99}$$

Using (8.74) we have

$$\boldsymbol{\mu}(t)\cdot\mathbf{E}_{RR}^{(+)}(t) = i\frac{2\pi}{V}\sum_{\mathbf{k}j}\sum_{m\ell pq}\omega_k\left(\boldsymbol{\mu}_{\ell m}\cdot\mathbf{F}_{\mathbf{k}j}^*(\mathbf{x}_A)\right)\left(\boldsymbol{\mu}_{pq}\cdot\mathbf{F}_{\mathbf{k}j}(\mathbf{x}_A)\right)$$
$$\times c_p^\dagger(t)c_q(t)\int_0^t e^{-i\omega_k(t-t')}c_\ell^\dagger(t')c_m(t')dt' \tag{8.100}$$

If we evaluate Δ_n at order e^2, we can approximate all atomic operators in the RHS of (8.100) with their free evolution

$$c_n(t) = e^{-iE_n t/\hbar}c_n(0) \tag{8.101}$$

and, after performing time integrations, Equation (8.100) becomes

$$\boldsymbol{\mu}(t)\cdot\mathbf{E}_{RR}^{(+)}(t) = \frac{2\pi}{V}\sum_{\mathbf{k}j}\sum_{m\ell pq}\omega_k\left(\boldsymbol{\mu}_{\ell m}\cdot\mathbf{F}_{\mathbf{k}j}^*(\mathbf{x}_A)\right)\left(\boldsymbol{\mu}_{pq}\cdot\mathbf{F}_{\mathbf{k}j}(\mathbf{x}_A)\right)$$
$$\times e^{i(\omega_{pq}-\omega_k)t}\frac{e^{i(\omega_{\ell m}+\omega_k)t}-1}{\omega_{\ell m}+\omega_k}c_p^\dagger(0)c_q(0)c_\ell^\dagger(0)c_m(0) \tag{8.102}$$

Finally

$$\langle\{0_{\mathbf{k}j}\},n \mid \boldsymbol{\mu}(t)\cdot\mathbf{E}^{(+)}_{RR}(t) \mid \{0_{\mathbf{k}j}\},n\rangle$$
$$= \frac{2\pi}{V}\sum_{\mathbf{k}j}\sum_{\ell}\omega_k \mid \boldsymbol{\mu}_{\ell n}\cdot\mathbf{F}_{\mathbf{k}j}(\mathbf{x}_A)\mid^2 \frac{1-e^{-i(\omega_k-\omega_{n\ell})t}}{\omega_k-\omega_{n\ell}} \tag{8.103}$$

and, taking the limit $t\to\infty$, we obtain Δ_n at large times as

$$\Delta_n = -\frac{2\pi}{V}\sum_{\mathbf{k}j}\sum_{\ell}\omega_k \mid \boldsymbol{\mu}_{\ell n}\cdot\mathbf{F}_{\mathbf{k}j}(\mathbf{x}_A)\mid^2 \frac{1}{\omega_k-\omega_{n\ell}} \tag{8.104}$$

The next step is the evaluation of the mode functions (8.94) at $\mathbf{r}_A$; retaining only corrections to the free-space modes that are linear in the polarizability α_B we have (Milonni 1982)

$$\mid \boldsymbol{\mu}_{\ell n}\cdot\mathbf{F}_{\mathbf{k}j}(\mathbf{x}_A)\mid^2 \simeq \frac{1}{V}(\boldsymbol{\mu}_{\ell m}\cdot\mathbf{e}_{\mathbf{k}j})^2$$
$$+\frac{2}{V}\mathrm{Re}\Big\{\alpha_B(\omega_k)k^3e^{ikR}e^{i\mathbf{k}\cdot\mathbf{R}}\Big[(\boldsymbol{\mu}_{\ell n}\cdot\mathbf{e}_{\mathbf{k}j})^2A^*(kR)$$
$$-(\mathbf{e}_{\mathbf{k}j}\cdot\boldsymbol{\mu}_{\ell n})(\mathbf{e}_{\mathbf{k}j}\cdot\hat{R})(\boldsymbol{\mu}_{\ell n}\cdot\hat{R})B^*(kR)\Big]\Big\} \tag{8.105}$$

where

$$A(kR)=\frac{1}{kR}+\frac{i}{(kR)^2}-\frac{1}{(kR)^3}\ ;\ B(kR)=\frac{1}{kR}+\frac{3i}{(kR)^2}-\frac{3}{(kR)^3} \tag{8.106}$$

Assuming that both atoms A and B are in their ground states g, after substitution of (8.105) into (8.104) and some lengthy calculation, we obtain the difference between the energy shift of the ground state of atom A when the two atoms are at distance R and when they are infinitely separated, as

$$\Delta_g(R)-\Delta_g(\infty)$$
$$=-\frac{4}{9\pi\hbar cR^2}\sum_{ij}\mid\mu_{ig}^{(A)}\mid^2\mid\mu_{jg}^{(B)}\mid^2\int_0^\infty \frac{k_{ig}^{(A)}k_{jg}^{(B)}u^4e^{-2uR}}{[k_{ig}^{(A)2}+u^2][k_{jg}^{(B)2}+u^2]}$$
$$\times\left(1+\frac{2}{uR}+\frac{5}{u^2R^2}+\frac{6}{u^3R^3}+\frac{3}{u^4R^4}\right) \tag{8.107}$$

where $k_{ig}^{(A,B)}=\omega_{ig}^{(A,B)}/c$ and the sums over i and j run over all excited levels of atoms A and B, respectively. Equation (8.107) coincides with the van der Waals potential (7.207). If (8.107) is approximated in the far zone we obviously get the same result as in (8.92).

This shows that radiation reaction arguments, without any mention of vacuum fluctuations, permit one to obtain the correct expression for the van der Waals forces.

We may conclude this section by stressing that, although the physical interpretation of the vacuum radiative processes that we have discussed can be given in terms of vacuum fluctuations as well as in terms of radiation reaction fields, both concepts are necessary for the internal consistency of quantum electrodynamics. In fact, it can be shown that both radiation reaction and free field are necessary to preserve the canonical commutation relations. A particular ordering of commuting operator can emphasize the role of one or the other in a particular physical phenomenon, but both are necessary for a self-consistent theory: they indeed represent "two sides of a coin" (Senitzky 1973, Milonni 1988).

Finally, we wish to spend a few words on the theory of dressing by zero-point fluctuations in the light of the arguments developed in this section. The physical interpretation of the virtual cloud proposed in Section 6.2 is remindful of the radiation reaction interpretation of radiative effects, because of the active role played by the atomic source in the formation of the cloud. This aspect is particularly evident in the case of van der Waals force in the far zone, which in Section 7.8 is presented as the interaction of one atom with the virtual field generated by the other. This is conceptually more similar to a radiation reaction than to a zero-point fluctuation point of view, also because in Section 7.8 no apparent recourse is made to zero-point fluctuations (that is, to field fluctuations in the absence of atoms). However, the dressing method that we have used throughout Chapters 6 to 8 is based on the structure of the complete ground state of the atom-field system, which is partitioned into a vacuum fluctuation and a radiation reaction part. This ground state, in our scheme, is perturbed by the atom-photon coupling which causes entanglement with the excited states of the atom-field system. Such an entanglement could also possibly be interpreted as due to a distortion of the unperturbed normal modes of the vacuum field. Thus the virtual cloud could also be visualized as due to the zero-point quanta populating the normal modes of the vacuum, distorted by the presence of the atom, rather than as a radiation reaction phenomenon. This second aspect of the theory of virtual clouds is perhaps particularly evident in the discussion of the radiative shift of a ground-state two-level atom presented in Section 6.4, where this shift is obtained as the quantum average of the complete Hamiltonian on the perturbed atom-photon ground state.

References

J.R. Ackerhalt, J.H. Eberly (1974). *Phys. Rev. D* **10,** 3350
J.R. Ackerhalt, P.L. Knight, J.H. Eberly (1973). *Phys. Rev. Lett.* **30,** 456
G. Barton (1987). *J. Phys. B* **20,** 879
R. Becker (1982). *Electromagnetic Fields and Interactions,* Vol. II (Dover Publications Inc., New York)
T.H. Boyer (1969). *Phys. Rev.* **180,** 19
H.B.G. Casimir (1949). *J. Chim. Phys.* (France) **46,** 407
C. Cohen-Tannoudji (1984). In *Les Houches, Session XXXVIII, 1982 - Tendances Actuelles en Physique Atomique,* G. Grynberg and R. Stora (eds.) (Elsevier Science Publishers B.V. 1984), p. 1
G. Compagno, R. Passante, F. Persico (1990). *Europhys. Lett.* **12,** 301
G. Compagno, R. Passante, F. Persico (1991). *Phys. Rev. A* **44,** 1956
J. Dalibard, J. Dupont-Roc, C. Cohen-Tannoudji (1982). *J. Physique* **43,** 1617
E.L. Feinberg (1980). *Usp. Fis. Nauk.* **132,** 255 [*Sov. Phys. Usp.* **23,** 629 (1980)]
R.P. Feynman (1961). *Solvay Institute Proceedings* (Interscience Publishers Inc., New York 1961), p. 76
H. Friedrich (1991). *Theoretical Atomic Physics* (Springer-Verlag, New York)
V.L. Ginzburg (1983). *Usp. Fiz. Nauk.* **140,** 687 [*Sov. Phys. Usp.* **26,** 713]
M. Grabowski (1990). *Ann. der Phys.* **47,** 391
J.D. Jackson (1962). *Classical Electrodynamics* (John Wiley and Sons, Inc., New York 1962)
L. Lo Cascio, F. Persico (1991). *J. Mod. Opt.* **39,** 87
E. Merzbacher (1961). *Quantum Mechanics* (John Wiley and Sons, New York)
P.W. Milonni (1976). *Phys. Rep.* **25,** 1
P.W. Milonni (1980). In *Foundations of Radiation Theory and Quantum Electrodynamics,* A.O. Barut (ed.) (Plenum Press, New York 1980), p. 1
P.W. Milonni (1982). *Phys. Rev. A* **25,** 1315
P.W. Milonni (1988). *Physica Scripta T* **21,** 102
P.W. Milonni, J.R. Ackerhalt, W.A. Smith (1973). *Phys. Rev. Lett.* **31,** 958
P.W. Milonni, M.L. Shih (1992). *Phys. Rev. A* **45,** 4241
P.W. Milonni, W.A. Smith (1975). *Phys. Rev. A* **11,** 814
A. Peres (1989). *Phys. Rev. D* **39,** 2943
A. Peres, W.K. Wootters (1985), *Phys. Rev. D* **32,** 1968
E.A. Power (1966). *Am. J. Phys.* **34,** 516
E.A. Power (1972). In *Magic Without Magic: John Archibald Wheeler,* J.R. Klauder (ed.) (Freeman, California)
E.A. Power, T. Thirunamachandran (1993). *Phys. Rev. A* **48,** 4761
F. Rohrlich (1965). *Classical Charged Particles* (Addison-Wesley Publ. Co., Redwood City 1965)
I.R. Senitzky (1973). *Phys. Rev. Lett.* **31,** 955
L. Spruch, E.J. Kelsey (1978). *Phys. Rev. A* **18,** 845
J. von Neumann (1955). *Mathematical Foundations of Quantum Mechanics* (Princeton University Press, Princeton)
V.F. Weisskopf (1981). *Physics Today,* Nov. 1981, p. 69
T. A. Welton (1948). *Phys. Rev.* **74,** 1157

Further reading

For further reading on the theory of partially dressed states:

F. Persico, E.A. Power, *Phys. Rev. A* **36,** 475 (1987)
G. Compagno, R. Passante, F. Persico, *Phys. Rev. A* **38,** 600 (1988)
G. Compagno, R. Passante, G.M. Palma, F. Persico, in *New Frontiers in Quantum Electrodynamics and Quantum Optics,* A.O. Barut (ed.) (Plenum Press, New York 1990), p. 129.

Appendix A

Multipolar expansion for the vector potential

We set out to obtain the expansion of the free e.m. field in terms of spherical waves. The starting point is the dyadic (Morse and Feshbach 1953)

$$\mathcal{I}e^{i\mathbf{k}\cdot\mathbf{x}} = 4\pi \sum_{\ell m} i^{\ell}\{-i\mathbf{P}^*_{\ell m}(u,v)\mathbf{L}_{\ell m}(\mathbf{x}) + \frac{1}{\sqrt{\ell(\ell+1)}} \times \left[\mathbf{C}^*_{\ell m}(u,v)\mathbf{M}_{\ell m}(\mathbf{x}) - i\mathbf{B}^*_{\ell m}(u,v)\mathbf{N}_{\ell m}(\mathbf{x})\right]\} \tag{A.1}$$

where ℓ is an integer which varies between 0 and ∞, m another integer which varies between $-\ell$ and $+\ell$, u and v are the angular polar coordinates of $\mathbf{k}$, whose corresponding unit vectors are $\hat{u}$ and $\hat{v}$ respectively, θ and φ are the angular polar coordinates of $\mathbf{x}$, whose corresponding unit vectors are $\hat{\theta}$ and $\hat{\varphi}$ respectively. The geometry of the system is described in Figure A.1. Moreover, in (A.1)

$$\mathbf{P}_{\ell m}(u,v) = \hat{\mathbf{k}} Y_\ell^m(u,v) \;;$$

$$\mathbf{B}_{\ell m}(u,v) = \frac{k}{\sqrt{\ell(\ell+1)}} \nabla Y_\ell^m(u,v) = \frac{\sqrt{\ell(\ell+1)}}{(2\ell+1)\sin u} \times \left\{ \hat{u}\left[\frac{\ell-m+1}{\ell+1} Y_{\ell+1}^m(u,v) - \frac{\ell+m}{\ell} Y_{\ell-1}^m(u,v)\right] + \hat{v}\frac{m(2\ell+1)}{\ell(\ell+1)} i Y_\ell^m(u,v)\right\} \;;$$

$$\mathbf{C}_{\ell m}(u,v) = -\hat{\mathbf{k}} \times \mathbf{B}_{\ell m}(u,v) = \frac{\sqrt{\ell(\ell+1)}}{(2\ell+1)\sin u} \times \left\{ \hat{u}\frac{m(2\ell+1)}{\ell(\ell+1)} i Y_\ell^m(u,v) - \hat{v}\left[\frac{\ell-m+1}{\ell+1} Y_{\ell+1}^m(u,v) - \frac{\ell+m}{\ell} Y_{\ell-1}^m(u,v)\right]\right\} \tag{A.2}$$

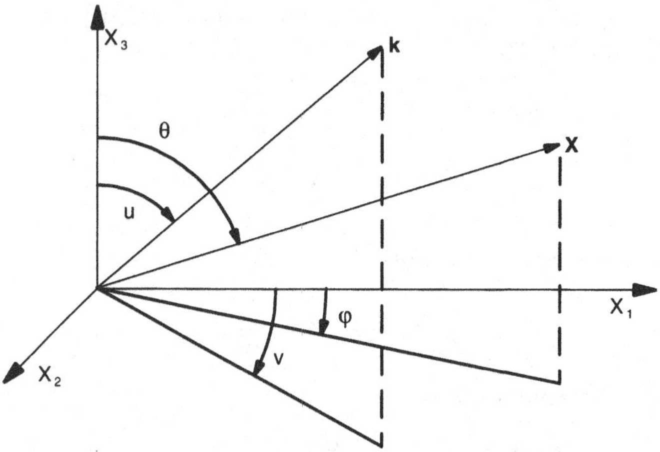

Fig. A.1. Polar coordinates for vectors **k** and **x**.

The spherical harmonics appearing in (A.2) are defined as (see e.g. Merzbacher 1961)

$$Y_\ell^m(u,v) = \sqrt{\frac{(2\ell+1)(\ell-m)!}{4\pi(\ell+m)!}} \times (-1)^m e^{imv} P_\ell^m(\cos u) \quad (m \geq 0) \; ;$$

$$Y_\ell^{-m}(u,v) = (-1)^m Y_\ell^{m^*}(u,v) \; ;$$

$$P_\ell^m(\xi) = \frac{1}{2^\ell \ell!}\left(1-\xi^2\right)^{m/2} \frac{d^{\ell+m}}{d\xi^{\ell+m}}\left(\xi^2-1\right)^\ell \; ;$$

$$\int Y_\ell^{m^*}(u,v)\, Y_{\ell'}^{m'}(u,v) d\Omega = \delta_{\ell\ell'}\delta_{mm'} \; ;$$

$$d\Omega = \sin u du dv \tag{A.3}$$

Finally

$$\mathbf{L}_{\ell m}(\mathbf{x}) = \frac{1}{2\ell+1}\left\{\ell j_{\ell-1}(kx)\left[\mathbf{P}_{\ell m}(\theta,\varphi) + \sqrt{\frac{\ell+1}{\ell}}\mathbf{B}_{\ell m}(\theta,\varphi)\right] - (\ell+1) j_{\ell+1}(kx)\left[\mathbf{P}_{\ell m}(\theta,\varphi) - \sqrt{\frac{\ell}{\ell+1}}\mathbf{B}_{\ell m}(\theta,\varphi)\right]\right\} \; ;$$

$$\mathbf{M}_{\ell m}(\mathbf{x}) = \sqrt{\ell(\ell+1)} j_\ell(kx) \mathbf{C}_{\ell m}(\theta,\varphi) \; ;$$

$$\mathbf{N}_{\ell m}(\mathbf{x}) = \frac{\ell(\ell+1)}{2\ell+1}\left\{j_{\ell-1}(kx)\left[\mathbf{P}_{\ell m}(\theta,\varphi) + \sqrt{\frac{\ell+1}{\ell}}\mathbf{B}_{\ell m}(\theta,\varphi)\right] + j_{\ell+1}(kx)\left[\mathbf{P}_{\ell m}(\theta,\varphi) - \sqrt{\frac{\ell}{\ell+1}}\mathbf{B}_{\ell m}(\theta,\varphi)\right]\right\} \tag{A.4}$$

where the spherical Bessel functions are defined as (see e.g. Cohen-Tannoudji *et al.* 1977)

$$j_\ell(\rho) = (-i)^\ell \rho^\ell \left(\frac{1}{\rho}\frac{d}{d\rho}\right)^\ell \frac{\sin\rho}{\rho} \; ;$$
$$\int_0^\infty x^2 j_\ell(kx) j_\ell(k'x) dx = \frac{\pi}{2k^2}\delta(k-k') \tag{A.5}$$

Introducing the new unit vectors

$$\mathbf{e}_1 = \frac{1}{\sqrt{2}}(\hat{u}+\hat{v}) \; ; \; \mathbf{e}_2 = \frac{1}{\sqrt{2}}(\hat{u}-\hat{v}) \tag{A.6}$$

one obtains after some algebra

$$\mathbf{e}_1 \cdot \mathcal{I} e^{i\mathbf{k}\cdot\mathbf{x}} = \mathbf{e}_1 e^{i\mathbf{k}\cdot\mathbf{x}} = 4\pi \sum_{\ell m} i^\ell \frac{1}{\sqrt{\ell(\ell+1)}}$$
$$\times \left[\mathbf{e}_1 \cdot \mathbf{C}^*_{\ell m}(u,v)\mathbf{M}_{\ell m}(\mathbf{x}) - i\mathbf{e}_2 \cdot \mathbf{C}^*_{\ell m}(u,v)\mathbf{N}_{\ell m}(\mathbf{x})\right] \; ;$$
$$\mathbf{e}_2 \cdot \mathcal{I} e^{i\mathbf{k}\cdot\mathbf{x}} = \mathbf{e}_2 e^{i\mathbf{k}\cdot\mathbf{x}} = 4\pi \sum_{\ell m} i^\ell \frac{1}{\sqrt{\ell(\ell+1)}}$$
$$\times \left[\mathbf{e}_2 \cdot \mathbf{C}^*_{\ell m}(u,v)\mathbf{M}_{\ell m}(\mathbf{x}) - i\mathbf{e}_1 \cdot \mathbf{C}^*_{\ell m}(u,v)\mathbf{N}_{\ell m}(\mathbf{x})\right] \tag{A.7}$$

Now $\hat{u}$ and $\hat{v}$ are mutually orthogonal and orthogonal to $\hat{\mathbf{k}}$. Therefore one can identify $\mathbf{e}_1$ and $\mathbf{e}_2$ in (A.6) with $\mathbf{e}_{\mathbf{k}1}$ and $\mathbf{e}_{\mathbf{k}2}$ in (1.84) for each $\mathbf{k}$. Thus, substituting (A.7) in (1.84) one has

$$\mathbf{A}(\mathbf{x},t) = 4\pi \sum_{\ell m} \frac{i^\ell}{\sqrt{\ell(\ell+1)}} \int_0^\infty \left\{ \int [\mathbf{e}_1 A_1(\mathbf{k}) + \mathbf{e}_2 A_2(\mathbf{k})] \right.$$
$$\cdot \mathbf{C}^*_{\ell m}(u,v) d\Omega \mathbf{M}_{\ell m}(\mathbf{x})$$
$$\left. - i \int [\mathbf{e}_2 A_1(\mathbf{k}) - \mathbf{e}_1 A_2(\mathbf{k})] \cdot \mathbf{C}^*_{\ell m}(u,v) d\Omega \mathbf{N}_{\ell m}(\mathbf{x}) \right\}$$
$$\times e^{-i\omega_k t} k^2 dk + \text{c.c.}$$
$$= 4\pi \sum_{\ell m} \frac{i^\ell}{\sqrt{\ell(\ell+1)}} \int_0^\infty \{A(k,\mathcal{M},\ell,m)\mathbf{M}_{\ell m}(\mathbf{x})$$
$$- iA(k,\mathcal{E},\ell,m)\mathbf{N}_{\ell m}(\mathbf{x})\} e^{-i\omega_k t} k dk + \text{c.c.} \tag{A.8}$$

where

$$A(k,\mathcal{M},\ell,m) = k \int [\mathbf{e}_1 A_1(\mathbf{k}) + \mathbf{e}_2 A_2(\mathbf{k})] \cdot \mathbf{C}^*_{\ell m}(u,v) d\Omega \; ;$$
$$A(k,\mathcal{E},\ell,m) = k \int [\mathbf{e}_2(\mathbf{k}) A_1(\mathbf{k}) - \mathbf{e}_1 A_2(\mathbf{k})] \cdot \mathbf{C}^*_{\ell m}(u,v) d\Omega \tag{A.9}$$

Finally, we introduce the vector spherical harmonics

$$\mathbf{Y}_{\ell m-} = i^{\ell+1}\sqrt{\frac{\ell}{2\ell+1}}\left(\mathbf{P}_{\ell m} + \sqrt{\frac{\ell+1}{\ell}}\mathbf{B}_{\ell m}\right) ;$$

$$\mathbf{Y}_{\ell m+} = i^{\ell-1}\sqrt{\frac{\ell+1}{2\ell+1}}\left(\mathbf{P}_{\ell m} - \sqrt{\frac{\ell}{\ell+1}}\mathbf{B}_{\ell m}\right) ;$$

$$\mathbf{Y}_{\ell m0} = i^{\ell}\mathbf{C}_{\ell m} \tag{A.10}$$

and recast (A.8) in the form

$$\mathbf{A}(\mathbf{x},t) = 4\pi\sum_{\ell m}\int_0^\infty \Big\{A(k,\mathcal{M},\ell,m)j_\ell(kx)\mathbf{Y}_{\ell m0}(\theta,\varphi)$$
$$+ A(k,\mathcal{E},\ell,m)\frac{1}{\sqrt{2\ell+1}}\Big[\sqrt{\ell}j_{\ell+1}(kx)\mathbf{Y}_{\ell m+}(\theta,\varphi)$$
$$- \sqrt{\ell+1}j_{\ell-1}(kx)\mathbf{Y}_{\ell m-}(\theta,\varphi)\Big]\Big\}e^{-i\omega_k t}kdk$$
$$+ \text{c.c.} \tag{A.11}$$

The vector spherical harmonics, in spite of their rather horrid appearance, are simple to work with. This is due to their rather simple algebraic properties, which we shall now briefly describe.

First we give a proof of the orthonormality relation

$$\int \mathbf{Y}^*_{\ell mi}\cdot\mathbf{Y}_{\ell' m' i'}d\Omega = \delta_{\ell\ell'}\delta_{mm'}\delta_{ii'} \quad (i = \pm, 0) \tag{A.12}$$

Since for any scalar ψ

$$\nabla^2\psi^*\psi = \psi^*\nabla^2\psi + \psi\nabla^2\psi^* + 2\nabla\psi^*\cdot\nabla\psi \tag{A.13}$$

from the definition (A.2) follows

$$\mathbf{B}^*_{\ell m}\cdot\mathbf{B}_{\ell' m'} = \frac{k^2}{\sqrt{\ell\ell'(\ell+1)(\ell'+1)}}\frac{1}{2}$$
$$\times\left(\nabla^2 Y_\ell^{m*}Y_{\ell'}^{m'} - Y_\ell^{m*}\nabla^2 Y_{\ell'}^{m'} - Y_{\ell'}^{m'}\nabla^2 Y_\ell^{m*}\right) \tag{A.14}$$

Now we use

$$\nabla^2 Y_\ell^m(u,v) = -\frac{\ell(\ell+1)}{k^2}Y_\ell^m(u,v) \tag{A.15}$$

in (A.14). (A.15) can be directly obtained from (A.3). Then we integrate (A.14) over Ω and k, and use Gauss's theorem to transform the first integration on the RHS from volume to surface. We easily obtain

$$\int\int \mathbf{B}^*_{\ell m}\cdot\mathbf{B}_{\ell' m'}d\Omega dk$$
$$=\frac{1}{\sqrt{\ell\ell'(\ell+1)(\ell'+1)}}\frac{1}{2}\int_S \nabla Y_\ell^{m*}Y_{\ell'}^{m'}\cdot\hat{\mathbf{k}}dS$$
$$+\frac{1}{2}\sqrt{\frac{\ell(\ell+1)}{\ell'(\ell'+1)}}\int\int Y_\ell^{m*}Y_{\ell'}^{m'}d\Omega dk$$
$$+\frac{1}{2}\sqrt{\frac{\ell'(\ell'+1)}{\ell(\ell+1)}}\int\int Y_\ell^{m*}Y_{\ell'}^{m'}d\Omega dk \tag{A.16}$$

where S is a spherical surface in k-space centred at the origin. Since in the surface integral

$$\nabla Y_\ell^{m*}Y_{\ell'}^{m'} = Y_\ell^{m*}\nabla Y_{\ell'}^{m'} + Y_{\ell'}^{m'}\nabla Y_\ell^{m*} \tag{A.17}$$

both contributions are proportional to **B**, which from definition (A.2) is perpendicular to $\hat{\mathbf{k}}$, the whole integrand is perpendicular to $\hat{\mathbf{k}}$. Therefore the surface integral vanishes. The other two integrals in (A.16) are both equal to $\delta_{\ell\ell'}\delta_{mm'}$. Hence

$$\int\int \mathbf{B}^*_{\ell m}\cdot\mathbf{B}_{\ell' m'}d\Omega dk = \int \delta_{\ell\ell'}\delta_{mm'}dk\,;$$
$$\text{or}\quad \int \mathbf{B}^*_{\ell m}\cdot\mathbf{B}_{\ell' m'}d\Omega = \delta_{\ell\ell'}\delta_{mm'} \tag{A.18}$$

Moreover from definition (A.2) $\mathbf{B}_{\ell m}$ is perpendicular to $\hat{\mathbf{k}}$, hence

$$\mathbf{C}^*_{\ell m}\cdot\mathbf{C}_{\ell' m'} = \hat{\mathbf{k}}\times\mathbf{B}^*_{\ell m}\cdot\hat{\mathbf{k}}\times\mathbf{B}_{\ell' m'} = \mathbf{B}^*_{\ell m}\cdot\mathbf{B}_{\ell' m'}\,;$$
$$\text{or}\quad \int \mathbf{C}^*_{\ell m}\cdot\mathbf{C}_{\ell' m'}d\Omega = \delta_{\ell\ell'}\delta_{mm'} \tag{A.19}$$

Further, in view of (A.3), we have

$$\int \mathbf{P}^*_{\ell m}\cdot\mathbf{P}_{\ell' m'}d\Omega = \delta_{\ell\ell'}\delta_{mm'} \tag{A.20}$$

Finally, from definition (A.2), **P**, **B** and **C** are mutually perpendicular. When this is taken into account and (A.18), (A.19) and (A.20) are used, the orthonormality property (A.12) follows easily.

A second useful property of vector spherical harmonics is the sum rule

$$\sum_{m=-\ell}^{\ell}\mathbf{Y}^*_{\ell mi}\cdot\mathbf{Y}_{\ell mi'} = \frac{2\ell+1}{4\pi}\delta_{ii'} \tag{A.21}$$

The proof of (A.21) is based on three intermediate steps:

i) generalization of (A.13) to the sum of N scalar functions ψ_i $(i = 1, \ldots N)$

$$\nabla^2 \sum_i \psi_i^* \psi_i = \sum_i \psi_i^* \nabla^2 \psi_i + \sum_i \psi_i \nabla^2 \psi_i^* + 2 \sum_i |\nabla \psi_i|^2 \qquad \text{(A.22)}$$

ii) the sum rule of Legendre polynomials in the form

$$\sum_{m=-\ell}^{\ell} Y_\ell^{m*} Y_\ell^m = \frac{2\ell+1}{4\pi} \qquad \text{(A.23)}$$

which yields

$$\nabla^2 \sum_m Y_\ell^{m*} Y_\ell^m = 0 \qquad \text{(A.24)}$$

For $\psi_i = Y_\ell^m$, $i = m$, $N = 2\ell + 1$, (A.24) implies vanishing of the LHS of (A.22).

iii) The analogue of (A.15) in real space

$$\nabla^2 Y_\ell^m(\theta, \varphi) = -\frac{\ell(\ell+1)}{x^2} Y_\ell^m(\theta, \varphi) \qquad \text{(A.25)}$$

From the three intermediate results (A.22), (A.24) and (A.25) it follows that

$$\sum_m |\nabla Y_\ell^m|^2 = \frac{\ell(\ell+1)}{x^2} \sum_m |Y_\ell^m|^2 = \frac{\ell(\ell+1)(2\ell+1)}{4\pi x^2} \qquad \text{(A.26)}$$

On the basis of (A.26) and (A.23)

$$\begin{aligned}
\sum_m |\mathbf{Y}_{\ell m0}|^2 &= \sum_m |\mathbf{C}_{\ell m}|^2 = \sum_m |\mathbf{B}_{\ell m}|^2 = \frac{x^2}{\ell(\ell+1)} \sum_m |\nabla Y_\ell^m|^2 \\
&= \frac{2\ell+1}{4\pi}\,; \\
\sum_m |\mathbf{Y}_{\ell m+}|^2 &= \frac{\ell+1}{2\ell+1}\left(\sum_m |\mathbf{P}_{\ell m}|^2 + \frac{\ell}{\ell+1}\sum_m |\mathbf{B}_{\ell m}|^2\right) \\
&= \frac{2\ell+1}{4\pi}\,; \\
\sum_m |\mathbf{Y}_{\ell m-}|^2 &= \frac{\ell}{2\ell+1}\left(\sum_m |\mathbf{P}_{\ell m}|^2 + \frac{\ell+1}{\ell}\sum_m |\mathbf{B}_{\ell m}|^2\right) \\
&= \frac{2\ell+1}{4\pi}
\end{aligned} \qquad \text{(A.27)}$$

Moreover

$$\sum_m \mathbf{Y}^*_{\ell m+} \cdot \mathbf{Y}_{\ell m-} = \sum_m \mathbf{Y}^*_{\ell m-} \cdot \mathbf{Y}_{\ell m+}$$

$$= -\frac{\sqrt{\ell(\ell+1)}}{2\ell+1} \sum_m (|\, \mathbf{P}_{\ell m} \,|^2 - |\, \mathbf{B}_{\ell m} \,|^2) = 0 \qquad \text{(A.28)}$$

and, in view of the orthogonality of $\mathbf{C}_{\ell m}$ to both $\mathbf{B}_{\ell m}$ and $\mathbf{P}_{\ell m}$,

$$\sum_m \mathbf{Y}^*_{\ell m0} \cdot \mathbf{Y}_{\ell m\pm} = \sum \mathbf{Y}^*_{\ell m\pm} \cdot \mathbf{Y}_{\ell m0} = 0 \qquad \text{(A.29)}$$

Results (A.27), (A.28) and (A.29) are summarized in the sum rule (A.21).

References

C. Cohen-Tannoudji, B. Diu, F. Lalöe (1977). *Quantum Mechanics*, vol. 2 (John Wiley and Sons, Paris)

E. Merzbacher (1961). *Quantum Mechanics* (John Wiley and Sons, New York)

P.M. Morse, H. Feshbach (1953). *Methods of Theoretical Physics*, part II (McGraw-Hill Co, New York)

Further reading

Other presentations of vector spherical harmonics can be found in

J.M. Blatt, V.F. Weisskopf, *Theoretical Nuclear Physics* (Dover Publications Inc., New York 1991).

V.B. Berestetskii, E.M. Lifshitz, L.P. Pitaevskii, *Quantum Electrodynamics* (Pergamon Press, London 1982).

Further information on Bessel functions and spherical harmonics can be found in

M. Abramovitz, I.A. Stegun (eds.), *Handbook of Mathematical Functions* (Dover Publications Inc., New York 1965).

Appendix B

Electric polarization and magnetization of the Schrödinger field

This appendix is largely based on the paper by Power and Thirunamachandran (1983) and on the book by Craig and Thirunamachandran (1984).

The electric charge density ρ and the current density $\mathbf{j}$ of the Schrödinger field in Section 4.1 are essential in determining the dynamics of the electromagnetic field, since they appear as sources in the equations of motion of the latter. We shall now introduce two quantities, the electric polarization field and the magnetization field, which are related to ρ and to $\mathbf{j}$ respectively and which play an important role in electrodynamics, as is plain in the development of Chapter 4.

The best way of introducing the electric polarization field is to express the electric charge density of the classical Schrödinger field at an observation point $\mathbf{r}$ in integral form as

$$\rho(\mathbf{r}) = q\psi^*(\mathbf{r})\psi(\mathbf{r}) = q\int \psi^*(\mathbf{x})\psi(\mathbf{x})\delta(\mathbf{r}-\mathbf{x})d^3\mathbf{x} \tag{B.1}$$

If the charge density distribution has a symmetry point at $\mathbf{R}$, it is convenient to refer the observation to this symmetry point. For example, in atomic physics a most obvious symmetry point is the atomic nucleus, and in this case $\mathbf{R}$ would be the nuclear position. Thus it is appropriate to express the δ-function appearing in (B.1) as

$$\begin{aligned}\delta(\mathbf{r}-\mathbf{x}) = \delta[(\mathbf{r}-\mathbf{R})-(\mathbf{x}-\mathbf{R})] &= \delta(\mathbf{r}-\mathbf{R}) \\ &+ (\mathbf{x}-\mathbf{R})_i\{\nabla_i\delta[(\mathbf{r}-\mathbf{R})-(\mathbf{x}-\mathbf{R})]\}_{\mathbf{x}-\mathbf{R}=0} \\ &+\frac{1}{2!}(\mathbf{x}-\mathbf{R})_i(\mathbf{x}-\mathbf{R})_j \\ &\times\{\nabla_i\nabla_j\delta[(\mathbf{r}-\mathbf{R})-(\mathbf{x}-\mathbf{R})]\}_{\mathbf{x}-\mathbf{R}=0}+\cdots\end{aligned} \tag{B.2}$$

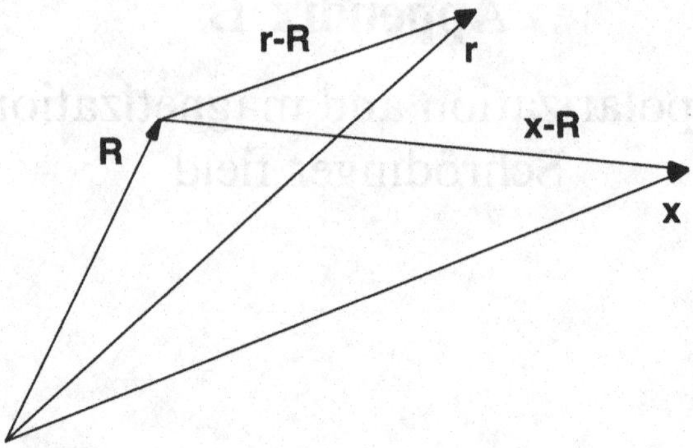

Fig. B.1 **R** is the symmetry point of the charge distribution, **r** is the observation point and **x** is the integration variable.

Expression (B.2) is a Taylor expansion performed over a distribution. Thus all derivatives of the δ function are to be taken in the sense of the theory of generalized functions. Moreover, in (B.2), $\nabla_i = \partial/\partial(\mathbf{x}-\mathbf{R})_i$. These may be transformed into $\nabla_i = \partial/\partial(\mathbf{r}-\mathbf{R})_i$ remembering that for a function $f(x-y)$ one has $\partial f/\partial x = -\partial f/\partial y$ (Jeffreys and Jeffreys 1972). This changes the sign of all odd terms in (B.2). Moreover, for fixed **R** one has $\partial/\partial(\mathbf{r}-\mathbf{R})_i = \partial/\partial \mathbf{r}_i$. Consequently,

$$\delta(\mathbf{r}-\mathbf{x}) = \delta(\mathbf{r}-\mathbf{R}) - (\mathbf{x}-\mathbf{R})_i \nabla_i \delta(\mathbf{r}-\mathbf{R}) + \frac{1}{2!}(\mathbf{x}-\mathbf{R})_i(\mathbf{x}-\mathbf{R})_j \nabla_i \nabla_j \delta(\mathbf{r}-\mathbf{R}) - \cdots \tag{B.3}$$

where $\nabla_i = \partial/\partial r_i$. Substitution of (B.3) into (B.1) yields

$$\begin{aligned}\rho(\mathbf{r}) &= q\int \psi^*(\mathbf{x})\psi(\mathbf{x})d^3\mathbf{x}\delta(\mathbf{r}-\mathbf{R}) \\ &\quad - \nabla_i\Big\{q\int \psi^*(\mathbf{x})\psi(\mathbf{x})(\mathbf{x}-\mathbf{R})_i d^3\mathbf{x}\delta(\mathbf{r}-\mathbf{R}) \\ &\quad - \frac{1}{2!}q\int \psi^*(\mathbf{x})\psi(\mathbf{x})(\mathbf{x}-\mathbf{R})_i(\mathbf{x}-\mathbf{R})_j d^3\mathbf{x}\nabla_j\delta(\mathbf{r}-\mathbf{R}) + \cdots\Big\} \\ &= q\int \psi^*(\mathbf{x})\psi(\mathbf{x})d^3\mathbf{x}\delta(\mathbf{r}-\mathbf{R}) \\ &\quad - \nabla_i\int \psi^*(\mathbf{x})\psi(\mathbf{x})\mu_i(\mathbf{x})d^3\mathbf{x}\delta(\mathbf{r}-\mathbf{R}) \\ &\quad + \nabla_i\int \psi^*(\mathbf{x})\psi(\mathbf{x})Q_{ij}(\mathbf{x})d^3\mathbf{x}\nabla_j\delta(\mathbf{r}-\mathbf{R}) - \ldots\end{aligned} \tag{B.4}$$

where $\mu_i(\mathbf{x})$ are the components of a vector and $Q_{ij}(\mathbf{x})$ those of a dyadic (Morse and Feshbach 1953) defined as

$$\boldsymbol{\mu}(\mathbf{x}) = q(\mathbf{x} - \mathbf{R}) \; ; \; \underline{\underline{Q}}(\mathbf{x}) = \frac{1}{2!} q(\mathbf{x} - \mathbf{R})(\mathbf{r} - \mathbf{R}) \tag{B.5}$$

$\boldsymbol{\mu}(\mathbf{x})$ is called the dipole moment and $\underline{\underline{Q}}(\mathbf{x})$ the quadrupole moment about $\mathbf{R}$. Higher order terms in the expansion (B.4) contain tensors of increasing rank and the expansion itself is called the multipolar expansion of $\rho(\mathbf{r})$ about $\mathbf{R}$. The advantage of such an expansion, if $\mathbf{R}$ is a point of symmetry for $\rho(\mathbf{r})$, is evident. For example, the first term equals the total charge of the Schrödinger field localized at point $\mathbf{R}$. For a neutral atom this cancels with the nuclear charge. Moreover, the integrals in the successive terms normally take a simple form. Furthermore, one can introduce the electric polarization field $\mathbf{p}(\mathbf{r})$ by putting

$$\rho(\mathbf{r}) = q \int \psi^*(\mathbf{x})\psi(\mathbf{x}) d^3\mathbf{x}\, \delta(\mathbf{r} - \mathbf{R}) - \nabla \cdot \mathbf{p}(\mathbf{r}) \tag{B.6}$$

where

$$\begin{aligned}
\mathbf{p}(\mathbf{r}) &= \int \psi^*(\mathbf{x}) \mathbf{p}(\mathbf{r}, \mathbf{x}) \psi(\mathbf{x}) d^3\mathbf{x} \; ; \\
\mathbf{p}(\mathbf{r}, \mathbf{x}) &= \boldsymbol{\mu}(\mathbf{x}) \delta(\mathbf{r} - \mathbf{R}) - \underline{\underline{Q}}(\mathbf{x}) \cdot \nabla \delta(\mathbf{r} - \mathbf{R}) + \cdots \\
&= q(\mathbf{x} - \mathbf{R}) \left\{ 1 - \frac{1}{2!} (\mathbf{x} - \mathbf{R}) \cdot \nabla + \frac{1}{3!} [(\mathbf{x} - \mathbf{R}) \cdot \nabla]^2 - \cdots \right\} \\
&\times \delta(\mathbf{r} - \mathbf{R}) \\
&= q(\mathbf{x} - \mathbf{R}) \int_0^1 \Big\{ 1 - \lambda(\mathbf{x} - \mathbf{R}) \cdot \nabla \\
&+ \frac{1}{2!} [\lambda(\mathbf{x} - \mathbf{R}) \cdot \nabla]^2 - \cdots \Big\} d\lambda\, \delta(\mathbf{r} - \mathbf{R})
\end{aligned} \tag{B.7}$$

From (B.6) it is obvious that $\mathbf{p}(\mathbf{r})$ indicates how much the charge distribution of the Schrödinger field differs from a point charge centred at $\mathbf{R}$. Furthermore, since for any f($\mathbf{r}$) one has

$$\left\{ 1 - \mathbf{a} \cdot \nabla + \frac{1}{2!} (\mathbf{a} \cdot \nabla)^2 - \cdots \right\} f(\mathbf{r}) = f(\mathbf{r} - \mathbf{a}) \tag{B.8}$$

where $\mathbf{a}$ is an arbitrary vector, one can express (B.7) in the compact form

$$\mathbf{p}(\mathbf{r}, \mathbf{x}) = q(\mathbf{x} - \mathbf{R}) \int_0^1 \delta[\mathbf{r} - \mathbf{R} - \lambda(\mathbf{x} - \mathbf{R})] d\lambda \tag{B.9}$$

We now turn to the magnetization field. For this, it is usual to express the electric current operator in terms of the velocity field $\dot{\mathbf{x}}$ of the

Schrödinger field. We proceed as follows. First we use expression (3.92) to obtain the velocity of the field as

$$\dot{\mathbf{X}} = \int \dot{\psi}^*(\mathbf{x})\mathbf{x}\psi(\mathbf{x})d^3\mathbf{x} + \int \psi^*(\mathbf{x})\mathbf{x}\dot{\psi}(\mathbf{x})d^3\mathbf{x} \tag{B.10}$$

It should be noted that here $\mathbf{x}$ is the space coordinate, an independent variable to be treated on the same footing as t: consequently, its time derivative must be taken to be zero. From (4.3) one obtains

$$\begin{aligned}\dot{\mathbf{X}} &= \frac{1}{i\hbar}\int \{-\psi^*(\mathbf{x})H_P\mathbf{x}\psi(\mathbf{x}) + \psi^*(\mathbf{x})\mathbf{x}H_P\psi(\mathbf{x})\}d^3\mathbf{x} \\ &= \frac{1}{i\hbar}\int \psi^*(\mathbf{x})[\mathbf{x}, H_P]\psi(\mathbf{x})d^3\mathbf{x}\end{aligned} \tag{B.11}$$

where we have introduced an operator H_P, corresponding to a single-particle Hamiltonian, acting on $\mathbf{x}$, as

$$\begin{aligned}H_P &= \frac{1}{2m}\left(\frac{\hbar}{i}\nabla - \frac{q}{c}\mathbf{A}\right)^2 - qV \equiv \frac{1}{2m}\left(\mathbf{p} - \frac{q}{c}\mathbf{A}\right)^2 - qV\,; \\ \dot{\psi} &= \frac{1}{i\hbar}H_P\psi\,;\ \dot{\psi}^* = -\frac{1}{i\hbar}\psi^* H_P\end{aligned} \tag{B.12}$$

and the single-particle momentum operator such that

$$[x_i, p_j] = i\hbar\delta_{ij}\,;\ [x_i, p_j^2] = 2i\hbar p_i \tag{B.13}$$

On the basis of (B.13), (B.11) yields

$$\dot{\mathbf{X}} = \frac{1}{m}\int \psi^*(\mathbf{x})\left(\mathbf{p} - \frac{q}{c}\mathbf{A}\right)\psi(\mathbf{x})d^3\mathbf{x} \tag{B.14}$$

which can be symmetrized using the fact that $\mathbf{p} = -i\hbar\nabla$ is a Hermitian operator, not to be confused with the polarization field, as

$$\begin{aligned}\dot{\mathbf{X}} = \frac{1}{2m}\int \Bigg\{&\psi^*(\mathbf{x})\left(\frac{\hbar}{i}\nabla - \frac{q}{c}\mathbf{A}\right)\psi(\mathbf{x}) \\ &+ \psi(\mathbf{x})\left(-\frac{\hbar}{i}\nabla - \frac{q}{c}\mathbf{A}\right)\psi^*(\mathbf{x})\Bigg\}d^3\mathbf{x}\end{aligned} \tag{B.15}$$

This suggests a way of defining the velocity $\dot{\mathbf{x}}$ of the Schrödinger field such that

$$\begin{aligned}\dot{\mathbf{X}} &= \frac{1}{2}\int \{\psi^*(\mathbf{x})\dot{\mathbf{x}}\psi(\mathbf{x}) + \psi(\mathbf{x})\dot{\mathbf{x}}^*\psi^*(\mathbf{x})\}d^3\mathbf{x}\,; \\ \dot{\mathbf{x}} &= \frac{1}{m}\left(\frac{\hbar}{i}\nabla - \frac{q}{c}\mathbf{A}\right) = \frac{1}{i\hbar}[\mathbf{x}, H_P]\end{aligned} \tag{B.16}$$

We note that this new field is real within the space of functions $\psi(\mathbf{x})$ which vanish sufficiently fast as $x \to \infty$. In fact, in this case one can use the definition (B.16) of $\dot{\mathbf{x}}$ and the second of (4.30) to obtain

$$\int \psi(\mathbf{x})\dot{\mathbf{x}}\psi^*(\mathbf{x})d^3\mathbf{x} = \int \psi^*(\mathbf{x})\dot{\mathbf{x}}^*\psi(\mathbf{x})d^3\mathbf{x} \tag{B.17}$$

Consequently, the current density (4.6) can be expressed in terms of the velocity field as

$$\mathbf{j}(\mathbf{x}) = q\psi^*(\mathbf{x})\dot{\mathbf{x}}\psi(\mathbf{x}) \tag{B.18}$$

We are now ready to discuss the magnetization field. In the same spirit of our treatment leading to the electric polarization field, we introduce the observation point $\mathbf{r}$, the symmetry point $\mathbf{R}$ and use (B.2) to obtain

$$\begin{aligned} j_i(\mathbf{r}) &= \frac{1}{2}q\int \psi^*(\mathbf{x})[\dot{x}_i\delta(\mathbf{r}-\mathbf{x}) + \delta(\mathbf{r}-\mathbf{x})\dot{x}_i]\psi(\mathbf{x})d^3\mathbf{x} \\ &= \frac{1}{2}q\int \psi^*(\mathbf{x})\Big\{\dot{x}_i\Big[1-(\mathbf{x}-\mathbf{R})_j\nabla_j. \\ &\quad + \frac{1}{2!}(\mathbf{x}-\mathbf{R})_j(\mathbf{x}-\mathbf{R})_k\nabla_j\nabla_k \\ &\quad - \frac{1}{3!}(\mathbf{x}-\mathbf{R})_j(\mathbf{x}-\mathbf{R})_k(\mathbf{x}-\mathbf{R})_\ell\nabla_j\nabla_k\nabla_\ell + \cdots] \\ &\quad + [1-(\mathbf{x}-\mathbf{R})_j\nabla_j + \frac{1}{2!}(\mathbf{x}-\mathbf{R})_j(\mathbf{x}-\mathbf{R})_k\nabla_j\nabla_k \\ &\quad - \frac{1}{3!}(\mathbf{x}-\mathbf{R})_j(\mathbf{x}-\mathbf{R})_k(\mathbf{x}-\mathbf{R})_\ell\nabla_j\nabla_k\nabla_\ell + \cdots]\dot{x}_i\Big\} \\ &\quad \times \psi(\mathbf{x})d^3\mathbf{x}\delta(\mathbf{r}-\mathbf{R}) \end{aligned} \tag{B.19}$$

The symmetric form of j_i is important for hermiticity, since $\dot{\mathbf{x}}$ is a differential operator which does not commute with $\delta(\mathbf{r}-\mathbf{x})$. We also use (B.7) in the form

$$\begin{aligned} p_i(\mathbf{r}) &= q\int \psi^*(\mathbf{x})(\mathbf{x}-\mathbf{R})_i\Big\{1 - \frac{1}{2!}(\mathbf{x}-\mathbf{R})_j\nabla_j \\ &\quad + \frac{1}{3!}(\mathbf{x}-\mathbf{R})_j(\mathbf{x}-\mathbf{R})_k\nabla_j\nabla_k \\ &\quad - \frac{1}{4!}(\mathbf{x}-\mathbf{R})_j(\mathbf{x}-\mathbf{R})_k(\mathbf{x}-\mathbf{R})_\ell\nabla_j\nabla_k\nabla_\ell\Big\} \\ &\quad \times \psi(\mathbf{x})d^3\mathbf{x}\delta(\mathbf{r}-\mathbf{R}) \end{aligned} \tag{B.20}$$

The time derivative of this expansion, using the same approach as that leading from (B.11) to (B.14), yields

$$\dot{p}_i(\mathbf{r}) = q\frac{1}{i\hbar}\int \psi^*(\mathbf{x})\left[(\mathbf{x}-\mathbf{R})_i\left\{1-\frac{1}{2!}(\mathbf{x}-\mathbf{R})_j\nabla_j+\cdots\right\}, H_P\right]$$
$$\times\, \psi(\mathbf{x})d^3\mathbf{x}\delta(\mathbf{r}-\mathbf{R}) \tag{B.21}$$

The commutators within the above integral can be performed assuming that $\mathbf{R}$ is a fixed parameter which commutes with the single-particle operator $\mathbf{p} = -i\hbar\nabla$ (where $\nabla = \partial/\partial\mathbf{x}$). Then it is an easy matter to obtain the array of commutators

$$\frac{1}{i\hbar}[(\mathbf{x}-\mathbf{R})_i, H_P] = (\dot{\overline{\mathbf{x}-\mathbf{R}}})_i = \dot{x}_i\,;$$

$$\frac{1}{i\hbar}[(\mathbf{x}-\mathbf{R})_i(\mathbf{x}-\mathbf{R})_j, H_P] = (\mathbf{x}-\mathbf{R})_i\dot{x}_j + \dot{x}_i(\mathbf{x}-\mathbf{R})_j$$

$$=\frac{1}{2}\Big\{\dot{x}_j(\mathbf{x}-\mathbf{R})_i$$

$$+\,\dot{x}_i(\mathbf{x}-\mathbf{R})_j + (\mathbf{x}-\mathbf{R})_i\dot{x}_j + (\mathbf{x}-\mathbf{R})_j\dot{x}_i\Big\}\,;$$

$$\frac{1}{i\hbar}[(\mathbf{x}-\mathbf{R})_i(\mathbf{x}-\mathbf{R})_j(\mathbf{x}-\mathbf{R})_k, H_P]$$

$$= (\mathbf{x}-\mathbf{R})_i(\mathbf{x}-\mathbf{R})_j\dot{x}_k + (\mathbf{x}-\mathbf{R})_i\dot{x}_j(\mathbf{x}-\mathbf{R})_k$$

$$+\,\dot{x}_i(\mathbf{x}-\mathbf{R})_j(\mathbf{x}-\mathbf{R})_k$$

$$=\frac{1}{2}\Big\{\dot{x}_k(\mathbf{x}-\mathbf{R})_i(\mathbf{x}-\mathbf{R})_j + \dot{x}_j(\mathbf{x}-\mathbf{R})_i(\mathbf{x}-\mathbf{R})_k$$

$$+\,\dot{x}_i(\mathbf{x}-\mathbf{R})_j(\mathbf{x}-\mathbf{R})_k + (\mathbf{x}-\mathbf{R})_i(\mathbf{x}-\mathbf{R})_j\dot{x}_k$$

$$+(\mathbf{x}-\mathbf{R})_i(\mathbf{x}-\mathbf{R})_k\dot{x}_j + (\mathbf{x}-\mathbf{R})_j(\mathbf{x}-\mathbf{R})_k\dot{x}_i\Big\} \tag{B.22}$$

and similarly for the others. In the last step of each expression in (B.22) the commutators have been taken to the symmetric form by use of the commutation relations

$$[(\mathbf{x}-\mathbf{R})_i, \dot{x}_j] = \frac{\hbar}{im}[x_i, \nabla_j]\delta_{ij} = \frac{i\hbar}{m}\delta_{ij} \tag{B.23}$$

based on expression (B.16) for $\dot{\mathbf{x}}$. Substituting (B.22) into (B.21), one gets after some algebra

$$\begin{aligned}\dot{p}_i(\mathbf{r}) = \frac{1}{2}q \int \psi^*(\mathbf{x})\Big\{ \dot{x}_i \Big[1 - \frac{1}{2!}(\mathbf{x}-\mathbf{R})_j\nabla_j. \\ + \frac{1}{3!}(\mathbf{x}-\mathbf{R})_j(\mathbf{x}-\mathbf{R})_k\nabla_j\nabla_k - \cdots\Big] \\ + \Big[1 - \frac{1}{2!}(\mathbf{x}-\mathbf{R})_j\nabla_j + \frac{1}{3!}(\mathbf{x}-\mathbf{R})_j(\mathbf{x}-\mathbf{R})_k\nabla_j\nabla_k - \cdots\Big]\dot{x}_i \\ + \nabla_j\dot{x}_j(\mathbf{x}-\mathbf{R})_i\Big[-\frac{1}{2!} + \frac{2}{3!}(\mathbf{x}-\mathbf{R})_k\nabla_k \\ - \frac{3}{4!}(\mathbf{x}-\mathbf{R})_k(\mathbf{x}-\mathbf{R})_\ell\nabla_k\nabla_\ell + \cdots\Big] \\ + \Big[-\frac{1}{2!} + \frac{2}{3!}(\mathbf{x}-\mathbf{R})_k\nabla_k - \frac{3}{4!}(\mathbf{x}-\mathbf{R})_k(\mathbf{x}-\mathbf{R})_\ell\nabla_k\nabla_\ell + \cdots\Big] \\ \times \nabla_j(\mathbf{x}-\mathbf{R})_i\dot{x}_j\Big\}\psi(\mathbf{x})d^3\mathbf{x}\delta(\mathbf{r}-\mathbf{R}) \qquad \text{(B.24)}\end{aligned}$$

Subtracting (B.24) from (B.19) and using

$$\frac{1}{n!} - \frac{1}{(n+1)!} = \frac{n}{(n+1)!} \qquad \text{(B.25)}$$

one obtains

$$\begin{aligned}j_i(\mathbf{r}) - \dot{p}_i(\mathbf{r}) = \frac{1}{2}q \int \psi^*(\mathbf{x})\Big\{\Big[\nabla_j\dot{x}_i(\mathbf{x}-\mathbf{R})_j - \nabla_j\dot{x}_j(\mathbf{x}-\mathbf{R})_i\Big] \\ \times \Big[-\frac{1}{2!} + \frac{2}{3!}(\mathbf{x}-\mathbf{R})_k\nabla_k - \frac{3}{4!}(\mathbf{x}-\mathbf{R})_k(\mathbf{x}-\mathbf{R})_\ell\nabla_k\nabla_\ell + \cdots\Big] \\ + \Big[-\frac{1}{2!} + \frac{2}{3!}(\mathbf{x}-\mathbf{R})_k\nabla_k - \frac{3}{4!}(\mathbf{x}-\mathbf{R})_k(\mathbf{x}-\mathbf{R})_\ell\nabla_k\nabla_\ell + \cdots\Big] \\ \times \Big[\nabla_j(\mathbf{x}-\mathbf{R})_j\dot{x}_i - \nabla_j(\mathbf{x}-\mathbf{R})_i\dot{x}_j\Big]\Big\}\psi(\mathbf{x})d^3\mathbf{x}\delta(\mathbf{r}-\mathbf{R}) \qquad \text{(B.26)}\end{aligned}$$

Remembering that the ∇ appearing explicitly in (B.26) is $\nabla = \partial/\partial\mathbf{r}$, using $[\mathbf{a}\times(\mathbf{b}\times\mathbf{c})] = a_j b_i c_j - a_j b_j c_i$ and defining

$$\begin{aligned}\mathbf{n}(\mathbf{r},\mathbf{x}) = q(\mathbf{x}-\mathbf{R})\Big\{\frac{1}{2!} - \frac{2}{3!}(\mathbf{x}-\mathbf{R})\cdot\nabla + \frac{3}{4!}[(\mathbf{x}-\mathbf{R})\cdot\nabla]^2 - \cdots\Big\} \\ \times \delta(\mathbf{r}-\mathbf{R}) \qquad \text{(B.27)}\end{aligned}$$

expression (B.26) can be cast in the form

$$\mathbf{j}(\mathbf{r}) - \dot{\mathbf{p}}(\mathbf{r}) = c \int \psi^*(\mathbf{x}) \nabla \times \mathbf{m}(\mathbf{r}, \mathbf{x}) \psi(\mathbf{x}) d^3\mathbf{x} \tag{B.28}$$

where

$$\mathbf{m}(\mathbf{r}, \mathbf{x}) = \frac{1}{2c} [\mathbf{n}(\mathbf{r}, \mathbf{x}) \times \dot{\mathbf{x}} - \dot{\mathbf{x}} \times \mathbf{n}(\mathbf{r}, \mathbf{x})] \tag{B.29}$$

The magnetization of the Schrödinger field is defined as

$$\mathbf{m}(\mathbf{r}) = \int \psi^*(\mathbf{x}) \mathbf{m}(\mathbf{r}, \mathbf{x}) \psi(\mathbf{x}) d^3\mathbf{x} \tag{B.30}$$

and, in view of (B.28), satisfies (Healy 1982)

$$\mathbf{j}(\mathbf{r}) - \dot{\mathbf{p}}(\mathbf{r}) = c \nabla \times \mathbf{m}(\mathbf{r}) \tag{B.31}$$

Expansion (B.27) can be used in turn to expand $\mathbf{m}(\mathbf{r})$ in a multipole series. For this purpose, the quantities

$$\mathbf{m}^{(1)}(\mathbf{r}, \mathbf{x}) = \frac{1}{2!2} \frac{q}{c} [(\mathbf{x} - \mathbf{R}) \times \dot{\mathbf{x}} - \dot{\mathbf{x}} \times (\mathbf{x} - \mathbf{R})] \delta(\mathbf{r} - \mathbf{R}) \; ;$$

$$\underline{\underline{m}}^{(2)}(\mathbf{r}, \mathbf{x}) = -\frac{1}{3!} \frac{q}{c} \{[(\mathbf{x} - \mathbf{R}) \cdot \nabla \delta(\mathbf{r} - \mathbf{R})](\mathbf{x} - \mathbf{R}) \times \dot{\mathbf{x}}$$
$$- \dot{\mathbf{x}} \times (\mathbf{x} - \mathbf{R})[(\mathbf{x} - \mathbf{R}) \cdot \nabla \delta(\mathbf{r} - \mathbf{R})]\} \tag{B.32}$$

are the leading terms of a multipole expansion of $\mathbf{m}(\mathbf{r}, \mathbf{x})$ which is the analogue of the multipole expansion for $\mathbf{p}(\mathbf{r}, \mathbf{x})$ given in (B.7). The quantities $\int \psi^*(\mathbf{x}) \mathbf{m}^{(1)} \psi(\mathbf{x}) d^3\mathbf{x}$ and $\int \psi^*(\mathbf{x}) \underline{\underline{m}}^{(2)} \psi(\mathbf{x}) d^3\mathbf{x}$ are the magnetic dipole and the magnetic quadrupole moment of the Schrödinger field about $\mathbf{R}$.

Finally, a procedure analogous to that followed to express $\mathbf{p}(\mathbf{r}, \mathbf{x})$ in closed form (B.9), yields from (B.27)

$$\mathbf{n}(\mathbf{r}, \mathbf{x}) = q(\mathbf{x} - \mathbf{R}) \int_0^1 \lambda \delta[\mathbf{r} - \mathbf{R} - \lambda(\mathbf{x} - \mathbf{R})] d\lambda \tag{B.33}$$

Moreover, we remark that on the basis of

$$\delta[\mathbf{r} - \mathbf{R} - \lambda(\mathbf{x} - \mathbf{R})] = \{1 - \lambda(\mathbf{x} - \mathbf{R}) \cdot \nabla$$
$$+ \frac{1}{2!} [\lambda(\mathbf{x} - \mathbf{R}) \cdot \nabla]^2 - \cdots\} \delta(\mathbf{r} - \mathbf{R}) \tag{B.34}$$

it is easy to obtain the following useful identity

$$[(\mathbf{x} - \mathbf{R}) \cdot \nabla]\delta[\mathbf{r} - \mathbf{R} - \lambda(\mathbf{x} - \mathbf{R})] = -\frac{d}{d\lambda}\delta[\mathbf{r} - \mathbf{R} - \lambda(\mathbf{x} - \mathbf{R})] \quad \text{(B.35)}$$

References

D.P. Craig, T. Thirunamachandran (1984). *Molecular Quantum Electrodynamics* (Academic Press Inc., London)

W.P. Healy (1982). *Non-relativistic Quantum Electrodynamics* (Academic Press, London)

H. Jeffreys, B. Jeffreys (1972). *Methods of Mathematical Physics* (Cambridge University Press, Cambridge)

P.M. Morse, H. Feshbach (1953). *Methods of Theoretical Physics* , vol. 1 (McGraw-Hill Book Co., New York)

E.A. Power, T. Thirunamachandran (1983). *Phys. Rev. A* **28,** 2649

Further reading

A short account on the theory of generalized functions can be found in the following book.

M.J. Lighthill, *Introduction to Fourier Analysis and Generalized Functions* (Cambridge University Press, London 1970).

Appendix C

Rayleigh-Schrödinger perturbation theory

Consider a system described by a Hamiltonian $H = H_0 + \eta h$, where the eigensolutions of H_0 are known as solutions of the eigenvalue equation

$$H_0 \mid \phi\rangle = E_0 \mid \phi\rangle \tag{C.1}$$

ηh is a perturbation, and η is a dimensionless parameter which we assume to be small enough for our purposes. We also assume that an approximation to the eigensolutions of the complete Hamiltonian H exists in the form of a series of powers of η. Thus we shall look for solutions of the eigenvalue equation

$$H \mid \psi\rangle = E \mid \psi\rangle \tag{C.2}$$

in the form (Brueckner 1959)

$$\begin{aligned} &\mid \psi\rangle = S \mid \phi\rangle \; ; S = 1 + \eta S_1 + \eta^2 S_2 + \cdots \; ; \\ &E = E_0 + \eta E_1 + \eta^2 E_2 + \cdots \end{aligned} \tag{C.3}$$

We finally assume that $\mid \psi\rangle$ is nondegenerate. Substituting (C.3) in (C.2) yields

$$\begin{aligned} &(H_0 + \eta h)\big(1 + \eta S_1 + \eta^2 S_2 + \cdots\big) \mid \phi\rangle \\ &= (E_0 + \eta E_1 + \eta^2 E_2 + \ldots)\big(1 + \eta S_1 + \eta^2 S_2 + \cdots\big) \mid \phi\rangle \end{aligned} \tag{C.4}$$

Equating the coefficients of equal powers in η in (C.4), one obtains

$$\begin{aligned} S_n \mid \phi\rangle = &\frac{1}{E_0 - H_0} h S_{n-1} \mid \phi\rangle - \sum_{m=1}^{n-1} E_m \frac{1}{E_0 - H_0} S_{n-m} \mid \phi\rangle \\ &- \frac{1}{E_0 - H_0} E_n \mid \phi\rangle \quad (n = 1, 2, \ldots) \end{aligned} \tag{C.5}$$

where $S_0 = 1$. One must, however, eliminate from the RHS of (C.5) the terms which diverge. These divergences arise when $(E_0 - H_0)^{-1}$ acts on $| \phi\rangle$, in view of (C.1). Thus we shall request that $S_n | \phi\rangle$ be orthogonal to $| \phi\rangle$ for any n. Then divergences can arise only from the first and the last term on the RHS of (C.5), and one can get rid of them by requiring that

$$\begin{aligned}
&\frac{1}{E_0 - H_0} h S_{n-1} | \phi\rangle - \frac{1}{E_0 - H_0} E_n | \phi\rangle \\
&= \frac{1}{E_0 - H_0}(1 - P_0) h S_{n-1} | \phi\rangle + \frac{1}{E_0 - H_0} P_0 h S_{n-1} | \phi\rangle \\
&- \frac{1}{E_0 - H_0} E_n | \phi\rangle \\
&= \frac{1}{E_0 - H_0}(1 - P_0) h S_{n-1} | \phi\rangle
\end{aligned} \tag{C.6}$$

Here P_0 is the projection operator on eigenstate $| \phi\rangle$ of H_0. Evidently, vector (C.6) is orthogonal to $| \phi\rangle$, and this makes $S_n | \phi\rangle$ in (C.5) orthogonal to $| \phi\rangle$ as required. Moreover, (C.6) is satisfied only if

$$P_0 h S_{n-1} | \phi\rangle = E_n | \phi\rangle \text{ or } E_n = \langle\phi | h S_{n-1} | \phi\rangle \tag{C.7}$$

and with this condition (C.5) reduces to

$$\begin{aligned}
S_n | \phi\rangle &= \frac{1}{E_0 - H_0}(1 - P_0) h S_{n-1} | \phi\rangle \\
&- \sum_{m=1}^{n-1} \langle\phi | h S_{m-1} | \phi\rangle \frac{1}{E_0 - H_0} S_{n-m} | \phi\rangle \quad (n = 1, 2, \ldots)
\end{aligned} \tag{C.8}$$

Using (C.8) it is a simple matter to evaluate

$$\begin{aligned}
| \psi^{(1)}\rangle &= \{1 + \eta S_1\} | \phi\rangle\,; \\
S_1 | \phi\rangle &= \frac{1}{E_0 - H_0}(1 - P_0) h | \phi\rangle\,; \\
E^{(1)} &= E_0 + \langle\phi | \eta h | \phi\rangle
\end{aligned} \tag{C.9}$$

$$\begin{aligned}
| \psi^{(2)}\rangle &= \{1 + \eta S_1 + \eta^2 S_2\} | \phi\rangle\,; \\
S_2 | \phi\rangle &= \left\{ \frac{1}{E_0 - H_0}(1 - P_0) h \frac{1}{E_0 - H_0}(1 - P_0) h. \right. \\
&\left. - \langle\phi | h | \phi\rangle \frac{1}{(E_0 - H_0)^2}(1 - P_0) h \right\} | \phi\rangle\,; \\
E^{(2)} &= E_0 + \langle\phi | \eta h | \phi\rangle + \langle\phi | \eta h \frac{1}{E_0 - H_0}(1 - P_0) \eta h | \phi\rangle
\end{aligned} \tag{C.10}$$

$$
\begin{aligned}
| \psi^{(3)}\rangle &= \{1+\eta S_1+\eta^2 S_2+\eta^3 S_3\} \mid \phi\rangle ; \\
S_3 \mid \phi\rangle &= \left\{\frac{1}{E_0-H_0}(1-P_0)h\frac{1}{E_0-H_0}(1-P_0)h\frac{1}{E_0-H_0}(1-P_0)h\right. \\
&\quad -\frac{1}{E_0-H_0}(1-P_0)h\langle\phi \mid h \mid \phi\rangle\frac{1}{(E_0-H_0)^2}(1-P_0)h \\
&\quad -\langle\phi \mid h \mid \phi\rangle\frac{1}{(E_0-H_0)^2}(1-P_0)h\frac{1}{E_0-H_0}(1-P_0)h \\
&\quad +\langle\phi \mid h \mid \phi\rangle^2\frac{1}{(E_0-H_0)^3}(1-P_0)h \\
&\quad \left.-\langle\phi \mid h\frac{1}{E_0-H_0}(1-P_0)h \mid \phi\rangle\frac{1}{(E_0-H_0)^2}(1-P_0)h\right\} \mid \phi\rangle ; \\
E^{(3)} &= E_0+\langle\phi \mid \eta h \mid \phi\rangle+\langle\phi \mid \eta h\frac{1}{E_0-H_0}(1-P_0)\eta h \mid \phi\rangle \\
&\quad +\langle\phi \mid \eta h\frac{1}{E_0-H_0}(1-P_0)\eta h\frac{1}{E_0-H_0}(1-P_0)\eta h \mid \phi\rangle \\
&\quad -\langle\phi \mid \eta h \mid \phi\rangle\langle\phi \mid \eta h\frac{1}{(E_0-H_0)^2}(1-P_0)\eta h \mid \phi\rangle \qquad \text{(C.11)}
\end{aligned}
$$

It should be noted that the various approximations $| \psi^{(n)}\rangle$ to $| \psi\rangle$ in (C.9) to (C.11) are not yet normalized. Assuming that $\langle\phi \mid \phi\rangle = 1$, in fact one has

$$
\begin{aligned}
Z_{(1)}^2 &= \langle\psi^{(1)} \mid \psi^{(1)}\rangle = 1+O(\eta^2) ; \\
Z_{(2)}^2 &= \langle\psi^{(2)} \mid \psi^{(2)}\rangle = 1+\langle\phi \mid \eta h(1-P_0) \\
&\quad \times\frac{1}{(E_0-H_0)^2}(1-P_0)\eta h \mid \phi\rangle+O(\eta^3) ; \\
Z_{(3)}^2 &= \langle\psi^{(3)} \mid \psi^{(3)}\rangle \\
&= 1+\eta^2\langle\phi \mid S_1^\dagger S_1 \mid \phi\rangle \\
&\quad +\eta^3\langle\phi \mid \left(S_1^\dagger S_2+S_2^\dagger S_1\right) \mid \phi\rangle+O(\eta^4) \qquad \text{(C.12)}
\end{aligned}
$$

Choosing to normalize $| \psi^{(n)}\rangle$ up to terms of order η^n, one has

$$
\begin{aligned}
Z_{(1)}^{-1} \mid \psi^{(1)}\rangle &= \left\{1+\frac{1}{E_0-H_0}(1-P_0)\eta h\right\} \mid \phi\rangle ; \\
Z_{(2)}^{-1} \mid \psi^{(2)}\rangle &= \left\{1+\frac{1}{E_0-H_0}(1-P_0)\eta h\right. \\
&\quad +\frac{1}{E_0-H_0}(1-P_0)\eta h\frac{1}{E_0-H_0}(1-P_0)\eta h \\
&\quad -\langle\phi \mid \eta h \mid \phi\rangle\frac{1}{(E_0-H_0)^2}(1-P_0)\eta h \\
&\quad \left.-\frac{1}{2}\langle\phi \mid \eta h(1-P_0)\frac{1}{(E_0-H_0)^2}(1-P_0)\eta h \mid \phi\rangle\right\} \mid \phi\rangle ;
\end{aligned}
$$

$$Z_{(3)}^{-1} \mid \psi^{(3)}\rangle = \left\{ 1 + \eta S_1 - \eta^2 \left[\frac{1}{2} \langle \phi \mid S_1^\dagger S_1 \mid \phi\rangle - S_2 \right] - \eta^3 \left[\frac{1}{2} \langle \phi \mid (S_1^\dagger S_2 + S_2^\dagger S_1) \mid \phi\rangle + \frac{1}{2} \langle \phi \mid S_1^\dagger S_1 \mid \phi\rangle S_1 - S_3 \right] \right\} \mid \phi\rangle \qquad \text{(C.13)}$$

Finally, it should be mentioned that the perturbation expansions developed in this appendix may retain part of their validity also in the presence of eigenstates $\mid \phi\rangle'$ of H_0 degenerate with $\mid \phi\rangle$. They fail at order n, however, if degeneracy is not removed at order $n - 1$ and h connects $\mid \phi\rangle$ and $\mid \phi\rangle'$ at order n (see e.g. Schiff 1955).

References

K.A. Brueckner (1959). in *The Many Body Problem* (Dunod, Paris) p. 47

L.I. Schiff (1955). *Quantum Mechanics* (McGraw-Hill and Kogakusha Company Ltd, Tokyo)

Further reading

A simple treatment of the Rayleigh-Schrödinger perturbation theory can also be found in

E. Merzbacher, *Quantum Mechanics* (John Wiley and Sons, New York 1961).

Appendix D

Sum rules for the nonrelativistic hydrogen atom

We shall derive in what follows some useful sum rules involving the radial matrix elements of the spinless hydrogen atom (Passante and Power 1987). It should be emphasized that summation over N includes integration over the continuous part of the spectrum of the hydrogen Hamiltonian H_0, whereas summation over M runs over $M = -1, 0, +1$ only. The starting point is the Thomas-Reiche-Kuhn sum rule (see e.g. Merzbacher 1961) specialized for the ground state

$$\sum_{NM} E_N \mid \langle u_{N1M} \mid \mathbf{x}_e \mid u_{100} \rangle \mid^2 = \frac{3\hbar^2}{2m} \tag{D.1}$$

where E_N is the difference between the energy of u_{N1M} and that of u_{100}. Extensive use will also be made of the following expressions for the matrix elements of the electron momentum

$$\begin{aligned} \langle u_{N1M} \mid \mathbf{p} \mid u_{100} \rangle &= \frac{\hbar}{i} \langle u_{N1M} \mid \nabla^{x_e} \mid u_{100} \rangle \\ &= \frac{i\hbar}{a} \langle u_{N1M} \mid \hat{x}_e \mid u_{100} \rangle \end{aligned} \tag{D.2}$$

$$\begin{aligned} \langle u_{N1M} \mid \mathbf{p} \mid u_{100} \rangle &= \frac{m}{i\hbar} \langle u_{N1M} \mid [\mathbf{x}_e, H_0] \mid u_{100} \rangle \\ &= \frac{im}{\hbar} E_N \langle u_{N1M} \mid \mathbf{x}_e \mid u_{100} \rangle \end{aligned} \tag{D.3}$$

i) Starting from (D.2) one has

$$\langle u_{N1M} \mid \mathbf{p} \mid u_{100} \rangle = \frac{1}{\sqrt{4\pi}} \frac{i\hbar}{a} \langle R_{N1} \mid R_{10} \rangle \int Y_1^{M*}(\theta_e, \varphi_e) \hat{x}_e d\Omega \tag{D.4}$$

where

$$\hat{x}_e = \hat{x}_1 \sin\theta_e \cos\varphi_e + \hat{x}_2 \sin\theta_e \sin\varphi_e + \hat{x}_3 \cos\theta_e \tag{D.5}$$

From (A.3) one easily obtains

$$\sin\theta_e \cos\varphi_e = \sqrt{\frac{2\pi}{3}}\left[Y_1^{-1}(\theta_e,\varphi_e) - Y_1^1(\theta_e,\varphi_e)\right]\;;$$
$$\sin\theta_e \sin\varphi_e = i\sqrt{\frac{2\pi}{3}}\left[Y_1^{-1}(\theta_e,\varphi_e) + Y_1^1(\theta_e,\varphi_e)\right]\;;$$
$$\cos\theta_e = \sqrt{\frac{4\pi}{3}}Y_1^0 \qquad \text{(D.6)}$$

Substitution of (D.5) and (D.6) in (D.4) leads to the following integrals which are immediately evaluated using (A.3)

$$\int Y_1^{M*}(\theta_e,\varphi_e)\sin\theta_e\cos\varphi_e d\Omega$$
$$= \sqrt{\frac{2\pi}{3}}\int Y_1^{M*}(\theta_e,\varphi_e)\left[Y_1^{-1}(\theta_e,\varphi_e) - Y_1^1(\theta_e,\varphi_e)\right]d\Omega$$
$$= \sqrt{\frac{2\pi}{3}}(\delta_{M,-1} - \delta_{M,1});$$
$$\int Y_1^{M*}(\theta_e,\varphi_e)\sin\theta_e\sin\varphi_e d\Omega$$
$$= i\sqrt{\frac{2\pi}{3}}\int Y_1^{M*}(\theta_e,\varphi_e)\left[Y_1^{-1}(\theta_e,\varphi_e) + Y_1^1(\theta_e,\varphi_e)\right]d\Omega$$
$$= i\sqrt{\frac{2\pi}{3}}(\delta_{M,-1} + \delta_{M,1});$$
$$\int Y_1^{M*}(\theta_e,\varphi_e)\cos\theta_e d\Omega$$
$$= \sqrt{\frac{4\pi}{3}}\int Y_1^{M*}(\theta_e,\varphi_e)Y_1^0(\theta_e,\varphi_e)d\Omega = \sqrt{\frac{4\pi}{3}}\delta_{M,0} \qquad \text{(D.7)}$$

Thus (D.4) yields the following equalities

$$\langle u_{N1M} \mid \mathbf{p} \mid u_{100}\rangle = \frac{i\hbar}{a}\langle R_{N1} \mid R_{10}\rangle\frac{1}{\sqrt{3}}\left[\hat{x}_1\frac{1}{\sqrt{2}}(\delta_{M,-1} - \delta_{M,1}) + \hat{x}_2\frac{1}{\sqrt{2}}(\delta_{M,-1} + \delta_{M,1}) + \hat{x}_3\delta_{M,0}\right]\;;$$
$$\mid \langle u_{N1M} \mid \mathbf{p} \mid u_{100}\rangle \mid^2 = \frac{\hbar^2}{3a^2}(\langle R_{N1} \mid R_{10}\rangle)^2\;;$$
$$\sum_M \mid \langle u_{N1M} \mid \mathbf{p} \mid u_{100}\rangle \mid^2 = \frac{\hbar^2}{a^2}(\langle R_{N1} \mid R_{10}\rangle)^2 \qquad \text{(D.8)}$$

The last equality in (D.8) can be used together with (D.3) and (D.1) to obtain

$$\sum_N \frac{1}{E_N}(\langle R_{N1} \mid R_{10}\rangle)^2 = \frac{a^2}{\hbar^2}\sum_{NM}\frac{1}{E_N} \mid \langle u_{N1M} \mid \mathbf{p} \mid u_{100}\rangle \mid^2$$
$$= \frac{m^2a^2}{\hbar^4}\sum_{NM} E_N \mid \langle u_{N1M} \mid \mathbf{x}_e \mid u_{100}\rangle \mid^2$$
$$= \frac{3ma^2}{2\hbar^2} \tag{D.9}$$

which is the first useful sum rule, and which can also be put in the form

$$\sum_N \frac{1}{\omega_N}(\langle R_{N1} \mid R_{10}\rangle)^2 = \frac{3ma^2}{2\hbar} \tag{D.10}$$

where $\omega_N = E_N/\hbar$.

ii) Use of the last equality in (D.8) and of (D.3) yields

$$\sum_N \frac{1}{\omega_N^2}(\langle R_{N1} \mid R_{10}\rangle)^2 = a^2\sum_{NM}\frac{1}{E_N^2} \mid \langle u_{N1M} \mid \mathbf{p} \mid u_{100}\rangle \mid^2$$
$$= \frac{m^2a^2}{\hbar^2}\sum_{NM} \mid \langle u_{N1M} \mid \mathbf{x}_e \mid u_{100}\rangle \mid^2$$
$$= \frac{m^2a^2}{\hbar^2}\langle u_{100} \mid \mathbf{x}_e^2 \mid u_{100}\rangle$$
$$= \frac{m^2a^2}{4\pi\hbar^2}\langle R_{10} \mid x_e^2 \mid R_{10}\rangle \tag{D.11}$$

iii) The same procedure as in ii) gives

$$\sum_N \frac{1}{\omega_N^3}(\langle R_{N1} \mid R_{10}\rangle)^2 = a^2\hbar\sum_{NM}\frac{1}{E_N^3} \mid \langle u_{N1M} \mid \mathbf{p} \mid u_{100}\rangle \mid^2$$
$$= \frac{m^2a^2}{\hbar}\sum_{NM}\frac{1}{E_N} \mid \langle u_{N1M} \mid \mathbf{x}_e \mid u_{100}\rangle \mid^2$$
$$= \frac{3m^2a^2}{2e^2\hbar}\alpha_H \tag{D.12}$$

where we have introduced the static polarizability of the nonrelativistic hydrogen atom in the ground state (see e.g. Bransden and Joachain 1986)

$$\alpha_H = \frac{2}{3}\sum_{NM}\frac{\mid \langle u_{N1M} \mid e\mathbf{x}_e \mid u_{100}\rangle \mid^2}{\hbar\omega_N} \tag{D.13}$$

iv) Consider preliminarily the equality

$$\sum_M \langle u_{100} \mid \mathbf{x}_e \mid u_{N1M}\rangle \cdot \langle u_{N1M} \mid \hat{x}_e \mid u_{100}\rangle$$
$$= \frac{1}{4\pi}\sum_M \langle R_{10} \mid x_e \mid R_{N1}\rangle\langle R_{N1} \mid R_{10}\rangle \left| \int Y_1^M(\theta_e, \varphi_e)\hat{x}_e d\Omega \right|^2$$
$$= \langle R_{10} \mid x_e \mid R_{N1}\rangle\langle R_{N1} \mid R_{10}\rangle \tag{D.14}$$

where (D.7) has been used. From (D.14), (D.2) and (D.3) one obtains

$$\sum_N \frac{1}{\omega_N}\langle R_{N1} \mid x_e \mid R_{10}\rangle\langle R_{N1} \mid R_{10}\rangle$$
$$= \sum_{NM} \frac{1}{\omega_N}\langle u_{100} \mid \mathbf{x}_e \mid u_{N1M}\rangle \cdot \langle u_{N1M} \mid \hat{x}_e \mid u_{100}\rangle$$
$$= -i\frac{a}{\hbar}\sum_{NM} \frac{1}{\omega_N}\langle u_{100} \mid \mathbf{x}_e \mid u_{N1M}\rangle \cdot \langle u_{N1M} \mid \mathbf{p} \mid u_{100}\rangle$$
$$= \frac{ma}{\hbar}\sum_{NM} \langle u_{100} \mid \mathbf{x}_e \mid u_{N1M}\rangle \cdot \langle u_{N1M} \mid \mathbf{x}_e \mid u_{100}\rangle$$
$$= \frac{ma}{\hbar}\langle u_{100} \mid x_e^2 \mid u_{100}\rangle = \frac{ma}{4\pi\hbar}\langle R_{10} \mid x_e^2 \mid R_{10}\rangle \tag{D.15}$$

v) The same procedure as in iv) gives

$$\sum_N \frac{1}{\omega_N^2}\langle R_{N1} \mid x_e \mid R_{10}\rangle\langle R_{N1} \mid R_{10}\rangle$$
$$= \sum_{NM} \frac{1}{\omega_N^2}\langle u_{100} \mid \mathbf{x}_e \mid u_{N1M}\rangle \cdot \langle u_{N1M} \mid \hat{x}_e \mid u_{100}\rangle$$
$$= -i\frac{a}{\hbar}\sum_{NM} \frac{1}{\omega_N^2}\langle u_{100} \mid \mathbf{x}_e \mid u_{N1M}\rangle \cdot \langle u_{N1M} \mid \mathbf{p} \mid u_{100}\rangle$$
$$= ma\sum_{NM} \frac{1}{E_N}\langle u_{100} \mid \mathbf{x}_e \mid u_{N1M}\rangle \cdot \langle u_{N1M} \mid \mathbf{x}_e \mid u_{100}\rangle$$
$$= \frac{3ma}{2e^2}\alpha_H \tag{D.16}$$

and

vi)

$$\sum_N \langle R_{N1} \mid x_e \mid R_{10}\rangle\langle R_{N1} \mid R_{10}\rangle = \frac{3a}{2} \tag{D.17}$$

References

B.H. Bransden, C.J. Joachain (1986). *Physics of Atoms and Molecules* (Longman Inc., New York)

E. Merzbacher (1961). *Quantum Mechanics* (John Wiley and Sons, New York)

R. Passante, E.A. Power (1987). *Phys. Rev. A* **35,** 188

Appendix E
From Gauss system to SI

The two most commonly used unit systems, when dealing with electromagnetic phenomena, are the system of Gauss units and the Système International (SI). This book has been written using Gauss units. An abundant literature exists on the relationship between the two systems (see e.g. Leroy 1985). Perhaps the situation is most clearly described in an Appendix to Jackson's book (1988), which is concluded by two very useful tables. The first can be used to convert formulae from one system to the other, and the second can be used to convert physical quantities from one system to the other.

In the present appendix we shall take the very practical attitude of reproducing these two tables, introducing only minor changes to adapt them to the needs of our book. In Table E.1 which follows, a correspondence is established between symbols in Gauss units and groups of symbols in SI units. To convert any expression or equation in this book from Gauss to SI, one has simply to substitute in both sides of the expression or equation the Gauss symbols with the corresponding SI groups of symbols. The procedure can be inverted in order to go from SI to Gauss system. Thus, for example, the atom-photon interaction Hamiltonian in (4.85) in SI takes the form

$$H_{AF}^{mul} = -\frac{1}{\sqrt{4\pi\epsilon_0}}\boldsymbol{\mu}\cdot\sqrt{\frac{4\pi}{\epsilon_0}}\mathbf{D}_\perp(\mathbf{R}) = -\frac{1}{\epsilon_0}\boldsymbol{\mu}\cdot\mathbf{D}_\perp(\mathbf{R}) \tag{E.1}$$

while the normal mode expansion for the vector potential in (2.9) becomes

$$\sqrt{\frac{4\pi}{\mu_0}}\mathbf{A}(\mathbf{x},t) = \sum_{\mathbf{k}j}\sqrt{\frac{2\pi\hbar c^2}{V\omega_k}}\left\{\mathbf{e}_{\mathbf{k}j}a_{\mathbf{k}j}(t)e^{i\mathbf{k}\cdot\mathbf{x}} + \mathbf{e}_{\mathbf{k}j}^{*}a_{\mathbf{k}j}^{\dagger}(t)e^{-i\mathbf{k}\cdot\mathbf{x}}\right\} \tag{E.2}$$

Physical quantity	Gauss	SI
Electric field	$\mathbf{E}$	$\sqrt{4\pi\epsilon_0}\mathbf{E}$
Magnetic field	$\mathbf{H}$	$\sqrt{4\pi/\mu_0}\mathbf{H}$
Vector potential	$\mathbf{A}$	$\sqrt{4\pi/\mu_0}\mathbf{A}$
Scalar potential	V	$\sqrt{4\pi\epsilon_0}V$
Momentum conjugate to $\mathbf{A}$	$\mathbf{\Pi}$	$\sqrt{\mu_0/4\pi}\mathbf{\Pi}$
Electric charge	$q\ (e)$	$\frac{1}{\sqrt{4\pi\epsilon_0}}q\ (e)$
Electric charge density	ρ	$\frac{1}{\sqrt{4\pi\epsilon_0}}\rho$
Electric current density	$\mathbf{j}$	$\frac{1}{\sqrt{4\pi\epsilon_0}}\mathbf{j}$
Electric polarization field	$\mathbf{p}$	$\frac{1}{\sqrt{4\pi\epsilon_0}}\mathbf{p}$
Electric displacement field	$\mathbf{D}$	$\sqrt{4\pi/\epsilon_0}\mathbf{D}$
Field defined in (B.33)	$\mathbf{n}$	$\frac{1}{\sqrt{4\pi\epsilon_0}}\mathbf{n}$
Magnetization field	$\mathbf{m}$	$\sqrt{\mu_0/4\pi}\ \mathbf{m}$
Electric dipole moment	$\boldsymbol{\mu}$	$\frac{1}{\sqrt{4\pi\epsilon_0}}\ \boldsymbol{\mu}$
Dielectric constant	ϵ	ϵ/ϵ_0
Atomic polarizability	α	$\frac{1}{4\pi\epsilon_0}\alpha$

Table E.1.

which yields the desired expansion as

$$\mathbf{A}(\mathbf{x},t) = \sum_{\mathbf{k}j}\sqrt{\frac{\mu_0\hbar c^2}{2V\omega_k}}\left\{\mathbf{e}_{\mathbf{k}j}a_{\mathbf{k}j}(t)e^{i\mathbf{k}\cdot\mathbf{x}} + \mathbf{e}^*_{\mathbf{k}j}a^\dagger_{\mathbf{k}j}(t)e^{-i\mathbf{k}\cdot\mathbf{x}}\right\} \tag{E.3}$$

Finally, the purely transverse contribution to the energy density of a free electron, as given by (7.110), becomes

$$\langle\frac{1}{8\pi}4\pi\epsilon_0\mathbf{E}^2_\perp(\mathbf{x})\rangle = \frac{1}{2}\frac{1}{V}\sum_{\mathbf{k}j}\frac{1}{2}\hbar\omega_k + \frac{5}{16\pi^2}\frac{1}{4\pi\epsilon_0}\frac{e^2\hbar}{mc}\frac{1}{\mid\mathbf{x}-\mathbf{x}_e\mid^5} \tag{E.4}$$

which gives

$$\langle\frac{1}{2}\epsilon_0\mathbf{E}^2_\perp(\mathbf{x})\rangle = \frac{1}{2}\frac{1}{V}\sum_{\mathbf{k}j}\frac{1}{2}\hbar\omega_k + \frac{5}{64\pi^3\epsilon_0}\frac{e^2\hbar}{mc}\frac{1}{\mid\mathbf{x}-\mathbf{x}_e\mid^5} \tag{E.5}$$

In Table E.2 which follows, the units of various physical quantities in the SI are expressed in Gauss units. The velocity of light is $c = 2.99792458 \cdot 10^{10}\text{cm s}^{-1}$

In Table E.3 which follows, we give the values of some useful fundamental constants expressed in SI units. The errors in ppm are given in the AIP Vademecum (1981). Finally, other frequently used miscellaneous units and their expressions in SI are given in Table E.4 below.

Physical quantity	Symbol	1 SI Unit	Value in Gauss units
Length	ℓ	1 metre (m)	10^2 centimetres (cm)
Mass	m	1 kilogram (kg)	10^3 grams (g)
Time	t	1 second (s)	1 second
Frequency	ν	1 hertz (Hz)	1 hertz
Energy	E	1 joule (J)	10^7 erg
Energy density	U	1 joule m^{-3}	10 erg cm^{-3}
Electric charge	q	1 coulomb (C)	$c \cdot 10^{-1}$ statcoulomb
Electric charge density	ρ	1 coulomb m^{-3}	$c \cdot 10^{-7}$ statcoulomb cm^{-3}
Electric current density	j	1 ampere (A) m^{-2}	$c \cdot 10^{-5}$ statampere cm^{-2}
Electric field	**E**	1 volt m^{-1}	$\frac{1}{c} \cdot 10^6$ statvolt cm^{-1}
Electric potential	V	1 volt (V)	$\frac{1}{c} \cdot 10^8$ statvolt
Electric polarizability field	**p**	1 coulomb m^{-2}	$c \cdot 10^{-5}$ gauss units
Electric displacement field	**D**	1 coulomb m^{-2}	$4\pi c \cdot 10^{-5}$ statvolt cm^{-1}
Magnetic field	**H**	1 tesla	10^4 gauss (g)

Table E.2.

Physical quantity	Symbol	Value in SI units
Speed of light *in vacuo*	c	$2.99792458 \cdot 10^8$ m s^{-1}
Electron charge	e	$-1.6021892 \cdot 10^{-19}$ C
Planck's constant	$\hbar$	$1.0545887 \cdot 10^{-34}$ J · s
Electron rest mass	m_e	$0.9109534 \cdot 10^{-30}$ kg
Boltzmann constant	K	$1.380662 \cdot 10^{-23}$ J · K^{-1}
Fine structure constant	α^{-1}	137.035963
Electron Compton wavelength	λ_c	$3.8615905 \cdot 10^{-13}$ m
Bohr radius	a	$0.52917706 \cdot 10^{-10}$ m
Proton rest mass	m_p	$1.6726485 \cdot 10^{-27}$ kg
Neutron rest mass	m_n	$1.6749543 \cdot 10^{-27}$ kg
$\pi^{\pm}$ rest mass	$m_{\pi^{\pm}}$	$2.48804 \cdot 10^{-28}$ kg
π^0 rest mass	m_{π^0}	$2.4059 \cdot 10^{-28}$ kg

Table E.3.

Physical quantity	Unit	Value in SI units
energy: electron-volt	1 eV	$1.6022 \cdot 10^{-19}$J
length: Ångstrom	1 Å	$1 \cdot 10^{-10}$ m
length: atomic unit	1 au	$0.52918 \cdot 10^{-10}$ m
length: fermi	1 fm	$1 \cdot 10^{-15}$ m
length: micron	1 μm	$1 \cdot 10^{-6}$ m
mass: megaelectron-volt	1 MeV	$1.78268 \cdot 10^{-30}$kg

Table E.4.

References

H.L. Anderson (ed.) (1981). *Physics Vademecum* (American Institute of Physics, New York)
J.D. Jackson (1988). *Elettrodinamica Classica* (Zanichelli, Bologna)
B. Leroy (1985). *Am. J. Phys.* **53**, 589

Further reading

A comprehensive review of high precision tests of QED is
T. Kinoshita (ed.), *Quantum Electrodynamics* (World Scientific Publishing Co., Singapore 1990).

Appendix F
Gauge invariance and field interactions

In Section 1.4 we have introduced the concept of gauge transformation for quantum electrodynamics and we have shown that Maxwell's equations are invariant under a combined transformation of the scalar and vector potentials. In Section 3.5 we have seen that the principle of gauge invariance can lead to a Lagrangian for the interacting Schrödinger and electromagnetic field. This possibility, as we will discuss in detail in this appendix, is not limited to nonrelativistic QED, but it applies also to relativistic QED as well as to other fields such as those of electroweak interactions and of quantum chromodynamics. The importance of gauge invariance stems from the fact that field theories that can be obtained by a gauge principle are "renormalizable", in the sense that all ultraviolet divergences can be removed at all orders of perturbation theory by introducing a finite number of renormalization constants. An extensive discussion of this point would lead us beyond the scope of this book; consequently, in this appendix we will only show how the Lagrangian of relativistic QED can be derived by a gauge principle, and we will extend this to the case of quantum chromodynamics (QCD), where it leads to new and unexpected features.

The Lagrangian of the free Dirac field is given in (3.59). Another Lagrangian leading also to the Dirac equation with $\hbar = c = 1$ (which is a system of units normally adopted in quantum field theory) is (Barut 1980)

$$\mathcal{L}_0 = \bar{\phi}(x)\big(i\gamma_\mu\partial_\mu - m\big)\phi(x) \tag{F.1}$$

It is easy to check that (F.1) is invariant under a phase transformation of the form

$$\begin{cases} \phi(x) & \rightarrow \phi'(x) = e^{-i\Lambda}\phi(x) \\ \bar{\phi}(x) & \rightarrow \bar{\phi}'(x) = e^{i\Lambda}\bar{\phi}(x) \end{cases} \tag{F.2}$$

provided the phase parameter Λ is space-time independent; the reader will recognize (F.2) as a global gauge transformation of the same kind as (3.102). In group theoretical language it is a U(1) transformation (unitary, dimension one).

Allowing Λ to depend on x we obtain a local gauge transformation as in Section 3.5. This leads us to extend the principle of invariance of the Lagrangian density to the Dirac Lagrangian (F.1). This can be achieved by substituting the derivative in (F.1) with the "covariant derivative" defined as

$$D_\mu \phi(\mathbf{x}) = \big(\partial_\mu + ieA_\mu(\mathbf{x})\big)\phi(\mathbf{x}) \tag{F.3}$$

and by assuming that the gauge field $A_\mu(\mathbf{x})$ transforms under the gauge transformation as

$$A_\mu(\mathbf{x}) \to A'_\mu(x) = A_\mu(x) + \frac{1}{e}\partial_\mu \Lambda(x) \tag{F.4}$$

Using (F.2) and (F.4), it is easily verified that the covariant derivative as defined in (F.3) transforms as

$$D_\mu \phi(\mathbf{x}) \to \big(D_\mu \phi(x)\big)' = e^{-i\Lambda(x)}\big(D_\mu \phi(x)\big) \tag{F.5}$$

After substitution of the derivative in (F.1) with the covariant derivative, we get the following Lagrangian

$$\mathcal{L}' = \bar{\phi}(x) i\gamma_\mu \big(\partial_\mu + ieA_\mu\big)\phi(x) - m\bar{\phi}(x)\phi(x) \tag{F.6}$$

Using (F.5) it is evident that (F.6) is invariant under the combined transformations (F.2) and (F.4). If the gauge field $A_\mu(x)$ is to be interpreted as a dynamical variable, the Lagrangian density of the free gauge field should be added to (F.6), and this must be done without destroying gauge invariance. The simplest way is to add the gauge-invariant term

$$\mathcal{L}'' = -\frac{1}{16\pi} F_{\mu\nu}(x) F_{\mu\nu}(x) \tag{F.7}$$

where

$$F_{\mu\nu}(x) = \partial_\mu A_\nu(x) - \partial_\nu A_\mu(x) \tag{F.8}$$

We finally obtain the following gauge-invariant Lagrangian

$$\mathcal{L}_{QED} = \bar{\phi}(x) i\gamma_\mu \big(\partial_\mu + ieA_\mu\big)\phi(x) - m\bar{\phi}(x)\phi(x) - \frac{1}{16\pi} F_{\mu\nu}(x) F_{\mu\nu}(x) \tag{F.9}$$

We note that the well-known relativistic electron-photon interaction in the minimal coupling form is contained in (F.9). It is also worth noting that no photon mass term in the form $-\frac{1}{2}m^2 A_\mu A_\mu$ is present in (F.9); a term of this kind would obviously violate gauge invariance. In other words, local gauge invariance requires a massless photon.

A fundamental point is that the principle of gauge invariance permits us to obtain the correct form of the Lagrangian for other (renormalizable) fundamental interactions such as electroweak interactions and strong interactions. We will discuss as an example only the case of the theory of strong interactions, that is, quantum chromodynamics (QCD). In this case, as we will see, the situation is more complicated because the gauge group is not Abelian, contrary to the QED case (see e.g. Mills 1989). QCD is the field theory describing the interaction between quarks (fermions with spin 1/2 and fractional electric charge, described by the Dirac equation) and gluons (quanta of the strong interaction). Analogously to (F.1), the Lagrangian density of the free quarks is

$$\mathcal{L}_{quarks} = \bar{\phi}(x)\left(i\gamma_\mu \partial_\mu - m\right)\phi(x) \tag{F.10}$$

In (F.10) a summation over the flavour quantum number of the quarks should be added, but for simplicity we will neglect here the existence of quarks with different flavours.

The quarks' property responsible for the strong force is called colour: with respect to strong forces, colour is the equivalent of electric charge with respect to electromagnetic forces. From high-energy physics experiments it has been inferred that quarks have three colour components, so that $\phi(x)$ is indeed a triplet of wave functions

$$\phi(x) = \begin{pmatrix} \phi_1(x) \\ \phi_2(x) \\ \phi_3(x) \end{pmatrix} \tag{F.11}$$

where indices $1, 2, 3$ specify the components of the quark wave function in colour space.

According to the gauge invariance principle, we must modify (F.10) in such a way as to make it invariant under local gauge transformation in colour space

$$\phi(x) \to \phi'(x) = U(x)\phi(x) \tag{F.12}$$

The 3×3 matrix $U(x)$ is a sort of generalization in colour space of the phase factor $e^{i\Lambda(x)}$ of transformation (F.1). A very natural generalization is obtained if $U(x)$ belongs to the SU(3) group (special unitary group of dimension 3), that is, if

$$U(x)U^{\dagger}(x) = U^{\dagger}(x)U(x) = 1$$
$$\det U(x) = 1 \tag{F.13}$$

where the elements of the matrix $U(x)$ can have an arbitrary x-dependence. Proceeding in complete analogy with the QED case, we note that the free quark Lagrangian is not invariant under the SU(3) gauge transformation (F.12), and that a gauge field must be introduced to restore gauge invariance. This can be performed by substitution of the derivative in (F.10), with the covariant derivative defined by

$$D_{\mu}\phi(x) = (\partial_{\mu} + igG_{\mu}(x))\phi(x) \tag{F.14}$$

where g is a constant and $G_{\mu}(x)$ are 3×3 traceless Hermitian matrices. These matrices can be expanded in terms of the eight linearly independent generators λ_a of the SU(3) colour group (Gell-Mann matrices)

$$G_{\mu}(x) = \sum_{a=1}^{8} G_{\mu}^{a}(x)\frac{\lambda_a}{2} \tag{F.15}$$

(for more on Gell-Mann matrices, see Nachtmann 1989). Therefore eight gauge fields are necessary in QCD.

The Lagrangian density becomes

$$\mathcal{L}'_{quarks}(x) = \bar{\phi}(x)(i\gamma_{\mu}D_{\mu} - m)\phi(x) = \mathcal{L}_{quarks} + \mathcal{L}_i \tag{F.16}$$

and it is invariant under transformation (F.12) provided the gluon field transforms as follows

$$G_{\mu}(x) \rightarrow G'_{\mu}(x) = U(x)G_{\mu}(x)U^{\dagger}(x) - \frac{i}{g}U(x)\partial_{\mu}U^{\dagger}(x) \tag{F.17}$$

The eight vector fields $G_{\mu}^{a}(x)$ should be interpreted as the dynamical variables corresponding to the quanta of the strong interaction, called gluons (consequently we have eight different kinds of gluons); we must therefore add to (F.16) the Lagrangian of the free gluon quanta. This Lagrangian should be gauge-invariant to maintain the gauge invariance of the total Lagrangian. At this point the analogy with the QED case breaks down because expressions analogous to (F.7) and (F.8) are not gauge invariant due to the non-Abelian character of the SU(3) gauge group. The simplest way to construct a gauge-invariant free field Lagrangian is

$$\mathcal{L}_{gluon} = -\frac{1}{4}\sum_{a=1}^{8} G^a_{\mu\nu}(x)G^a_{\mu\nu}(x) \tag{F.18}$$

where

$$G^a_{\mu\nu}(x) = \partial_\mu G^a_\nu(x) - \partial_\nu G^a_\mu(x) - gf_{abc}G^b_\mu(x)G^c_\nu(x) \tag{F.19}$$

and f_{abc} are the structure constants of the SU(3) group. The presence of a quadratic term in (F.19) should be noted; no quadratic term is present in the Abelian gauge theory of QED. This term is characteristic of all non-Abelian gauge theories, and leads to self-interactions of the field quanta.

Summation of (F.16) and (F.19) yields the total (gauge-invariant) Lagrangian density of QCD

$$\mathcal{L}_{QCD} = -\frac{1}{4}\sum_{a=1}^{8} G^a_{\mu\nu}(x)G^a_{\mu\nu}(x)$$

$$+ \bar{\phi}(x)\left(i\gamma_\mu\left(\partial_\mu - ig\sum_{a=1}^{8} G^a_\mu \frac{\lambda^a}{2}\right) - m\right)\phi(x) \tag{F.20}$$

Comparing Equations (F.20) and (F.19) with the corresponding QED Equation (F.9), a striking difference becomes evident. In the first term of (F.20) self-interactions among gluons are present, while photons do not interact among themselves. From a physical point of view, these self-interactions stem from the fact that gluons themselves carry a colour charge; on the contrary, in QED the photon does not carry any electric charge. Moreover, the coupling constant g of these self-interactions is the same one that governs the quark-gluon interaction. In other words, the gluon-gluon interaction is not independent of the quark-gluon interaction. This is a consequence of the non-Abelian character of the gauge group of QCD and, in fact, no photon-photon fundamental interaction exists in the Abelian gauge theory of QED. We mention only that gluon-gluon self-interaction is responsible for a peculiar property of QCD named asymptotic freedom, that is, the fact that the effective coupling constant of QCD decreases as the energy of the particles increases or, equivalently, as their separation decreases.

References

E.S. Abers, B.W. Lee (1973). *Phys. Rep.* **9** C, 1

A.O. Barut (1980). *Electrodynamics and Classical Theory of Fields and Particles* (Dover Publications Inc., New York)

T.P. Cheng, L.F. Li (1984). *Gauge Theory of Elementary Particle Physics* (Clarendon Press, Oxford)

R. Mills (1989). *Am J. Phys.* **57**, 493

O. Nachtmann (1989). *Elementary Particle Physics* (Springer-Verlag, Berlin)

Appendix G
Dressed sources in relativistic QED and in QCD

In this appendix we will discuss qualitatively some examples of dressed sources in the framework of relativistic QED and of quantum chromodynamics (QCD), focusing our interest on some aspects that are not present in the nonrelativistic case.

a) Relativistic QED: vacuum polarization

In Section 3.5 using the gauge principle discussed also in Appendix F, we derived the interaction Hamiltonian of relativistic QED (Equations (3.123) and (F.9)). The matter-field interaction Hamiltonian is (see also Mandl and Shaw 1984)

$$\mathcal{H}_i(x) = -e\bar{\phi}(x)\gamma^{\mu}\phi(x)A_{\mu}(x) \tag{G.1}$$

where $\phi(x) = \phi(\mathbf{x}, t)$ is the electron field operator (see (3.75) with $\hbar = c = 1$ as in Appendix F)

$$\phi(\mathbf{x}, t) = \frac{1}{\sqrt{V}}\sum_{s=1}^{2}\sum_{\mathbf{p}}\sqrt{\frac{m}{\mid E_p \mid}}\Big\{b_{\mathbf{p}}^{(s)}u^{(s)}(\mathbf{p})e^{i\mathbf{p}\cdot\mathbf{x}} + d_{\mathbf{p}}^{(s)\dagger}v^{(s)}(\mathbf{p})e^{-i\mathbf{p}\cdot\mathbf{x}}\Big\} \tag{G.2}$$

where we have omitted implicit time-dependence. The creation and annihilation operators $b_{\mathbf{p}}^{(s)}, b_{\mathbf{p}}^{(s)\dagger}, d_{\mathbf{p}}^{(s)}, d_{\mathbf{p}}^{(s)\dagger}$ contain all the time dependence of the field operator and satisfy the equal-time anticommutation relations (3.80).

The electromagnetic four-potential, in terms of its positive and negative frequency components, can be expanded in plane waves as

$$A^{\mu}(x) = A^{\mu+}(x) + A^{\mu-}(x) \; ;$$

$$A^{\mu+}(x) = \sum_{r\mathbf{k}} \left(\frac{2\pi}{\omega_k V} \right)^{1/2} \epsilon_r^{\mu}(k) a_r(k) e^{ikx} \; ;$$

$$A^{\mu-}(x) = \sum_{r\mathbf{k}} \left(\frac{2\pi}{\omega_k V} \right)^{1/2} \epsilon_r^{\mu}(k) a_r^{\dagger}(k) e^{-ikx} \qquad \text{(G.3)}$$

where $r = 1, 2, 3, 4$ indicate the four polarization states of the photon when a covariant quantization procedure is used; however, only two of them are independent (for more detail on covariant electromagnetic field quantization, see e.g. Mandl and Shaw 1984, Bjorken and Drell 1965). k and x in the exponent are four-vectors, with $k_4 = i\omega_k$ and $x_4 = it$.

Substitution of (G.2) and (G.3) into (G.1) leads to eight terms with different combinations of creation and annihilation operators in the interaction Hamiltonian, representing the eight fundamental vertices of QED. Two of these terms contain the following combinations of creation and annihilation operators (for simplicity we omit spin and momentum labels)

$$bda^{\dagger} \; , \; ab^{\dagger}d^{\dagger} \qquad \text{(G.4)}$$

The first represents a process in which an electron and a positron disappear and a photon is created (pair annihilation); the second represents a process in which a photon disappears and an electron-positron pair is created (pair creation). These processes are not allowed in the nonrelativistic case and are peculiar to the relativistic theory. They are usually negligible at low energies, but can become relevant when energies larger than the electron rest mass are involved. The processes of pair creation and pair annihilation can also be virtual. For example, a virtual photon present in the photon cloud around an electron (see Chapter 6) can in turn generate a virtual electron-positron pair; successively, the pair will obviously annihilate generating a virtual photon. This process is shown by the Feynman diagram in Figure G.1. These processes generate a virtual electron-positron cloud around the (real) electron.

The virtual pair cloud around the relativistic electron affects the physical properties of the electron itself. The main effect is the so-called vacuum polarization. It is due to the fact that the electron attracts the positron of the virtual pair, while it repels the electron. Thus, on the

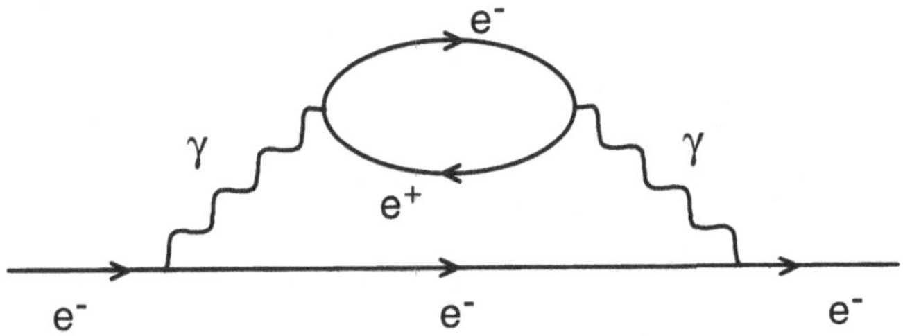

Fig. G.1 γ is a virtual photon contributing to the cloud of energy density around an electron. The diagram represents a process where γ gives rise to a virtual electron-positron pair which then annihilates, regenerating γ.

average, the virtual positron cloud is closer to the electron than the virtual electron cloud, yielding a polarization effect similar to that induced by a charged particle in a dielectric medium. This polarization is usually called vacuum polarization (see e.g. Gottfried and Weisskopf 1986, Griffiths 1987). If we consider a sphere around the electron the effective charge inside it is smaller than the charge of the electron, because of the vacuum polarization (considering that the positron cloud on the average is closer to the electron than the electron cloud): the "effective charge" of the electron is reduced by the vacuum polarization. In other words, the vacuum acts as a dielectric with a dielectric constant larger than unity. The nearer we get to the electron, the more the screening is reduced and the charge we can measure increases. One can also say that vacuum polarization effects generate an induced charge density around the electron.

The size of the relativistic virtual electron-positron cloud can be estimated qualitatively using arguments similar to those in Chapters 6 and 7 for the nonrelativistic photon cloud. The creation of a virtual electron-positron cloud by a photon, as in the process shown in Figure G.1, requires an energy of at least $2m_ec^2$, where m_e is the electron mass, and energy conservation is violated by an amount $\Delta E \geq 2m_ec^2$. The lifetime of the fluctuation is $\tau \sim \hbar/\Delta E \sim \hbar/mc^2$ and the size of the cloud is $\lambda \sim c\tau \sim \hbar/mc = \lambda_c$, where λ_c is the Compton wavelength of the electron. The electron-positron cloud is therefore localized within a distance λ_c around the electron; this distance is very small and consequently the effects of the cloud are expected to be negligible on a spatial scale (much larger than λ_c) typical of atomic physics and quantum optics.

The "unscreened" charge that we can observe at very small distances from the electron is the "bare" electron charge, while the "screened" charge observed at large distances is the charge of the electron "dressed"

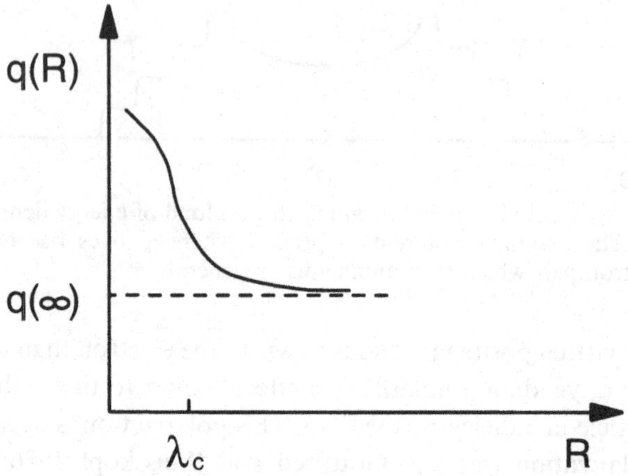

Fig. G.2 Qualitative representation of the value of the electric charge observed at distance R from the electron. At distances larger than the electron Compton wavelength λ_c the charge observed is pratically $q(\infty) \equiv -e$.

by the electron-positron cloud. It should be noted that what is normally called "the charge of the electron" is actually the fully screened charge that we can observe at very large distance (see Figure G.2). This consideration is very important in the framework of renormalization theory.

It is quite obvious that the polarization of the vacuum determines a change in the electric potential created by an electric charge. Let Q be an effective electric charge; the electric potential at distance r from the charge turns out to be (see e.g. Lifshitz and Pitaevskii 1974)

$$\phi(r) = \frac{Q}{r}(1 + C(r)) \tag{G.5}$$

where $C(r)$ is a correction due to the electron-positron vacuum fluctuations. In units where $\hbar = c = 1$, the limiting expressions of $C(r)$ are

$$C(r) \sim \begin{cases} \frac{\alpha}{4\sqrt{\pi}} \frac{e^{-2mr}}{(mr)^{3/2}} & \text{for } r \gg m^{-1} \\ \frac{2\alpha}{3\pi}\left(\log\left(\frac{1}{mr}\right) - \gamma - \frac{5}{6}\right) & \text{for } r \ll m^{-1} \end{cases} \tag{G.6}$$

where α is the fine-structure constant and γ is Euler's constant. As expected by our qualitative considerations, $C(r)$ grows for $r \ll m^{-1}$ and decreases exponentially for $r \gg m^{-1}$.

b) Quantum chromodynamics (QCD)

Dressing processes play a fundamental role also in QCD. In Appendix F we derived the Lagrangian of QCD using the gauge invariance principle; it is given by (F.20), where a summation over all flavours of quarks should be added to the second term on the RHS. A quantitative study of any vacuum process is much more complicated in QCD than in QED: in fact, not only the quark colour can appear in three different states and gluons can have eight different colour states (this should be compared with the single state of the electric charge of the electron and with the fact that the photon has no electric charge), but also the coupling constant of QCD is large to the point that perturbative methods are not generally applicable.

Besides vacuum polarization, which is present in QCD with qualitative features similar to those discussed in the QED case, other vacuum processes are present in QCD because of the non-Abelian character of the SU(3) gauge group. As mentioned after (F.20), gluons, the quanta of QCD, carry colour charge and consequently are self-interacting. The gluon self-interactions appear in the form of three-gluon and four-gluon vertices, and they lead to new vacuum processes. For example, a virtual gluon can emit another gluon (both of them carry colour and therefore interact between themselves and with other quarks and gluons) at a three-gluon vertex, and the consequent colour fluctuations can change the structure of the colour vacuum around quarks and/or gluons.

Vacuum polarization processes in QCD occur to a larger extent than in QED (a possible diagram is similar to that in Figure G.1, after substitution of electrons and photons with quarks and gluons respectively) because there are many more particles that can appear as virtual particles: eight gluons with different colour charges and three quarks' colours for each flavour. As in QED, vacuum polarization by virtual quark-antiquark pairs leads to screening of colour charges. It turns out, however, that the other vacuum processes already mentioned, due to three- and four-gluon vertices, give rise to the opposite effect, i.e. antiscreening. This has been proved rigorously only by perturbation methods, assuming a small coupling constant. There are, however, indications, based on qualitative models, that speak for a general validity of this result (for more details on qualitative models of antiscreening of the QCD vacuum, see e.g. Gottfried and Weisskopf 1986, Lee 1988). The prevailing effect in the competition between antiscreening and screening can be decided only by explicit calculations. The result is that antiscreening dominates over screening if the number of quark flavours is less than or equal to 16 (this is understandable considering that the screening due to virtual quark-antiquark pairs grows proportionally to the number of quarks). Because the (known) quark flavours are six, we can conclude that the QCD vacuum has antiscreening properties. This leads to a decrease of the "effective" or "dressed" coupling constant of QCD as the distance decreases or, equivalently, as the energy involved increases. This is exactly the opposite of the QED behaviour. This property of the effective coupling constant is called "asymptotic freedom" and it can be proved by group renormalization methods to hold for quite a wide range of energies (see e.g. Reya 1981). Indeed, the effective dimensionless coupling constant $\alpha_S(q^2) = g^2(q^2)/4\pi$ at the squared momentum scale q^2 is given by

$$\alpha_S(q^2) = \frac{\alpha_S(\Lambda^2)}{1 + b\alpha_S(\Lambda^2)\log\frac{q^2}{\Lambda^2}} \tag{G.7}$$

where Λ is a reference momentum scale of QCD and b is positive if the number of quark flavours is ≤ 16. Equation (G.7) shows that $\alpha_S(q^2) < \alpha_S(q'^2)$ if $q^2 > q'^2$, and also that $\alpha_S(q^2) \to 0$ if $q^2 \to \infty$. If the energy is sufficiently large, the effective coupling constant (also called the running coupling constant) becomes sufficiently small to permit use of perturbative calculations (for reviews of perturbative QCD, see e.g. Reya 1981 and Altarelli 1982).

As energy becomes smaller (or distance becomes larger) the running coupling constant grows and perturbative calculations fail. Many important physical situations, such as bound states of quarks and gluons forming hadrons, fall in this region, and no analytical technique can be successfully applied.

Although it has not yet been proved, there are reasons for believing that the QCD effective coupling constant grows indefinitely with increasing distance. This fact, if correct, leads to the idea of "confinement", that is to the fact that quarks and gluons cannot be separated indefinitely. This might explain why free quarks or free gluons have never been observed. Only particular combinations of quarks and gluons, yielding "neutral" colour composite particles can exist as separate entities; these combinations are supposed to form the observed hadrons. If this is correct, an infinite energy is required to separate completely the quarks constituting the hadrons or, in any case, enough energy to create from the vacuum new quark-antiquark pairs that then give rise to other colourless hadrons, without leaving any free isolated quark.

In conclusion, we wish to stress once more that many striking features of QCD such as asymptotic freedom and confinement are consequences of dressing processes by the vacuum; therefore the structure of the QCD vacuum plays a fundamental role in all phenomena involving strong interactions.

References

G. Altarelli (1982). *Phys. Rep.* **81,** 1

J.D. Bjorken, S.D. Drell (1965). *Relativistic Quantum Fields* (McGraw-Hill Book Company, New York)

K. Gottfried, V.F. Weisskopf (1986). *Concepts of Particle Physics,* vol. 2 (Oxford University Press, New York)

D. Griffiths (1987). *Introduction to Elementary Particles* (John Wiley and Sons, New York)

T.D. Lee (1988). *Particle Physics and Introduction to Field Theory* (Harwood Academic Publishers, London)

E.M. Lifshitz, L.P. Pitaevskii (1974). *Relativistic Quantum Theory* (Pergamon Press, London)

F. Mandl, G. Shaw (1984). *Quantum Field Theory* (John Wiley and Sons, Chichester)

E. Reya (1981). *Phys. Rep.* **69,** 195

Further reading

A non-technical introduction to the concept of vacuum in different areas of quantum field theory can be found in

I.J.R. Aitchison, *Contemp. Phys.* **26**, 333 (1985).

Appendix H
The energy-momentum tensor and Lagrangian density

The symmetric energy-momentum tensors for the electromagnetic field (Section 1.6), for the Klein-Gordon field (Section 3.2) and for the Dirac field (Section 3.3) have been given in the text without derivation. In this Appendix we wish to sketch briefly the procedure leading from the Lagrangian density to a symmetric energy-momentum tensor for an N-component field $\phi_r(x_\mu)$ with $r = 1, 2, \dots N$. Here each field component is taken as real and the Lagrangian density is taken to be of the form $\mathcal{L}(\phi_r, \partial\phi_r/\partial x_\mu)$.

The canonical stress tensor (see e.g. Bjorken and Drell 1964) $\Theta_{\mu\nu}$ is defined as

$$\Theta_{\mu\nu} = \frac{\partial \mathcal{L}}{\partial(\partial\phi_r/\partial x_\mu)} \frac{\partial\phi_r}{\partial x_\nu} - \delta_{\mu\nu}\mathcal{L} \tag{H.1}$$

The field equations of motion, analogous to (1.9) for the electromagnetic case, are

$$\frac{\partial \mathcal{L}}{\partial\phi_r} = \frac{\partial}{\partial x_\mu} \frac{\partial \mathcal{L}}{\partial(\partial\phi_r/\partial x_\mu)} \tag{H.2}$$

Using (H.2) it is easy to show that

$$\frac{\partial}{\partial x_\mu}\Theta_{\mu\nu} = 0 \tag{H.3}$$

Thus $\Theta_{\mu\nu}$ satisfies a continuity equation similarly to the energy-momentum tensor for the three fields previously mentioned. Definition (H.1), however, does not ensure symmetry in the sense that in general it is $\Theta_{\mu\nu} \neq \Theta_{\nu\mu}$. On the other hand, symmetry is a very important feature of $T_{\mu\nu}$ in Sections 1.6, 3.2 and 3.3 since it should be noted that $T_{\mu\nu} = T_{\nu\mu}$ was essential in ensuring conservation of angular momentum of the free field, a "must" for any reasonable field theory.

The question then arises if it is possible to derive from $\mathcal{L}$ an energy-momentum tensor which is symmetrical with respect to interchange of the indices. The first possibility is to exploit the fact, remarked on in Section 1.2, that $\mathcal{L}$ is undefined up to the divergence of a four-vector $\Gamma_\mu(\phi_r)$ (see e.g. Schweber 1964) in the sense that $\tilde{\mathcal{L}} = \mathcal{L} + \partial\Gamma_\mu/\partial x_\mu$ yields the same equations of motion of $\mathcal{L}$. Naturally this leads to a corresponding multiplicity of canonical stress tensors, since from (H.1)

$$\tilde{\Theta}_{\mu\nu} = \Theta_{\mu\nu} + \frac{\partial}{\partial x_\lambda}(\delta_{\lambda\nu}\Gamma_\mu - \delta_{\mu\nu}\Gamma_\lambda)$$
$$= \Theta_{\mu\nu} + \frac{\partial}{\partial x_\lambda}\psi_{\lambda\mu\nu} \tag{H.4}$$

where the tensor

$$\psi_{\lambda\mu\nu}(\phi_r) = \delta_{\lambda\nu}\Gamma_\mu - \delta_{\mu\nu}\Gamma_\lambda \tag{H.5}$$

is antisymmetric with respect to the first pair of indices because $\psi_{\lambda\mu\nu} = -\psi_{\mu\lambda\nu}$. One might hope that the antisymmetry of the ψ term in (H.4) would compensate for the antisymmetric part of $\Theta_{\mu\nu}$, yielding a symmetric $\tilde{\Theta}_{\mu\nu}$. This is unfortunately unlikely, since (H.4) and (H.5) are based on the assumption that Γ_μ, and consequently $\psi_{\lambda\mu\nu}$, are functions of only ϕ_r, while the antisymmetric part of $\Theta_{\mu\nu}$ might well be a function of $\partial\phi_r/\partial x_\mu$, as is evident from (H.1). In this case the compensation would not take place.

The idea of compensating the antisymmetric part of $\Theta_{\mu\nu}$ with an appropriate antisymmetric tensor seems however good and worth pursuing, even if this tensor cannot be given in general by $\psi_{\lambda\mu\nu}$ in (H.5). One proceeds in the following way (Belinfante 1940). Consider the infinitesimal coordinate transformation

$$x'_\mu = x_\mu + \epsilon_{\mu\nu}x_\nu + \delta_\mu \tag{H.6}$$

with $\epsilon_{\mu\nu} = -\epsilon_{\nu\mu}$ and δ_μ an infinitesimal displacement. (H.6) is called a Poincaré transformation (see e.g. Goldstein 1980) and it reduces to the infinitesimal pure Lorentz transformation for $\delta_\mu = 0$ and for an appropriate choice of $\epsilon_{\mu\nu}$. Pure Lorentz transformations have been introduced in Section 1.3. Under the Poincaré transformation (H.6) the field transforms as (Mandl and Shaw 1984)

$$\phi'_r(x'_\mu) = \phi_r(x_\mu) + \frac{1}{2}\epsilon_{\mu\nu}S^{rs}_{\mu\nu}\phi_s(x_\mu)\ ;$$
$$S^{rs}_{\mu\nu} = -S^{rs}_{\nu\mu} \tag{H.7}$$

The tensor $S^{rs}_{\mu\nu}$ is determined by the transformation properties of the field. Consider now the tensor

$$\Sigma_{\lambda\mu\nu} = \frac{\partial \mathcal{L}}{\partial(\partial\phi_r/\partial_\lambda)} S^{rs}_{\mu\nu}\phi_s = -\Sigma_{\lambda\nu\mu} \tag{H.8}$$

which, in view of (H.7), is antisymmetric with respect to the last pair of indices, and which is uniquely related to $\mathcal{L}$. In order to bring the antisymmetry to the first pair of indices, we shall construct the other tensor

$$S_{\lambda\mu\nu} = \Sigma_{\lambda\mu\nu} + \Sigma_{\mu\nu\lambda} + \Sigma_{\nu\mu\lambda} = -S_{\mu\lambda\nu} \tag{H.9}$$

which has the desired antisymmetry features. Moreover, it can be shown that

$$\frac{\partial}{\partial x_\lambda}\Sigma_{\lambda\mu\nu} = \Theta_{\nu\mu} - \Theta_{\mu\nu} \tag{H.10}$$

Clearly, the tensor $S_{\lambda\mu\nu}$ seems to possess all the necessary ingredients to play the compensating role which could not be played by $\psi_{\lambda\mu\nu}$ in an expression of the kind of (H.4). Consider in fact

$$T_{\mu\nu} = \Theta_{\mu\nu} + \frac{1}{2}\frac{\partial}{\partial x_\lambda} S_{\lambda\mu\nu} \tag{H.11}$$

It is very easy to show, using (H.9) and (H.10), that $T_{\mu\nu} = T_{\nu\mu}$. In addition, it is possible to show that, with respect to any transformation of the kind (H.4),

$$\tilde{T}_{\mu\nu} = T_{\mu\nu}\,; \quad \frac{\tilde{T}_{\mu\nu}}{\partial x_\mu} = \frac{T_{\mu\nu}}{\partial x_\mu} = 0 \tag{H.12}$$

Finally, the second of (H.12) can be shown to yield all the expected conservation laws, including angular momentum conservation as for example, in the electromagnetic case discussed in Section 1.6. Hence $T_{\mu\nu}$ is uniquely defined and it can be taken as the appropriate definition of the energy-momentum tensor of a field. The symmetric energy-momentum tensors defined in Sections 1.6, 3.2 and 3.3 can indeed be obtained from (H.11).

References

F.J. Belinfante (1940). *Physica* **7**, 449

J.D. Bjorken, S. Drell (1964). *Relativistic Quantum Fields* (McGraw-Hill Book Co., New York)

H. Goldstein (1980). *Classical Mechanics* (Addison-Wesley Publishing Co., Reading, Ma.)

F. Mandl, G. Shaw (1984). *Quantum Field Theory* (John Wiley and Sons, Norwich)

S.S. Schweber (1964). *An Introduction to Relativistic Quantum Field Theory* (John Weatherhill Inc., Tokyo)

Appendix I
The dressed relativistic hydrogen atom

The problem of the virtual cloud surrounding a relativistic hydrogen atom has been discussed by Radożycki (1990). This appendix gives a short account of this work and will follow closely the ideas discussed in Radożycki's papers. Thus the interest will be focused on the energy distribution of the virtual cloud surrounding a hydrogen atom in the ground state of the atom-field system. The main ideas are the same as in the nonrelativistic theory discussed in some detail in Sections 7.5 and 7.6, but the procedure exploits quantum field methods typical of relativistic QED and takes naturally into account the effects stemming from the electron spin, which have been completely neglected in the text. Thus the free matter field is described by the Dirac Lagrangian density

$$\mathcal{L}_D = -\hbar c\bar{\phi}\left(\gamma_\mu \frac{\partial}{\partial x_\mu} + \kappa\right)\phi \tag{I.1}$$

which differs from the form in Section 3.3, symmetric in ϕ and $\bar{\phi}$, only by a four-divergence. On the other hand, the electromagnetic field is partitioned as $A_\mu + A_{\mu c}$ where

$$A_{\mu c} = \left(0, i\frac{Ze}{r}\right) \tag{I.2}$$

is the field of the nucleus of charge Ze and r is the electron-nucleus distance. Consequently, $A_{\mu c}$ is not a field variable, unlike A_μ which can be described as the electromagnetic field amplitude. The Lagrangian density for this free field is taken to be of the form

$$\mathcal{L}_F = \frac{1}{8\pi} A_\mu \left(\frac{\partial^2 A_\mu}{\partial x_\nu^2} - (1-\lambda)\frac{\partial^2}{\partial x_\mu \partial x_\nu} A_\nu\right) \tag{I.3}$$

This expression differs from

$$\mathcal{L}_F = -\frac{1}{16\pi} F_{\mu\nu} F_{\mu\nu} - \frac{\lambda}{8\pi} \left(\frac{\partial A_\mu}{\partial x_\mu} \right)^2 \qquad \text{(I.4)}$$

by a four-divergence. Expression (I.4), in turn, can be obtained from the more familiar expression (1.8) by the addition of the term containing the Lagrange multiplier λ which fixes the gauge (Itzykson and Zuber 1985). Thus the total Lagrangian density of the atom-field system, assuming an infinitely heavy nucleus, can be written as

$$\begin{aligned} \mathcal{L} &= \mathcal{L}_D + \mathcal{L}_i + \mathcal{L}_F + \mathcal{L}_C \; ; \\ \mathcal{L}_i &= -e\bar{\phi}\gamma_\mu \phi A_\mu \; ; \\ \mathcal{L}_C &= -\bar{\phi}\gamma_4 \frac{Ze}{r} \phi \end{aligned} \qquad \text{(I.5)}$$

where the interaction Lagrangian density $\mathcal{L}_i$ is of the form (3.123).

The following states will be particularly useful in the present context.

- the vacuum $| \, \Omega \rangle$, which is the ground state of the system associated with the Lagrangian $\mathcal{L}_D + \mathcal{L}_F$;

- the state $| \, \tilde{\Omega} \rangle$, which is the ground state of the system associated with the total Lagrangian $\mathcal{L}$.

The procedure analogous to the one developed in Chapter 7 for a nonrelativistic system would be to determine $| \, \tilde{\Omega} \rangle$ by perturbation theory and then to evaluate the quantum average on this state of the energy density of the field expressed in terms of E^2 and H^2. For the relativistic case, however, it turns out to be more natural to determine first the quadratic correlation functions of the field amplitude A_μ, for which task the scattering methods of field theory are best suited (see e.g. Bjorken and Drell 1965), and to derive from these correlation functions the desired energy densities by appropriate differentiation.

Thus the basic object of the relativistic theory is the quantity

$$\begin{aligned} I_{\mu\nu}(x, x') = {} & \langle \tilde{\Omega} \mid \left[A_\mu(x) + A_{\mu c}(x) \right] \left[A_\nu(x') + A_{\nu c}(x') \right] \mid \tilde{\Omega} \rangle \\ & - \langle \Omega \mid A_\mu(x) A_\nu(x') \mid \Omega \rangle \end{aligned} \qquad \text{(I.6)}$$

whose evaluation is relatively straightforward using standard techniques of S-matrix theory which transform (I.6) into a linear conbination of vacuum expectation values. These expectation values are then evaluated using Feynman diagrams up to the desired order in the coupling constant e. Once this has been done, taking the appropriate space and time derivatives of $I_{\mu\nu}(x, x')$, and successively the limit $x' \to x$, yields not just the energy density, but all the elements of the energy-momentum tensor. Limiting ourselves to the quantum averages of the energy density on state $|\tilde{\Omega}\rangle$, the procedure outlined above yields at order e^2 rather involved expressions for the electric and magnetic parts which, however, in the far zone of the atom simplify as

$$\begin{aligned}\frac{1}{8\pi}\langle E^2(\mathbf{r})\rangle &= \frac{13}{16\pi^2}e^2\sum_n\int\frac{1}{E_n-E_1}\langle 1\mid x_i\mid n\rangle\langle n\mid x_i\mid 1\rangle\frac{1}{r^7}\\ &\quad+\frac{7}{16\pi^2}e^2\sum_n\int\frac{1}{E_n-E_1}\langle 1\mid x_i\mid n\rangle\langle n\mid x_j\mid 1\rangle\frac{\hat{r}_i\hat{r}_j}{r^7}\,;\\ \frac{1}{8\pi}\langle H^2(\mathbf{r})\rangle &= -\frac{7}{16\pi^2}e^2\sum_n\int\frac{1}{E_n-E_1}\langle 1\mid x_i\mid n\rangle\langle n\mid x_j\mid 1\rangle\\ &\quad\times\left(\delta_{ij}-\hat{r}_i\hat{r}_j\right)\frac{1}{r^7}\end{aligned}\tag{I.7}$$

These expressions constitute the relativistic generalization of the far-zone results for the nonrelativistic hydrogen atom in Section 7.6. It should be noted that the states $|n\rangle$ and the corresponding eigenenergies E_n in (I.7) are eigensolutions of the Dirac equation for the relativistic hydrogen atom.

We refer the reader to the second paper by Radożycki (1990) for a discussion of the properties of the electromagnetic field around an excited hydrogen atom.

References

J.D. Bjorken, S. Drell (1965). *Relativistic Quantum Fields* (McGraw-Hill Book Co., New York)

C. Itzykson, J.B. Zuber (1985). *Quantum Field Theory* (McGraw-Hill Book Co., Singapore)

T. Radożycki (1990). *J. Phys. A* **23**, 4911, 4925

Appendix J
The nonrelativistic Lamb shift in a hydrogenic atom

The interaction with the transverse modes of the vacuum electromagnetic field has been shown to yield a radiative (or self-energy) shift of the ground state of a two-level atom in Section 6.4. In this appendix we will show that the same effect occurs also in the energy of the states of a multilevel atom of the hydrogenic kind, which can be modelled by an electron bound to a fixed nucleus of charge Ze. This self-energy shift will turn out to depend on the form of the wavefunction of the state of the electron, leading to the possibility of lifting some of the accidental degeneracies which occur in hydrogenic atoms. Indeed the first experimental observation of this effect is related to the lifting of the well-known 2s-2p degeneracy in atomic hydrogen, and it is due to Lamb and Retherford (1947). Its nonrelativistic QED explanation, on the other hand, is due to Bethe (1947), and this appendix is simply a short account of his theory.

Consider a bare one-electron atom, whose energy levels and corresponding wavefunctions are denoted by E_n and $u_n(\mathbf{x}) = \langle \mathbf{x} \mid n \rangle$. Here n indicates the triplet of quantum numbers N, L, M and the electron spin is disregarded. Adopting the Coulomb gauge and the minimal coupling scheme in the electric dipole approximation, the Hamiltonian of the system can be partitioned as

$$H = H_A^{min} + H_F^{min} + H_{AF}^{min} \tag{J.1}$$

where H_A^{min}, H_F^{min} and H_{AF}^{min} are given by (4.42), (4.43) and (4.49) respectively. Using the results of Appendix C, it is easy to see that the energy shift of $\mid n \rangle$, due to H_{AF}^{min}, at order e^2 and in the absence of real photons, is given by

$$\begin{aligned}
\Delta_n &= \Delta_{1n} + \Delta_{2n} \, ; \\
\Delta_{1n} &= \langle \{0_{\mathbf{k}j}\}, n \mid H_1 \frac{1 - P_n}{E_n - H_0} H_1 \mid \{0_{\mathbf{k}j}\}, n \rangle \, ; \\
\Delta_{2n} &= \langle \{0_{\mathbf{k}j}\}, n \mid H_2 \mid \{0_{\mathbf{k}j}\}, n \rangle
\end{aligned} \tag{J.2}$$

where $\{0_{\mathbf{k}j}\}$ is the vacuum of the free field, $H_0 = H_A^{min} + H_F^{min}$ and where H_1 and H_2 correspond to the first and second terms of (4.49). We also remark that (J.2), in view of the arguments developed in Section 6.5, is based, strictly speaking, on the assumption that $\mid n \rangle$ is stable in the sense that it should not decay quickly to lower states via direct spontaneous emission. Although this condition is not satisfied for most $\mid n \rangle$, this conceptual difficulty will be disregarded here in order to follow Bethe's procedure, which involves evaluation of Δ_n without any restriction on $\mid n \rangle$.

Evaluation of Δ_{2n} is trivial, since taking $\mathbf{R} = 0$ one has from (4.49)

$$\Delta_{2n} = \frac{e^2}{2mc^2} \langle \{0_{\mathbf{k}j}\}, n \mid A^2(0) \mid \{0_{\mathbf{k}j}\}, n \rangle \tag{J.3}$$

which does not depend on n in view of the electric dipole approximation. Thus Δ_{2n} is a shift common to all the eigenstates of H_0. As such, it cannot influence the dynamics of the system and it can be disregarded. On the other hand, from (J.2) one has

$$\Delta_{1n} = -\frac{2}{3\pi} \alpha \frac{\hbar^2}{m^2 c^2} \sum_{n'} \int_0^{\omega_M} \frac{\mid \mathbf{p}_{nn'} \mid^2}{E_{n'} - E_n + \hbar\omega} \omega d\omega \tag{J.4}$$

where $\alpha = e^2/\hbar c$ is the fine structure constant and ω_M is a cut-off frequency, similar in nature to that defined in Section 5.4. Here one can assume $\omega_M = mc^2/\hbar$, since one expects the nonrelativistic scheme to provide a natural cut-off at photon energy of the order of mc^2.

It should be emphasized that the m appearing in (J.4) is the bare electron mass. Renormalization can be performed by the same prescription leading to (6.54). Neglecting terms $O(e^3)$, this consists of adding to (J.1) the counterterm

$$H_{MR} = \frac{4}{3\pi} \alpha \frac{p^2}{2m} \tag{J.5}$$

and of regarding m as the physical electron mass. Thus the contribution of mass renormalization to Δ_n is taken into account simply by adding the quantity

$$\langle\{0_{\mathbf{k}j}\}, n \mid H_{MR} \mid \{0_{\mathbf{k}j}\}, n\rangle = \frac{4}{3\pi}\alpha\frac{\mid \mathbf{p}_{nn} \mid^2}{2m}$$
$$= \frac{4}{3\pi}\alpha\frac{1}{2m}\sum_{n'} \mid \mathbf{p}_{nn'} \mid^2 \qquad \text{(J.6)}$$

to expression (J.4). Thus, for all practical purposes and within the approximations adopted, the self-energy shift of state $\mid n\rangle$ turns out to be

$$\Delta_n^{(MR)} = \Delta_{2n} + \langle\{0_{\mathbf{k}j}\}, n \mid H_{MR} \mid \{0_{\mathbf{k}j}\}, n\rangle$$
$$= \frac{2}{3\pi}\alpha\frac{\hbar}{m^2c^2}\sum_{n'}\int_0^{\omega_M} \frac{\mid \mathbf{p}_{nn'} \mid^2 (E_{n'} - E_n)}{E_{n'} - E_n + \hbar\omega} d\omega \qquad \text{(J.7)}$$

The integration in (J.7) can be performed by taking the principal value where necessary. Exploiting the nonrelativistic assumption $\mid E_{n'} - E_n \mid \ll \hbar\omega_M$, the result is

$$\Delta_n^{(MR)} = \frac{2}{3\pi}\alpha\frac{1}{m^2c^2}\sum_{n'} \mid \mathbf{p}_{nn'} \mid^2 (E_{n'} - E_n)\ln\frac{\hbar\omega_M}{\mid E_{n'} - E_n \mid} \qquad \text{(J.8)}$$

For a two-level atom and taking properly into account mass renormalization, this expression reduces to (6.71) in the case of the ground state. The sum in (J.8) can be approximated by substituting the slow-varying logarithm with an appropriate average over all possible n'

$$\ln\frac{\hbar\omega_M}{\langle E_{n'} - E_n\rangle_{\mathrm{Av}}} \qquad \text{(J.9)}$$

Moreover, as shown by Bethe (1947), one also has for any $\mid n\rangle$ belonging to the hydrogenic spectrum

$$\sum_{n'} \mid \mathbf{p}_{nn'} \mid^2 (E_{n'} - E_n) = 2\pi\hbar^2 e^2 Z \mid u_n(0) \mid^2 \qquad \text{(J.10)}$$

Thus $\Delta_n^{(MR)}$ vanishes unless $u_n(0) \neq 0$. This is true only for the s-states of the hydrogenic spectrum with $L = 0$, and consequently only these states display radiative shifts at the level of the approximations adopted. This yields a small energy difference between the 2s and 2p states which was experimentally detected by Lamb and Retherford (1947) using spectroscopic techniques and which is called the Lamb shift. Substitution of (J.9) and (J.10) in (J.8) gives a Lamb shift of $\sim$ 1040 MHz, which compares favourably with the experimental value of 1057 MHz.

References

H.A. Bethe (1947). *Phys. Rev.* **72**, 339
W.E. Lamb, R.C. Retherford (1947). *Phys. Rev.* **72**, 241

Index